materials science for engineers

LAWRENCE H. VAN VLACK

The University of Michigan

ADDISON-WESLEY PUBLISHING COMPANY

Reading, Massachusetts
Menlo Park, California · London · Amsterdam · Don Mills, Ontario · Sydney

This book is in the

ADDISON-WESLEY SERIES IN METALLURGY AND MATERIALS

Consulting Editor
MORRIS COHEN

ISBN 0-201-08074-5
PQRSTUVW-HA-8987654

preface

Many readers of this preface are aware of the significant changes which occurred in Materials Science and Materials Engineering within the decade preceding 1970. Concurrent with the nucleation and growth of these new disciplines was the establishment of a course in Materials Science as one of the Engineering Sciences, not only in this country but worldwide. With the changes just cited, it is necessary that a materials science textbook for undergraduate engineering students draw as much as possible upon the basic science background of those students—a background that commonly includes a working knowledge of calculus, an introduction to structural chemistry, and an initial familiarity with the principles of electricity in physics.

The starting point of the text which follows is the mathematics, chemistry, and physics just described. This review, with an emphasis on those introductory principles that specifically pertain to materials, is the content of the first three chapters. Chapters 4 through 8 present the structures of solid phases with emphasis on crystals, imperfections, molecular phases, and solid solutions. These chapters provide a framework for considering, first, those properties which arise from the atomic processes, such as diffusion and deformation (Chapters 6 through 12); and second, those properties which originate from electrical processes (Chapters 13 through 16). The latter area is expanded and closely integrated with the preceding topics of materials science. This close integration is a new venture for a second- or third-year text and may place many instructors in a new teaching situation. However, the author is confident that the long-term trend is in the direction taken by this book, and that it is only a matter of time before all materials courses will include such an arrangement. Following a more conventional presentation of multiphase equilibria, reactions, and microstructures (Chapters 17 through 19), there is an expanded focus on materials utilization in Chapters 20 through 23. Within this area, strengthening processes and materials failure receive separate attention, and corrosion is viewed with a modern presentation. Materials systems, including composites as well as bonding and induced stresses, constitute the final topic.

This textbook builds upon problem solving as a teaching tool. Not only is this a logical procedure for engineers, it also provides an analytical approach to the study of materials. To implement this teaching method, approximately 150 example problems have been included within the chapters, together with nearly 400 study problems which are placed at the ends of the chapters. It is to be hoped that the instructor will have the student use the former both as an aid to understanding the assigned home problems and also for the purpose of emphasizing new concepts which cannot be presented as effectively in prose.

Many faculty hours have been spent in discussions and conferences on the merits of teaching a general undergraduate materials course. The most common question may be phrased as "Does not such a course simply become a smattering of many topics?"—or with different phrasing, "I cannot really cover dislocations (or some other topic) thoroughly within a general materials course." These concerns have an apparent credibility until one looks at the analogous pedagogical situation in chemistry (or physics or biology). Few biochemistry instructors, for example, would consider teaching their specialty to students who have not had a basic course in chemistry. The author likes to consider that the course in which this book is used will serve a similar basic purpose in the materials area, and looks forward to the prospect that a reasonable number of specialized courses will be developed across the country as sequels in the applied engineering areas. However, even though the author has attempted to produce a textbook for a general course, individual instructors will find it necessary and often desirable to pick and choose topics out of the chapters which follow. It is anticipated that a one-semester course will be able to assign only about 75 percent of the subject matter in the book. Which 75 percent is used depends, of course, on the available prerequisites, the student cross section, subsequent materials and design courses, as well as many local factors.

Every author should feel considerable humility when he recalls the help which he has received from others. It is with regret that I cannot give individual recognition to all the students and colleagues who have given either direct or indirect help. Though space does not permit this, I am especially desirous to acknowledge the counsel of certain people. These include Professor W. F. Hosford of The University of Michigan and Professor M. Cohen of the Massachusetts Institute of Technology, both of whom have read the entire manuscript and offered numerous valuable suggestions. Equally appreciated is the assistance of Professors B. Cullity of the University of Notre Dame and F. Donahue and G. Yeh of The University of Michigan, who constructively criticized selected chapters. Dr. C. J. Osborn of the University of Melbourne and Prof. W. C. Bigelow, plus others among my colleagues at The University of Michigan, have given critical advice of a constructive nature on many occasions. Their help, along with that of the Addison-Wesley staff at Reading, is sincerely acknowledged. Finally, I must admit some feeling of guilt that only my name appears as author, since Fran, Laura, and Bruce spent many hundreds of hours doing typing, problem checking, editing, proofreading, and indexing in our third such family project.

Ann Arbor   L. H. V. V.
August 1969

Dedicated to
C. H. V. V. and R. A. R.
by L., F., L., and B.

contents

part I □ introductory principles

1 □ materials and properties

1-1 MATERIALS IN ENGINEERING

All engineering products utilize materials. The concrete of highways and the steel of automobiles are obvious materials of engineering. Less apparent but equally important are the materials of electronic circuits (Fig. 1-1), or the carbide bits on oil-well drills. Each of these products requires materials with specific characteristics such that (1) the materials can be processed into the final products satisfactorily and economically and (2) the products will behave appropriately in service.

1-1 Electronic microcircuit (made by IBM). Each component material has an active function in the circuit. The performance depends on the internal structure. The background is a 1-mm grid. (R. Landauer and J. J. Hall, *Science*, **160**, 1968, p. 738.)

Since the engineer must specify the materials for TV sets, computers, suspension bridges, oil refineries, rocket motors, nuclear reactors, or supersonic transports, he must have sufficient knowledge to select the optimum material for each application—taking into account material availability, processing requirements, service demands, and the ever-important cost factor, in addition to the design specifications

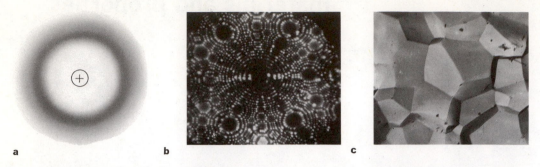

1-2 Internal structures. (a) Atomic level. A hydrogen atom has a central proton, ⊕, and an electron which moves continuously but has an average orbit radius of about 0.5 A (shaded). Other elements have additional electrons and protons (and neutrons) in their structures. (b) Crystal level [×1,300,000 (linear)]. Each bright spot reveals the location of an atom in a crystal on a very sharply pointed tungsten cathode. Crystals possess specific atom arrangements. (Courtesy of H. F. Ryan and J. W. Suiter, University of Melbourne.) (c) Microstructure level. Each area of this photomicrograph, magnified ×5000 (linear), is a single crystal of Al_2O_3 containing about 10^{12} oxygen atoms plus two-thirds that number of aluminum atoms. Properties may be altered by changing the size, shape, orientation, or composition of the crystalline grains which comprise a microstructure. (Courtesy of E. Scala, Cornell University.)

which match the part to the system as a whole. Although experience provides the engineer with a starting point for selection of materials, and the property data found in handbooks will aid him, the skill of the engineer will be limited unless he understands the factors which contribute to the properties of materials. Without this knowledge, a material is simply a "black box" and the engineer will have no conception of the material's limitations, or of possible modifications in materials selection and design.

1-2 INTERNAL STRUCTURE OF MATERIALS

Every material has an internal structure which involves the *atoms* and their electronic and nuclear modifications (Fig. 1-2a). The internal structure also includes the coordination of atoms with their neighbors into distinct *phases*. Commonly, but by no means always, this coordination is sufficiently regular to produce crystals (Fig. 1-2b). Finally, the *microstructure* of a material comprises the fitting together of individual crystals and phases into a coherent whole (Fig. 1-2c).

Each of these structural levels—atoms, phases, and microstructures—affects the properties of a material. The internal structure of a material may be modified, and any such change will alter the properties accordingly. Conversely, if we desire a specific set of properties for a material, we must develop an internal structure with the necessary characteristics.

This text contains a review of atomic characteristics (Chapters 2 and 3) as a basis for studying the *internal structure* of solid phases (Chapters 4 through 8). These structures are then related to *properties* and *service behavior* through atomic processes (Chapters 9 through 12) and electrical processes (Chapters 13 through 16). Finally, attention is given to *multiphase* materials (Chapters 17 through 19) and to the factors affecting *utilization* of materials in service (Chapters 20 through 23).

1-3 PROPERTIES OF MATERIALS

Our initial discussion of the properties of materials will be introductory only. Its purpose will be to identify a few of the more common properties so that later we can make comparisons between different materials or between different internal structures of the same material.

Three of the more general types of properties are mechanical, thermal, and electrical. By mechanical properties, which will be discussed in the following paragraphs, we mean those properties which a material has in response to mechanical forces. Mechanical properties will receive first attention because they are most easily comprehended in a discussion of properties versus internal structure. Thermal properties arise from internal energies that introduce atomic and electronic movements. Also thermal energy can influence mechanical behavior and mechanical properties. Electrical properties have two prime sources: electron

TABLE 1-1
PRINCIPAL MECHANICAL PROPERTIES

Property	Symbol	Definition (or comments)	Common units		
			English	Metric	Conversion
Stress	σ	force/unit area $(=F/A)$	psi	dynes/cm^2	1 psi = 69,000 dynes/cm^2
Strain	ϵ	fractional deformation $(=\Delta l/l)$		dimensionless	
Elastic modulus	Y*	$\dfrac{\text{elastic stress}}{\text{unit strain}}$	psi	dynes/cm^2	1 psi = 69,000 dynes/cm^2
Strength		stress at failure	psi	dynes/cm^2	1 psi = 69,000 dynes/cm^2
Yield	YS	strength for initial plastic deformation			
Tensile	TS	maximum strength (based on original dimensions)			
Ductility		plastic strain at fracture			
Elongation	El	$(l_f - l_o)/l_o$		dimensionless	
Reduction of area	R of A	$(A_o - A_f)/A_o$		dimensionless	
Hardness		resistance to plastic indentation			
Brinell	BHN	large indentor	—	kg$_f$/mm^2†	—
Rockwell	R	small indentor		arbitrary units	
Knoop	KHN	very small indentor	—	kg$_f$/mm^2	—
Toughness	—	energy for fracture	ft-lb	—	—

* Often expressed as E in other texts.
† The symbol kg$_f$ stands for kilogram as a force, in contrast to kg, which is the symbol for kilogram as a mass.

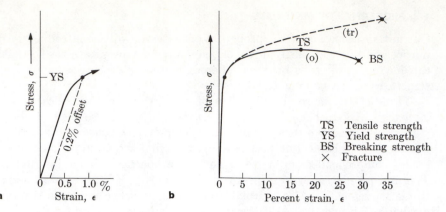

1-3 Strain versus stress. (a) Enlarged strain axis of (b). Initially strain (deformation) is proportional to stress (force per unit area). This strain is reversible (elastic) when the stress is removed. Stresses beyond the elastic limit, or yield point, introduce a permanent, or plastic, deformation, and finally fracture, ×. (o) Stress-strain curve based on original, or design, dimensions. (tr) Stress-strain curve based on minimum cross-sectional, or true, dimensions during deformation. (Unless stated otherwise, strengths are based on original areas.)

movements and charge displacements. Because of the growing importance of electrical materials, our treatment will give greater emphasis to electrical properties than does the normal introductory text.

Mechanical properties. Modulus of elasticity, strength, ductility, hardness, and toughness are useful properties to know in order to anticipate the mechanical behavior of various materials. They are summarized in Table 1-1 and Fig. 1-3. This figure shows a stress-strain curve which initially is linear and reversible. The proportionality coefficient relating the reversible strain to stress is called an *elastic modulus* and will subsequently reveal several characteristics of atomic bonding.

With sufficiently high stresses, ductile materials undergo a nonreversible yielding, or permanent *plastic deformation*. In practice, such plastic deformation produces a reduction in cross-sectional area. Thus stress-strain curves can be plotted on the basis of *original* dimensions or on the basis of instantaneous, or *true*, dimensions. The former are important for specifying the materials to be used in engineering products; the latter permit calculations of the actual stress and deformation, and are important for interpreting what is happening within the material itself.

Example 1-1*

An aluminum bar has a reduction of area (see Table 1-1) of 62% and a breaking strength of 28,000 psi based on the original design area, A_o. Calculate (a) the true stress, σ_{tr}, and (b) the true strain, ϵ_{tr}, just prior to fracture.

---

* Examples serve two purposes in this text. First, they provide a basis for the solution of study problems at the end of each chapter. Second, and equally important, they introduce certain concepts which cannot be presented as efficiently in the text proper. For this reason, the reader will need to examine the examples whether or not he expects to solve comparable study problems later as class assignments.

1-4 Hardness (Al + 13% Si) [×125 (linear)]. The indentations made on two different components of a microstructure reveal that the lighter, high-silicon area is harder than the dark, lower-silicon area. For ductile materials, hardness values may be used as an indication of strength. (Courtesy of J. Muscara and C. A. Siebert, The University of Michigan.)

Answer

a) $28,000 \text{ psi} = F/A_o, \quad F = 28,000 A_o.$
Since $A_f = A_o - 0.62 A_o = 0.38 A_o,$

$$\sigma_{\text{tr}} = F/A_f = 28,000 A_o / 0.38 A_o = 74,000 \text{ psi}.$$

b) Since the instantaneous strain $d\epsilon$ is equal to $dl/l,$

$$\epsilon_{\text{tr}} = \int_{l_o}^{l_f} \frac{dl}{l} = \ln \frac{l_f}{l_o}.$$

Assuming constant volume,

$$A_f l_f = A_o l_o,$$
$$\epsilon_{\text{tr}} = \ln(A_o/A_f) \tag{1-1}$$
$$= \ln \left[\frac{A_o}{0.38 A_o} \right] = 0.97 = 97\%. \; \blacktriangleleft$$

Hardness, which is the resistance of a material to plastic indentation, is not a simple property because complex stress patterns develop during the testing (Fig. 1-4). However, hardness measurements can be made relatively easily and consistently to give an index of strength and structural coherence. Therefore we will make extensive use of these values to develop a qualitative understanding of properties-versus-structure relationships in later chapters.

Toughness is important in fracture and materials failure (Chapter 21). Simply stated, toughness is the energy required for fracture. Hence, toughness is the integrated product of the stress and strain per unit volume required to produce fracture. Consequently, the units should be in-lb per in³. However, the fracture process is complex in that the strains are not uniformly distributed during fracture, particularly in the region of a crack. Accordingly, as indicated in Table 1-1, toughness is represented as the number of inch-pounds (or foot-pounds) of force necessary to break a standard test specimen. Toughness values are of major importance to design engineers and can be related to the internal structure of a material.

TABLE 1-2
PRINCIPAL THERMAL PROPERTIES

Property	Symbol	Definition	Common units English	Common units Metric	Conversion
Temperature	°	Thermal potential	°F	°K	$(°F + 460)/1.8 = °K$
Heat	H	Thermal energy	Btu	cal	$1 \text{ Btu} = 252 \text{ cal}$
Thermal expansion	α	Fractional dimension change per degree change	$°F^{-1}$	$°K^{-1}$	$1/°F = 1.8/°C$
Heat capacity	C	Heat per mass per degree change	$\dfrac{\text{Btu}}{\text{lb} \cdot °F}$	$\dfrac{\text{cal}}{\text{gm} \cdot °K}$	equal
Thermal conductivity	k	Thermal energy transport	$\dfrac{\text{Btu} \cdot \text{in.}}{°F \cdot \text{ft}^2 \cdot \text{sec}}$	$\dfrac{\text{cal} \cdot \text{cm}}{°K \cdot \text{cm}^2 \cdot \text{sec}}$	$0.806 \dfrac{\text{Btu} \cdot \text{in.}}{°F \cdot \text{ft}^2 \cdot \text{sec}} = 1 \dfrac{\text{cal} \cdot \text{cm}}{°K \cdot \text{cm}^2 \cdot \text{sec}}$

TABLE 1-3
PRINCIPAL ELECTRICAL PROPERTIES

Property	Symbol	Definition	Metric units*
Conductivity	σ	Charge flux per unit voltage gradient	$\text{ohm}^{-1} \cdot \text{m}^{-1}$
Resistivity	ρ	Reciprocal of conductivity	$\text{ohm} \cdot \text{m}$
Polarization	$\mathscr{P}$	Dipole moment per unit volume	$\text{coul} \cdot \text{m/m}^3$
Relative dielectric constant	κ	Ratio of dielectric constant to that of free space	dimensionless
Dielectric constant	ϵ	Electric charge density per unit field strength	farads/m (or coul/volt·m)

* Mks units are becoming the most common metric units for electrical properties; however, cgs units still persist, particularly for resistivity (ohm · cm) and for magnetic properties.

Thermal properties. Three thermal properties will receive specific attention in this text: (1) *thermal expansion*, (2) *heat capacity*, and (3) *thermal conductivity* (Table 1-2). In general, the reader is aware of these as bulk properties, encountered in physics courses and listed as handbook data. In subsequent chapters, attention will be given to the origins of these properties and how they are related to the internal structure of materials.

Electrical properties. (See Table 1-3.) The reader is familiar with the fact that electrical resistance, expressed in ohms, is a consequence of geometry as well as of the material itself. In this text, we shall be primarily concerned with the *resistivity* of materials. Resistance and resistivity are related as follows:

$$\rho = (R/l)(A), \tag{1-2}$$
$$\text{resistivity} = (\text{resistance/length})(\text{area})$$
$$= (\text{ohms/cm})(\text{cm}^2)$$
$$= \text{ohm} \cdot \text{cm}.$$

Example 1-2

Copper has a resistivity, ρ, of 1.54×10^{-6} ohm $\cdot$ cm at 0°C, and 2.22×10^{-6} ohm $\cdot$ cm at 100°C. What is the value of dR/dT per foot for a 0.022-in. wire if one assumes that the resistance, R, varies linearly with temperature?

Answer

$$\Delta R/\Delta T = (R_{100°} - R_{0°})/100°C$$
$$= \frac{(\rho_{100°} - \rho_{0°})(12 \text{ in./ft})(2.54 \text{ cm/in.})}{(100°C)(\pi/4)(0.022 \text{ in.})^2(2.54 \text{ cm/in.})^2}$$
$$= \frac{(0.68)(10^{-6} \text{ ohm} \cdot \text{cm})(12 \text{ in./ft})}{(100°C)(0.00038 \text{ in}^2)(2.54 \text{ cm/in.})}$$
$$= 0.84 \times 10^{-4} \text{ ohm/°C} \cdot \text{ft.} \quad \blacktriangleleft$$

Electrical *conductivity* is the reciprocal of electrical resistivity and is expressed in mho/cm (mho = ohm^{-1}). Sometimes it is convenient to express conductivity as the electrical charge flux per unit voltage gradient, V/d:

$$\sigma = 1/\rho = \text{flux}/(V/d); \tag{1-3}$$
$$\text{conductivity} = \frac{\text{coul}/(\text{cm}^2 \cdot \text{sec})}{\text{volt/cm}}.$$

Since

$$\text{coul} = \text{amp} \cdot \text{sec},$$
$$\text{conductivity} = \text{ohm}^{-1} \cdot \text{cm}^{-1}.$$

The relationship of conductivity to the number of carriers, the charge per carrier, and their mobility will be discussed in Chapters 14 and 15 along with the relationships between electrical conductivity and (1) temperature, (2) strain, (3) composition, and (4) thermal conductivity.

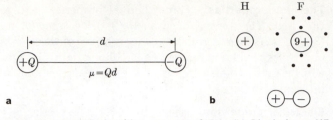

1-5 Electrical dipoles. (a) Example of a large dipole with a moment of Qd. (b) Dipole in an HF molecule. Since all the electrons cluster around the fluorine, leaving the hydrogen as an exposed proton, the centers of positive and negative charges are not coincident, and thus an electrical dipole is created.

Example 1-3

If a potential of 0.10 volt is placed across a 6-meter length of the copper wire described in Example 1-2, how many electrons will enter the wire per second at 0°C? (The charge on each electron is 1.6×10^{-19} coul.)

Answer. From Eq. (1-3),

$$
\begin{aligned}
\text{coul/sec} &= (\text{conductivity})(\text{volt/cm})(\text{cm}^2) \\
&= (\text{volt/cm})(\text{cm}^2)/(\text{resistivity}) \\
&= (\text{volt/cm})(\text{cm}^2)/(\text{ohm} \cdot \text{cm}) \\
&= \frac{(0.10 \text{ volt}/600 \text{ cm})(\pi/4)(0.022 \text{ in.})^2(2.54 \text{ cm/in.})^2}{(1.54 \times 10^{-6} \text{ ohm} \cdot \text{cm})} \\
&= 0.266 \text{ coul/sec};
\end{aligned}
$$

$$
\begin{aligned}
\text{electrons/sec} &= \frac{0.266 \text{ coul/sec}}{1.6 \times 10^{-19} \text{ coul/electron}} \\
&= 1.66 \times 10^{18} \text{ electrons/sec.} \blacktriangleleft
\end{aligned}
$$

Polarization, an important concept for understanding the structures and properties of certain materials, is commonly defined as the dipole moment per unit volume. A *dipole moment* is developed within a material when the centers of positive and negative charges are not coincident. Assume, for example, that one coulomb of negative charge, $-Q$, has been introduced on the right end (and a one-coulomb positive charge, Q, on the left end) of the one-meter dipole (literally, 2 poles) shown in Fig. 1-5(a). Within an electric field at right angles to the dipole, a moment, or torque, of 100 coul · cm is developed.

The various types of dipole moments present within a material are associated with atoms, ions, or molecules [see Fig. 1-5(b) and Chapter 13]. Although their individual magnitudes are very small, the numbers of atomic dipoles per unit volume are large and provide a significant polarization. This will be incorporated into the calculation of Example 2-5.

A third electrical characteristic of significance is the *dielectric constant*. In Chapter 13 we shall relate this constant to polarization. In the meantime, it is sufficient to indicate that the dielectric constant is the proportionality constant between the electric charge density, $\mathfrak{D}$, in coul/cm^2 and the field strength or voltage gradient, $\mathcal{E}$, in volts/cm. This proportionality constant is usually factored into a dielectric constant for a vacuum, ϵ_0, and the *relative dielectric constant*, κ: thus,

$$\mathfrak{D} = \epsilon_0 \kappa \mathcal{E}, \tag{1-4}$$

where $\epsilon_0 = 8.854 \times 10^{-14}$ coul/volt · cm when centimeters are used (or $\epsilon_0 = 8.854 \times 10^{-12}$ farad/m, since a farad equals 1 coul/volt). The relative dielectric constant is a useful property for comparing insulators. For example, a capacitor with an insulating spacer of κ equal to 10 will hold 10 times the charge of a comparable capacitor with no spacer; i.e.,

$$\kappa = C/C_0, \tag{1-5}$$

where C and C_0 are the capacitances with and without dielectric material present.

Example 1-4

A capacitor designed to use wax-paper spacers (relative dielectric constant, $\kappa = 1.75$) between aluminum-foil electrodes has a capacitance of 0.013 microfarad. A plastic film ($\kappa = 2.10$) with the same thickness is being considered as a substitute for the paper. If other factors are equal, what would be the new capacitance of the capacitor?

Answer. From Eq. (1-5),

$$\frac{\kappa_{\text{plastic}}}{\kappa_{\text{paper}}} = \frac{C_{\text{plastic}}/C_0}{C_{\text{paper}}/C_0} ;$$

$$C_{\text{plastic}} = \left(\frac{2.10}{1.75}\right)(0.013) = 0.0156 \text{ microfarad.} \quad \blacktriangleleft$$

Consideration of other electrical properties will be deferred until later chapters.

1-4 ENERGIES OF MATERIALS

An important feature of our physical universe is the fact that the stability of matter increases when its energy is reduced. In many situations we assume this automatically; e.g., a pencil has greater potential energy and less stability when standing on end than when lying horizontally on a table. In other cases the relationship may be deduced: natural gas burns with air because it releases energy (heat). The resulting products, CO_2 and H_2O, have less chemical energy than the original hydrocarbons and oxygen.

Energy considerations are also important in the internal structures of materials. Bonds are stronger between those atoms which require a large amount of energy for separation (Chapter 3). Conversely stated, the most stable bonds are those which release the most energy during formation. Likewise electrons tend to assume low-energy orbits except when they are specifically energized by external sources (Chapter 2). With time, however, even the energized electrons return to available lower-energy, more stable states.

Potential and kinetic energies. A material possesses potential energy by virtue of its location. The reader is familiar with potential energy which arises from gravitational forces. Another source of potential energy is the coulombic forces which arise from electric fields; specifically, unlike charges attract and like charges repel. Energy changes which occur when these forces are integrated over a distance will be the subject of Chapter 3.

Kinetic energy, or energy of motion, is as important within materials as within machines. For example, gas pressures arise from the kinetic energy possessed by the atoms or molecules. Also, atoms within solids are not static but are in continuous oscillation as a result of thermal agitation. The consequences of such motion on properties are significant and will be the basis of considerable study in later chapters.

Selected thermodynamic relationships. The *enthalpy*, H (loosely called the *heat content*), of an unaltered material increases with temperature:

$$H_2 = H_1 + \int_{T_1}^{T_2} C_p \, dT, \tag{1-6}$$

where H_1 is the enthalpy at the initial temperature and C_p is the *heat capacity* at constant pressure. Enthalpy may be divided in two manners. One is as a combination of the internal energy of the material, E, and the PV-work by the material against its surroundings:

$$H = E + PV \tag{1-7a}$$

(P is pressure and V is the volume of the material). The other is as a combination of the *free energy*, F, of the material and the TS-energy associated with internal disorder:

$$H = F + TS \tag{1-7b}$$

(T is absolute temperature, and S is the entropy of the material). Entropy is a measure of internal disorder, and like enthalpy and free energy, values of it are widely available in thermodynamic tables.

The relationships of Eqs. (1-6) and (1-7b) are summarized in Fig. 1-6. Since the concept of free energy will be very helpful to us in subsequent discussions of materials and their stability, several observations from this figure will be useful. (1) At absolute zero, $H = F$. (2) Free energy, the energy which must be considered for structural changes within materials, decreases with increased temperature. (3) The rate of decrease may be related to the above entropy as

$$dF/dT = -S, \tag{1-8}*$$

and further, since entropy always has a positive value and always increases with

---

* To show this we must differentiate Eq. (1–7b) with temperature,
$$dH/dT = dF/dT + T \, dS/dT + S, \tag{1–7c}$$
and substitute the two basic thermodynamic relationships
$$dH/dT = C_p \tag{1–9a}$$
and
$$dS/dT = C_p/T. \tag{1–9b}$$

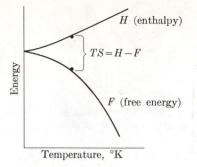

Energy

H (enthalpy)

$TS = H - F$

F (free energy)

Temperature, °K

1-6 Enthalpy and free energy. The heat content (i.e., enthalpy) increases at higher temperatures. However, the amount of energy required for thermal disorder (that is, *TS*) increases more markedly. Therefore, less free energy is available for reactions as the temperature is raised. Equation (1-7b) applies.

temperature, the slope of the free-energy curve gets ever steeper. These free-energy values will become important to us whenever phase changes occur, because the phase with the lower free energy will become the stable phase (Sections 3-8, 4-6, and 17-5).

Example 1-5

The constant-volume heat capacity, C_v, of quartz (silica sand) between 298°K and 848°K is

$$C_v = (11.22 + 8.20 \times 10^{-3}T - 2.70 \times 10^5/T^2) \text{ cal/mole} \cdot {}°\text{K}.$$

Since the volume changes in a solid are usually small, the constant-pressure heat capacity, C_p, is essentially identical to C_v.

What change in enthalpy occurs when the temperature of 10 lb of quartz is reduced from 145°C to 75°C?

Answer

$$\text{SiO}_2 = 60.1 \text{ gm/mole,}$$

$$H_{348°\text{K}} - H_{418°\text{K}} = \int_{418}^{348} (11.22 + 8.20 \times 10^{-3}T - 2.70 \times 10^5/T^2) \, dT$$

$$= 11.22(348 - 418) + 4.10 \times 10^{-3}(348^2 - 418^2)$$

$$+ (2.70 \times 10^5)\left(\frac{1}{348} - \frac{1}{418}\right)$$

$$= -866 \frac{\text{cal}}{60.1 \text{ gm}}$$

$$\Delta H_{10 \text{ lb}} = -(14.4 \text{ cal/gm})(10 \text{ lb})(454 \text{ gm/lb})$$

$$= -66,000 \text{ cal.}$$

In other words, 66,000 cal of energy are removed in the cooling process. ◀

Metastability. Although the lowest-energy condition (or state) of a material is the equilibrium state, we find situations where higher-energy states remain for extended periods of time in metastable equilibrium. Consider, for example, the 5-lb blocks of Fig. 1-7 in two possible positions. Block (c) is more stable than block (a), since its center of gravity is lower by two inches. The potential energy

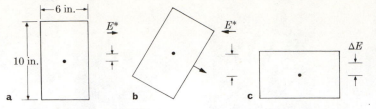

1-7 Activation energy. Although state (c) has less energy than state (a), the block cannot spontaneously move from (a) to (c) without an initial input of energy, E^*, to activate the change.

of block (a) would be reduced 10 in-lb if it could be changed to this lower position. We find, however, that if undisturbed, block (a) will remain indefinitely in its present position; it is metastable. To have its potential energy reduced by 10 in-lb, block (a) has to be raised and tipped with the input of some energy (Fig. 1-7b). Eventually this additional energy is released, but an energy input is necessary before the "reaction" can occur. This supplemental energy is called *activation energy*, E^*, and is widely encountered in atomic processes. We shall see in the next chapter that within a material this activation energy is commonly supplied by thermal energy.

Example 1-6

a) Calculate the minimum activation energy, E^*, required for the change in states for a 5-lb block from that of Fig. 1-7(a) to that of Fig. 1-7(c).
b) Calculate the minimum activation energy required for a change in the reverse direction.

Answer

a) The center of gravity in Fig. 1-7(a) is 5 in. above the base. The center of gravity in Fig. 1-7(b) is $\sqrt{34}$ in. above the base. Thus

$$\underrightarrow{E^*} = (5.83 \text{ in.} - 5.00 \text{ in.})(5 \text{ lb}) = 4.15 \text{ in-lb.}$$

b) The center of gravity in Fig. 1-7(c) is 3 in. above the base. Thus

$$\underleftarrow{E^*} = (5.83 \text{ in.} - 3.00 \text{ in.})(5 \text{ lb}) = 14.15 \text{ in-lb.}$$

Note. $\underrightarrow{\Delta E} = -10$ in-lb; $\underleftarrow{\Delta E} = +10$ in-lb. The minus sign indicates a decrease in energy of the system, whereas the plus sign indicates an increase in energy. ◄

REFERENCES FOR FURTHER READING

Appendix C. "Properties of Selected Engineering Materials."

Avner, S. H., *Introduction to Physical Metallurgy*. New York: McGraw-Hill, 1964. The laboratory tools used by materials engineers are discussed in Chapter 1. Freshman level.

Baker, W. O., "Solid State Science and Materials Development," *Journal of Materials*, **2** [4], 1967, pp. 917–963. An excellent presentation of properties versus structure for the instructor or other reader who already has some familiarity with materials.

Brady, G. S., *Materials Handbook*. New York: McGraw-Hill, 1951. This book has a paragraph or two describing each of a thousand or more types of materials.

Keyser, C. A., *Materials Science in Engineering*. Columbus, O.: Merrill, 1968. Chapter 2 provides a discussion of mechanical properties pertinent to the behavior of materials. Introductory undergraduate level.

Keyser, C. A., *Materials Testing*. Columbus, O.: Merrill, 1968. A laboratory text for materials testing. Twelve experiments with emphasis on mechanical properties and heat treatment.

McCabe, C. L., and C. L. Bauer, *Metals, Atoms, and Alloys*. Washington: National Science Teachers Association, 1964. Introductory background for metals. Nontechnical high school level.

Mitchell, L., *Ceramics: Stone Age to Space Age*. Washington: National Science Teachers Association, 1963. Introductory review for ceramics. Nontechnical high school level.

Richards, C. W., *Engineering Materials Science*. San Francisco: Wadsworth, 1961. A thorough discussion of mechanical properties. Advanced undergraduate level.

Rogers, B. A., *The Nature of Metals*, second edition. Metals Park, O.: American Society for Metals, 1964. An excellent introductory book for the prospective student.

Smith, C. S., "Materials," *Scientific American*, **217** [3], September 1967, pp. 68–79. Lead article of a special issue featuring materials. Reviews the history of materials. Introductory level.

Van Vlack, L. H., *Elements of Materials Science*, second edition. Reading, Mass.: Addison-Wesley, 1964. The engineering requirements of materials are presented in Chapter 1. Freshman-sophomore level.

PROBLEMS*

1-1 A brass has a modulus of elasticity (Young's modulus) of 1.6×10^7 psi. A 42.5-in. rod must not elongate more than 0.04 in. What diameter must the rod be if it has to carry a tensile load of 4000 lb?

Answer. 0.58 in.

1-2 A copper alloy has the following properties:

Modulus of elasticity	17,000,000 psi
0.2% yield strength (Fig. 1–3a)	17,000 psi
Breaking strength (original area)	17,000 psi
Elongation (2-in. gage)	17%
Reduction in area	17%

a) What is the true breaking strength (in psi)?
b) What cross-sectional area (in²) is required to support a load of 1700 lb with an elastic strain of 0.00017?
c) What is the total strain at the 0.2% yield stress?

---

* Answers cannot be more precise than the data.

1-3 a) A vertical rod of iron is cooled from 50°C to —5°C. What is the percent decrease in length?

   b) It is loaded at the lower temperature until it has returned to its original length. What stress is required? (See Appendix C for data.)

Answer. a) 0.065%   b) 20,000 psi

1-4 An aluminum bar 0.505 in. in diameter has 2-in. gage marks. The following data are obtained:

Load, lb	Gage length, in.
2000	2.0020
4000	2.0039
6000	2.0061
8000	2.193

   a) Draw a stress-strain curve.

   b) What is the bar's modulus of elasticity?

1-5 The maximum load in Problem 1-4 is 8230 lb and the breaking load is 3630 lb, with diameter of 0.261 in.

   a) What is the tensile strength?

   b) The true tensile strength (= true breaking strength)?

   c) The true strain at the point of fracture?

   d) The elongation, if the final gage length is 2.63 in.?

Answer. a) 41,000 psi   b) 68,000 psi   c) 132%   d) 31.5%

1-6 Eight thousand feet of wire are to be purchased and cut into shorter lengths. The wire must carry loads of 12 lb without permanent deformation, and have a resistance of less than 0.01 ohm/ft. Choose between annealed copper wire and annealed 70-30 brass wire.

	Annealed copper	Annealed 70-30 brass
Density, gm/cm³	8.9	8.5
Yield strength, psi	4,100	10,000
Tensile strength, psi	32,000	46,000
Hardness, R_b	40	61
Cost, cents/lb	c	0.9c
Resistivity, ohm · cm	1.7×10^{-6}	6.2×10^{-6}
Modulus of elasticity, psi	16×10^6	16×10^6
Resistivity, ohm · in.	0.67×10^{-6}	2.44×10^{-6}

1-7 An electric oven 2.0 ft × 2.0 ft × 2.0 ft has 0.5 in. of glass wool insulation on all sides.

   a) If the oven is maintained at 450°F, at what rate will heat be lost to the surrounding atmosphere (70°F) as a result of conduction through the walls?

   b) What length of heater wire 0.06 in. in diameter, having an electrical resistivity of 80×10^{-6} ohm · cm, and heated continuously by 110-volt ac current would be required to make a heater that would just balance this heat loss?

Note. See Appendix C for useful data; 1 watt = 1 joule/sec = 0.0143 Kcal/min.

Answer. a) 2220 cal/sec   b) 10 ft.

1-8 a) If a pure copper wire (resistivity $= 1.7 \times 10^{-6}$ ohm · cm) 0.04 in. in diameter is used for an electrical circuit carrying 10 amp, how many watts of heat will be lost per foot?

b) How many more watts would be lost if the copper wire were replaced by a brass wire of the same size (resistivity $= 3.2 \times 10^{-6}$ ohm · cm)?

1-9 The average coefficient of thermal expansion of a steel rod is 6.0×10^{-6} in./in · °F.

a) How much temperature change is required to provide the same linear change as a stress of 90,000 psi?

b) What volume change does this temperature change produce?

Answer. a) 500°F   b) 0.9%

1-10 A wall is 5 in. thick and has a thermal conductivity of 1.44 Btu · in./°F · hr · ft².

a) What is the rate of heat loss through the wall in Btu if the inside is at 80°F and the outside is at 20°F?

b) What is the average temperature gradient in the wall in metric units?

1-11 The dielectric constant of a glass ribbon is 5.1. Would a capacitor using such a glass ribbon 0.01 in. thick have greater or less capacitance than another similar capacitor using a 0.005-in. plastic with a dielectric constant of 2.1? (Capacitance is inversely proportional to the spacing between the electrodes.)

Answer. Greater by 20%.

1-12 A file case which weighs 160 lb is 52 in. high, sitting on a 15 in. × 27 in. base.

a) Assuming its center of gravity is at the center of volume, how much activation energy is required to tip it over in the two principal directions?

b) In the reverse directions, to return it to the upright position?

2 □ structure and energy of atoms

2-1 SINGLE ATOMS

This section on single atoms is primarily a review of previously encountered chemistry and physics. It is included here because certain features of single atoms have special significance when more than 10^{22} atoms are associated in each cubic centimeter of an engineering material.

Atomic numbers and atomic weights. The nucleus of each element has neutrons and a unique number of protons ranging from one for hydrogen to more than 100 for the transuranium elements (Fig. 2-1). The atomic number is equal to the number of protons, and this in turn is equal to the number of surrounding electrons in an electrically neutral atom. These electrons have a major effect on electrical and chemical properties (see Chapters 13 through 16 and 22).

Since the mass of an electron is only about 0.0005 times that of either the proton or the neutron, the mass of an atom is nearly proportional to the total number of protons plus neutrons. Carbon-12, with six neutrons and six protons, has been chosen as the mass reference for atoms. It is taken as 12.0000 atomic mass units (amu). More commonly we refer to the atomic weight of C^{12} as being 12.0000 gm/mole, where a mole has *Avogadro's number* of atoms, specifically, 0.602×10^{24}. Thus one amu is equal to 1.66×10^{-24} gm.

The atomic weight of an element with more than one isotope is proportional to the fraction of each isotope present.

Example 2-1

Natural carbon contains 98.89 a/o* C^{12} and 1.11 a/o C^{13}. The masses of these two isotopes are 12.0000 and 13.0033 amu, respectively. What is the atomic weight of natural carbon?

Answer. Basis: 10,000 atoms (9889 C^{12} and 111 C^{13}).

$$9889 \times 12.0000 = 118{,}668 \text{ amu}$$
$$111 \times 13.0033 = \underline{1{,}443 \text{ amu}}$$
$$120{,}111 \text{ amu}/10{,}000 \text{ atoms}$$
$$\text{Atomic weight} = 12.0111 \text{ amu.} \blacktriangleleft$$

---

* The symbol a/o will be used for atom percent; w/o for weight percent; m/o for mole percent; v/o for volume percent; etc.

2-1 Periodic table of elements. The atomic number (upper) of each element equals the number of protons or electrons. The atomic weight (lower) of each element is based on $C_{12} = 12.0000$.

— Metals —   — Nonmetals —

IA	IIA	IIIB	IVB	VB	VIB	VIIB	VIII	VIII	VIII	IB	IIB	IIIA	IVA	VA	VIA	VIIA	0
1 H 1.00797																	2 He 4.0026
3 Li 6.939	4 Be 9.012											5 B 10.811	6 C 12.011	7 N 14.007	8 O 15.9994	9 F 18.998	10 Ne 20.183
11 Na 22.990	12 Mg 24.312											13 Al 26.98	14 Si 28.086	15 P 30.97	16 S 32.064	17 Cl 35.453	18 Ar 39.95
19 K 39.102	20 Ca 40.08	21 Sc 44.96	22 Ti 47.90	23 V 50.94	24 Cr 52.00	25 Mn 54.94	26 Fe 55.85	27 Co 58.93	28 Ni 58.71	29 Cu 63.54	30 Zn 65.37	31 Ga 69.72	32 Ge 72.59	33 As 74.92	34 Se 78.96	35 Br 79.91	36 Kr 83.80
37 Rb 85.47	38 Sr 87.62	39 Y 88.91	40 Zr 91.22	41 Nb 92.91	42 Mo 95.94	43 Tc 99	44 Ru 101.07	45 Rh 102.91	46 Pd 106.4	47 Ag 107.87	48 Cd 112.40	49 In 114.82	50 Sn 118.69	51 Sb 121.75	52 Te 127.60	53 I 126.90	54 Xe 131.30
55 Cs 132.90	56 Ba 137.34	57–71 La series*	72 Hf 178.49	73 Ta 180.95	74 W 183.85	75 Re 186.2	76 Os 190.2	77 Ir 192.2	78 Pt 195.1	79 Au 196.97	80 Hg 200.59	81 Tl 204.37	82 Pb 207.19	83 Bi 208.98	84 Po 210	85 At 210	86 Rn 222
87 Fr 223	88 Ra 226	89– Ac series†															

*La series (lanthanides):

58 Ce 140.12	59 Pr 140.91	60 Nd 144.24	61 Pm 147	62 Sm 150.35	63 Eu 151.96	64 Gd 157.25	65 Tb 158.92	66 Dy 162.50	67 Ho 164.93	68 Er 167.26	69 Tm 168.93	70 Yb 173.04	71 Lu 174.97

†Ac series (actinides):

90 Th 232.04	91 Pa 231	92 U 238.03	93 Np 237	94 Pu 239	95 Am 241	96 Cm 242	97 Bk 249	98 Cf 252	99 Es 254	100 Fm 253	101 Md	102 No	103 Lw

We note here that C^{13}, with seven neutrons and six protons, has slightly more than 13 amu. This apparent discrepancy arises from the energy of nuclear binding and belongs to the subject matter of nuclear physics. Its significance to us is simply that we cannot automatically calculate the atomic weight of an element on the basis of the number of neutrons and protons in the nucleus.

Atomic weights have a major effect on the density of solids, and some effect on heat capacity. Otherwise, atomic weights have relatively little direct influence on engineering properties.

Example 2-2

Argon has a density of 1.78 gm/liter at a pressure of 1 atm and a temperature of 0°C. How many atoms are there per cm^3?

Answer. From Appendix B, the atomic weight of argon is 39.95 gm/mole. Therefore,

$$\left(\frac{1.78 \text{ gm/liter}}{10^3 \text{ cm}^3/\text{liter}}\right)\left(\frac{0.602 \times 10^{24} \text{ atoms/mole}}{39.95 \text{ gm/mole}}\right) = 2.7 \times 10^{19} \frac{\text{atoms}}{\text{cm}^3}. \quad \blacktriangleleft$$

2-2 ELECTRONIC STRUCTURE OF ATOMS

Much of our discussion of atoms will be concerned mainly with the electron. Although its exact mass (9.107×10^{-28} gm) is not critical for our purposes, its charge of 1.6×10^{-19} coul is important, since electrical conductivity and electronic polarization are related to the number of electrons and their charge (Example 1-3). The charges on ions are integral multiples of the electron charge, because ions differ from atoms only through the gain or loss of electrons. This in turn determines the strength of ionic bonds (Chapter 3) and a number of related properties. A third property of electrons, the magnetic moment, will be important when we consider magnetism. Each electron has the characteristic of a spinning charge, with the consequence that it develops a magnetic moment, called a *Bohr magneton*, of 9.27×10^{-24} amp $\cdot$ m^2. This may be calculated, as indicated in the footnote to Table 2-1, from the charge, mass, and Planck's constant, h.

The electrons associated with each atom are not static but are in continuous wavelike motion. On a mathematical basis, this motion may be compared with standing waves which have characteristic frequencies (and harmonics). The mathematics of wave mechanics are more complex than we want to handle here; in brief, however, two conclusions are important to us: (1) electrons around atoms have only specific energy values, and (2) we can only indicate the probability of electron locations around and between atoms.

Quantum numbers. The electrons which surround an atomic nucleus do not all possess the same energy; therefore, we divide the electrons into shells, or groups with different energy levels. The first or lowest quantum shell contains a maximum of 2 electrons; the second shell contains a maximum of 8; the third, 18; and the fourth, 32. Thus the maximum number of electrons in a given shell is $2n^2$, where n is the principal quantum number of the shell.

TABLE 2-1

ELECTRON PROPERTIES

Property	Symbol	Common units
Mass (at rest)	m	9.107×10^{-28} gm
Charge	q	1.602×10^{-19} coul or 4.8×10^{-10} esu*
Magnetic moment (Bohr magneton†)	p_β	9.27×10^{-24} amp · m² or 9.27×10^{-21} erg/gauss

* 1 esu² = 1 erg · cm

† $p_\beta = qh/4\pi m = \dfrac{(1.60 \times 10^{-19} \text{ coul})(6.62 \times 10^{-34} \text{ joule · sec})}{4\pi(9.11 \times 10^{-31} \text{ kg})}$

joule = kg · m²/sec²
coul = amp · sec

Although the concept of the quantum shell is a convenient one and will be used frequently in succeeding sections, it is an oversimplification, because it implies that all electrons within a shell are equivalent. A more complete discussion of electrons is necessary if the properties of materials are to be understood. Without rigorous explanation, we can approach the subject by way of the *Pauli exclusion principle*, which states that no more than two interacting electrons in an atom can have the same orbital quantum number. Even these two electrons are not identical, because they possess opposite magnetic behaviors, or "spins." Since the electrons in a quantum shell do not all possess the same energy, as mentioned above, it is desirable to subdivide the shells into subshells. Here again, subshells with the lowest energy are filled before those with higher energy; and, as with the main quantum shells, the upper energy levels within a given subshell may be in a higher energy state than the lowest energy level in a succeeding subshell.

Electron notations. Experimental verification of the electronic groupings and sub-groupings was initially obtained through spectrometric studies, where it was concluded that a quantum of energy is required to move an electron from a lower energy level to the next higher energy level. Conversely, a quantum of energy (a photon) is released when an electron drops into a lower level. The energy, E, possessed by the photon may be calculated directly from the photon wavelength, λ:

$$E = hc/\lambda = h\nu, \qquad (2\text{-}1)$$

where h is Planck's constant, 6.62×10^{-34} joule · sec or 6.62×10^{-27} erg · sec, and c is the velocity of light, 2.998×10^{10} cm/sec. The wave frequency is c/λ or ν.

The sharpest spectral lines are produced by the electrons which drop to the lowest energy level within a given quantum shell. Early spectrographic experimenters used the letter "s" to describe the electrons that are at the lowest energy levels within each shell. A notation such as 1s² indicates that two electrons (of opposite magnetic spins) are located in the low energy position of the *first* quantum

TABLE 2-2
ELECTRON QUANTUM NUMBERS (H to Kr)

Element Symbol	Number	K (n = 1) 1s	L (n = 2) 2s	2p	M (n = 3) 3s	3p	3d	N (n = 4) 4s	4p	4d	4f	O (n = 5) 5s	5p	5d	5f
H	1	1													
He	2	2													
Li	3	2	1												
Be	4	2	2												
B	5	2	2	1											
C	6	2	2	2											
N	7	2	2	3											
O	8	2	2	4											
F	9	2	2	5											
Ne	10	2	2	6											
Na	11	2	2	6	1										
Mg	12	2	2	6	2										
Al	13	2	2	6	2	1									
Si	14	2	2	6	2	2									
P	15	2	2	6	2	3									
S	16	2	2	6	2	4									
Cl	17	2	2	6	2	5									
Ar	18	2	2	6	2	6									
K	19	2	2	6	2	6		1							
Ca	20	2	2	6	2	6		2							
Sc	21	2	2	6	2	6	1	2							
Ti	22	2	2	6	2	6	2	2							
V	23	2	2	6	2	6	3	2							
Cr	24	2	2	6	2	6	5	1							
Mn	25	2	2	6	2	6	5	2							
Fe	26	2	2	6	2	6	6	2							
Co	27	2	2	6	2	6	7	2							
Ni	28	2	2	6	2	6	8	2							
Cu	29	2	2	6	2	6	10	1							
Zn	30	2	2	6	2	6	10	2							
Ga	31	2	2	6	2	6	10	2	1						
Ge	32	2	2	6	2	6	10	2	2						
As	33	2	2	6	2	6	10	2	3						
Se	34	2	2	6	2	6	10	2	4						
Br	35	2	2	6	2	6	10	2	5						
Kr	36	2	2	6	2	6	10	2	6						

shell (i.e., the K-shell). Similarly, $2s^2$ indicates that two electrons are located in the lowest energy position, i.e., *orbital*, of the *second* quantum shell (L-shell). Two is the maximum number of electrons that can exist in the s-subshell.

Each succeeding shell has additional subshells, which are designated "p," "d," and "f." The maximum numbers of electrons in these subshells are 6, 10, and 14, respectively. Neon, which has its L-shell (or second quantum shell) completely filled, possesses the electron notation $1s^2 2s^2 2p^6$, which indicates that two electrons are in the first shell and eight in the second, the latter containing two in its s-subshell and six in its p-subshell. The extension of this notation scheme may be deduced from Table 2-2.

Example 2-3

With neutral atoms, two electrons go into the 4s-subshell before the 3d-subshell is filled; however, with ionization, the 4s-electrons are removed first. Show the electron notation for a single iron atom, and for ferrous and ferric iron ions.

Answer

$$Fe = 1s^2 2s^2 2p^6 3s^2 3p^6 3d^6 4s^2, \qquad (2\text{-}2)$$

$$Fe^{2+} = 1s^2 2s^2 2p^6 3s^2 3p^6 3d^6, \qquad (2\text{-}3)$$

$$Fe^{3+} = 1s^2 2s^2 2p^6 3s^2 3p^6 3d^5. \quad \blacktriangleleft \qquad (2\text{-}4)$$

We might pause at this point to compare the electronic orbitals of an uncharged vanadium atom and of an Fe^{3+} ion. Although each has 23 electrons, the ferric ion has 26 protons against only 23 for vanadium; therefore,

$$Fe^{3+} = 1s^2 2s^2 2p^6 3s^2 3p^6 3d^5,$$

$$V^0 = 1s^2 2s^2 2p^6 3s^2 3p^6 3d^3 4s^2. \qquad (2\text{-}5)$$

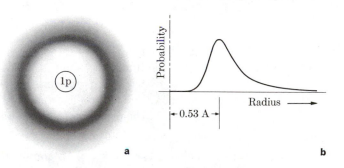

a

b

2-2 Hydrogen. The electron has the highest probability of being 0.53 A from the proton in a hydrogen atom. It is not in an orbit with a fixed radius.

Location probability. The exact location of an electron cannot be ascertained because any attempt to do so by light, x-rays, magnetic measurements, or other conceivable means would modify the wave characteristics of the electron. This means that we can only talk about probabilities of electron locations. Calculations which involve electron locations may be summarized as follows. For hydrogen, with one electron, the probability is greatest that the electron will be located at a radius of 0.53 A from the nucleus. The probability is smaller, but still finite, that the moving electron will be nearer, or more remote, at any instant of time (Fig. 2-2b). The two electrons of helium have the same type of distribution as the electron of hydrogen because they differ only in their spin orientations.

Lithium has an atomic number of 3. Its third electron goes into the second quantum shell, with a probable location as shown in Fig. 2-3. Next in order are beryllium, boron, carbon, nitrogen, oxygen, fluorine, and neon, each of which adds one more electron to the second quantum shell. However, as we noted earlier,

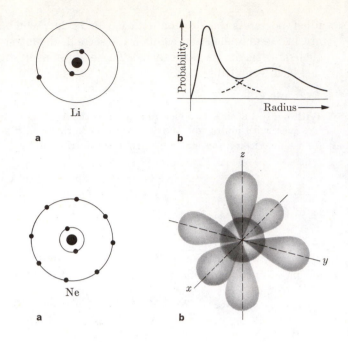

2-3 Lithium. (a) This simplified sketch shows a second quantum group because the first one has its maximum of two electrons. (b) The radial probability for electron location.

Li

a

Probability

Radius

b

2-4 Neon. (a) In this simplified sketch, the second quantum shell is filled. (b) The distribution probabilities for the second-shell electrons are indicated. The distribution probability is spherical for the first two electrons only. The remaining three pairs of electrons are in quantum subshells which have a high probability of being located along the three axes. This is a very stable electron arrangement.

Ne

a

b

only *two* electrons can have the same energy state, and hence the same location probability. Additional location or distribution probabilities therefore arise, and suborbitals are established. Figure 2-4 shows the distribution probabilities for the eight valence electrons of the second shell of neon. (The two electrons of the first quantum shell are subvalent.) This distribution of the eight electrons around an atom, whenever it occurs, is very stable.

3d-electrons. The 3d-electrons of the transition elements (Table 2-2) are important because they influence magnetic properties. In this chapter we will consider only single atoms, but in Chapter 16 these considerations will be adapted to solids with more than 10^{22} atoms per cm^3.

The electron distribution probabilities for the s- and p-electrons may be visualized as shown in the previous figures. Such visualization is more difficult for the 3d-electrons and will not be attempted; however, an important consequence of their distribution is expressed by Hund's rule, which states that there is a tendency for the 3d-electrons to align their magnetic spins. Figure 2-5 shows schematically the spin alignments within the first transition series. The limit of ten 3d-electrons permits as many as five electrons to align in one direction before the alignment starts in the opposite direction.

Magnetic moments. Since the major observable magnetic behavior* of an atom results from electron spins, the unbalance of spin orientation in the atoms of

---

* Electron orbits and nuclear spins make minor contributions to ferromagnetic behavior.

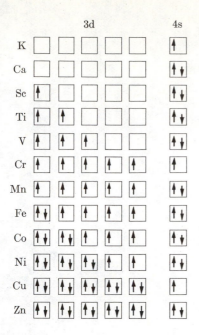

2-5 Spin alignment. The 3d-electrons of atoms will orient their spins in the same direction as far as possible. The second electron in each state assumes the opposite orientation.

Fig. 2-5 permits us to estimate the magnetic moment of an atom. In brief, an isolated titanium atom has a magnetic moment of

$$(2\uparrow - 0\downarrow)(9.27 \times 10^{-24} \text{ amp} \cdot \text{m}^2/\text{electron}), \quad \text{or} \quad 18.54 \times 10^{-24} \text{ amp} \cdot \text{m}^2/\text{atom}.$$

Likewise,

$$(5\uparrow - 2\downarrow)(9.27 \times 10^{-24} \text{ amp} \cdot \text{m}^2/\text{electron}), \quad \text{or} \quad 2.78 \times 10^{-23} \text{ amp} \cdot \text{m}^2/\text{atom}$$

is the magnetic moment arising from the electron spins of an isolated cobalt atom.

Although the reasons are complex, it is interesting to note that the full magnetic alignment of five electrons represents a reduced-energy (i.e., more stable) situation. Thus we find a $3d^54s^1$ electronic configuration rather than $3d^44s^2$ for chromium (Table 2-2). Likewise, copper has a $3d^{10}4s^1$ configuration, rather than $3d^94s^2$ as might be predicted on a casual basis.

2-3 IONS

The reader will observe that throughout this text we will first present an idealized structure and then give attention to nonideal or imperfect structures. Although the idealized structure may represent a lower-energy (i.e., more stable) structure, the altered structure is often more common. For example, a metal ion in which the numbers of protons and electrons are not equal is produced if energy is supplied to remove one or more electrons. Later (in Chapter 6) we shall see that a crystal structure may contain imperfections if a limited number of atoms are displaced

TABLE 2-3

IONIZATION POTENTIALS, eV*

	H		He
I	13.6	I	24.5
		II	54.1

	Li		Be
I	5.4	I	9.3
		II	18.2

	Na		Mg
I	5.1	I	7.6
		II	15.0

	K		Ca
I	4.3	I	6.1
		II	11.8

	Rb		Sr
I	4.2	I	5.7
		II	11.0

	Cs		Ba
I	3.9	I	5.2
		II	9.9

* $1 \text{ eV} = 1.602 \times 10^{-12}$ erg
$= 1.602 \times 10^{-19}$ joule

I First electron removed.
II Second electron removed.

or removed. Almost universally these imperfections have significant consequences for properties.

Metallic ions (cations). The major portion of the elements in Fig. 2-1 consists of metals. Metallic atoms have the characteristic of easy ionization; i.e., they give up electrons:

$$M \rightarrow M^{m+} + me^-, \qquad (2\text{-}6)$$

where m is a small integer. The energy required to remove electrons is called the *ionization potential*. Selected values are listed in Table 2-3. Those elements in the lower left corner of the periodic table are most readily ionized, requiring the least energy for electron removal. In every case, more energy is required to remove the second electron because its removal leaves a doubly charged ion.

For the sake of simplicity we can state that the electrical conductivity of liquid and solid metals arises from the easy separation of valence electrons from their parent atom. There are, however, additional considerations which must be incorporated into the explanation when the individual atom is modified by many neighboring atoms (Chapter 14).

TABLE 2-4

ELECTRONEGATIVITIES OF THE
REPRESENTATIVE ELEMENTS (From Mahan)

H 2.1						
Li 0.97	Be 1.5	B 2.0	C 2.5	N 3.1	O 3.5	F 4.1
Na 1.0	Mg 1.2	Al 1.5	Si 1.7	P 2.1	S 2.4	Cl 2.8
K 0.90	Ca 1.0	Ga 1.8	Ge 2.0	As 2.2	Se 2.5	Br 2.7
Rb 0.89	Sr 1.0	In 1.5	Sn 1.72	Sb 1.82	Te 2.0	I 2.2
Cs 0.86	Ba 0.97	Tl 1.4	Pb 1.5	Bi 1.7	Po 1.8	At 1.9

Anions. The elements in the upper right corner of the periodic table (Fig. 2-1) have a high attraction for electrons either through the formation of negative ions or through the covalent sharing of electrons. A qualitative index of this attraction is presented in Table 2-4 as *electronegativity*. Thus the reaction

$$X + me^- \rightarrow X^{m-} \tag{2-7}$$

is more readily obtained for halides and adjacent elements than for metals.

Because the anion receives additional electrons, its radius is increased; likewise, the cation radius is reduced. As a consequence, all but the heaviest of the cations are smaller than anions.

2-4 ELECTRON EXCITATION

An atom may have its electrons excited to higher energies without ionization. Specifically the electrons may be raised from their *ground state* of lowest energy to a higher level. Perhaps this idea is conveyed most simply by indicating the permissible energy states for an electron which is associated with the single proton of a hydrogen atom. When brought together after separation, the reaction

$$p^+ + e^- \rightarrow H \tag{2-8}$$

releases energy, E_n:

$$E_n = \frac{-13.6 \text{ eV}}{n^2}, \tag{2-9}$$

where n is an integer and may be related to the principal quantum numbers of

Table 2-2. Figure 2-6 shows these energy levels for different values of n. Note first that the energy is given off as the electron joins the atom. Second, the maximum amount of energy released is -13.6 eV and is equal to the energy required for ionization (Table 2–3). Third, the energy differences, ΔE, between levels are specific. Thus, as indicated in Eq. (2-1), a photon with a unique wavelength is released whenever an electron drops from one level to another. This is the basis of the spectrographic data cited in Section 2-2.

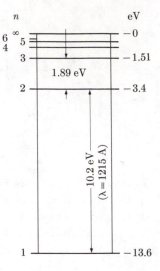

2-6 Excitation energy levels (hydrogen). Electrons can possess only the energies indicated (based on the energy at infinite separation). Photons are released when an electron drops from one level to another.

Example 2-4

A wavelength of 6562 A is observed in the hydrogen spectrum. Does this emission involve the energy level $n = 3$ in Fig. 2-6?

Answer. From Eq. (2-1),

$$E = hc/\lambda$$

$$= \frac{-13.6 \text{ eV}}{3^2} - \frac{-13.6 \text{ eV}}{n^2}.$$

Thus

$$\frac{(6.62 \times 10^{-27} \text{ erg} \cdot \text{sec})(2.998 \times 10^{18} \text{ A/sec})}{6562 \text{ A}}$$

$$= -(13.6 \text{ eV})(1.6 \times 10^{-12} \text{ erg/eV}) \left(\frac{1}{3^2} - \frac{1}{n^2} \right),$$

$$\frac{1}{3^2} - \frac{1}{n^2} = -0.1389,$$

$$n = 2.$$

The photon released when an electron moves from $n = 3$ to $n = 2$ has a wavelength of 6562 A.

Note. We assumed initially that $n < 3$. Had this assumption been incorrect, our answer would have appeared negative—an impossibility. In such a case, we would be required to reverse our assumption. ◄

2-5 GASES

Gas law. Noble gases such as He, Ne, and Ar, and other gases at sufficiently high temperatures or sufficiently low pressures, exhibit the relationship

$$PV = nRT, \tag{2-10}$$

where the letters stand for pressure, volume, number of moles, the gas constant, and absolute temperature, respectively. The gas constant, R, may be expressed in different units as listed in Table 2-5.

Example 2-5

An electric dipole moment of 1.43×10^{-36} coul · cm is induced in each argon atom when it is placed in an electric field of 100 volts/cm. What is the polarization (coul · cm per cm^3) when argon is compressed to 100 atm at 0°C within that electric field?

Answer. From Eq. (2-10),

$$n/V = P/RT = (100 \text{ atm})/(82.06 \text{ cm}^3 \cdot \text{atm/mole} \cdot °\text{K})(273°\text{K})$$
$$= 0.00447 \text{ mole/cm}^3;$$
$$\text{polarization} = (0.00447 \text{ mole/cm}^3)(0.602 \times 10^{24} \text{ atoms/mole})$$
$$\times (1.43 \times 10^{-36} \text{ coul} \cdot \text{cm/atom})$$
$$= 3.87 \times 10^{-15} \text{ coul} \cdot \text{cm/cm}^3. \quad ◄$$

Energy distributions. The thermal energy of atoms in a monatomic gas is entirely in the form of kinetic energy, KE, and is proportional to temperature. This statement may be formalized through the use of the gas constant, R, as

$$\text{KE} = \tfrac{3}{2}RT. \tag{2-11}$$

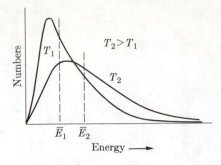

2-7 Energy distribution in a gas. Both the average energy and the fraction of atoms having energies in excess of a specified level increase as the temperature, T, is increased.

As such, the energy is usually expressed in cal/mole. Alternatively, we may express kinetic energy per atom as

$$\text{KE} = \tfrac{3}{2}kT, \tag{2-12}$$

where k is *Boltzmann's constant*, 1.38×10^{-16} erg/atom $\cdot$ °K.

The above equation does not imply that all the atoms within a gas have exactly the same energy. In fact, Boltzmann has shown that there is a statistical distribution of velocities, and therefore energies (since $\text{KE} = \tfrac{1}{2}mv^2$), as shown in Fig. 2-7. At any particular instant, very few atoms have nearly zero velocity and negligible energy; many have energies near the average value, $\overline{E}$; and some have extremely high energies as a result of high velocities. As the temperature increases, (1) both the average velocity and the average kinetic energy increase, and (2) the total number of atoms with energies in excess of any specified value increases. Of course, the integrated totals under each of the curves in Fig. 2-7 are equal and represent the sum of all atoms present.*

Our interest will be directed toward those atoms with very high energies, because they will be the ones which may possess the activation energy (Section 1-4) for a reaction, whereas the average energy is much less. Boltzmann showed that the fraction of atoms (n atoms out of a total number, N_{tot}) which exceed an energy E (Fig. 2-8) may be calculated as

$$\frac{n}{N_{\text{tot}}} = Me^{-(E-\overline{E})/kT}, \tag{2-13}$$

where M is a proportionality constant which may be determined by experiment, and k is Boltzmann's constant, already cited in Eq. (2-12). If $E \gg \overline{E}$, Eq. (2-13) approximates

$$\frac{n}{N_{\text{tot}}} \cong M/e^{E/kT}. \tag{2-14}$$

---

* In a gas where the atoms are independent, a continuous range of energies is possible, and nothing prevents several atoms out of the approximately 10^{20} per cm^3 from having identical energies. The significance of this footnote will become apparent in later chapters, where interacting atoms and electrons are discussed.

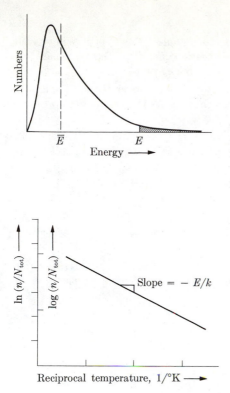

2-8 Molecular energies. The ratio of the number of high-energy atoms (shaded) to the total number of atoms is an exponential function of $-E/kT$ when $E \gg \bar{E}$.

2-9 Arrhenius plot. Many rate processes exhibit the relationship given in Eq. (2-16) because activation energies are exponential functions of temperature.

From this we may compare the value of n_1 at T_1 with that of n_2 at T_2:

$$\frac{n_1}{n_2} = \frac{e^{E/kT_2}}{e^{E/kT_1}}, \tag{2-15}$$

or

$$\ln \frac{n_1}{n_2} = \frac{E}{k}\left(\frac{1}{T_2} - \frac{1}{T_1}\right). \tag{2-16}$$

This is shown in Fig. 2-9 as an *Arrhenius plot*, a type of diagram we shall encounter with several modifications elsewhere in the text.

Example 2-6

The average energy, $\bar{E}$, at 100°C of helium gas is 7.7×10^{-14} erg per atom. Compare the numbers of helium atoms with energies in excess of 10^{-12} erg at 70°C and 100°C.

Answer. Consider $10^{-12} \gg 7.7 \times 10^{-14}$ erg. From Eq. (2-16),

$$\ln \frac{n_{100°C}}{n_{70°C}} = \frac{10^{-12} \text{ erg}}{1.38(10^{-16}) \text{ erg/°K}}\left[\frac{1}{343°K} - \frac{1}{373°K}\right] = 1.7,$$

$$\frac{n_{100°C}}{n_{70°C}} = 5.5. \blacktriangleleft$$

REFERENCES FOR FURTHER READING

Addison, W. E., *Structural Properties of Inorganic Materials*. New York: Wiley, 1961. Paperback. Chapter 1 has a nonmathematical presentation of electron theory of the atom. Recommended for the undergraduate who wants more background.

Hoffman, B., *The Strange Story of the Quantum*. New York: Dover, 1959. Paperback. Historical background of present-day quantum mechanics. Nonmathematical presentation for the educated layman.

Mahan, B., *University Chemistry*, second edition. Reading, Mass.: Addison-Wesley, 1969. This book provides the chemistry background assumed in this text. Freshman level.

Moffatt, W. G., G. W. Pearsall, and J. Wulff, *Structure and Properties of Materials: I. Structure*. New York: Wiley, 1964. Paperback. Electrons and bonding are discussed on an introductory level.

Ryschkewitsch, G. E., *Chemical Bonding and the Geometry of Molecules*. New York: Reinhold, 1963. Paperback. Chapter 2 gives an introductory background to the electronic structure of atoms.

Sisler, H. H., *Electronic Structure, Properties, and the Periodic Law*. New York: Reinhold, 1963. Paperback. Recommended as supplementary reading. General chemistry background is required.

PROBLEMS

2-1 a) What is the weight of an aluminum atom?
  b) The density of aluminum is 2.70 gm/cm^3. How many atoms are there per cm^3?

Answer. a) 4.48×10^{-23} gm/atom   b) 6.02×10^{22} atoms/cm^3

2-2 a) How many iron atoms are there per gram?
  b) What is the volume of a grain of metal containing 10^{20} iron atoms?

2-3 MgO has a density of 3.65 gm/cm^3. How many atoms will there be per cm^3?

Answer. 1.1×10^{23} atoms/cm^3

2-4 How many Bohr magnetons does a single Fe atom have? An Fe^{2+} ion? An Fe^{4+} ion?

2-5 Give the notation for the electronic structure of (a) silver, (b) iodine. (5p-electrons are formed before 4f-electrons.)

Answer. a) $1s^2 2s^2 2p^6 3s^2 3p^6 3d^{10} 4s^2 4p^6 4d^{10} 4f^0 5s^1$
  b) $1s^2 2s^2 2p^6 3s^2 3p^6 3d^{10} 4s^2 4p^6 4d^{10} 4f^0 5s^2 5p^5$

[*Note.* The $4f^0$ notation is commonly omitted.]

2-6 What is the maximum wavelength of a photon which can ionize cesium?

2-7 The temperature for Example 2-6 is lowered to 40°C. Compare the numbers of helium atoms with energies in excess of 10^{-12} erg at 40°C and 70°C.

Answer. $n_{70°C}/n_{40°C} = 7.6$

2-8 Indicate the number of 3d-electrons in each of the following: (a) V^{3+}; (b) V^{5+}; (c) Cr^{3+}; (d) Fe^{3+}; (e) Fe^{2+}; (f) Mn^{2+}; (g) Mn^{4+}; (h) Ni^{2+}; (i) Co^{2+}; (j) Cu^+; (k) Cu^{2+}.

2-9 It takes approximately 10^{-19} cal to break the covalent bond between carbon and nitrogen. What wavelength would be required of a photon to supply this energy? (See Appendix A for constants.)

Answer. 4750 A

2-10 An electron absorbs the energy from a photon of ultraviolet light ($\lambda = 2768$ A). How many eV are absorbed?

2-11 One gram of copper is ionized to Cu^{2+}. How many amperes are required to do this in one hour?

Answer. 0.84 amp

2-12 Solid neon has a density of 1.204 gm/cm^3. If the ideal gas law applied, what pressure would be required to give a gas the same density at 0°C?

3 □ atomic bonding and coordination

3-1 INTERATOMIC ATTRACTIONS

Engineering materials possess interatomic attractions which lead to atomic bonding. If there were no such attractions, each atom would behave independently. Furthermore, materials would have neither coherence nor resistance to externally applied forces.

It is convenient to identify four general types of interatomic bonds. The first three—ionic, covalent, and metallic—are called *primary bonds* because they are relatively strong. Each is a direct consequence of the exchange or sharing of valence electrons which lie in the s- and p-orbitals (Table 2-2 and Section 2-2). The fourth bond category—van der Waals—includes several variants of weaker, but important, attractive forces.

3-2 IONIC BONDS

Ions of unlike charges are attracted to one another by *coulombic forces*. The energy change, E_c, as two ions approach each other is given by the expression

$$E_c = \frac{(Z_1 q)(Z_2 q)}{a}. \qquad (3\text{-}1)$$

Here, a is the interionic distance; q is the electronic charge of 1.6×10^{-19} coul, or 4.8×10^{-10} esu; and Z_1 and Z_2 are the charges of the two ions and may be either positive or negative. As shown by Eq. (3-1), like charges require energy as they approach each other ($E_c > 0$), and unlike charges release energy as they approach each other.

We know from experiment that the interionic distances in ionic solids are seldom less than 2 A, yet Eq. (3-1) indicates that more energy is released as the value of a is reduced. Why can't the centers of ions be brought even closer together into a still more stable, lower-energy position? Since experiments show that the nucleus has a diameter of only about 10^{-13} cm (i.e., 10^{-5} A) and that electrons occupy no space *per se*, lack of space is not the problem. It is a matter of the repulsion of the electrons of the adjacent atoms at close distances. Another way to look at the matter would be to cite the Pauli exclusion principle, which states that no more than two electrons (with opposite spins) may occupy the same orbital (Section 2-2), as would have to occur if the interatomic distance were reduced still further. With either explanation, it may be shown that an *electron-repulsion energy* becomes significant as atoms approach one another. This energy, E_R, is inversely proportional to a high power ($6 < n < 12$) of the interatomic distance:

$$E_R = b/a^n. \tag{3-2}$$

Thus the total energy change as a pair of *unlike monovalent* ions are brought together is

$$E = -q^2/a + b/a^n, \tag{3-3a}$$

or, for the more general case,

$$E = Z_1Z_2q^2/a + b/a^n. \tag{3-3b}$$

In Eqs. (3-2) and (3-3), b is a constant which may be determined experimentally, and Z_1 and Z_2 are the respective charges.

The curves for Eqs. (3-1), (3-2), and (3-3) are shown in Fig. 3-1. The total-energy curve possesses a sharp minimum which locates the most stable *interatomic distance*, a', and energy is required for either a reduction or an increase in this distance. The *energy trough* of Fig. 3-1 will be used in a later discussion on compressibility and thermal expansion. We might note here, however, that the value $E_\infty - E_{min}$ represents the energy (heat) of sublimation, because that much energy must be supplied to separate two atoms or ions completely.

Example 3-1

A pair of oppositely charged divalent ions has an equilibrium distance, a', of 2.40 Å. Assuming that n of Eq. (3-3b) is 9, what energy is required to separate the ions?

Answer. From Eq. (3-3b),

$$dE/da = -Z_1Z_2q^2/a^2 - nb/a^{n+1},$$

and at $a' = 2.4$ A,

$$dE/da = 0.$$

Therefore,

$$4q^2/a^2 = 9b/a^{10},$$

$$b = \tfrac{4}{9}a^8q^2,$$

$$E_\infty - E_{min} = 0 - [-4q^2/a + (4a^8q^2/9a^9)]$$

$$= \frac{4q^2}{a}\left(1 - \frac{1}{9}\right) = \frac{32}{9}\frac{(4.8 \times 10^{-10})^2 \text{ esu}^2}{(2.4 \times 10^{-8} \text{ cm})}$$

$$= 34.3 \times 10^{-12} \text{ erg}$$

$$= 8.2 \text{ eV. } \blacktriangleleft$$

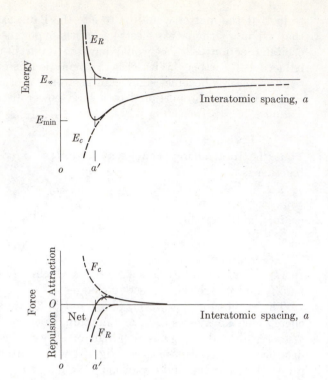

3-1 Potential energy trough. The solid line gives the sum of the energies: E_c due to the net coulombic forces (attraction) and E_R due to electronic overlap (repulsion). The equilibrium spacing is oa'.

3-2 Interatomic forces. At the equilibrium spacing, oa', the coulombic attractive forces, F_c, equal the electronic repulsive forces, F_R.

Interatomic forces. Since $F = dE/da$, the interatomic forces responsible for the energies between the *unlike monovalent* ions of Fig. 3-1 may be expressed as

$$F = q^2/a^2 - nb/a^{n+1}. \tag{3-4a}$$

Again, the more general case gives

$$F = -Z_1Z_2q^2/a^2 - nb/a^{n+1}. \tag{3-4b}$$

The two terms on the right represent the coulombic attraction force and the electron-repulsion force, respectively. These and the net force are presented as a function of interatomic distances in Fig. 3-2. The net force, of course, is zero at a' (and again at $a = \infty$). The slope, dF/da, of the net curve at a' is related to the modulus of elasticity, since the latter is the ratio, stress/strain. We shall return to this point in Chapter 10 when we consider elasticity.

Hard-ball model of atoms. The energy trough is very narrow for strongly bonded atoms. As a result, the slope, dF/da, of the force curve at a' in Fig. 3-2 is steep. This produces an interatomic distance which is constant to within 0.001 A at a given temperature, so that it is convenient to speak of atoms as hard balls with definite radii (Fig 3-3). We will do this often in succeeding chapters because such a model provides a very usable concept.

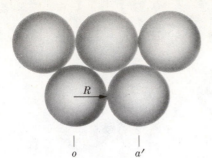

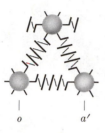

3-3 Hard-ball model. Since the average interatomic spacing between atoms normally varies by only a fraction of a percent, we find the hard-ball model of atoms to be useful for many concepts.

3-4 Spring model. Although the *average* interatomic spacing is constant, the atoms are in continual vibration; thus a spring model is useful in certain cases. The spring constant is nonlinear. The interatomic distance, *oa'*, corresponds to that in Figs. 3-1 and 3-2.

Spring model of atomic forces. A more accurate (but still not perfect) model of interatomic dimensions is provided in Fig. 3-4 by a spring. If the equilibrium distance is exceeded, the spring is under tension and the atoms are subject to a net attractive force. Conversely, a net repulsive force is developed if the atoms are forced together. It will be noted that the spring is nonlinear because the "spring constant" under compression increases rapidly, whereas it decreases under tension. (Observe dF/da of Fig. 3-2.) The spring model has the disadvantage of greater complexity than the hard-ball model, but it does permit us to depict the effect of atomic vibrations, thermal expansion, activation energies, and several other properties.

Coordination numbers. Coulombic forces are nonspecific; i.e., a positive (or negative) ion will produce attraction for *all* oppositely charged ions in the vicinity according to Eq. (3-1). This is in contrast to covalent bonds, which will be discussed in the next section, and this contrast produces important differences in the mechanical and electrical properties of the two kinds of bonds.

Because coulombic attractions are nonspecific, ions tend to maximize their coordination with neighboring ions. As a result, a large positive ion can have more negative neighbors than a small positive ion. We see evidence of this in the chlorides. A cesium ion with a radius of more than 1.6 A (hard-ball model) is large enough to have eight chlorine ion neighbors without chlorine ion contact,

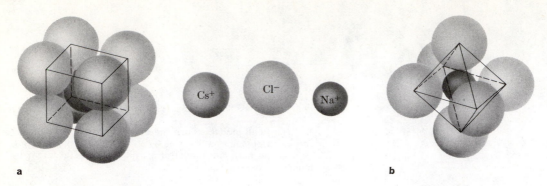

a

Cs⁺   Cl⁻   Na⁺

b

3-5 Coordination numbers. (a) Eight Cl⁻ ions can be coordinated with a Cs⁺ ion. (b) A maximum of only six Cl⁻ ions can surround each Na⁺ ion.

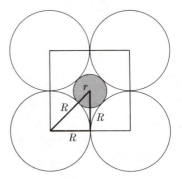

3-6 Coordination limit. If the ratio r/R is less than 0.414, it is impossible to coordinate six or more large ions with each small ion. (See Example 3-2 and Table 3-1.)

i.e., without major electronic repulsions (Fig. 3-5a). In contrast, a sodium ion with a radius of slightly less than 1 A can be coordinated with only six chlorine ions, as shown in Fig. 3-5(b). The smaller sodium ion could not have contact with eight Cl⁻ neighbors without concurrent contact, i.e., electron repulsion, between the negatively charged chlorine ions. We speak of the number of contacting neighbors as the *coordination number*.

Example 3-2

Ionic fluorine (F⁻) has a radius of 1.33 A. What is the radius of the smallest positive ion which can be coordinated with six neighboring fluorine ions?

Answer. View Fig. 3-6 as the limiting case of Fig. 3-5(b) in which the anions "touch":

$$2(r + R)^2 = (2R)^2;$$

therefore,

$$r = 0.414R,$$

and since $R_{F-} = 1.33$ A,

$$r = 0.55 \text{ A.}$$

Note. Related calculations can be made to provide the other values of Table 3-1. ◄

Geometric calculations relating the coordination number (CN) to the radii ratios are summarized in Table 3-1. As minimum values, these ratios are ex-

3-7 Hydrogen molecule. (a) The two protons share two electrons of opposite spins. (b) Simplified presentation (generally used in this text). (c) Probability contours for electron distributions. The electrons are between the hydrogen atoms the major part of the time.

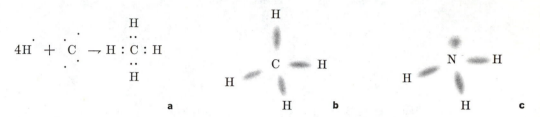

$$4H \;+\; \overset{\displaystyle \cdot}{\underset{\displaystyle \cdot}{C}} \;\rightarrow\; \overset{\displaystyle H}{\underset{\displaystyle H}{H : \overset{\displaystyle \cdot\cdot}{\underset{\displaystyle \cdot\cdot}{C}} : H}}$$

a b c

3-8 Covalent bonds in molecules. (a and b) Two presentations of methane, CH_4. (c) Ammonia, NH_3. (Cf. Table 3-2.)

TABLE 3-1
IONIC COORDINATION VERSUS SIZE RATIO

Coordination number	Minimum ratio of ionic radii, r/R*
3-fold	0.155
4	0.225
6	0.414
8	0.732
12	1.0

* Smaller ion—r; larger ion—R.

tremely limiting; e.g., an ion ratio, r/R, of 0.40 would not permit CN = 6. By contrast, values of CN = 4 can occur when $r/R > 0.414$, particularly when bonds are not wholly ionic but partially covalent.

3-3 COVALENT BONDS

The term "covalent" describes the sharing of valence electrons by adjacent atoms. This is exemplified most readily by one of several presentations of the hydrogen molecule in Fig. 3-7. Consistent with the Pauli exclusion principle (Section 2-2), two electrons may have the same energy values and wave behavior if they have opposite spins. Two such electrons have a high probability of being located between the atoms, as shown in Fig. 3-7(c) for hydrogen and again in Fig. 3-8 (a and b) for the methane molecule. As an oversimplification, the covalent bond in a hydrogen molecule may be considered to arise from the attraction of positive ions to the intervening pair of electrons with opposite spins ($+\; \rightleftharpoons \;+$).

TABLE 3-2
COVALENT COORDINATION VERSUS PERIODIC GROUP

Elements	Coordination number	Periodic group
H	1	—
He	0	—
C, Si, Ge	4	4
N, P, As, Sb	3	5
O, S, Se, Te	2	6
F, Cl, Br, I	1	7
Ne, Ar, Kr	0	8

TABLE 3-3
BOND ANGLES

Methane	CH_4	H—C—H	109.5°
Ethane	C_2H_6	H—C—H	109.3°
Chloromethane	CH_3Cl	H—C—H	110. °
Ammonia	NH_3	H—N—H	107.3°
Water	H_2O	H—O—H	104.5°
Hydrogen peroxide	H_2O_2	O—O—H	100. °
Hydrogen sulfide	H_2S	H—S—H	93.3°
Diamond	C	C—C—C	109.5°
Propylene	$H_2C{=}C\begin{smallmatrix}CH_3\\\\H\end{smallmatrix}$	C=C—C	124.7°
Carbonate ion	$CO_3^=$	O—C—O	120. °

The number of bonds per atom is reviewed for the reader in Table 3-2. We will use the convenient presentations of Fig. 3-9 (a and b) to show double bonds, but the reader must realize that a description of the electron probabilities within a double bond is appreciably more complex.*

Directionality of bonds. The covalent bond involves a pair of shared electrons between specific atoms. Thus while coulombic attractions arise from an electric field in *all* directions, a covalent bond is *stereospecific*; i.e., it has a specific direction in space. One consequence of stereospecificity is shown in Fig. 3-8(b), where carbon has a coordination of only 4 in spite of the fact that the surrounding space could contain many more of the small hydrogen atoms (protons). *The coordination numbers for covalently bonded atoms are not controlled by radii ratios.*

A second consequence of stereospecificity arises because strong covalent attractions occur only between first neighbors. Thus a carbon atom of one methane

---

* The chemist refers to σ and π bonds in compounds. See References for Further Reading.

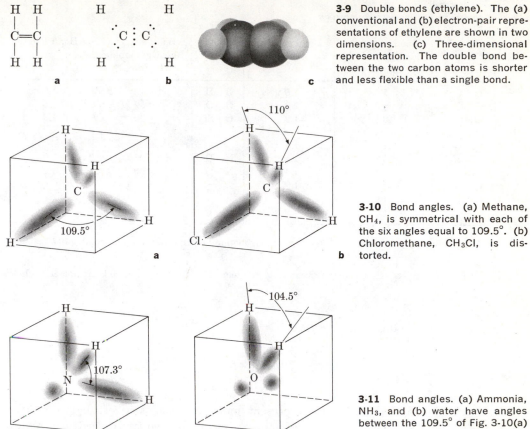

3-9 Double bonds (ethylene). The (a) conventional and (b) electron-pair representations of ethylene are shown in two dimensions. (c) Three-dimensional representation. The double bond between the two carbon atoms is shorter and less flexible than a single bond.

3-10 Bond angles. (a) Methane, CH_4, is symmetrical with each of the six angles equal to 109.5°. (b) Chloromethane, CH_3Cl, is distorted.

3-11 Bond angles. (a) Ammonia, NH_3, and (b) water have angles between the 109.5° of Fig. 3-10(a) and the 90° of Fig. 2-4(b).

molecule has only minor attraction to atoms of other molecules. The low melting and boiling temperatures of methane give evidence of this (-183°C and -151°C, respectively).

Bond angles. The directionality of covalent bonds also produces characteristic bond angles (Table 3-3). In methane, for example, the average H—C—H bond angle is 109.5°, as shown in Fig. 3-10(a). (The word "average" is included because the atoms of a molecule are in a continuous state of vibration, with frequencies of approximately 10^{13}/sec; hence, at any one instant of time, the true angle may be different.)

This tetrahedral bond angle of 109.5° is encountered in engineering materials whenever four equal atoms are bonded to a central atom (e.g., CH_4, CCl_4, SiF_4, diamond, silicon metal, SiC, etc.). Often, however, the molecules are not symmetrical. For example, chloromethane (CH_3Cl) is distorted slightly, as shown in Fig. 3-10(b), because one of the hydrogen atoms of methane is replaced by a chlorine atom with more electrons. The H—C—Cl angle decreases slightly, and the H—C—H bond angles are increased to 110°.

TABLE 3-4

BOND ENERGIES IN MOLECULES

Bond	Bond energy, kcal/gm · mole (approx.)*	Bond length, A	Bond	Bond energy, kcal/gm · mole (approx.)*	Bond length, A
C—C	88	1.5	O—H	119	1.0
C=C	162	1.3	O—O	52	1.5
C≡C	213	1.2	O—Si	90	1.8
C—H	104	1.1			
C—N	73	1.5	N—H	103	1.0
C—O	86	1.4	N—O	60	1.2
C=O	128	1.2			
C—F	108	1.4			
C—Cl	81	1.8	H—H	104	0.74

* The values vary somewhat with the type of neighboring bonds; e.g., the C—H bond is 104 kcal/mole

in methane,
$$\text{H—}\overset{\displaystyle H}{\underset{\displaystyle H}{\text{C}}}\text{—H},$$
but only 98 kcal/mole in ethane,
$$\text{H—}\overset{\displaystyle H}{\underset{\displaystyle H}{\text{C}}}\text{—CH}_3,$$
and 90 kcal/mole in trichloro-

methane,
$$\text{H—}\overset{\displaystyle Cl}{\underset{\displaystyle Cl}{\text{C}}}\text{—Cl}.$$

Ammonia, NH_3, and water present more extreme cases of angle distortion (Fig. 3-11), because the nitrogen and oxygen, with their extra electrons, require only three and two covalent bonds each, respectively.

Example 3-3

An O—H pair has a dipole moment of 1.52 debyes (1 debye $= 3.33 \times 10^{-28}$ coul · cm). What is the dipole moment, p_{H_2O}, of a water molecule where the H—O—H bond is 104.5°?

Answer. The moment of the molecule is the vector sum of the two components:

$$
\begin{aligned}
p_{H_2O} &= 2(\cos 104.5°/2)(p_{OH}) \\
&= 2(0.61)(1.52 \text{ debyes}) \\
&= 1.85 \text{ debyes} \\
&= 6.2 \times 10^{-28} \text{ coul · cm.} \quad \blacktriangleleft
\end{aligned}
$$

Bond energies. As with ionic bonds, energy is required to separate covalently bonded atoms. Table 3-4 presents data for commonly encountered atom pairs. Two generalizations are noteworthy. (1) Multiple bonds have shorter bond lengths. (2) Double bonds are stronger than single bonds (but not twice as strong). The latter situation is of major importance in the formation of polymers (i.e., plastics), as will be discussed in Chapter 7.

Example 3-4

Hydrogen peroxide, H_2O_2, and ethylene, C_2H_4, react to form ethylene glycol, $C_2H_4(OH)_2$:

$$H\!-\!O\!-\!O\!-\!H + \begin{array}{c} H \quad H \\ | \quad\quad | \\ C\!=\!C \\ | \quad\quad | \\ H \quad H \end{array} \rightarrow H\!-\!O\!-\!\begin{array}{c} H \quad H \\ | \quad\quad | \\ C\!-\!C \\ | \quad\quad | \\ H \quad H \end{array}\!-\!O\!-\!H. \qquad (3\text{-}5)$$

What is the net change in bond energy?

Answer. Energy must be supplied to break one O—O bond and one C=C bond per molecule. However, energy is released when two C—O bonds and one C—C bond are formed. (The O—H and the C—H bonds are only slightly affected.)

Basis: 1 mole, or 0.602×10^{24} molecules of each. From Table 3-4,

$$\Delta E = +(52 + 162) - [2(86) + 88]$$
$$= -46 \text{ kcal.}$$

Note. Since the result is negative, energy is released in the process. Note, however, that energy must be supplied as an activation energy (Section 1-4) to alter the previous bonds before the reaction can proceed to the more stable product. ◄

3-4 METALLIC BONDS

The third type of primary bond is the metallic bond. Although this bond is sometimes considered to be intermediate between ionic and covalent, it is convenient to categorize it separately because metals comprise a major class of materials. Metallic atoms, i.e., those which possess loosely held valence electrons (Section 2-3), may be bonded together into very stable structures. Within a metal, many of the valence electrons are "free" to move throughout the structure, thus imparting the familiar metallic conductivity and optical properties such as luster and opacity.

The metallic bond may be compared with the covalent bond in that the valence electrons are shared by adjacent atoms. The metallic bond may also be compared to the ionic bond if one envisions the negative electrons as holding positive ions together. Depending on the metal, one or the other of the above comparisons may be preferred.

Because of the mobility of electrons within a metal, the terms "electron gas" and "electron cloud" are useful. Since the metallic bond is nondirectional, the coordination number is generally high. A large number of metals (e.g., Be, Mg, Al, Ca, Ti, Co, Ni, Cu, Zn, Sr, Ag, Cd, and Fe at high temperature) have 12 neighbors for each atom. Others (Li, Na, K, V, Cr, Mo, W, and Fe at room temperature) have eight neighbors for each atom. Only those "metals" with a major degree of covalency (Si, Ge, and low-temperature Sn) have as few as four neighbors.*

---

* The coordination number of eight for the alkali and transition metals, rather than the maximum of 12, is attributed to partial covalency.

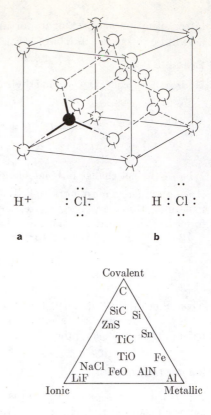

3-12 Diamond structure. Each carbon atom is covalently bonded to four adjacent atoms. The bond angle is 109.5°. Silicon and germanium have the same structure but have a finite number of electrons which are able to leave the covalent bond location.

$$H^+ \qquad : \overset{\cdot\cdot}{Cl} : ^{-} \qquad\qquad H : \overset{\cdot\cdot}{\underset{\cdot\cdot}{Cl}} :$$

3-13 HCl molecules. Both (a) ionic and (b) covalent bonding occur. The former is more common in aqueous solutions.

a   b

Covalent
C
SiC Si
ZnS
TiC Sn
TiO Fe
NaCl
LiF FeO AlN Al
Ionic   Metallic

3-14 Bonding mixtures. Most materials possess a combination of bonds.

3-5 INTERMEDIATE PRIMARY BONDS

Although we have categorized primary bonds into three major types, we seldom encounter "pure" bonds. Silicon, like the carbon of diamond (Fig. 3-12), covalently shares electrons. However, in silicon a few electrons are able to leave the covalent location between adjacent atoms to permit limited conductivity. This occurs more extensively in germanium and tin, serving as the basis of semiconductivity (Chapter 15) and providing a more metallic behavior.

A second example of bond combinations can be deduced from the last sketch of hydrogen in Fig. 3-7. During short intervals of time ($\sim 10^{-17}$ sec) when both electrons may be located near the same nucleus, the molecule is ionic; i.e., H^+ and H^- ions exist with mutual attraction. Since this extreme exists only a small fraction of the time, we commonly identify the H_2 molecule as covalent. HCl has greater ionic and lesser covalent tendency than H_2 (Fig. 3-13a). The choice between these two states—ionic or covalent—often depends on the environment. For example, (a) of Fig. 3-13 predominates in an aqueous solution, (b) predominates in a gas.

Ionic and metallic bond combinations are found in a number of compounds. For instance, TiO is sufficiently metallic to be opaque and conductive; it may also be envisioned in chemical reactions as $Ti^{2+}O^{2-}$. These intermediate bond combinations are illustrated schematically in Fig. 3-14.

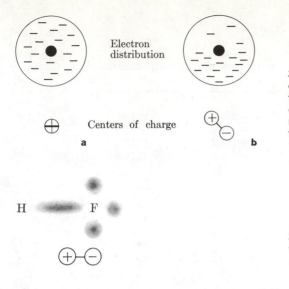

Electron distribution

Centers of charge

a

b

3-15 Dispersion effects. (a) Time-averaged, the electrons are uniformly distributed, with the result that the centers of positive and negative charge are coincident. (b) Momentarily, however, electrons may be off-center, to give a small dipole moment to the molecule.

H   F

3-16 Permanent dipoles, HF. The center of positive charge, ⊕, and the center of negative charge, ⊖, are not coincident.

3-6 SECONDARY (van der Waals) BONDS

The forces responsible for weaker, secondary bonding arise from internal dipoles. These dipoles may result from *dispersion effects*, which are simply the statistical irregularities of electron distribution in atoms or molecules; or they may be the *permanent dipoles* in asymmetric molecules. The former (Fig. 3-15) produce very weak forces and are evident only as the attractive forces which, at low temperatures, condense noble gases (He, Ne, Ar, etc.) or symmetrical molecules such as CH_4, F_2, H_2, CO_2, and N_2. Although masked by other stronger bonds, the dispersion effects are present in all materials. Permanent dipoles are found in asymmetric molecules where the centers of positive and negative charges are not coincident (Fig. 3-16). Because they are not coincident, the positive end of one dipole will have a preferential attraction to the negative pole of the next molecule to give intermolecular forces of attraction.

The bonding energies from atomic and molecular dipoles may be compared qualitatively by means of the boiling points shown in Table 3-5. They range from small fractions of a kcal/mole in the noble gases up to about 10 kcal/mole for the hydrogen bridge, which is most apparent in HF and H_2O.

The *hydrogen bridge* deserves special mention. Within a molecule, the hydrogen atom can be easily pictured as a proton on the end of a covalent bond (Fig. 3-16). Unlike other atoms bonded with covalent bonds, the positive charge of the proton is not shielded by surrounding electrons. Therefore, it may be attracted closely and rather strongly to the electrons of atoms in other molecules. This attraction accounts for the fact that water and HF have higher boiling points than other molecules of comparable weights. It also accounts for the high specific heat of water and is a factor in the extensive surface adsorptions possible by organic molecules.

TABLE 3-5
MELTING AND BOILING POINTS

	Melting temperature, °C	Boiling temperature, °C
Noble gases		
He	−272.2*	−268.9
Ne	−248.7	−245.9
Ar	−189.2	−185.7
Kr	−157.0	−152.9
Symmetric molecules		
H_2	−259	−252
N_2	−209	−195
O_2	−218	−183
Cl_2	−102	−34
CH_4	−183	−161
CF_4	−185	−128
CCl_4	−23	+26
Asymmetric molecules		
HF	−83	+20
HCl	−115	−85
H_2O	0	+100
CH_3Cl	−160	−14
CH_3OH	−98	+65
C_3H_7OH	−90	+117

* Melting temperature with 26 atm pressure. At 1 atm pressure, helium remains as a liquid at absolute zero (−273.16°C).

Example 3-5

When covalently bonded, an HCl molecule has a permanent electric dipole moment of 0.35×10^{-27} coul·cm. If all of the HCl dipoles are aligned with the field, what HCl pressure would be required to provide the same polarization as was induced in argon at 100 atm pressure and 0°C by an electric field of 100 volts/cm? (Compare Example 2-5.)

Answer. From Example 2-5, for 100 atm pressure,

$$\text{polarization} = 3.87 \times 10^{-15} \text{ coul·cm/cm}^3$$

$$= \left(n \frac{\text{moles}}{\text{cm}^3} \right) \left(0.602 \times 10^{24} \frac{\text{HCl}}{\text{mole}} \right) \left(0.35 \times 10^{-27} \frac{\text{coul·cm}}{\text{HCl}} \right),$$

$$n = 1.8 \times 10^{-11} \text{ mole/cm}^3;$$

$$1.8 \times 10^{-11} \text{ mole} = (P \text{ atm})(1 \text{ cm}^3)/(82.06 \text{ cm}^3 \cdot \text{atm/mole} \cdot °K)(273°K),$$

$$P = 4 \times 10^{-7} \text{ atm} = 0.0003 \text{ mm Hg}.$$

Note. The induced dipole moment of an argon atom and the permanent dipole moment of an HCl molecule differ by several orders of magnitude. We shall observe later, however, that thermal agitation prevents perfect alignment of all molecules with the field (Section 13-3). ◄

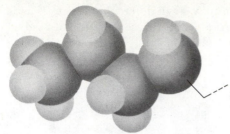
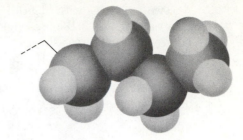

3-17 Macromolecule (polyethylene, C_nH_{2n+2}). Each molecule has strong internal bonds but only weak van der Waals bonds to other molecules.

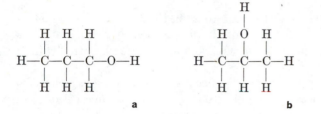

a

b

3-18 Isomers. (a) Normal propyl alcohol. (b) Isopropyl alcohol. The molecules have the same composition but different structures. Consequently, the properties are different. Compare with polymorphism of crystalline materials (Section 4-6).

3-7 POLYATOMIC UNITS

It is convenient at this point to identify three types of coordination units for later discussions—molecules, polyatomic ions, and free radicals. Each will have significance as we develop our concepts of the internal structures in engineering materials.

Molecules. We shall define a molecule as a neutrally charged group of atoms which are strongly bonded together but whose bonds to other, similar groups of atoms are relatively weak. The *intra*molecular forces provide primary bonds; the weaker *inter*molecular forces result in secondary bonds.* Compounds such as H_2O, CO_2, CCl_4, O_2, and HNO_3 produce small molecules. Size is not, however, a determining factor in the definition. In fact, large molecules are the basis for polymeric materials like polyethylene, C_nH_{2n+2}, where there may be hundreds or thousands of atoms (Fig. 3-17). Yet even in polyethylene and related vinyl compounds (Chapter 7) each molecule is still distinct, with strong intramolecular bonds and weak intermolecular bonds.

Other materials such as metals, MgO, SiO_2, and phenol-formaldehyde plastics have continuing structures of primary bonds. The difference between the structures of molecular materials and those with primary bonds throughout produces a major difference in properties. These differences will be considered in subsequent chapters.

---

* We include the hydrogen bridge among the secondary bonds because its 1- to 10-kcal/mole bond energy is still low as compared to ionic, covalent, or metallic bonds. (Compare Table 3-4.)

Two molecules with identical composition may have different structures, as is illustrated in Fig. 3-18 for propyl and isopropyl alcohol. These two variants, called *isomers*, have different shapes and different dipole moments. Therefore they have different properties; e.g., the melting points are $-127°C$ and $-89°C$, respectively.

Polyatomic ions. Although a molecule is electrically neutral, other polyatomic units must carry a charge. For example:

$$\begin{array}{ccccc}
 & \ddot{O}\, \cdot & & & \ddot{O}\, \dot{\bar{\,}} \\
\ddot{O} : \ddot{S} : \ddot{O}\, \cdot & + 2e^- \rightarrow & \ddot{O} : \ddot{S} : \ddot{O}\, \dot{\bar{\,}} \\
 & \ddot{O} : & & & \ddot{O} :
\end{array} \qquad (3\text{-}6)$$

Since the addition of two electrons to the SO_4 group on the left would complete a full outer shell of eight electrons for each of the five atoms, it readily attracts electrons (from metal atoms) to produce an SO_4^{2-} anion. The bonds within sulfate ions are covalent.

Most polyatomic ions are negative, chiefly because those atoms which commonly share electrons readily accept extra electrons also. Polyatomic cations are encountered, however, as evidenced by the ammonium ion NH_4^+.

Anions, like molecules, can be large. An example is the stearate ion $H_3C(CH_2)_{16}COO^-$, which has a weight of 283 amu. In practice, however, the ionic charge does not increase in proportion to the mass. As a result, large anions have less effective coulombic attractions than those found in small ions.

Free radicals. The reaction

$$H_2O \rightarrow H^+ + OH^- \qquad (3\text{-}7)$$

produces a negative hydroxyl ion because an electron is removed from the hydrogen ion. The dissociation of hydrogen peroxide, H_2O_2, also gives oxygen-hydrogen combinations,

$$H_2O_2 \rightarrow 2(OH), \qquad (3\text{-}8)$$

although in this case the atom pairs do not produce charged ions. Instead the product is a *free radical* and is very reactive. Equation (3-8) can be written differently as

$$H : \ddot{O} : \ddot{O} : H \rightarrow H : \ddot{O} \bullet + \bullet \ddot{O} : H \qquad (3\text{-}9)$$

or

$$H_2O_2 \rightarrow 2HO \bullet , \qquad (3\text{-}10)$$

where the large dot implies a *reactive site*. Because the free radical combines readily with ions, molecules, or other free radicals, its lifetime is normally short (except at very low temperatures). Other common free radicals are $H_3C\bullet$, $H_5C_2\bullet$, and $H_2N\bullet$.

Since free radicals will have some importance in polymer processing, to be discussed in Chapter 7, a short paragraph is warranted here as a preview. If hydrogen peroxide, H_2O_2, dissociates in the presence of ethylene, C_2H_4, the HO• radicals react as follows:

$$
\begin{array}{ccc}
& \text{H} \quad \text{H} & \text{H} \quad \text{H} \\
\text{H} : \overset{..}{\underset{..}{\text{O}}} \bullet + \text{C} :: \text{C} & \rightarrow \text{H} : \overset{..}{\underset{..}{\text{O}}} : \text{C} : \text{C} \bullet, \\
& \text{H} \quad \text{H} & \text{H} \quad \text{H}
\end{array} \tag{3-11a}
$$

or

$$
\begin{array}{ccc}
& \text{H} \quad \text{H} & \text{H} \quad \text{H} \\
\text{H}-\text{O} \bullet + \; \text{C}{=}\text{C} & \rightarrow \text{H}-\text{O}-\text{C}-\text{C} \bullet. \\
& \text{H} \quad \text{H} & \text{H} \quad \text{H}
\end{array} \tag{3-11b}
$$

A *chain reaction* continues with other ethylene molecules:

$$
\text{HO}(C_2H_4)_n^\bullet + C_2H_4 \rightarrow \text{HO}(C_2H_4)_{n+1}^\bullet. \tag{3-12}
$$

Large polyethylene molecules can thus be generated from small ethylene molecules. Obviously, considering that ethylene is a gas and polyethylene is a commercially useful plastic, *polymerization* causes a major change in properties. The same reaction is applicable to other vinyl compounds such as vinyl chloride, CH_3Cl.

Example 3-6

How much energy is released when a mole (i.e., 0.6×10^{24}) of ethylene molecules reacts according to Eq. (3–12)?

Answer. A mole of ethylene molecules has 0.6×10^{24} double-bonded carbon pairs. After reaction there are 1.2×10^{24}, or 2 moles of single bonds. From Table 3-4,

$$
\begin{aligned}
\Delta E &= +162 \text{ kcal} + 2(-88 \text{ kcal}) \\
&= -14 \text{ kcal}.
\end{aligned}
$$

Note. Energy must be supplied (+) to break the double bonds, and energy is released (−) when subsequent single bonds are formed. ◄

3-8 CONDENSED PHASES

Except for air and a few fuels such as natural gas, most engineering materials are condensed; i.e., they are liquids or solids with unlimited numbers of atoms and molecules. Several distinctions between gases, on the one hand, and liquids and solids, on the other, are worthy of note.

Molecules of a gas (or atoms of a noble gas) have only kinetic energy and minimal potential energy as a result of their location among neighbors. Thus each molecule acts independently, except that two may collide in their flight path. As observed in Section 2-5, the total pressure exerted by a gas at any temperature is a function of the number of moles per unit volume (Eq. 2-10):

$$P = R(n/V)T.$$

Likewise, the average kinetic energy, $\overline{\mathrm{KE}}$, of each molecule depends only on the temperature (Eq. 2-12):

$$\overline{\mathrm{KE}} = \tfrac{3}{2}kT,$$

where the constant, k, is 1.38×10^{-16} erg/°K (Boltzmann's constant). A *continuous* range of energies is possible for molecules within a gas, as indicated by the Boltzmann distribution in Fig. 2-8. Within this distribution it is not impossible for several molecules out of the approximately $10^{20}/\mathrm{cm}^3$ to have identical energies.

In a liquid or solid, atoms (or molecules) can have up to 12 immediate neighbors. Because of their mutual attraction, energy is given off as atoms are brought together, or energy is required to separate them as will be recalled from the energy trough of Fig. 3-1. A similar energy relationship exists for covalent, metallic, or van der Waals bonds. The energy released during condensation, or required during volatilization, is greater than for a single pair of ions or atoms because the coordination number involves several adjacent neighbors.

Example 3-7

Each C—C bond in diamond has an energy of $(85{,}400 \text{ cal}/0.602 \times 10^{24})$.* What is the energy required to volatilize 0.01 gm of diamond?

Answer. Each carbon atom *shares* four bonds. Therefore *twice* as many bonds as carbon atoms are present in diamond. [*Note.* To view this another way, focus on one carbon atom of Fig. 3-12 and assign *one-half* of each bond to that atom (and the other half to the adjacent atom). Thus there are 4/2 bonds per atom.]

$$\text{Carbon atoms} = \left(\frac{0.01 \text{ gm}}{12.01 \text{ gm/mole}}\right)(0.602 \times 10^{24} \text{ atoms/mole}).$$

$$\text{Energy required} = 2(\text{carbon atoms})(85{,}400 \text{ cal}/0.602 \times 10^{24})$$
$$= 2(0.00083 \text{ mole})(85{,}400 \text{ cal/mole})$$
$$= 142 \text{ cal.} \ \blacktriangleleft$$

Heat of vaporization. When a liquid is vaporized, energy must be supplied to break bonds. This change in energy is called the heat of vaporization, ΔH_v.† The enthalpy change for the vaporization of water to steam at 1 atm pressure is shown in Fig. 3-19(a). Since vaporization reduces atomic and molecular coordination, there is a considerable increase in disorder, or entropy, of the system. The thermodynamicist, consequently, finds it advantageous to factor the enthalpy change

---

* This value does not match that in Table 3-4 exactly. As indicated in the footnote to Table 3-4, the energy varies with the type of neighboring bonds. In diamond, each carbon atom has four C—C bonds. In most molecules, other types of bonds are also present around the carbon atom (e.g., C—H and C—O).

† Vaporization from a solid, as in Example 3-7, is comparable, but is called the *heat of sublimation*.

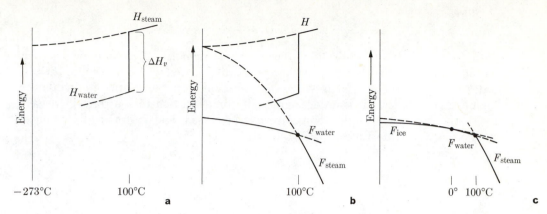

3-19 Energies of transformation (water). (a) Enthalpy and heat of vaporization, ΔH_v. (b) The free energies of steam (1 atm) and water are equal at 100°C. (c) Free energies of stable phases.

into temperature, T, and the change in entropy during vaporization, ΔS_v:

$$\Delta H_v = T\,\Delta S_v. \tag{3-13}$$

Equilibrium temperatures. Free energy was identified in Eq. (1-7b) and discussed in Section 1-4. Since this is the energy available for reaction, *the free energy of the reactant and product must be equal at equilibrium.* This is shown in Fig. 3-19(b) for a free-energy plot of water and steam at 1 atm pressure*:

$$H_2O_{water} \rightleftarrows H_2O_{steam}. \tag{3-14}$$

Note that above 100°C, steam has lower free energy than water, so the reaction of Eq. (3-14) progresses to the right to reduce the free energy. Below 100°C, the stable, lower-energy phase is liquid and the above reaction progresses to the left. According to thermodynamic laws, *the phases with the minimum free energy are the stable phases.*

The familiar reaction at 0°C,

$$H_2O_{ice} \rightleftarrows H_2O_{water}, \tag{3-15}$$

also has a change in enthalpy called the *heat of fusion*, ΔH_f, with $T\,\Delta S_f$ as an analog to the $T\,\Delta S_v$ shown in Eq. (3-13):

$$\Delta H_f = T\,\Delta S_f. \tag{3-16}$$

The free energies of ice and water are equal at 0°C. Thus Fig. 3-19(c) represents the F-versus-T curves for the three forms of H_2O.

---

* Since the effect of pressure is not considered here, the interested student is referred to standard physical chemistry or thermodynamics textbooks for further elaboration.

Since

$$dF/dT = -S$$

(Eq. 1-8), the slope of the solid curve reveals the change of *entropy*, or disorder, as the temperature is changed. At low temperatures and within the solid, the disorder is low. In the next chapter we shall see that a characteristic of crystals is a high degree of *order*. This order is gradually reduced, and the entropy increased, by thermal agitation as the temperature is increased, and is drastically changed by melting and vaporization.

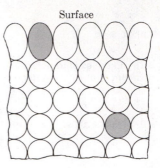
Surface

3-20 Surface coordination. The surface atoms have higher potential energy because they have fewer neighbors than the interior atoms.

3-9 SURFACE ENERGIES

Unlike the atoms or molecules within a condensed phase, which are coordinated on all sides, those on a surface have a lower coordination number because they lack neighbors on one side (Fig. 3-20). Consequently those on the surface have a higher energy than those in the interior. We speak of this extra energy as surface energy, with units commonly expressed in $ergs/cm^2$.

Example 3-8

The surface energy of glass at 650°C is 300 $ergs/cm^2$. How much energy, $\Delta\Gamma$, is released if a 10-cm length of a 0.02-mm fiber of glass spheroidizes?

Answer. Volume $= (\pi)(0.001 \text{ cm})^2(10 \text{ cm}) = 10^{-5}\pi \text{ cm}^3$,
$$d = \sqrt[3]{(6/\pi)(10^{-5}\pi)} = 0.039 \text{ cm}.$$

Fiber	*Sphere*
$\Gamma_{\text{fib}} = (300 \text{ ergs/cm}^2)(0.002\pi \text{ cm})(10 \text{ cm})$	$\Gamma_{\text{sph}} = (300 \text{ ergs/cm}^2)(4\pi)(0.039 \text{ cm})^2$
$= 18.8 \text{ ergs}$	$= 6.0 \text{ ergs}$

$$\Delta\Gamma = -12.8 \text{ ergs}$$

Note. Spheroidization of both liquids and solids is a natural phenomenon when only one type of boundary is present, because surface energy is reduced. The spheroidization rate is slow for solids. *Wetting*, in which there are several types of boundaries, will be considered in Fig. 18-10 and Section 18-3. ◀

Example 3-9

The surface energy of water is 70 ergs/cm^2 at 25°C. If we assume that all the surface energy is concentrated in a monomolecular surface layer,* what is the surface energy per molecule?

Answer. Basis: 1 cm^3 = 1 gm.

$$\text{Volume per molecule} = \left(\frac{1 \text{ cm}^3}{1 \text{ gm}}\right)\left(\frac{18 \text{ gm/mole}}{0.602 \times 10^{24} \text{ molecules/mole}}\right)$$
$$= 30 \times 10^{-24} \text{ cm}^3/\text{molecule}.$$

Now assume that a cubic volume is associated with each molecule in order to fill space:

$$\text{volume per molecule} = (3.1 \times 10^{-8}\text{cm})^3/\text{molecule}.$$

On the monomolecular surface,

$$\text{molecules/cm}^2 = (1 \text{ molecule})/(3.1 \times 10^{-8}\text{cm})^2,$$
$$\text{energy/molecule} = \frac{70 \text{ ergs/cm}^2}{(1 \text{ molecule})/(3.1 \times 10^{-8} \text{ cm})^2}$$
$$= 7 \times 10^{-14} \text{ erg/molecule}$$
$$(= 0.04 \text{ eV/molecule}).$$

Note. Since the heat of vaporization of water is 540 cal/gm, or an average of 0.42 eV/molecule, the surface molecules already have a significant fraction of the energy necessary for volatilization. ◀

REFERENCES FOR FURTHER READING

Addison, W. E., *Structural Principles in Inorganic Compounds.* New York: Wiley, 1961. Paperback. Recommended as supplementary reading on chemical bonds. Introductory level.

DiBenedetto, A. T., *The Structure and Properties of Materials.* New York: McGraw-Hill, 1967. Bonding of atoms is presented at an advanced level in Chapter 3.

Mahan, B., *University Chemistry,* second edition. Reading, Mass.: Addison-Wesley, 1969. Chapter 11 provides supplementary reading on the subject of chemical bonds. Freshman-sophomore level.

Parker, E. R., F. R. Davis, and E. L. Langer, *Solid State Structures and Reactions.* Metals Park, O.: American Society for Metals, 1968. Bonding in solids is presented on the basis of high school science.

Ryschkewitsch, G. E., *Chemical Bonding and the Geometry of Molecules.* New York: Reinhold, 1963. Paperback. Excellent background reading for this chapter.

Van Vlack, L. H., *Elements of Materials Science,* second edition. Reading, Mass.: Addison-Wesley, 1964. Chapter 2 introduces bonds at the freshman-sophomore level.

---

* This is a better assumption for stereospecific bonds (Section 3-2) than for ionic or metallic bonds. The hydrogen bridge in H$_2$O is stereospecific.

PROBLEMS

3-1 The equilibrium distance between Mg^{2+} and O^{2-} ions in MgO is 2.10 A.

a) What is the attractive force between an Mg^{2+} ion and a neighboring O^{2-} ion at this equilibrium distance? What is the repulsive force at this equilibrium distance? [*Note.* To simplify units, solve for dynes (i.e., ergs/cm) by using the appropriate values for the charge on an electron from Appendix A. $Esu = (erg \cdot cm)^{1/2}$.]

b) What are the comparable values for NaCl where the radii of Na^+ and Cl^- are 0.98 A and 1.81 A, respectively?

Answer. a) 21×10^{-4} dyne   b) 3×10^{-4} dyne

3-2 Noting that energy is the integration of force and distance, plot the coulombic energy given off versus distance as a Cl^- ion is brought in from ∞ to within 3 A from an Na^+ ion. ($E_\infty \equiv 0$; assume that $n = 9$.)

3-3 If the repulsion exponent, n, is equal to 9, what is the energy of bonding between a Ca^{2+} ion and *one* of its O^{2-} neighbors in CaO?

Answer. -34.5×10^{-12} erg/bond (-21.2 eV). [*Note.* Each ion reacts with neighboring ions, both positive and negative. There is less energy involved at greater distances, however.]

3-4 a) Refer to Appendix C-1. Plot thermal expansion coefficients versus moduli of elasticity.

b) Explain the correlation. [*Note.* This correlation is reasonably good. However, the metals of Appendix C-1 all have comparable structures. Our correlation would break down if we added data from Appendix C-2 for materials with distinctly different bonds and structures.]

3-5 Show the origin of 0.732 in Table 3-1.

Answer. $2(r + R) = \sqrt{3}\,(2R)$

3-6 Show the origin of 0.225 in Table 3-1. [*Hint.* The height of a tetrahedron is (0.817)(edge), and the center of the tetrahedron is 25% of the height.]

3-7 a) What is the radius of the smallest cation that can have a 6-fold coordination with O^{2-} ions?

b) 8-fold coordination?

Answer. a) 0.545 A   b) 0.965 A

3-8 a) From Appendix B, cite three divalent cations which can have $CN = 6$ with S^{2-} but not $CN = 8$.

b) Cite two divalent cations which may have $CN \geq 8$ with F^-.

3-9 An N—H pair has a dipole moment of 1.7 debyes. Determine the dipole moment of NH_3.

Answer. 1.95 debyes

3-10 Sketch the structure of the various isomers of C_7H_{16}.

3-11 From Problem 3-9, the dipole moment of NH_3 was determined to be 1.95 debyes (or 6.5×10^{-27} coul $\cdot$ cm). *If* all of the molecules could be aligned with the field, what would the polarization be at 0°C and 760 mm pressure?

Answer. 1.7×10^{-7} coul $\cdot$ cm/cm^3. [*Note.* Thermal agitation will prevent all of the molecules being aligned with the field at one time. (See Section 13-3.)]

3-12 An organic compound contains 62.1 w/o* carbon, 10.3 w/o hydrogen, and 27.6 w/o oxygen. Name a possible compound.

3-13 A molecule built up according to Eq. (3-12) has a molecular weight of 17,380. What is the value of n?

Answer. 620

3-14 Sulfur dichloride has a molecular weight of 103 and a boiling point of 59°C. Draw a diagram showing the valence electron structure of this compound.

3-15 Sketch the valence-shell electron structure of an SiO_4^{4-} ion.

3-16 Show the centers of positive and negative charges in (a) CCl_4, (b) $C_2H_2Cl_2$, and (c) CH_3Cl.

3-17 The heat of vaporization of water at 25°C is 580 cal/gm. Compare this figure to the surface energy calculated in Example 3-9.

Answer. 7.3×10^{-13} versus 7×10^{-14} erg/molecule

---

* Recall that w/o = weight percent, v/o = volume percent, l/o = linear percent, etc.

part II □ structure of solid phases

4 □ crystalline phases

4-1 PHASES

A *phase*, as the term pertains to materials, is a *structurally homogeneous part of a system*. In introductory science courses, frequent references are made to gas, liquid, and solid phases. This is natural because the structure of a gas is markedly different from that of a liquid, and as we shall see, the structure of a liquid is different from that of a *crystalline* solid.

A material system may include several phases; an example is water in a glass fiber-reinforced plastic tank. The water, the glass, and the plastic (and any air) all have different structures. Multiphase systems always have *phase boundaries* which are *structural and/or compositional discontinuities.*

Only one gas phase is encountered in a given system. The atomic and molecular packing is very low, and each molecule is sufficiently independent so that there is no order. The structure above the molecular level is random.

Several liquid phases may coexist; oil, water, and liquid mercury provide three familiar examples. The oil has long molecules with weak van der Waals forces between neighbors; water has small molecules with hydrogen bridges; and mercury has atoms held together by metallic bonds.* By separating into three distinct liquid phases, the total free energy of a mixture of these components is minimized, and phase boundaries develop to give structural and compositional discontinuities.

Numerous solid phases with high degrees of order are encountered in engineering materials. For example, each atom in solid aluminum not only has 12 immediate neighbors at 2.862 A, but also has a specific number of second- and third-nearest neighbors, all in specific directions and at specific distances. Such a solid is called a *crystal*. In contrast, although every silicon atom in silica *glass* is coordinated with four neighboring oxygen atoms at a distance of 1.6 A and every oxygen atom is coordinated with two silicon atoms, the second and third neighbors are *not* specific. Such a solid is *noncrystalline*. The structures of solids will be the topic of this and the next four chapters. Emphasis will be placed first on crystals which have the highest degree of order (lowest entropy), and then on various types of disorder. After this we will discuss molecular solids and solid solutions.

---

* The metallic bond in mercury is much weaker than the typical metallic bond; hence its low melting point.

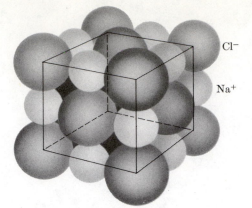

Cl⁻

Na⁺

4-1 Crystal structure (NaCl). If this structure is extended in three dimensions, every Na⁺ ion is found to be coordinated with six Cl⁻ ions, and every Cl⁻ ion with six Na⁺ ions. The resulting long-range order introduces characteristic crystal planes and directional vectors.

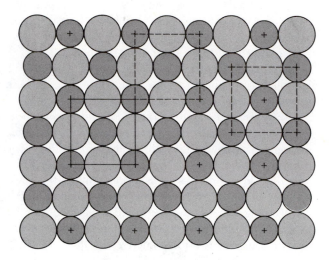

4-2 Repeating patterns. (Cf. Fig. 4-1.) The description of one "unit square" would be sufficient to describe the total area. The center atom of each square of this pattern has the same coordination as the corner atoms.

4-2 CRYSTALLINE SOLIDS

Many solid phases are crystals; i.e., they have a long-range order. Under normal conditions, all solid metals fall in this category as well as most natural minerals. We will also find that a number of molecular phases may be crystalline; an example is polyethylene. Other plastics and all glasses are amorphous, i.e., noncrystalline solids.

Short-range order. We apply the term "short-range order" when a regular pattern of first-neighbor coordination is encountered. Such order exists either (1) because there is a specific number of covalent bonds to be satisfied, as with silicon and oxygen atoms in a silica glass, or (2) because the size ratio of the ions permits a certain coordination number, as with ionic bonding.

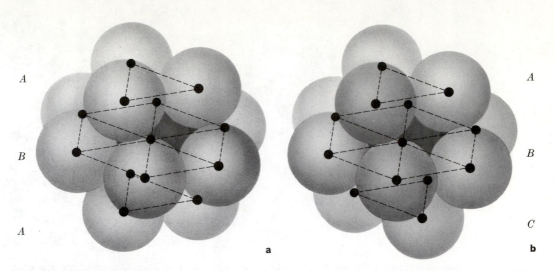

4-3 Close-packed coordinations. In elemental solids as many as 12 immediate neighbors are possible. (a) Hexagonal close-packed. (b) Cubic close-packed (or face-centered cubic). (See Fig. 4-4.)

Long-range order. In NaCl, each sodium ion is surrounded by six first-neighbor chlorine ions, and each chlorine ion is coordinated with six sodium ions. If this occurs for *all ions*, a long-range structure develops in which there is an order not only of first neighbors, but also of more distant neighbors. Refer to Fig. 4-1 and visualize the repetition of this structure in three dimensions. In addition to its nearest-neighbor Na^+ ions, each chlorine ion has 12 second-neighbor Cl^- ions which are $\sqrt{2}$ times as far away, and 8* third-neighbor Na^+ ions at $\sqrt{3}$ times the distance of the nearest neighbors. These and more distant neighbors can be counted and located to define a long-range order which is a distinguishing characteristic of crystalline materials.

When long-range order exists, a repeating pattern is present. This is illustrated in Fig. 4-2, which shows one plane of Fig. 4-1. Each "unit square" is identical and the whole plane could be filled with repeating unit squares.

Unit cells. The unit repeating volume in three dimensions is called a unit cell. It is the *simplest* volume which completely fills space,† and has all the characteristics of the whole crystal. We shall focus our attention on the unit cell in the next chapter when we look at the geometric features of crystals.

4-3 ELEMENTAL CRYSTALS

The simplest crystals contain atoms of only one kind of element, so the radius ratio for neighboring atoms is 1.0. Unless covalent bonds are present, a maximum of

---

* Each corner Cl^- ion in Fig. 4-1 has a third neighbor at the center of the illustrated cube. Other third neighbors are to be found in the centers of the seven other cubes which corner on each Cl^- ion. The 12 second neighbors can be located through similar analysis.

† Often this is the smallest repeating volume. Exceptions occur when a higher degree of symmetry is possible through the use of a larger volume (see Section 5-2).

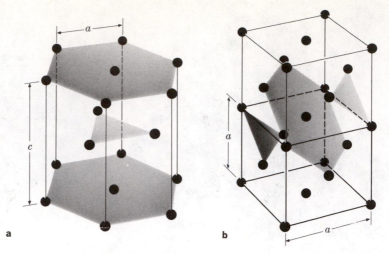

4-4 Close-packed unit cells. (a) Hexagonal close-packed. (b) Face-centered cubic. (Cf. Fig. 4-3.)

12 first neighbors is possible. Two possible arrangements of atoms with 12-fold coordination are shown in Fig. 4-3 (a and b). Only second- and third-neighbor differences exist. [In (a), the third layer is directly over the first layer. In (b), the fourth layer is directly over the first layer.] The atom coordinations of Fig. 4-3 are resketched in Fig. 4-4 to show the corresponding unit cells. They are hexagonal and cubic, respectively.

Close-packed crystals (hcp and fcc). Since 12 nearest neighbors are the maximum possible when all atoms are identical, the arrangements in Figs. 4-3 and 4-4 have the greatest possible number of atoms per cm³. The two packings are called hexagonal close-packed (hcp) and cubic close-packed, respectively. [The term "face-centered cubic" (fcc) is also applicable to Fig. 4-4(b) and will be employed in this book since it has wider use.]

An important concept in studying crystal structures is that of the *atomic packing factor* (APF), the fraction of the space occupied by atoms. The hard-ball model and spherical atoms are assumed.

Example 4-1

Calculate the atomic packing factor for solid argon which has an fcc structure.

Answer. There is an equivalent of 4 atoms per unit cell (Fig. 4-5). Furthermore, $4R = a\sqrt{2}$. As shown below, it is not necessary to know the radius of the argon atom.*

$$\text{APF} = \frac{\text{atom volume}}{\text{total volume}}$$

$$= \frac{(4 \text{ atoms/unit cell})(4\pi/3)(a^3\sqrt{2}/32 \text{ cm}^3/\text{atom})}{(a^3 \text{ cm}^3/\text{unit cell})}$$

$$= 0.74. \quad \blacktriangleleft$$

---

* The radii must be known in comparable problems for bielemental crystals, e.g., NaCl in Fig. 4-1.

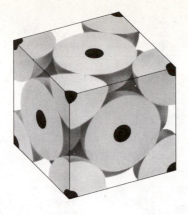

4-5 Face-centered cubic unit cell (argon). An fcc structure has four atoms per unit cell and an atomic packing factor of 0.74 in most elemental crystals. The lattice constant, a, equals $4R/\sqrt{2}$.

Both fcc and hcp have a packing factor of 0.74. Because they provide efficient packing and therefore relatively low energy, these two structures are encountered in approximately half of the elemental solids. The solids of the noble gas elements are all fcc. Of the metals, Al, Ca, Sc, Fe (at elevated temperatures), Ni, Cu, Sr, Rh, Pd, Ag, Ir, Pt, Au, Pb, and Th are fcc; and Be, Mg, Ti (at low temperatures), Co, Zn, Y, Zr, Ru, Hf, Re, Os, and Tl are hcp.* All close-packed crystals exhibit expansion on melting because the increase in disorder decreases the atomic packing factor.

Covalent crystals. The structure of covalently bonded diamond was shown in Fig. 3-12. Since it has a coordination number of 4, its calculated atomic packing factor is only 0.34. White tin, which is partially metallic and partially covalent, has a packing factor of 0.55. These crystals are stable despite their low packing factor, because covalent bonds are strong.

Example 4-2

Silicon has the same structure as diamond (Fig. 3-12) and a bond length of 2.351 A. Estimate its density.

Answer. Examine Fig. 3-12 and observe that the interatomic distance is 1/4 of the body diagonal, the direction along which atoms are in contact. Therefore,

$$3(a/4)^2 = (2R)^2,$$
$$a = 4(2.351 \times 10^{-8}\text{ cm})/\sqrt{3}$$
$$= 5.43 \times 10^{-8}\text{ cm}.$$

There are

4 atoms totally within the unit cell,

6/2 atoms on the six faces,

8/8 atoms on the eight corners,

---

* "Rare earth" metals have not been listed. The majority of these are hcp.

TABLE 4-1
INTERATOMIC DISTANCES IN SELECTED ELEMENTAL CRYSTALS (20°C)

Element	Structure	Distance, A*	Element	Structure	Distance, A*
Ag	fcc	2.888	Nb	bcc	2.859
Al	fcc	2.862	Ne	fcc	3.21
Ar	fcc	3.84	Ni	fcc	2.491
Au	fcc	2.882	Os	hcp	2.734*
Ba	bcc	4.348	Pb	fcc	3.499
Be	hcp	2.286*	Pd	fcc	2.750
C	covalent	1.545	Pt	fcc	2.775
Ca	fcc	3.94	Rb	bcc	4.88
Cd	hcp	2.979*	Re	hcp	2.76*
Co	hcp	2.507*	Rh	fcc	2.689
Cr	bcc	2.498	Ru	hcp	2.704*
Cs	bcc	5.30	Sc	fcc	3.21
Cu	fcc	2.556	Si	covalent	2.351
Fe	bcc	2.4824	Sr	fcc	4.31
Fe	fcc	2.540	Ta	bcc	2.859
Ge	covalent	2.449	Th	fcc	3.60
Hf	hcp	3.188*	Ti	hcp	2.9503*
Ir	fcc	2.714	Ti	bcc	2.85
K	bcc	4.624	Tl	hcp	3.408*
Kr	fcc	4.03	V	bcc	2.632
Li	bcc	3.038	W	bcc	2.734
Mg	hcp	3.209*	Y	hcp	3.65*
Mo	bcc	2.725	Zn	hcp	2.665*
Na	bcc	3.714	Zr	hcp	3.231*

* The a-dimension is given for hcp. These values may differ slightly from the values in Appendix B because of anisotropies in the noncubic unit cells.

or

$$8 \text{ atoms/unit cell.}$$

$$\text{Density} = \frac{\text{mass/unit cell}}{\text{volume/unit cell}}$$

$$= \frac{(8 \text{ atoms/unit cell})(28.09 \text{ gm/mole})/(0.602 \times 10^{24} \text{ atoms/mole})}{(5.43 \times 10^{-8} \text{ cm})^3/\text{unit cell}}$$

$$= \frac{(8)(28.09)}{(0.602)(160.1)} \frac{\text{gm}}{\text{cm}^3} = 2.33 \text{ gm/cm}^3.$$

Note. Experiments give a value of 2.4 gm/cm³. ◄

Body-centered cubic (bcc). Eight-fold coordination (Fig. 4-6) can exist in metals, The slight inefficiency in packing (0.68 versus a maximum of 0.74) is tolerated in favor of some covalency, magnetic coupling, and other secondary factors. Bcc metals include Li, Na, K, Ti (at elevated temperatures), V, Cr, Fe (at low temperatures), Rb, Nb, Mo, Cs, Ba, Ta, and W.

4-6 Body-centered cubic unit cell (sodium). A bcc metal has two atoms per unit cell and an atomic packing factor of 0.68 in elemental crystals. The lattice constant, *a*, equals $4R/\sqrt{3}$.

4-7 CsCl structure. Every Cs^+ ion is coordinated with eight Cl^- ions, and (with the structure continued) every Cl^- ion is coordinated with eight Cs^+ ions.

Atomic radii. Through x-ray diffraction measurements (Section 5-6) it is possible to determine *interatomic distances* with a high degree of precision. Such data are listed in Table 4-1 for those elemental solids cited in this section. In elemental crystals, it is convenient to consider the atoms as being spherical with a radius equal to one-half the interatomic distance. (These radius values cannot be used for ionic crystals of the next section because ions have electrons added or removed.)

When the atomic weights are comparable, the atomic radii are largest for noble gases and alkali metals, and smallest for covalently bonded atoms. The interatomic distances are 2% or 3% smaller when the coordination number is equal to 8 (bcc) than when it is equal to 12 (fcc and hcp). (See iron and titanium in Table 4-1.) This closer approach is possible because there is less electronic repulsion with lower atomic packing factors.

4-4 IONIC CRYSTALS

The prototype for ionic crystals is NaCl (Section 4-2), but a large number of other crystals (CaO, FeO, LiF, MgO, MnS, PbS, ThC, and ZrN, to name but a few) possess the same structure. In each of these, the number of metal atoms equals the number of nonmetal atoms. Furthermore, the radius ratio of the ions is such that a maximum of only six anions may surround each cation (Table 3-1).

Compounds with larger cations usually have an 8-fold coordination. If equal numbers of cations and anions are required to balance charges, a CsCl-type structure results (Fig. 4-7). In certain AX_2-type compounds (e.g., CaF_2, ZrO_2, UO_2),

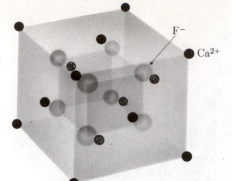

4-8 CaF$_2$ structure. Every Ca^{2+} ion is coordinated with eight F$^-$ ions, but every F$^-$ ion is coordinated with only four Ca^{2+} ions.

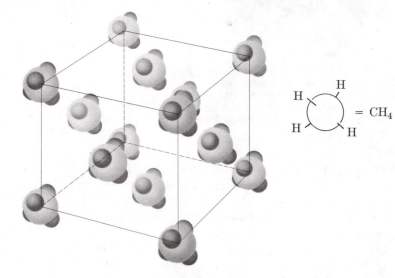

$$\begin{array}{c} H \\ H - \bigcirc - H \\ H \end{array} = CH_4$$

4-9 Molecular crystals (methane). Since intermolecular attractions between nonpolar molecules are weak, methane does not crystallize into an fcc structure until −183°C (90°K). Even then, the CH$_4$ molecules are able to rotate within their fcc locations. Finally, at 20°K, the thermal agitation is sufficiently reduced so that all molecules assume identical orientations and provide the highly ordered (low-entropy) structure shown above.

each anion is coordinated with only four cations (Fig. 4-8). In effect, only half of the possible cation sites are occupied so that the charge may remain balanced.

Later we shall refer to additional crystals by noting the coordination number and the locations of unoccupied sites. However, no attempt can be made here to systematize all possible ionic crystals.

4-5 MOLECULAR CRYSTALS

Since molecules, by our definition (Section 3-7), are self-contained units, they possess a possibility of forming crystals as do the atoms themselves. Methane, CH$_4$, for example, freezes to an fcc crystal (Fig. 4-9). Like fcc crystals of argon

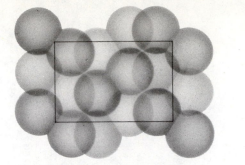

4-10 Molecular crystal (iodine). The molecule, I_2, acts as a unit in the repetitive crystal structure. As a result of the molecular shape, the unit cell cannot simultaneously be cubic and have a high packing factor.

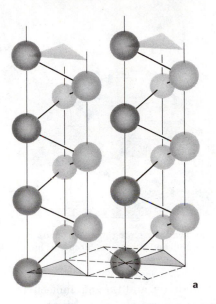

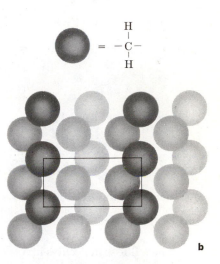

a   b

4-11 Crystals of linear molecules. (a) Tellurium. (b) Polyethylene (cf. Fig. 7-6). Because of the molecule shape, the three-dimensional structures which form do not have cubic unit cells.

and other materials with van der Waals bonds, the melting temperature of CH_4 is low. One carbon and four hydrogen atoms associate with each fcc position of the unit cell.

Diatomic molecules such as O_2, N_2, and the halides cannot coordinate as spheres because of their shape. As a result, they do not form cubic crystals. This is illustrated for iodine in Fig. 4-10, where the observed unit-cell dimensions are not equal.

Linear molecules produce crystals which vary markedly in the structure for their three dimensions. As shown in Fig. 4-11(a), tellurium forms a chain-like molecule because there are two covalent bonds per atom. These chains coordinate into a crystal by aligning themselves in one direction, each chain assuming a helical conformation. As a second example (Fig. 4-11b), polyethylene may crystallize by aligning its chain-like molecules of CH_2 units into a very regular three-dimensional pattern. Another presentation of this structure is given in Fig. 7-6.

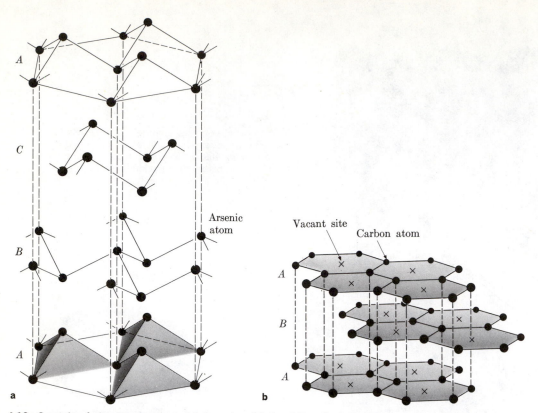

4-12 Crystals of sheet molecules. (a). Arsenic. (b) Graphite. Each layer has strong internal bonds but weak bonds between layers which stack into crystals. The properties parallel and perpendicular to the sheets are markedly different.

Since linear molecules involve many atoms, often hundreds, and the *inter*molecular forces are only weak van der Waals attractions, it is not surprising that crystallization is not always as perfect over great distances in these materials as in elemental solids. This will be the subject of discussion in later chapters.

Sheet molecules are also large. Arsenic, even though uncommon, will be used as our prototype (Fig. 4-12a). These "puckered" sheets stack one above another with the sequence repeating every third layer. Graphite also possesses a sheet structure, as shown in Fig. 4-12(b). The contrast between this structure and that of diamond (Fig. 3-12) reveals why the properties of these two carbon crystals differ so markedly. Diamond, with strong covalent bonds in three dimensions, is the hardest of all naturally occurring materials. Graphite, on the other hand, has strong bonds within its layers but only weak bonds between layers; hence, it may be used as a lubricant! Its properties are highly anisotropic, i.e., directional, with thermal (and electrical) conductivities varying by a factor of 100 (0.006 versus $0.6 \, \text{cal} \cdot \text{cm}/\text{°C} \cdot \text{cm}^2 \cdot \text{sec}$ in directions perpendicular and parallel to the sheets).

Ice deserves special mention (1) because it is so familiar to us and (2) because it has the uncommon property of contraction during melting. The hexagonal

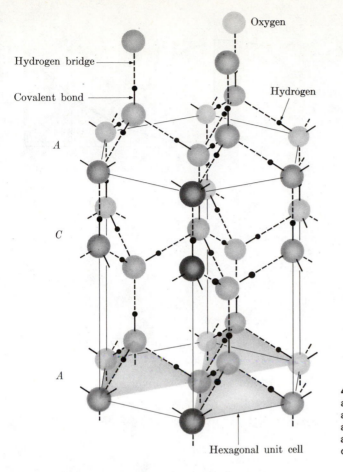

Oxygen

Hydrogen bridge

Covalent bond

Hydrogen

A

C

A

Hexagonal unit cell

4-13 Structure of ice. The hydrogen atoms form a bridge to oxygen atoms of adjacent molecules. This structure has a low packing factor; hence, the volume actually contracts when the long-range order is destroyed during melting.

structure shown in Fig. 4-13 accounts for the shape of snowflakes. Recall from Example 3-3 that the water molecule has a bond angle of 104.5°; also, from Section 3-6, that a hydrogen bridge is formed between an unshielded proton and the unshared electrons of oxygen in adjacent water molecules. These factors dictate the structure which is shown, even though the packing factor is low.* When this long-range order is destroyed during melting, the structure collapses into a less open phase and contraction occurs. Observe from Fig. 4-14, however, that the hydrogen bridge still affects the density of water for several degrees above the melting point. The maximum density occurs at 4°C, and the full liquid expansion coefficient is not achieved until nearly 40°C.

Example 4-3

Determine the unit-cell volume in ice at 0°C. (Use the hexagonal unit cell sketched in Fig. 4-13, and data from Fig. 4-14.)

---

* If the *molecules* were spheres, the packing factor would be 0.34. Since they are not, the packing factor is even less.

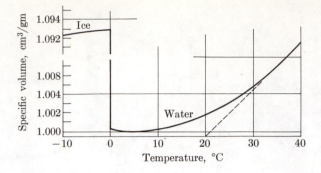

4-14 Specific volume of ice and water. The structure of ice collapses during melting. However, the minimum volume is not reached until 4°C, indicating that the water retains some of the order found in ice.

Answer. The number of molecules per unit cell is most readily determined by an oxygen count:

a) oxygen atoms within the hexagonal prism   7
b) oxygen atoms along prism edges   6/3
c) oxygen atoms centered on the prism end   2/2
d) oxygen atoms at the prism corners   <u>12/6</u>
                 Oxygen/unit cell   12

Therefore,

$$\text{cm}^3/\text{unit cell} = \left(1.093\ \frac{\text{cm}^3}{\text{gm}}\right)\left(12\ \frac{\text{molecules}}{\text{unit cell}}\right)\left(\frac{18.01\ \text{gm/mole}}{0.602 \times 10^{24}\ \text{molecules/mole}}\right)$$
$$= 393 \times 10^{-24}\ \text{cm}^3 \quad (=393\ \text{A}^3).$$

Note. Experimental data obtained by x-ray diffraction (Chapter 5) give values of 7.367 A for the height of the hexagonal prism and 53.15 A^2 for its base, or 392 A^3 for the volume of the unit cell. ◄

4-6 POLYMORPHISM

Within Section 3-7 reference was made to isomeric molecules, i.e., compositions with two distinct molecular structures. Subsequently it was noted, in Section 4-3, that iron has an fcc structure at high temperatures and a bcc structure at low temperatures. Finally, a contrast was made in the previous section between graphite and diamond. The latter form is the stable crystal structure of carbon at very high pressures, the former under ambient conditions.* Multiple crystal structures of the same composition are generally called *polymorphs.* When this happens in the chemical elements, such as in metals, we often call the different phases *allotropes.*

Polymorphism is widespread among engineering materials and becomes very important in their processing and behavior. For example, the heat treatment of steel is designed to make use of the stability of fcc iron at high temperatures and

---

* Diamond, the "gem to eternity," is not stable in our surroundings. Given an opportunity, it would change to graphite, the form with lower free energy. Fortunately, the activation energy (Section 1-4) necessary for graphitization is high because strong bonds must be broken. As a result, the transformation is almost infinitely slow.

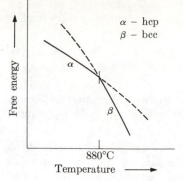

α – hcp
β – bcc

4-15 Polymorphic equilibrium (titanium). At 880°C the free energies of the hcp and bcc structures of titanium are equal. Below this temperature α-Ti is stable because it has the minimum free energy; above 880°C, β-Ti is stable. (Cf. Fig. 3-19.)

its change to bcc on cooling to normal service temperatures (Chapters 17 through 20). Like melting or boiling reactions, polymorphic phase changes arise from free-energy differences. This is shown in Fig. 4-15 for metallic titanium,

$$\text{Ti}_{\text{hcp}} \xrightleftharpoons[\quad]{880°C} \text{Ti}_{\text{bcc}}. \tag{4-1}$$

A volume (and density) change occurs here, since a bcc structure has a lower packing factor than an hcp structure.

Example 4-4

Calculate the volume percent (v/o) change as iron transforms from fcc to bcc.

Answer. Basis: 4 atoms = 1 fcc unit cell = 2 bcc unit cells. From Table 4-1, the atomic radii are 1.270 A and 1.241 A, respectively. Thus

$$a_{\text{fcc}} = 4(1.270 \text{ A})/\sqrt{2} \qquad a_{\text{bcc}} = 4(1.241 \text{ A}/\sqrt{3})$$
$$= 3.591 \text{ A}, \qquad\qquad = 2.863 \text{ A},$$
$$\text{vol}_{\text{fcc}} = (3.591 \text{ A})^3 \qquad \text{vol}_{\text{bcc}} = 2(2.863 \text{ A})^3$$
$$= 46.20 \text{ A}^3; \qquad\qquad = 46.94 \text{ A}^3;$$

and

$$\Delta V/V = (46.94 - 46.20)/46.20 = +1.6 \text{ v/o}.$$

Note. This calculation is based on dimensions at 20°C. ◄

REFERENCES FOR FURTHER READING

Addison, W. E., *Structural Principles in Inorganic Compounds*. New York: Wiley, 1961. Paperback. Supplementary reading for atomic coordination. Introductory level.

Mahan, B., *University Chemistry*, second edition. Reading, Mass.: Addison-Wesley, 1969. Chapter 3 of Mahan gives a helpful background to this chapter. Freshman level.

Moffatt, W. G., G. W. Pearsall, and J. Wulff, *The Structure and Properties of Materials: I. Structure*. New York. Wiley, 1964. Paperback. Atomic packing is presented in Chapter 2. Introductory level. Alternate reading for part of this chapter.

Mott, N., "The Solid State," *Scientific American*, **217** [3], September 1967, pp. 80–89. Presents the internal structure of solids in an easily read manner. Supplementary reading for this and several following chapters.

Parker, E. R., F. R. Davis, and E. L. Langer, *Solid State Structures and Reactions*. Metals Park, O.: American Society for Metals, 1968. The nature of solids is presented at an elementary level.

Rogers, B. A., *The Nature of Metals*, second edition. Metals Park, O.: American Society for Metals, 1964. Supplementary reading for an introduction to crystal structures. Freshman level.

Sisler, H. H., *Electronic Structure, Properties and the Periodic Law*. New York: Reinhold, 1963. Paperback. A nonmathematical presentation of the subject. Bonding in crystals is given under properties. Recommended as brief supplementary reading for the student.

Van Vlack, L. H., *Elements of Materials Science*, second edition. Reading, Mass.: Addison-Wesley, 1964. Bcc, fcc, and hcp structures are presented in Chapter 3.

PROBLEMS

4-1 Aluminum is fcc and has a density of 2.699 gm/cm^3. Calculate (a) its unit-cell dimension and (b) its interatomic distance.

Answer.   a) 4.05 A   b) 2.86 A

4-2 Verify the value of 0.68 as the atomic packing factor for the bcc crystal of Fig. 4-6.

4-3 White tin is tetragonal with $\rho = 7.3 \text{ gm/cm}^3$, $a = 5.83$ A, $c = 3.18$ A. How many atoms are there per unit cell?

Answer. 4

4-4 Show in tabular form the relationship between atom radii and unit-cell dimensions for face-centered, body-centered, and simple cubic metals:

	fcc	bcc	sc
Side of unit cell			
Face diagonal			
Body diagonal			

4-5 Refer to Fig. 4-1. The radius of Na^+ is 0.98 A; that of Cl^- is 1.81 A.

a) What is the volume of the unit cell?

b) The ion-size ratio, r/R, is such that $0.414 < (r/R) < 0.73$. Therefore, CN = 6 and the Cl^- ions do not touch. How much space is there between the surfaces of the nearest Cl^- ions? Between the nearest Na^+ ions?

c) How many Na^+ ions are there per unit cell? Cl^- ions?

Answer. a) $174 \times 10^{-24} \text{ cm}^3$ b) 0.3 A, 2.0 A   c) 4, 4

4-6 Calculate the atomic packing factor of diamond.

4-7 Is there an expansion or contraction as titanium transforms from hcp to bcc? How much? ($R_{hcp} \approx 1.458$ A; $R_{bcc} = 1.425$ A.)

Answer. Expansion of 1.6 v/o. [*Note.* The volume change arises from two counteracting sources: (a) lower packing factor and (b) smaller radii (Table 4-1).]

4-8 Estimate the density of sodium on the basis of its interatomic distance.

4-9 MgO has the same structure as NaCl. Its density is 3.65 gm/cm^3. Use these data to calculate the length of the edge of the unit cell. (Do not use the radii of the ions, $Mg^{2+} = 0.78$ A or $O^{2-} = 1.32$ A, to arrive at your answer.)

Answer. 4.19 A

4-10 Silica glass has a density of 2.3 gm/cm^3 and contains only silicon and oxygen (SiO_2). Assume radii of 0.4 A and 1.3 A, respectively.

    a) What is the atomic packing factor? [*Hint.* How many atoms are there per cm^3?]

    b) Why is it so low?

4-11 The common polymorph of manganese has a complex cubic structure that is neither bcc nor fcc. Its lattice parameter is 8.93 A. From data in Appendix B, determine the number of atoms per unit cell.

Answer. 58

4-12 Verify 0.74 as the atomic packing factor for an hcp metal.

5 □ crystal geometry

5-1 CRYSTAL SYSTEMS

To date, our discussion of crystals has not included any reference to the regular shapes that are commonly identified as crystals (Fig. 5-1a). These planar surfaces are simply external boundaries of the internal atomic and molecular structure. Although the external shape can be altered (Fig. 5-1b), the internal structures remain. While fascinating, the external shape will have relatively little importance for us except to help reveal the internal structure.

Crystals can be categorized into seven *systems* based on their internal structure (Table 5-1). It is important to realize that Table 5-1 refers to axial dimensions and angles and not to external geometry. Thus each example of Fig. 5-2 belongs to the cubic system.

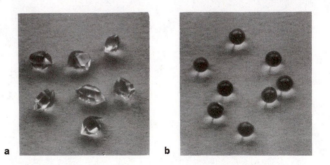

5-1 Crystals (quartz). (a) Euhedral crystals, which have faces determined by the internal structure. (b) Rounded crystals. The *internal structure* remains, although the external shape has been ground to a spherical shape and polished. **a**   **b**

5-2 BRAVAIS LATTICES

External crystal shapes provide us with relatively little information about the internal arrangement of atoms. To discover that information, it is necessary to resort to x-ray diffraction analyses (Section 5-5). In the meantime, we will look at repetitive space-filling arrangements. There are only 14 possibilities. These are called the *Bravais lattices* and will be quite useful to us in subsequent studies.

The Bravais lattices are shown in Fig. 5-3. The characteristic of a lattice is the equivalency of points (Table 5-2). For example, each Bravais point in a face-centered cubic lattice has identical surroundings—the same number of neighbors

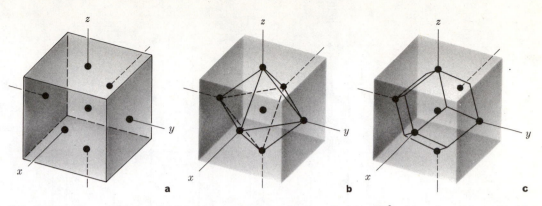

5-2 Cubic system. (a) Cube. (b) Octahedron. (c) Dodecahedron. Each has 90° axial angles and equal axial dimensions; therefore each belongs to the cubic system. (See Table 5-1.)

TABLE 5-1
GEOMETRY OF CRYSTAL SYSTEMS (Compare with Fig. 5-3)

System	Axes	Axial angles
Cubic	$a_1 = a_2 = a_3$	All angles = 90°
Tetragonal	$a_1 = a_2 \neq c$	All angles = 90°
Orthorhombic	$a \neq b \neq c$	All angles = 90°
Monoclinic	$a \neq b \neq c$	2 angles = 90°; 1 angle ≠ 90°
Triclinic	$a \neq b \neq c$	All angles different; none equal 90°
Hexagonal	$a_1 = a_2 \,(= a_3) \neq c$	Angles = 90° and 120° (or 60°)
Rhombohedral	$a_1 = a_2 = a_3$	All angles equal, but not 90°

TABLE 5-2
EQUIVALENT POINTS

Bravais lattices*	Equivalent points
sc, st, so, sm, tri, h, rh	0, 0, 0
bcc, bct, bco	0, 0, 0 $\frac{1}{2}, \frac{1}{2}, \frac{1}{2}$
eco, ecm	0, 0, 0 $\frac{1}{2}, \frac{1}{2}, 0$
fcc, fco	0, 0, 0 $\frac{1}{2}, \frac{1}{2}, 0$ $\frac{1}{2}, 0, \frac{1}{2}$ $0, \frac{1}{2}, \frac{1}{2}$
hcp†	0, 0, 0 $\frac{1}{3}, \frac{2}{3}, \frac{1}{2}$

* See Fig. 5-3 for identification.
† This is a special case of the hexagonal (h) lattice. Refer to Section 5-4 for additional information.

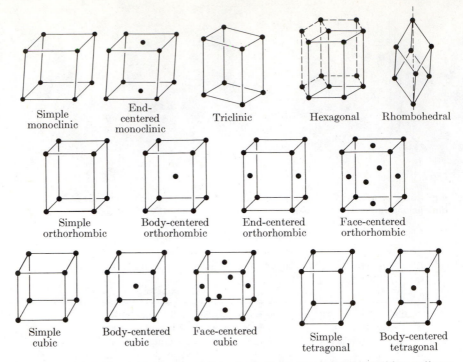

5-3 Space lattices. These 14 *Bravais lattices* represent the basic three-dimensional repetition patterns which fill space. When extended into space, each Bravais point of a lattice has equivalent surroundings. (See Table 5-2).

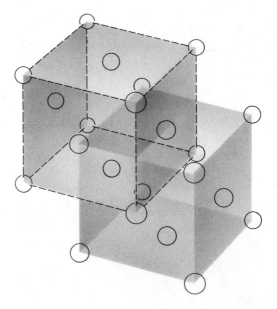

5-4 Face-centered cubic lattice. Each point is equivalent, except for our arbitrary choice of an origin for reference.

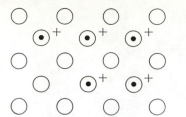

5-5 Equivalency of points in a lattice. All points within a lattice (+) are repeated, even though they may not be directly associated with an atom.

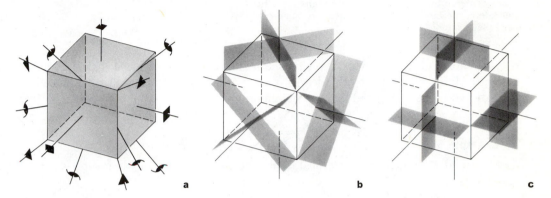

a   b   c

5-6 Symmetry. (a) Axes of symmetry. Three 4-fold axes repeat the structure every 90°; four 3-fold axes, every 120°; six 2-fold axes, every 180°. (b and c) Planes of symmetry. Most cubic crystals also have a center of symmetry.

in the same directions and at the same distances (Fig. 5-4). The *unit cells* of Fig. 5-3 are repetitive within the crystal.

The points of a Bravais lattice may represent a single atom, as in many metals, or they may represent a group of atoms, as they do in the methane crystal of Fig. 4-9. Furthermore, the points of a Bravais lattice need not coincide with any atoms. This is best illustrated with a two-dimensional sketch, as in Fig. 5-5. All points have equivalent points even though they are outside the spherical limits of an atom.

Several *crystal structures* may belong to a given Bravais *lattice*. For example, (1) the diamond structure of Fig. 3-12, (2) the NaCl structure of Fig. 4-1, (3) the CaF$_2$ structure of Fig. 4-8, (4) the CH$_4$ structure of Fig. 4-9, and (5) the elemental fcc crystal structures, such as aluminum, nickel, or argon, of Fig. 4-5 have face-centered lattices. Although the atom arrangements around the equivalent points are not identical in the above five examples, the *repetition pattern* of the equivalent points is that shown for fcc in Fig. 5-3.

Symmetry. Several types of symmetry are possible within a lattice. An fcc crystal, for example, may have as many as 13 axes of symmetry, as shown in Fig. 5-6 (plus nine planes of symmetry and a center of symmetry). In contrast, an ortho-rhombic lattice can have a maximum of only three axes of symmetry. When there

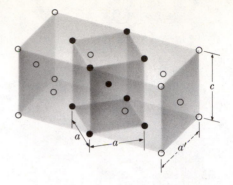

5-7 Body-centered tetragonal. Although bct and fct lattices are equivalent, the former is used because it is smaller and simpler. (See Example 5-1.)

is a choice of lattices, the one with the higher symmetry is used. To illustrate, an fcc lattice can also be divided into body-centered tetragonal (bct) unit cells.* However, this would reduce the maximum symmetry to only five axes and five planes.

Example 5-1

Show that a face-centered tetragonal lattice is equivalent to a body-centered tetragonal lattice.

Answer. See Fig. 5-7. The a of bct equals $a'/\sqrt{2}$ of fct.

Note. Both of the above lattices have the same maximum symmetry; however, the bct lattice is smaller. Therefore, only the bct lattice is recognized among the 14 Bravais lattices. ◄

Example 5-2

a) Determine the Bravais lattice of NaCl (Fig. 4-1).
b) Determine the Bravais lattice of CsCl (Fig. 4-7).

Answer. a) Fcc. [*Note.* This is best appreciated by considering the locations of the Cl^- ions, all of which have identical surroundings.]

b) Simple cubic. [*Note.* This is *not* bcc! Unlike the corner point with a Cl^- ion, the center point has a Cs^+ ion. The two points are *not* equivalent.] ◄

Unit-cell geometry. The unit cell has a-, b-, and c- dimensions parallel to the x-, y-, and z-axes, respectively. As shown in Table 5-1, $a = b = c$ in the cubic system; therefore, it is necessary to identify only one *lattice constant*, a. (If we had chosen to use a bct lattice rather than an fcc lattice, as suggested in the footnote, it would have been necessary to identify *two* lattice constants. Hence, we have another reason for preferring the more symmetrical fcc lattice.)

Any point within a crystal may be chosen as the axial origin for a unit cell (see Fig. 5-5). Once chosen, the origin provides a reference point for other unit-cell

---

* To envision this, use the face-centered point on the lower side of the fcc Bravais lattice as an origin, and rotate the x- and y-axes $45°$. The new lattice has $c_{bct} = a_{fcc}$ and $a_{bct} = a_{fcc}/\sqrt{2}$.

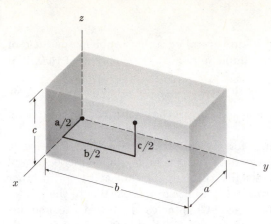

5-8 Unit-cell locations (bco). Positions are measured in terms of axial vectors, **a**, **b**, and **c**.

locations. This is done most advantageously as vectorial translations along axial directions. Thus the center position of *any* unit cell is

$$u, v, w = \tfrac{1}{2}, \tfrac{1}{2}, \tfrac{1}{2}, \tag{5-1}$$

regardless of the Bravais lattice (Fig. 5-8). Here the terms u, v, and w are the respective coefficients of the vectors **a**, **b**, and **c**. Likewise, the center of the base face is

$$u, v, w = \tfrac{1}{2}, \tfrac{1}{2}, 0 \tag{5-2}$$

when the axes have the commonly used orientation shown in Fig. 5-8.

A body-centered lattice (bcc, bct, or bco) has two equivalent points per unit cell. These are 0, 0, 0 and $\tfrac{1}{2}, \tfrac{1}{2}, \tfrac{1}{2}$, as indicated in Table 5-2. In this connection, note that we do not include 1, 0, 0 or other full unit-cell translations, because the 1, 0, 0 site is also the 0, 0, 0 site in the next unit cell. Furthermore, *all* points within a unit cell have equivalent points at translations consistent with the Bravais lattice. Thus, in a body-centered lattice, points u, v, w and $(u + \tfrac{1}{2})$, $(v + \tfrac{1}{2})$, $(w + \tfrac{1}{2})$ are equivalent, regardless of the values of u, v, and w.

Example 5-3

What points within an fcc unit cell are equivalent to $\tfrac{1}{2}, 0, 0$?

Answer. Based on Table 5-2. Translations of 0, 0, 0; $\tfrac{1}{2}, \tfrac{1}{2}, 0$; $\tfrac{1}{2}, 0, \tfrac{1}{2}$; and 0, $\tfrac{1}{2}, \tfrac{1}{2}$ give equivalent points.

u, v, w	$\tfrac{1}{2}, 0, 0$
$(u + \tfrac{1}{2})$, $(v + \tfrac{1}{2})$, $(w + 0)$	$1, \tfrac{1}{2}, 0$
$(u + \tfrac{1}{2})$, $(v + 0)$, $(w + \tfrac{1}{2})$	$1, 0, \tfrac{1}{2}$
$(u + 0)$, $(v + \tfrac{1}{2})$, $(w + \tfrac{1}{2})$	$\tfrac{1}{2}, \tfrac{1}{2}, \tfrac{1}{2}$

Note. To check this, look at the locations of the sodium ions in Fig. 4-1. They all have identical environments when the pattern is repeated in three dimensions. ◄

5-3 LATTICE DIRECTIONS

Vector relationships. Vectors are useful in crystallographic problems because of the repetitive nature of crystals. For example, a point u, v, w is related to 0, 0, 0 by the vector translation

$$\mathbf{r} = u\mathbf{a} + v\mathbf{b} + w\mathbf{c}, \tag{5-3}$$

where $\mathbf{a}$, $\mathbf{b}$, $\mathbf{c}$ are the lattice constants and u, v, w are the vector coefficients along x, y, and z.

A characteristic of major importance for crystal properties is their *anisotropy*, i.e., the fact that the properties are directional. For example, an fcc nickel crystal has greater magnetic permeability diagonally through the unit cell than in other directions; in contrast, the greatest permeability of bcc iron is parallel to the crystal axes. Effects of these differences will be cited in Chapter 16. Likewise, the modulus of elasticity, Y, varies with crystal direction (19,000,000 to 41,000,000 psi for iron). For these reasons it is advantageous to provide a simple index for directions.

Directions are shown as a *ray* originating at the origin and paralleling the vector $u\mathbf{a} + v\mathbf{b} + w\mathbf{c}$. The index $[hkl]$ is used for this direction.* The square brackets are used to differentiate these indices from others to be defined later. (Commas are omitted here.) A quick examination will reveal that $[\frac{1}{2}\frac{1}{2}\frac{1}{2}]$, $[111]$, $[333]$, etc., all have the same direction. Therefore, the index is reduced to the smallest appropriate integers.

Example 5-4

Show the following directions for an orthorhombic lattice with dimensions $a = 2.53$ A, $b = 5.8$ A, and $c = 4.27$ A:

[100]	[112]
[110]	[$\bar{1}$00]
[111]	[2$\bar{1}$0]

[*Note.* A negative direction is indexed by a superbar, or more simply, a *bar*.]

Answer. See Fig. 5-9.

Note. Parallel directions have the same index, since our choice of an origin is arbitrary. ◄

Lattice vectors. A vector passing through identical lattice points is called a *lattice vector*, $\mathbf{r}_{hkl}$. There are as many lattice vectors as there are equivalent points, but in practice only the shorter ones will interest us. The length of the lattice vector is the shortest *repetitive distance*. Thus $\mathbf{r}_{100}$ is preferred to higher multiples, such as $\mathbf{r}_{200}$ and $\mathbf{r}_{300}$.

Example 5-5

a) Identify the following lattice vectors for an fcc lattice:

$\mathbf{r}_{111}$	$\mathbf{r}_{0\bar{1}0}$
$\mathbf{r}_{110}$	$\mathbf{r}_{112}$

---

* Since a direction lacks magnitude, it is not a vector.

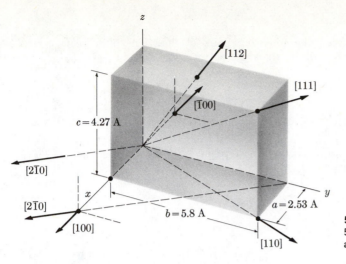

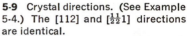

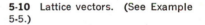

5-9 Crystal directions. (See Example 5-4.) The [112] and $[\frac{1}{2}\frac{1}{2}1]$ directions are identical.

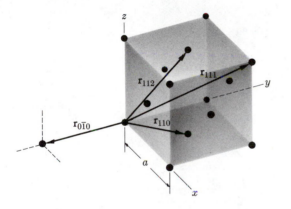

5-10 Lattice vectors. (See Example 5-5.)

b) What are their lengths in terms of the lattice constant, a?

Answer. a) *See* Fig. 5-10.

b) If Eq. (5-3) is expressed in terms of the Pythagorean Theorem,

$$\mathbf{r}_{111} = \sqrt{a^2 + a^2 + a^2} = a\sqrt{3},$$
$$\mathbf{r}_{110} = (1/2)\sqrt{a^2 + a^2 + 0} = a/\sqrt{2},$$
$$\mathbf{r}_{0\bar{1}0} = \sqrt{0 + (-a)^2 + 0} = a,$$
$$\mathbf{r}_{112} = (1/2)\sqrt{a^2 + a^2 + (2a)^2} = (a\sqrt{6})/2$$
$$= \sqrt{(a/2)^2 + (a/2)^2 + a^2} = (a\sqrt{6})/2.$$

Note. These calculations cannot be generalized to other lattices. Each lattice must be considered individually. ◄

Example 5-6

What are the angles between the following pairs of directions of a *cubic* crystal? ($[h_1k_1l_1]$ ⤨ $[h_2k_2l_2]$ is used to indicate the angle between $[h_1k_1l_1]$ and $[h_2k_2l_2]$.)

a) $\theta_a = [001]$ ⤨ $[011]$
b) $\theta_b = [001]$ ⤨ $[111]$
c) $\theta_c = [011]$ ⤨ $[101]$

Answer. Refer to Fig. 5-11.

a) By inspection, $\cos \theta_a = a/a\sqrt{2} = 0.707$, and thus $\theta_a = 45°$.
b) By inspection, $\cos \theta_b = a/a\sqrt{3} = 0.577$, and thus $\theta_b = 54.7°$, or one-half the tetrahedral angle of $109.5°$ (Fig. 3–10a).

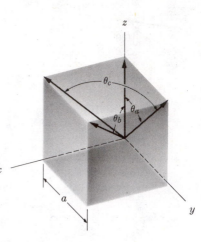

5-11   Angles between directions. (See Example 5-6.)

c) Insert the [112] direction into Fig. 5-11:

$$\theta_c = 2([112] ⤨ [011]),$$
$$\cos (\theta_c/2) = (a\sqrt{6}/2)/(a\sqrt{2}) = 0.866,$$
$$\theta_c/2 = 30°,$$
$$\theta_c = 60°. \quad ◄$$

Angles between directions (cubic crystals). The example just presented shows how the angles between directions may be determined by inspection. This will suffice for most of the calculations in this book. However, with vectors, a *dot product* may be utilized (*for cubic lattices only*):

$$\cos \phi = \frac{\mathbf{A} \cdot \mathbf{B}}{|A| \cdot |B|}. \tag{5-4}$$

When $\phi = [h_1k_1l_1]$ ⤨ $[h_2k_2l_2]$,

$$\cos \phi = \frac{h_1h_2 + k_1k_2 + l_1l_2}{\sqrt{h_1^2 + k_1^2 + l_1^2} \cdot \sqrt{h_2^2 + k_2^2 + l_2^2}}. \tag{5-5}$$

Example 5-7

Calculate the tetrahedral angle of a CH_4 molecule.

Answer. Refer to Fig. 3-10. One of the six equal tetrahedral angles may be identified in terms of *cubic* directions as

$$\phi = [111] \not\angle [\overline{1}\overline{1}1].$$

From Eq. (5-5),

$$\cos \phi = \frac{-1 - 1 + 1}{\sqrt{3}\,\sqrt{3}} = -\frac{1}{3},$$

$$\phi = 109.5° \quad \text{(or } 250.5°\text{)}.$$

This procedure is valid because the CH_4 molecule has *cubic* symmetry. ◀

A family of directions. A number of lattice vectors within a cubic crystal have identical lengths but different directions. For convenience, we group their directions together into a *family* of directions, $\langle hkl \rangle$ (note the use of the caret symbol). All six $\langle 100 \rangle$ directions* possess the same high magnetic permeability in a bcc iron crystal; likewise the elastic moduli in the two $\langle 001 \rangle$ directions† of a simple tetragonal lattice are equal, but different from the elastic moduli in the four $\langle 100 \rangle$ directions‡ of the same lattice.

Linear densities of equivalent points. It is sometimes convenient to speak of linear densities of *equivalent lattice points per unit length* along a given direction:

$$\text{(linear density)}_{hkl} = \mathbf{r}_{hkl}^{-1}. \tag{5-6}$$

Although this is done most commonly when there is only one atom per lattice point, the following example will serve to illustrate the more general situation.

Example 5-8

White tin (stable above 13°C) is bct with equivalent points at $0, 0, 0$ and $\frac{1}{2}, \frac{1}{2}, \frac{1}{2}$ and lattice constants of $a = 5.820$ A and $c = 3.175$ A. The unit cell, however, has four atoms (at $0, 0, 0$; $\frac{1}{2}, \frac{1}{2}, \frac{1}{2}$; $\frac{1}{2}, 0, \frac{1}{4}$; and $0, \frac{1}{2}, \frac{3}{4}$). What are the linear densities of equivalent points in the (a) [111] and (b) [201] directions? [The student is asked to ascertain for himself that the points $0, 0, 0$ and $\frac{1}{2}, \frac{1}{2}, \frac{1}{2}$ are equivalent in a white tin crystal, while $0, 0, 0$ and $\frac{1}{2}, 0, \frac{1}{4}$ are not.]

Answer

a) $\mathbf{r}_{111} = [(5.820 \text{ A}/2)^2 + (5.820 \text{ A}/2)^2 + (3.175 \text{ A}/2)^2]^{1/2}$

$\qquad = 4.4$ A.

Thus the linear density of equivalent points is

$$1/(4.4 \text{ A}) = 2.26 \times 10^7/\text{cm}.$$

---

* [100], [010], [001], [\overline{1}00], [0\overline{1}0], and [00\overline{1}].
† [001] and [00\overline{1}].
‡ [100], [010], [\overline{1}00], and [0\overline{1}0].

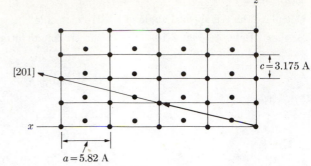

5-12 Linear density [$\mathbf{r}_{201}$ of tin in the (010) plane]. (See Example 5-8.) There are two atoms per repetitive distance.

b) Since $0, 0, 0$ and $\frac{1}{2}, 0, \frac{1}{4}$ are not equivalent, the lattice vector extends from $0, 0, 0$ to $2, 0, 1$ and includes two atoms (Fig. 5-12):

$$\begin{aligned}\mathbf{r}_{201} &= [(2 \times 5.820 \text{ A})^2 + 0 + (3.175 \text{ A})^2]^{1/2} \\ &= 12.1 \text{ A};\end{aligned}$$

thus the linear density of *equivalent points* is

$$1/(12.1 \text{ A}) = 8.25 \times 10^6/\text{cm}.$$

In this direction there are 16.5×10^6 atoms/cm. ◄

5-4 LATTICE PLANES

A lattice contains numerous identifiable planes (Figs. 5-13, 5-14, and 5-15) which will be important in future discussions of crystal properties. For example, cleavage occurs along specific lattice planes. This is most vivid in mica but also occurs in the brittle fracture of many other materials (Chapter 21). Likewise, plastic deformation proceeds most readily along favored directions within specific planes.

Miller indices. A plane may be described either in terms of perpendicular [hkl] direction or in terms of its three axial intercepts. The intercept method will be used here because it will be more pertinent in subsequent calculations.* This method produces *Miller indices*, (hkl). The parentheses are used to avoid confusion with other indices and *must* be used for all single planes.

As we know from analytical geometry,

$$\frac{x}{u} + \frac{y}{v} + \frac{z}{w} = 1 \tag{5-7a}$$

for a plane in space intercepting the x-, y-, and z-coordinates at u, v, and w. We may write

$$hx + ky + lz = 1, \tag{5-7b}$$

where h, k, and l are the reciprocals of the corresponding axial intercepts. This

---

* We shall observe shortly that the two methods give identical indices in cubic lattices, but not in others.

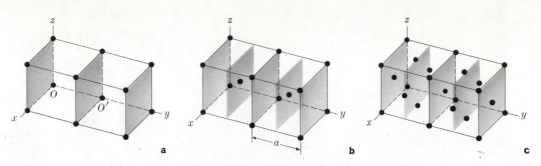

5-13 (010) planes in cubic structures. (a) Simple cubic. (b) bcc. (c) fcc. [Note that the 020 planes included for bcc and fcc are equivalent to (010) planes.]

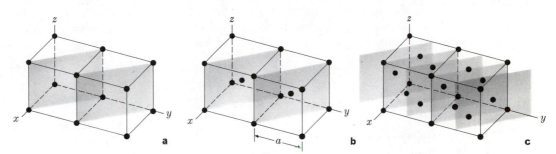

5-14 (110) planes in cubic structures. (a) Simple cubic. (b) bcc. (c) fcc. [The 220 planes included for fcc are equivalent to (110) planes.]

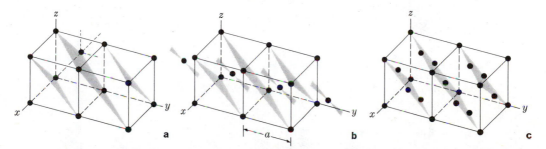

5-15 ($\bar{1}$11) planes in cubic structures. (a) Simple cubic. (b) bcc. (c) fcc. Negative intercepts are indicated by bars above the index. [The $\bar{2}$22 planes included for bcc are equivalent to ($\bar{1}$11) planes.]

permits the use of (*hkl*) indices to identify planes by the following procedure: (1) Note the axial intercepts on the *x*-, *y*-, and *z*-axes in terms of *u*, *v*, and *w*. (2) Write their reciprocals $1/u$, $1/v$, and $1/w$ as *hkl*. (3) Enclose these reciprocals in parentheses and reduce them to the lowest possible set of integers. For example, the lower shaded plane of Fig. 5-16 intercepts each of the three axes at unit-cell distances of 1, 1, and $\frac{1}{2}$; so the (*hkl*) index is (112). A (111) plane intercepts all three axes at one unit distance of each axial dimension (Fig. 5-15); a (110) plane intercepts the *c*-axis at ∞ (Fig. 5-14); and a (010) plane is parallel to two axes (Fig. 5-13).

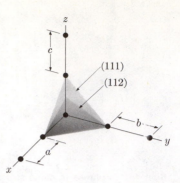

5-16 Miller indices. The (112) plane cuts the three axes at 1, 1, and $\frac{1}{2}$ unit distances.

Since the Miller index of parallel planes is based on an arbitrary choice of origin, we can multiply the index by a constant to avoid fractions or unnecessarily large numbers; for example, $(\frac{1}{2}\frac{1}{3}1)$ corresponds to (326), and (224) corresponds to (112).* Negative intercepts and indices are indicated by a bar.

Plane normals (cubic crystals). Because of the symmetry of *cubic* crystals, where all axial angles equal 90° and $a_1 = a_2 = a_3$, the planes of this system and their normals have the same indices. Thus [312] $\perp$ (312) for *cubic crystals only*. When $a \neq b$, $[hkl] \not\perp (hkl)$.

Planar density of equivalent points. Analogous to linear density, planar density of equivalent points refers to the number of equivalent lattice points per unit area within a plane. This is significant because slip occurs with least stress on those planes with the highest planar density (other factors being equal).

5-17 Planar density [(100) of Pb (fcc)]. A (100) plane of an fcc structure has two lattice points per a². There is one lead atom per lattice point.

Example 5-9

How many atoms per mm² are there on the (100) and (111) planes of lead (fcc)?

Answer. In an fcc metal, each lattice point has only one atom. On the basis of the interatomic data of Table 4-1,

$$a_{\text{Pb}} = \frac{2(3.499 \text{ A})}{1.414} = 4.95 \text{ A}.$$

Figure 5-17 shows that the (100) plane contains two atoms per unit-cell face:

$$(100): \text{ atoms/mm}^2 = \frac{2 \text{ atoms}}{(4.95 \times 10^{-7} \text{ mm})^2} = 8.2 \times 10^{12} \text{ atoms/mm}^2.$$

---

* In general, we remove all common factors, so that (224) corresponds to (112). An exception will be encountered in the next section, where we deal with interplanar spacings, e.g., d_{224}.

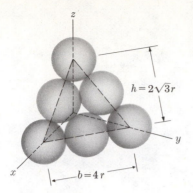

5-18 Planar density [(111) of Pb (fcc)]. A (111) plane of an fcc structure has one lattice point per $2r^2\sqrt{3}$.

Figure 5-18 shows that the (111) plane contains $\frac{3}{6} + \frac{3}{2} = 2$ atoms for the triangular area shown:

$$\text{(111):} \quad \text{atoms/A}^2 = \frac{2}{\frac{1}{2}bh} = \frac{2}{\frac{1}{2}(4)(1.750\ \text{A})(\sqrt{3})(2)(1.750\ \text{A})}$$

$$= 0.095\ \text{atoms/A}^2$$

$$= 9.5 \times 10^{12}\ \text{atoms/mm}^2. \quad \blacktriangleleft$$

Crystal forms. Depending on the symmetry, two or more planes may possess the same geometry, differing only by our arbitrary choice of axes. In the cubic system this is true for the following planes, which constitute a *form*:

$$\begin{Bmatrix} (100) & (\bar{1}00) \\ (010) & (0\bar{1}0) \\ (001) & (00\bar{1}) \end{Bmatrix} = \{100\}.$$

The best verification for this statement is provided by Fig. 4-1 for NaCl. For simplification, the notation $\{hkl\}$ is used to indicate the *form*, or family of planes, rather than the individual planes.

Example 5-10

Cite the planes which fall in (a) the $\{111\}$ form of a cubic crystal, (b) the $\{0001\}$ form of a hexagonal crystal, and (c) the $\{100\}$ form of a tetragonal crystal.

Answer

a) (111), ($\bar{1}$11), (1$\bar{1}$1), (11$\bar{1}$), ($\bar{1}\bar{1}\bar{1}$), (1$\bar{1}\bar{1}$), ($\bar{1}$1$\bar{1}$), ($\bar{1}\bar{1}$1). See Fig. 5-2(b). The latter four are not distinct, since they parallel the first four.
b) Only (0001) and (000$\bar{1}$). [See the next subsection for $(hkil)$ notations.]
c) (100), (010), ($\bar{1}$00), (0$\bar{1}$0). [*Note.* (001) and (00$\bar{1}$) cannot be included in part (c) because $1/c \neq 1/a$ in a tetragonal lattice.] $\blacktriangleleft$

Miller-Bravais indices (hexagonal crystals). Any plane can be identified by the three Miller indices. In the hexagonal systems, however, it is possible and frequently desirable to use four axes, three of them coplanar (Fig. 5-19). This leads to four

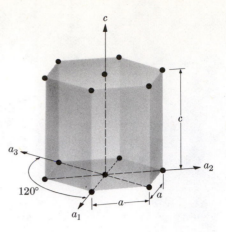

5-19 Hexagonal axes. The a_1-, a_2-, and a_3-axes are coplanar. Although the resulting i-index is not independent (Eq. 5-8), ($hkil$) indices are often useful.

intercepts and ($hkil$) indices (called Miller-Bravais indices), where i is the additional, nonindependent, index. It can be shown that

$$h + k = -i. \tag{5-8}$$

The Miller-Bravais system utilizes a hexagonal prism as a unit cell rather than the heavily outlined rhombic prism of Fig. 5-3. The ratio of volumes is 3-to-1; thus the ratio of atoms present will also be 3-to-1.

The Miller-Bravais indices are often favored because they reveal hexagonal symmetry more clearly. Alternatively, the three-axis Miller index is preferred for the relationships shown in the next two subsections. Fortunately, Eq. (5-8) permits immediate conversion.

Example 5-11

Sketch the $\{10\bar{1}1\}$ and the $\{11\bar{2}0\}$ planes of a hexagonal lattice.

Answer. See Fig. 5-20.

Note. Equivalent planes such as $(1\bar{2}10)$ or $(\bar{2}110)$ in Fig. 5-20(b) might escape detection if we simply used the rhombic prism of Fig. 5-3 and the Miller indices $\{110\}$. ◄

Intersection of planes. The line of intersection $[hkl]$ of two planes, $(h_1k_1l_1)$ and $(h_2k_2l_2)$, can be determined by taking a *cross product* of Miller indices of the two planes. We will not derive this relationship, but merely note its simplicity by using the above indices:

$$\begin{bmatrix} h \\ k \\ l \end{bmatrix} = \begin{matrix} k_1l_2 - l_1k_2 \\ l_1h_2 - h_1l_2 \\ h_1k_2 - k_1h_2 \end{matrix} \tag{5-9}$$

Thus the direction of intersection between a (112) plane and a (111) plane is $[\bar{1}10]$, or alternatively $[1\bar{1}0]$. (For verification, see Fig. 5-16.)

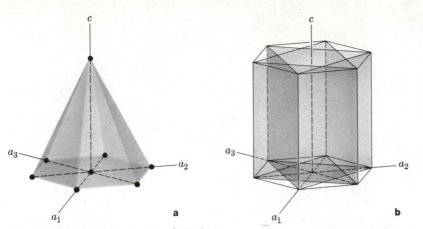

5-20 Miller-Bravais indices. (a) $\{10\bar{1}1\}$ form. This form also includes six more planes which produce an inverted hexagonal pyramid below the basal plane. (b) $\{11\bar{2}0\}$ form. The form includes $(11\bar{2}0)$, $(1\bar{2}10)$, $(\bar{2}110)$, $(\bar{1}\bar{1}20)$, $(\bar{1}\bar{2}\bar{1}0)$, and $(2\bar{1}\bar{1}0)$ planes to complete a hexagonal prism. The unit cell is enclosed by $\{10\bar{1}0\}$ and $\{0001\}$ planes as shown in Fig. 5-19.

Directions within a plane. There is an unlimited number of directions within a given plane. The *dot product* of the index for the direction $[h_1k_1l_1]$ and the plane normal $[h_2k_2l_2]$ is zero when the direction lies within the plane $(h_2k_2l_2)$. Consequently,

$$h_1 \cdot h_2 + k_1 \cdot k_2 + l_1 \cdot l_2 = 0. \tag{5-10}$$

To illustrate, a (111) plane contains the following directions, among many others: $[11\bar{2}]$, $[\bar{1}01]$, $[5\bar{6}1]$, $[3\bar{1}\bar{2}]$, and $[\bar{9}72]$.

The above relationship can be proven simply for a cubic crystal by noting that the angle between an included direction $[h_1k_1l_1]$ and a plane normal $[h_2k_2l_2]$ is 90°, and hence $\cos \phi$ of Eq. (5-5) is zero. Proof for other systems is more complex.

The above relationship is useful as a double check on calculations concerning plastic deformation (Chapter 11), where the slip direction must lie within the slip plane.

5-5 DIFFRACTION

Bragg's law. The structures of crystals are analyzed most readily by x-ray (or electron) diffraction procedures. As radiation passes through a periodic structure, the wavefronts are diffracted.

The physical explanation of diffraction may be summarized by stating that the electrons associated with each atom are polarized as the wavefront passes (Fig. 5-21). The induced dipoles are reversed at the frequency of the radiation, and in turn emit their own radiations in *all* directions at that same frequency. This new radiation is coherent only in directions perpendicular to the tangents shown. At other angles, the wavefronts are lost by destructive interference.

A geometric explanation of diffraction leads to the formulation of Bragg's law:

$$\lambda = 2d_{hkl} \sin \theta, \tag{5-11}$$

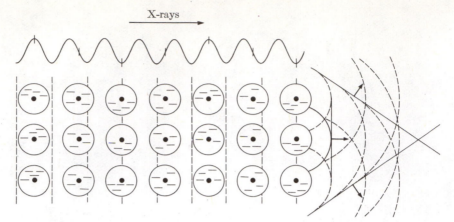

X-rays

5-21 Diffraction. A coherent wavefront emerges from the individual wave tangents.

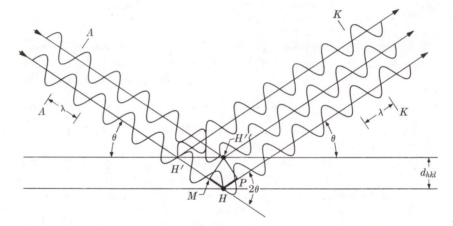

5-22 Bragg's law of diffraction. Diffraction occurs only when the conditions of Eq. (5-11) are met. Thus the diffraction angles for x-rays of a known wavelength provide a means of calculating interplanar dimensions.

where λ is the wavelength of the radiation, d_{hkl} is the spacing between (hkl) planes, and θ is the diffraction angle. As shown in Fig. 5-22, the rays of the incident wavefront, AA, are in phase. The diffracted wavefront, KK, can be coherent (i.e., in phase) only if the distance MHP is equal to one or more integral wavelengths. By calculation, this must equal $2d_{hkl} \sin \theta$.

Interplanar spacings. The interplanar spacings referred to in Bragg's law (Eq. 5-11) are revealed in Figs. 5-13, 5-14, and 5-15 as the perpendicular distances between parallel planes. Since two planes must be involved for interplanar spacings, our calculations will be simplified if we let one plane pass through the origin of the lattice, and measure the perpendicular distance, d_{hkl}, to the second, parallel plane, (hkl). Thus, in Fig. 5-13(a), d_{010} of a simple cubic lattice is the edge of the unit cell, a. In Fig. 5-13(b), d_{010} of a bcc lattice is also a, while d_{020} is $a/2$. Further,

TABLE 5-3

INTERPLANAR SPACINGS

System	Spacings (d_{hkl})
Cubic	$\left[\dfrac{h^2 + k^2 + l^2}{a^2}\right]^{-1/2}$
Tetragonal	$\left[\dfrac{h^2 + k^2}{a^2} + \dfrac{l^2}{c^2}\right]^{-1/2}$
Orthorhombic	$\left[\dfrac{h^2}{a^2} + \dfrac{k^2}{b^2} + \dfrac{l^2}{c^2}\right]^{-1/2}$
Hexagonal	$\left[\dfrac{4(h^2 + k^2 + kh)}{3a^2} + \dfrac{l^2}{c^2}\right]^{-1/2}$
	or $\left[\dfrac{2(h^2 + k^2 + i^2)}{3a^2} + \dfrac{l^2}{c^2}\right]^{-1/2}$

d_{110} of a bcc lattice (Fig. 5-14b) is $a/\sqrt{2}$, and d_{111} of the fcc lattice in Fig. 5-15(c) is $a/\sqrt{3}$.

For cubic lattices,

$$d_{hkl} = \frac{a}{\sqrt{h^2 + k^2 + l^2}}. \tag{5-12}$$

Interplanar spacings for several other systems are listed in Table 5-3.

[For Miller indices, we rationalize the index to the lowest integer set. Thus, except for our arbitrary choice of an origin, the (020) plane of a bcc lattice is identical to the (010) plane, and we use (010) to indicate all such planes. Since an interplanar spacing involves a distance as well as a plane, we must be more specific than for plane identification only. In Fig. 5-13(c), for example, d_{020} is the perpendicular distance from the origin to the plane going through intercepts ∞, $\frac{1}{2}$, ∞; while d_{010} is the perpendicular distance from the origin to the plane going through intercepts ∞, 1, ∞.]

5-6 DIFFRACTION PATTERNS

Since Bragg's law (Eq. 5-11) is adhered to strictly, it is possible to use diffraction procedures to determine crystal structures and lattice dimensions for identification purposes. Two of several diffraction techniques will be described briefly: (1) Laue patterns and (2) powder patterns.

The two experimental variables of Bragg's law are the wavelength, λ, and the diffraction angle, θ. If we chose both randomly, our chances of observing diffraction would be low. However, if "white" x-rays are used (i.e., a spectrum rather than a single wavelength), diffraction occurs from those x-rays which have the correct wavelength to obey Bragg's law. This procedure gives the *Laue* pattern (Fig. 5-23). It may be used to identify the crystal orientation and the lattice, since each spot represents a given crystal plane.

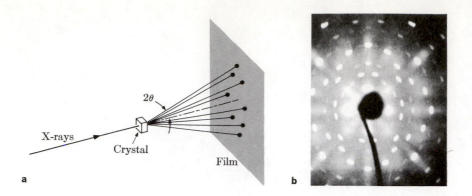

5-23 Laue pattern. (a) Procedure. (b) Results. Each point on the film arises from a set of parallel crystal planes. (Courtesy of L. Thomassen, The University of Michigan).

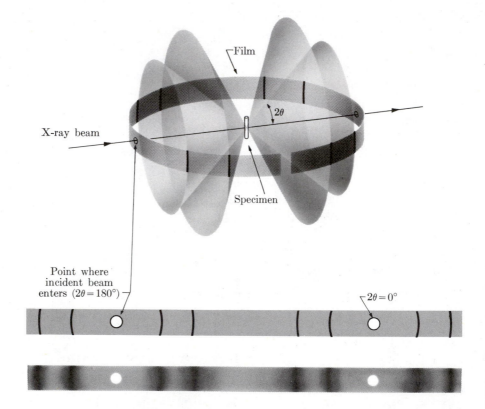

5-24 Powder pattern. Each diffraction line arises from a specific interplanar spacing. Through the use of monochromatic x-rays, the value of d_{hkl} may be calculated (Eq. 5-11). (Based on B. D. Cullity, *Elements of X-ray Diffraction*. Reading, Mass.: Addison-Wesley, 1956).

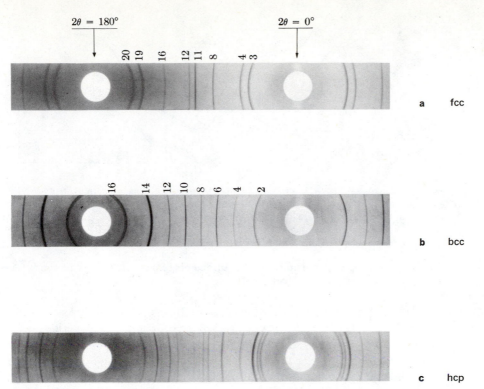

5-25 X-ray diffraction patterns for (a) copper, fcc; (b) tungsten, bcc; and (c) zinc, hcp. The numbers correspond to the values of $(h^2 + k^2 + l^2)$ in Table 5-4. Values of 2θ may be measured directly from the film arc. (Cf. Fig. 5-24.) (B. D. Cullity, *Elements of X-ray Diffraction*, Reading, Mass.: Addison-Wesley, 1956.)

The *powder pattern* utilizes a monochromatic wavelength directed at many particles of a fine powder. As a result, diffraction occurs from those particles which have the correct orientation to obey Bragg's law (Fig. 5-24). A strip of film (or a Geiger counter) records the 2θ-angles as diffraction lines (Fig. 5-25). These angles, together with the known values of λ, permit a direct calculation of interplanar spacings.

Example 5-12

The first line (from 0°) in Fig. 5-25(a) is for the d_{111} diffraction of copper.

a) What is the lattice constant, a, of copper if the wavelength of the x-rays used is 1.54 A?
b) What is the atomic radius of copper?

Answer

a) Interpolating between $2\theta = 0°$ and $2\theta = 180°$, we find that the first lines in Fig. 5-25(a) are at $2\theta \cong 43°$. Therefore,

$$d_{111} = \frac{1.54 \text{ A}}{2(\sin 21.5°)} = 2.1 \text{ A},$$

$$a = d_{111}\sqrt{1^2 + 1^2 + 1^2} = 0.0 \text{ A},$$

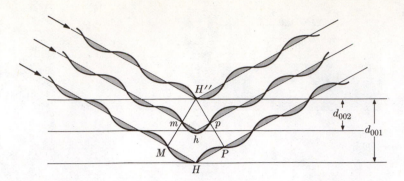

5-26 Diffraction cancellation. No d_{001} diffraction occurs from one extra wavelength in *MHP* because *mhp* is only one-half wavelength and cancellation results. Second-order ($n = 2$) diffraction from d_{001} is the same as first-order ($n = 1$) diffraction from d_{002}.

b) $r = \sqrt{2}(a/4) = 1.28$ A.

Note. Within the accuracy of our angular measurements, this corresponds to the inter-atomic distance listed as 2.556 A in Table 4-1. ◀

Diffraction lines. Not all interplanar spacings produce diffraction. Assume, for example, that Eq. (5-11) (Bragg's law) is to be used for a d_{001} diffraction in fcc copper (compare Fig. 5-13c). As shown in Fig. 5-26, the extra wavelength in the distance *MHP* corresponds to only one-half wavelength, *mhp*, for the d_{002} diffraction. The two rays emerge 180° out of phase and are destructively canceled. No d_{001} lines appear in an fcc diffraction pattern.

TABLE 5–4

DIFFRACTION LINES IN CUBIC CRYSTALS

Diffraction line, d_{hkl}	$(h^2 + k^2 + l^2)$		
	sc	bcc	fcc
100	1	—†	—‡
110	2	2	—
111	3	—	3
200	4	4	4
210	5	—	—
211	6	6	—
—	*	*	*
220	8	8	8
221	9	—	—
310	10	10	—
311	11	—	11
222	12	12	12

* Exclude 7, 15, and 23 because they do not correspond to any index combination.
† Exclude all lines with $(h^2 + k^2 + l^2) =$ odd.
‡ Exclude all lines with mixed odd and even in h, k, and l.

A detailed inspection of Figs. 5-13, 5-14, and 5-15 would reveal that d_{010} diffraction is canceled in both bcc and fcc; d_{110} lines will appear in simple cubic and bcc, but not in fcc; and d_{111} lines will be canceled in bcc lattices. Other diffraction lines can be similarly examined, with the result that a list of observed (or canceled) lines can be made. With appropriate proofs, it can be shown that the only lines observed in *cubic* crystals are those which have the values of $(h^2 + k^2 + l^2)$ that are listed in Table 5-4.

The significance of Table 5-4 is that it permits us to identify the Bravais lattice of a cubic crystal. Figure 5-25(a and b) illustrates that fcc copper and bcc tungsten have a different sequence of lines. Each gives the "fingerprint" of the specific lattice.

Example 5-13

Verify that the diffraction lines for copper in Fig. 5-25(a) conform with the fcc listing of Table 5-4. (Assume that the first line is for d_{111}.)

Answer. Interpolating in Fig. 5-25(a), we find that the diffraction lines have the positions indicated in column A below.

Also, from Eqs. (5-11) and (5-12) and the above assumption,

$$\frac{\lambda}{2a} = \frac{\sin\theta}{\sqrt{h^2 + k^2 + l^2}} = \frac{0.367}{\sqrt{3}} = 0.212.$$

A	B	C	D
2θ, deg	$\sin^2\theta$	$h^2 + k^2 + l^2 =$ $[\sin\theta/(\lambda/2a)]^2$	d_{hkl}
43	0.135	3	111
50	0.18	4	200
74	0.36	8	220
90	0.49	11	311
94	0.54	12	222
116	0.72	16	400
136	0.85	19	331
146	0.90	20	420 ◀

Second-order diffraction. In copper and other fcc metals, there is no diffraction from d_{010} spacings, because of the previously discussed cancellation. Diffraction is possible from d_{020} spacings if the wavelength of Fig. 5-26 is changed to equal *mhp*. Under these conditions *MHP* is equal to *two* wavelengths and we can speak of a *second-order diffraction* for d_{010}. Likewise, a first-order (one-wavelength) diffraction of d_{220} is a second-order diffraction for d_{110}. This has significance for the fifth line, d_{222}, of Example 5-13, since copper has no lattice points on a (222) plane; however, it does have a second-order d_{111} diffraction.

The more general form of Eq. (5-11) is

$$n\lambda = 2d \sin\theta, \tag{5-13}$$

where n is an integer indicating the diffraction "order." In practice, the integer n seldom has significance in calculations, so we always will use Eq. (5-11) even though the intervening plane may pass through hypothetical lattice points.*

Example 5-14

Will the CsCl structure (Fig. 4-7) give a d_{010} diffraction line? Explain.

Answer. Since CsCl is simple cubic (Example 5-2b), it can have a d_{010} line (Table 5-4). More specifically, diffraction from d_{020} spacing does not fully cancel that from d_{010} spacing, because in CsCl, d_{020} involves planes of unlike ions while d_{010} is from one kind only. The two atoms have different numbers of electrons, and different polarizations for diffraction. ◄

5-7 ANISOTROPY AND PROPERTIES

Properties such as density and specific heat are affected by the crystal structure primarily through the atomic packing factor. Such properties do not have direction. However, other properties, such as magnetic permeability, elastic moduli, and thermal conductivity, are highly directional, i.e., anisotropic. In present-day materials science, it is important to recognize the anisotropy of materials, as opposed to an isotropic continuum. Wherever possible in this text, we shall point out the consequences of anisotropy.

REFERENCES FOR FURTHER READING

Barrett, C. S., and T. B. Massalski, *Structure of Metals*, third edition. New York: McGraw-Hill, 1966. The classic text in the crystallography of metals. Advanced undergraduate and graduate level.

Cullity, B. D., *Elements of X-ray Diffraction*. Reading, Mass.: Addison-Wesley, 1956. The standard undergraduate textbook in x-ray diffraction.

Kittel, C., *Introduction to Solid State Physics*, third edition. New York: Wiley, 1966. Chapter 1 is good supplementary reading on crystal structure at the level of this text. Chapter 2 presents the details of crystal diffraction at a more advanced level.

Kroeber, G. G., *Properties of Solids*. Englewood Cliffs, N. J.: Prentice-Hall, 1962. Chapter 4 gives an introductory presentation of the structure of solids. Chapter 7 follows up on a more rigorous basis through the use of transformation matrices. A useful supplement for the student who has studied tensors.

Moffatt, W. G., G. W. Pearsall, and J. Wulff, *The Structure and Properties of Materials: I. Structure*. New York: Wiley, 1964. Paperback. Appendix III introduces vector notations for crystallographic indices. Appendix IV elaborates on x-ray diffraction.

Nutt, M., *Principles of Metallurgy*. Columbus, O.: Merrill, 1968. Chapter 3 provides background reading for this chapter. Introductory level.

---

* Lattice points, rather than atoms, receive attention because the former provide equivalent diffraction power.

Parker, E. R., F. R. Davis, and E. L. Langer, *Solid State Structure and Reactions*. Metals Park, O.: American Society for Metals, 1968. The crystalline nature of solids is introduced at the precollege level.

Richman, M. H., *Science of Metals*. Waltham, Mass.: Blaisdell, 1967. Crystal structure is introduced in Chapter 2. Alternate reading for this topic. Chapter 3 presents x-ray diffraction.

Rogers, B. A., *The Nature of Metals*, second edition. Metals, Park, O.: American Society for Metals, 1964. Self-education text. An excellent book for the student who needs additional background for crystal structure.

Rosenthal, D., *Introduction to Properties of Materials*. Princeton, N. J.: Van Nostrand, 1964. The determination of the structure of solids is presented in an introductory manner in Chapter 4.

Van Vlack, L. H., *Elements of Materials Science*, second edition. Reading, Mass.: Addison-Wesley, 1964. Chapter 3 gives another but parallel presentation of crystal structure. In general, it is more elementary than this chapter.

Van Vlack, L. H., *Physical Ceramics for Engineers*. Reading, Mass.: Addison-Wesley, 1964. The crystal structure of compounds is presented in Chapter 3. More detail is given than in this book. Undergraduate level.

Wood, E. A., *Crystals and Light*. Princeton, N. J.: Van Nostrand, 1964. Paperback. The first chapters discuss symmetry, crystal lattices, and planes. Alternate reading for some of this chapter. Introductory level.

Wyckoff, R. W. G., *Crystal Structures*, Vol. 1, second edition. New York: Wiley, 1963. This first volume of several includes elements and the simpler compounds. Prime reference book for crystal structures.

PROBLEMS

5-1 a) Give the unit-cell locations of all atoms in diamond (Fig. 3-12).
b) Give the unit-cell locations of all atoms in NaCl (Fig. 4-1).

Answer. a) $0, 0, 0$ $\quad 0, \frac{1}{2}, \frac{1}{2}$ $\quad \frac{1}{2}, 0, \frac{1}{2}$ $\quad \frac{1}{2}, \frac{1}{2}, 0$
$\quad\quad\quad \frac{3}{4}, \frac{3}{4}, \frac{3}{4}$ $\quad \frac{3}{4}, \frac{1}{4}, \frac{1}{4}$ $\quad \frac{1}{4}, \frac{3}{4}, \frac{1}{4}$ $\quad \frac{1}{4}, \frac{1}{4}, \frac{3}{4}$

b) Na^+ $\quad \frac{1}{2}, 0, 0$ $\quad 0, \frac{1}{2}, 0$ $\quad 0, 0, \frac{1}{2}$ $\quad \frac{1}{2}, \frac{1}{2}, \frac{1}{2}$
$\quad\; Cl^-$ $\quad 0, 0, 0$ $\quad 0, \frac{1}{2}, \frac{1}{2}$ $\quad \frac{1}{2}, 0, \frac{1}{2}$ $\quad \frac{1}{2}, \frac{1}{2}, 0$

5-2 A cubic metal has atoms *only* at $\frac{1}{4}, \frac{1}{4}, \frac{1}{4}$ and $\frac{3}{4}, \frac{3}{4}, \frac{3}{4}$.

a) Sketch such a unit cell.
b) What is the common name for this lattice? [*Hint*. Translate the corner of the unit cell to one of these positions.]
c) Sketch the bcc unit cell of vanadium, with an atom placed at $0, \frac{1}{2}, 0$.

5-3 In one form of SiO_2 (Fig. 18-1) the crystal structure is cubic with a silicon atom at $0, 0, 0$ and the closest oxygen at $\frac{1}{8}, \frac{1}{8}, \frac{1}{8}$. If the radii of the two ions are 0.39 A and 1.32 A, respectively, what is the volume of the unit cell?

Answer. 493 A^3

5-4 Sketch an fcc unit cell which has been translated so that the *origin* rests at the $\frac{1}{2}, \frac{1}{2}, \frac{1}{2}$ position of Fig. 4-5.

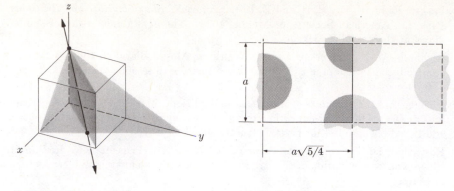

5-27 See Problem 5-15.　　　　　　**5-28** See Problem 5-17.

5-5 A unit cell of fcc nickel is positioned so that the cell origin is in the interstitial position between 4 adjacent atoms. What are the locations of the atoms in the unit cell?

Answer. $\frac{1}{4}, \frac{1}{4}, \frac{1}{4}$　　$\frac{1}{4}, \frac{3}{4}, \frac{3}{4}$　　$\frac{3}{4}, \frac{1}{4}, \frac{3}{4}$　　$\frac{3}{4}, \frac{3}{4}, \frac{1}{4}$

5-6 What is the direction between the $\frac{1}{2}, \frac{1}{4}, 0$ location and the $\frac{1}{4}, 0, \frac{1}{2}$ location?

5-7 A [112] direction passes into a unit cell $\frac{1}{3}, \frac{1}{5}, 0$. Where does that particular ray leave the unit cell?

Answer. $\frac{5}{6}, \frac{7}{10}, 1$

5-8 The $\langle 110 \rangle$ family of equivalent directions includes 12 directions in a cubic crystal (half of which are simply the negative of the other six). Identify the six distinct directions.

5-9 Identify the four distinct directions of the $\langle 101 \rangle$ family in a tetragonal crystal.

Answer. [101], [011], [10$\bar{1}$], [01$\bar{1}$]. (*Note.* [$\bar{1}$0$\bar{1}$], [0$\bar{1}$$\bar{1}$], [$\bar{1}$01], and [0$\bar{1}$1] are the negative directions of the above set. Also, [110] and [1$\bar{1}$0] are not included because $c \neq a = b$.)

5-10 What is the linear density of equivalent points in the [112] direction of silver?

5-11 What is the lattice vector in (a) the [120] direction of gold? (b) the [111] direction? (c) the [211] direction?

Answer. a) 9.1 A　　b) 7.1 A　　c) 5.0 A

5-12 What is the angle between [111] and [110] in a cubic crystal?

5-13 A force of 1000 dynes is applied in the [221] direction of a cubic crystal. What is the resolved force in the [201] direction?

Answer. 745 dynes

5-14 a) Show the (102) and (231) planes of a cubic unit cell.
   b) Show the same planes for an orthorhombic unit cell with $a/b/c = 0.9/1.0/3.2$.

5-15 a) What are the Miller indices of the two shaded planes sketched in Fig. 5-27?
   b) What is the direction of the line of intersection?

Answer. a) (1$\bar{1}$0) and (212)　　b) [22$\bar{3}$]

5-16 Refer to Fig. 5-17. Make a similar sketch to scale of the (101) plane (a) of sodium (bcc); (b) of nickel (fcc).

5-17 Refer to Fig. 5-17. Figure 5-28 is a comparable sketch of another set of fcc planes in lead. What are they?

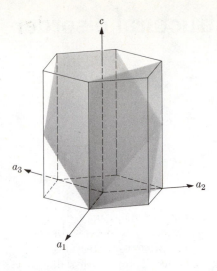

5-29 See Problem 5-19.

Answer. {012}

5-18 a) Sketch the $(11\bar{2}0)$ plane of a hexagonal unit cell.

b) What other planes belong to the same form?

5-19 What planes are shaded in Fig. 5-29?

Answer. $(10\bar{1}2)$ and $(11\bar{2}0)$

5-20 Show that $[\bar{2}201]$ is the direction of the line of intersection between the two planes sketched in Fig. 5-29.

5-21 What is the planar density of equivalent points on the (021) plane of copper?

Answer. $6.9 \times 10^{14}/\text{cm}^2$

5-22 How many atoms are there per cm^2 on the (111) plane of vanadium?

5-23 What are the values of 2θ for the first three diffraction lines of bcc iron when the x-ray wavelength is 0.58 A?

Answer. 16.4°, 23.6°, 28.7°

5-24 Refer to Fig. 5-25(b). The x-rays had a wavelength of 1.54 A.

a) Determine the lattice constant for tungsten.

b) Determine the interatomic distance.

5-25 An iron wire 1 mm in diameter and 2 cm long contains only one crystal. The [111] direction is parallel to the length of the wire.

a) What angle does the normal to the (100) plane make to the length of the wire?

b) What crystal direction in the (100) plane is in the radial direction of the wire?

c) A force of 7000 dynes is applied along the wire. What is the resolved force in the [100] direction?

Answer. a) 55°   b) $[0\bar{1}1]$ c) 4050 dynes

5-26 A sodium chloride crystal is used to measure the wavelength of some x-rays. The diffraction angle is 5.2° for the d_{111} spacing of the chloride ions. What is the wavelength? (The lattice constant is 5.63 A.)

6 □ structural disorder

6-1 IMPERFECTIONS IN CRYSTALS

Fortunately crystals are seldom, if ever, perfect. This statement may come as a surprise after spending a full chapter on the high degree of order exhibited by crystals. Furthermore, our esthetic sense urges us to appreciate perfection.

However, were it not for imperfections, materials could not possess ductility and could not be mechanically formed into engineering products. The heat treatment and enhancement of properties of steels depend on structural changes and their accompanying imperfect transition states. Plastics, ceramics, and glasses can be strengthened by the incorporation of impurities. Semiconductors have controlled conductivities because of electronic imperfections that accompany crystal imperfections. This listing could be continued at great length. The point to be made is that structural imperfections are important aspects of many properties of engineering materials.

Our starting place for a study of *structural disorder* is the perfect crystal, because it permits us to classify the type and extent of disorder. Within this chapter we shall pay attention first to thermal disorder; then we shall categorize imperfections into point defects, dislocations (linear defects), and boundaries (two-dimensional discontinuities) and their microstructural consequences; finally we shall look at amorphous, or noncrystalline, solids which in effect possess three-dimensional defects. In Chapter 8 we shall consider the role of impurities.

6-2 THERMAL DISORDER

Atomic motion is reduced to a minimum at absolute zero. Above that temperature, there is a gradual increase in the *amplitude* of the vibration.* According to our spring model (Fig. 3-4), these vibrations introduce both potential energy and kinetic energy into the structure. At the end point of the vibration, the kinetic energy is zero because the atoms come to a momentary stop; however, there is high potential energy because the interatomic distances are not at an equilibrium spacing (Fig. 6-1). Near the midpoint of the vibration, the potential energy decreases toward the minimum observed at absolute zero, but the velocity and the kinetic energy reach a maximum.

---

* For a given atom within a structure, the frequency of vibration remains essentially constant.

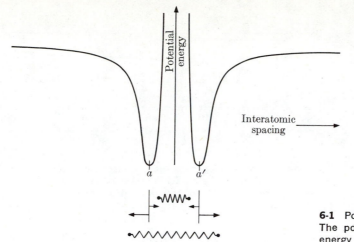

Heat capacity. The mean thermal energy per atom, $\overline{E}$, of a monatomic *gas* was expressed in Eq. (2-12) as

$$\overline{E} = \tfrac{3}{2}kT.$$

In a monatomic gas, all of the thermal energy is kinetic (KE), while in a solid there is also potential energy (PE). In fact, as described in the previous paragraph,

TABLE 6-1

HEAT CAPACITIES OF CLOSE-PACKED SOLIDS

Solid	C_v, 300°K, cal/mole · °K	C_v, 800°K, cal/mole · °K
Ag	6.1	6.7
Al	5.8	7.3
Cr	5.8	6.9
Cu	5.8	6.6
Fe	5.9	8.9
Mg	5.7	7.3
P	5.7	—
Pb	6.5	—
Zn	6.0	—
CuO	11.0	12.9
FeO	12.1	13.0
MgO	9.0*	11.5
ZnO	10.7	12.1
Ag$_2$S	18.0	21.6
SiO$_2$	13.4*	17.1
Al$_2$O$_3$	20.0*	28.8
Cr$_2$O$_3$	26.4	29.9

* The heat capacities of compounds of light elements do not approximate 6 times the number of atoms per "molecule" until above room temperature.

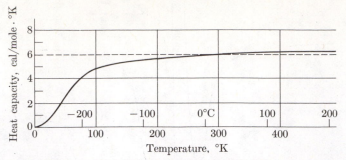

6-2 Heat capacity (silver). At room temperature and above, the specific heat of most ionic and metallic solids is approximately 6 cal/mole·°K. (See Eq. 6-1b.) These data are for constant-volume heat capacity, C_v. Since the energy change with volume changes is very small in solids at normal pressures, these data may be used for constant-pressure heat capacities, C_p.

the two energies must be equal, on the average. Therefore, in a simple *solid* at elevated temperatures,

$$\overline{E} = \text{KE} + \text{PE} \cong 3kT. \tag{6-1a}$$

Since $R = k(0.602 \times 10^{24} \text{ atoms/mole})$, the energy per mole is

$$E \cong 3RT, \tag{6-1b}$$

where R is the "gas constant" of 1.987 cal/mole · °K. The coefficient $3R$ is the heat capacity and has a value of approximately 6 cal/mole · °K. This is illustrated in Table 6-1 for a number of close-packed solids.

The sentence containing Eq. (6-1a) was carefully phrased to include *simple* solids and *elevated* temperatures. In the solids of Table 6-1, the bulk of the thermal energy is absorbed by atomic vibration as described in Fig. 6-1. More complex solids such as polymers can possess rotational as well as the above translational movements, so that proportionally more energy is required per mole.

The fact that Eq. (6-1) is applicable only at elevated temperatures is best illustrated by Fig. 6-2. Except for solids of light elements (which have a high vibrational frequency because of their low mass), the heat capacity is reasonably constant at room temperature and above. Heat capacities always drop significantly at reduced temperatures, finally approaching 0°K as a function of T^3. A quantum-mechanical model is required to explain this low-temperature relationship. The interested reader may consult the references at the end of the chapter for further detail.

Example 6-1

Estimate the heat capacity (per gram) at room temperature (a) of chromium metal, (b) of Cr_2O_3.

Answer

a) The atomic weight of chromium is 52.00 gm. Thus,

$$\frac{\text{cal}}{\text{gm} \cdot °\text{K}} = \frac{(\sim 6 \text{ cal/mole} \cdot °\text{K})}{(52.00 \text{ gm/mole})}$$

$$\cong 0.12 \text{ cal/gm} \cdot °\text{K}.$$

b) The molecular weight of Cr_2O_3 is 152.00 gm; however, this includes $5(0.602 \times 10^{24})$ atoms. Thus,

$$\frac{cal}{gm \cdot °K} = \frac{5(\sim 6 \text{ cal/mole} \cdot °K)}{(152.00 \text{ gm/mole})}$$

$$\cong 0.20 \text{ cal/gm} \cdot °K.$$

Note. Experimental data give 0.11 cal/gm $\cdot$ °K and 0.17 cal/gm $\cdot$ °K, respectively, at 20°C. ◄

6-3 THERMAL EXPANSION

Increased thermal agitation expands the volume of a solid. We can discover the reason for this expansion by recalling, from our discussion of the spring model in Fig. 3-4, that the spring is nonlinear. Compression introduces abnormally high repulsive forces, while tensile forces result in lower dF/da values (Fig. 3-2). Consequently, the energy trough is asymmetric and the mean interatomic distance increases at higher energy (and temperature) levels.

The *thermal expansion coefficient* is an inverse function of the slope of the curve for the mean vibrational position versus distance. This permits a direct comparison between the expansion coefficient and the bonding energy. Other factors being equal, those three-dimensional structures with high attractive forces also have deep, more symmetric energy troughs than do weakly bonded materials (Fig. 6-3).

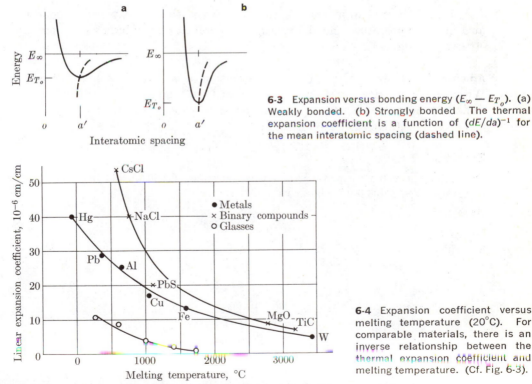

6-3 Expansion versus bonding energy ($E_\infty - E_{T_o}$). (a) Weakly bonded. (b) Strongly bonded The thermal expansion coefficient is a function of $(dE/da)^{-1}$ for the mean interatomic spacing (dashed line).

6-4 Expansion coefficient versus melting temperature (20°C). For comparable materials, there is an inverse relationship between the thermal expansion coefficient and melting temperature. (Cf. Fig. 6-3).

TABLE 6-2

THERMAL EXPANSION OF CRYSTALLINE SOLIDS

Solid	α_l, cm/cm $\cdot$ °C at 20°C
Tungsten	4×10^{-6}
Iron	12
Copper	17
Aluminum	23
Lead	29
TiC	7
MgO	9
PbS	20

	$\alpha \perp [001]$ or $[0001]$	$\alpha \parallel [001]$ or $[0001]$
Magnesium	24×10^{-6}	27×10^{-6}
Zinc	61	15
Tin	15	30
ZnO (zincite)	6	5
SiO$_2$ (quartz)	14	9
CaCO$_3$ (calcite)	-6	25
ZrSiO$_4$ (zircon)	4	6
TiO$_2$ (rutile)	7	8

As a result, there is an inverse correlation between thermal expansion coefficients at a given temperature and the melting temperatures of materials of comparable structures (Fig. 6-4).

Anisotropy of expansion. Thermal expansion arises from atomic vibrations. Within a symmetrical structure, the vibrations are comparable in all three dimensions. In noncubic structures, the thermal expansion is anisotropic and in some cases even negative (Table 6-2).

Example 6-2

Determine the volume expansions of tetragonal zircon (ZrSiO$_4$) and hexagonal calcite (CaCO$_3$). (See Table 6-2 for linear values.)

Answer

$$\alpha_v = (dV/V)/dT. \tag{6-2a}$$

Since $V = l_x l_y l_z$,

$$dV \cong l_y l_z \, dl_x + l_x l_z \, dl_y + l_x l_y \, dl_z.$$

Therefore,

$$\alpha_v = \frac{dl_x}{l_x \, dT} + \frac{dl_y}{l_y \, dT} + \frac{dl_z}{l_z \, dT},$$

or

$$\alpha_v = \alpha_x + \alpha_y + \alpha_z. \tag{6-2b}$$

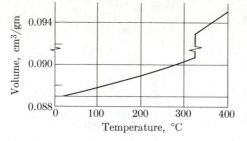

6-5 Volume expansion versus temperature (lead). The thermal expansion coefficient, dV/dT, of a close-packed crystalline solid is always less than the expansion coefficient for the liquid.

In tetragonal and hexagonal crystals, $\alpha_x = \alpha_y \neq \alpha_z$. For zircon,

$$\alpha_v = (4 + 4 + 6) \times 10^{-6}/°C$$
$$= 14 \times 10^{-6}/°C.$$

For calcite,

$$\alpha_v = (-6 - 6 + 25) \times 10^{-6}/°C$$
$$= 13 \times 10^{-6}/°C. \blacktriangleleft$$

6-4 MELTING

A liquid lacks the repetitive order of a crystal. Since most ionic and metallic crystals have high packing factors, the atomic disorder which accompanies melting usually introduces a volume expansion (Fig. 6-5).* This added volume arising from the decreased packing factor of the liquid is sometimes called "free volume," to indicate that there is space in addition to that normally associated with each atom.

Above the melting point, the added thermal energy not only increases the amplitude of atomic and molecular vibrations, but also introduces still more "free volume" through atom rearrangements. Thus liquids normally have a greater thermal expansion coefficient than solids, as is illustrated by the slopes of the curves in Fig. 6-5.†

6-5 POINT DEFECTS

Local imperfections are encountered within crystals. These commonly appear as lattice *vacancies* or as *interstitial* atoms (Fig. 6-6a and b). Envisioned most simply, they are accidents in crystallization; i.e., there was not a one-to-one correspondence between available atoms and possible lattice sites. In practice, they can also occur as a result of thermal disorders which displace atoms from their ideal locations (Fig. 6-6c and d).

---

* Ice, with its hydrogen bond, is the most notable exception (Fig. 4-14). Others include silicon, germanium, and bismuth.
† Even atypical water exhibits this feature at higher temperatures (Fig. 4-14). The V-versus-T curve of water is modified, however, as the freezing temperature is approached, because the hydrogen bridge starts to induce the ordering found in ice (Section 4-5).

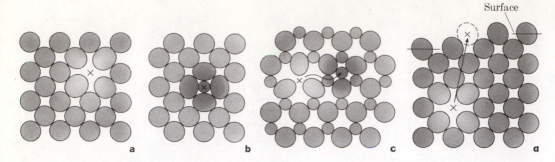

6-6 Point defects. (a) Vacancy. (b) Interstitial. (c) Displacement to interstitial site (Frenkel defect). (d) Displacement to surface. An imperfection distorts the crystal lattice.

The energy required to displace an atom from its regular site is illustrated schematically in Fig. 6-7, where $\underrightarrow{E^*}$ is the activation energy necessary to break original bonds and move the atom to an energy crest before descending into a metastable position. In general, very few atoms have enough energy to cross that barrier; however, as indicated in Fig. 2-8 and Eq. (2-13), a small though statistically important fraction have the necessary energy at any specific increment of time. Of those which have sufficient energy, a fraction will have their momentum in the right direction to move across an adjoining energy barrier. This number per unit time, $\underrightarrow{n}$, will be

$$\underrightarrow{n} = \underrightarrow{C} N_s e^{-\underrightarrow{E^*}/kT}, \qquad (6\text{-}3)$$

where $\underrightarrow{C}$ is a proportionality constant for each type of atom and structure, N_s is the number of atoms in stable positions, and k is Boltzmann's constant (1.38×10^{-16} erg/atom·°K). Temperature, T, is in °K, as was the case in Eq. (2-13).

The reverse step, from the metastable position to the stable position, requires less activation energy, $\underleftarrow{E^*}$. Thus,

$$\underleftarrow{n} = \underleftarrow{C} N_m e^{-\underleftarrow{E^*}/kT}, \qquad (6\text{-}4)$$

where N_m is the number of atoms in metastable positions.

Equilibrium is established when

$$\underrightarrow{n} = \underleftarrow{n};$$

or, from Eqs. (6-3) and (6-4),

$$N_m/N_s = Ce^{-\Delta E/kT}, \qquad (6\text{-}5\text{a})$$

where ΔE is the difference between $\underleftarrow{E^*}$ and $\underrightarrow{E^*}$, or the energy of formation per defect, and C is still a constant. Equation (6-5a) presents a conclusion which may appear surprising—under *equilibrium* conditions, vacancies and interstitials are present! Since $N_s \cong N_{\text{tot}}$, the total number of sites, the fraction of sites which

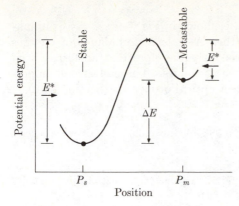

6-7 Energy barrier. An activation energy is required to move an atom from its present position. (ΔE = formation energy per defect; $E*$ = activation energy.)

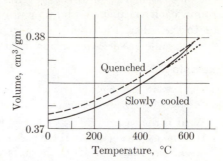

6-8 Trapped vacancies (aluminum). Vacancies formed at elevated temperatures cannot diffuse out of the structure during a quench; therefore, the quenched sample of aluminum is about 0.2% less dense than a slowly cooled sample.

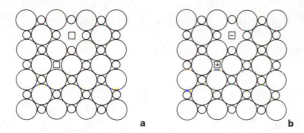

6-9 Schottky pair (LiF). (a) Vacancies (□) must occur as pairs in ionic solids to ensure a charge balance within the crystal as a whole. (b) Ionization of vacancies. If an electron leaves a fluorine ion which is adjacent to the cation vacancy and locates with a lithium ion which is adjacent to the anion vacancy, a local charge balance is maintained. Both the electron hole and the excess electron can move among the several ions adjacent to each vacancy.

have vacancies (or interstitials) can be expressed as

$$\ln \frac{N_m}{N_{\text{tot}}} = -\frac{\Delta E}{kT} + \ln C. \tag{6-5b}$$

Direct evidence of vacancy formation in crystals is shown by taking two identical samples of aluminum to within a few degrees of the melting point (660°C). If one is quenched and the other slowly cooled, the former will have a greater volume at room temperature (about 0.2%) than the slowly cooled sample (Fig. 6-8). At high temperatures, vacancies form because some atoms diffuse to the surface (Fig. 6-6d). These vacancies can be trapped within the structure of the quenched sample, whereas slower cooling allows time for most of the atoms to return to equilibrium positions in accordance with Eq. (6-5b). Furthermore, if the quenched sample is reheated and then slowly cooled, the volume difference can be eliminated.

The energy difference between the stable atom site and the metastable site appears partly as *strain energy* around the point defects. The strain distorts the lattice, and these distortions, plus those from linear and boundary defects, influence atom movements and other properties (Chapters 9 and 18).

In the absence of interstitials, vacancies must occur in pairs within an ionic solid such as LiF, in order to maintain a charge balance within a crystal as a whole.

Such a defect, called a *Schottky imperfection*, is shown in Fig. 6-9(a). An examination of that figure, however, reveals that there is a local imbalance of charges around each vacancy of the pair; an electron may therefore be transferred from one to the other (Fig. 6-9b), thus ionizing the vacancies. These electronic imperfections, called *color centers*, significantly affect optical properties (Section 13-6).

Example 6-3

At 10°C below the melting point of aluminum, 0.08% of the atom sites are vacant. At 484°C, only 0.01% are vacant.

a) What is the energy, E, associated with each vacancy?
b) How many vacancies are there per cm^3 at 527°C? Assume that most of the vacancies arise from displacements to the surface (Fig. 6-6d).

Answer. See Eq. (6-5).

a) At 650°C (923°K),

$$\ln (0.0008) = \ln C - \Delta E/(1.38 \times 10^{-16} \text{ erg/vacancy} \cdot °K)(923°K).$$

At 484°C (757°K),

$$\ln (0.0001) = \ln C - \Delta E/(1.38 \times 10^{-16} \text{ erg/vacancy} \cdot °K)(757°K).$$

Solving simultaneously, we obtain

$$\Delta E = 1.2 \times 10^{-12} \text{ erg/vacancy}$$
$$= 0.75 \text{ eV/vacancy},$$
$$\ln C = 2.3.$$

b) At 527°C (800°K),

$$\ln N_m = 2.3 - (1.2 \times 10^{-12} \text{ erg/vacancy})/(1.38 \times 10^{-16} \text{ erg/vacancy} \cdot °K)(800°K)$$
$$= -8.6,$$
$$N_m = 1.8 \times 10^{-4};$$

$$\left(1.8 \times 10^{-4} \frac{\text{vacancies}}{\text{atom}}\right)\left(\frac{2.699 \text{ gm/cm}^3}{26.98 \text{ gm/mole}}\right)\left(0.602 \times 10^{24} \frac{\text{atoms}}{\text{mole}}\right) = 1.1 \times 10^{19} \frac{\text{vacancies}}{\text{cm}^3}.$$

◄

6-6 DISLOCATIONS (Linear defects)

The most important imperfection affecting mechanical properties is the dislocation. This is a one-dimensional defect which is best visualized by imagining a crystal undergoing shear (Fig. 6-10). As shear proceeds, the unit cells of the upper part of the crystal are pushed ahead one full step, where they again register with most of the cells in the lower part of the crystal. However, a line of mismatch, D, called a *dislocation*, remains between the slipped and unslipped portions. As the shear continues, the dislocation will move through the crystal, causing the shear displacement to expand. Plastic deformation (Chapter 11) is the result of the continued generation and movements of such dislocations.

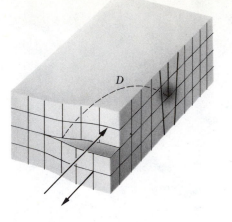

6-10 Shear dislocation. The dislocation loop, *D,* is a line of disregistry. All other unit cells match their neighbors.

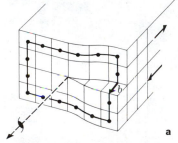

a

b

6-11 Dislocations. (a) Screw. (b) Edge. The displacement is parallel to the screw dislocation but perpendicular to the edge dislocation.

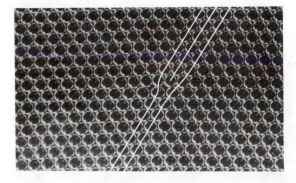

6-12 Edge dislocation. "Bubble-raft" model of an imperfection in a crystal structure. Note the extra row of atoms. [Bragg and Nye, *Proc. Roy. Soc. (London),* **A190**, 1947, p. 474.]

Part of the dislocation *loop* is parallel to, and part is perpendicular to, the displacement (Fig. 6-11). We call the *perpendicular* part an *edge dislocation,* because if we were to look along this part of the loop we would see the edge of an extra half-plane of atoms (Fig. 6-12). The part of the dislocation loop which is *parallel* to the displacement is called a *screw dislocation,* for reasons best visualized by looking at Fig. 6-11(a). The rest of the loop length in Fig. 6-10 involves combinations of these two types of dislocations.

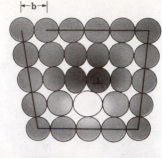

6-13 Dislocation strains. Atoms are under compression (darker) and tension (lighter) adjacent to the dislocation. The displacement vector (Burgers vector) is perpendicular to the dislocation line.

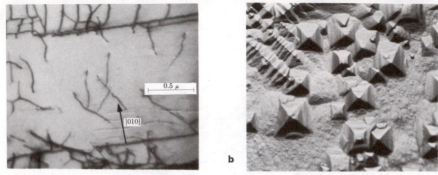

a   b

6-14 Dislocation observations. (a) Dislocation network (stainless-steel foil). Dislocations are revealed by an electron microscope because the lattice distortion diffracts the electrons. (× 25,000.) [Courtesy of M. J. Whelan, *Proc. Roy. Soc.* (*London*), **A249**, 1959, p. 114.] (b) Etch pits of dislocations (MnS). The atoms adjacent to the point where the dislocation emerges from the surface are easily removed because of their higher energy. (× 300.) (Courtesy of H. C. Chao.)

The *Burgers vector*, **b**, may be identified as the "error of closure" for a circuit around a dislocation (Fig. 6-13). In order for the atoms to have their regular co-ordinations, its length must be equal to the distance between equivalent lattice sites in the direction of shear. Usually, the Burgers vector corresponds to a unit shear in the plastic straining process (Chapter 11).

Observation of dislocations. Dislocations were predicted long before they were verified. Subsequently their existence has been amply supported by several means, the most direct being the electron microscope (Fig. 6-14a), which reveals the dislocation through the accompanying *strain* or lattice distortion. As shown in Fig. 6-13, the atoms above an edge dislocation are under compression; those below are under tension. Thus an electron beam is diffracted as it passes through this lattice distortion.* A second means of observing dislocations is through etch pits (Fig. 6-14b). Chemical etching occurs where the dislocation emerges through a polished surface, because those atoms around a dislocation possess strain energy and are thus more readily taken into solution by the etchant.

---

* The lattice distortion around a point defect will also diffract an electron beam. However, detection is difficult because only a few unit cells are involved, whereas a dislocation distorts literally thousands of unit cells along its path.

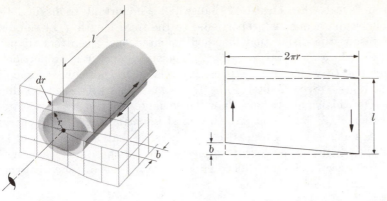

6-15 Energy of a screw dislocation. The shear strain introduces strain energy proportional to *l*, *G*, and b^2, as described in the text.

Energy of dislocations. Because of the importance of dislocations in plastic deformation (Chapter 11), it is desirable to be able to calculate the energy of this linear defect. For simplicity, some reasonable approximations will be required.

In Fig. 6-15, a thin cylinder is identified and removed from the distorted zone around a *screw dislocation*. The elastic shear strain, γ, will be shown in Section 10-1 to be

$$\gamma = b/2\pi r.$$

Since elastic strain energy per unit volume is equal to $\tau\gamma/2$, and τ, the shear stress, is equal to the product of the shear modulus, G, and the shear strain, γ,*

$$\frac{dE}{dV} = \tfrac{1}{2} G\gamma^2 = \frac{G}{2}\left[\frac{b}{2\pi r}\right]^2. \qquad (6\text{-}6)$$

Furthermore, the increment volume, dV, in Fig. 6-15 equals $2\pi rl\, dr$, and therefore

$$dE = \frac{lGb^2}{4\pi}\frac{dr}{r}. \qquad (6\text{-}7)$$

To simplify we will integrate from $r = b$ to $r = be^{4\pi}$. Therefore, the strain energy, E, is approximately

$$E \cong lGb^2. \qquad (6\text{-}8)$$

This simplification is logical. First, the outer limit, $be^{4\pi}$, which is selected for mathematical convenience, is well beyond the zone of significant distortion; further, the inner limit may be shown by more elaborate means to exclude only a small fraction of the total strain energy which accompanies a dislocation line.

An *edge dislocation* has a somewhat higher (25% to 40%) energy because of geometric factors; however, for our purposes, we will simply consider that Eq. (6-8) applies to the total dislocation loop.

---

* See Eq. (10-6) in Section 10-1.

Equation (6-8) reveals three features that will be important to us in later chapters. (1) The strain energy is proportional to the total length, l, of the dislocation. As a result, the dislocation loop will spontaneously contract, if permitted to do so, to reduce the total energy. (2) The strain energy is proportional to the shear modulus, G, which in turn is closely related to the more commonly used Young's modulus. (See Chapter 10.) Those materials with high bonding energies, and therefore higher elastic moduli, will have more strain energy associated with their dislocations. (3) Last, and probably most important, the strain energy is a function of the square of the Burgers vector, **b**. As we shall see in Chapter 11, this means that shear deformation occurs preferentially in those directions with the shortest Burgers vector, i.e., with the shortest repetitive distance between equivalent points in the Bravais lattices.

Example 6-4

Microscopic examination of a polished and etched copper surface reveals 0.2×10^{11} dislocation etch pits per cm^2. *Estimate* the amount of strain energy per cm^3. (On the basis of the note in Example 6-5, assume a dislocation length of $2 \ cm/cm^3$ for each surface interception per cm^2. $G_{Cu} = 6,000,000 \ psi \cong 4 \times 10^{11} \ dynes/cm^2$.)

Answer. Since there are 0.2×10^{11} dislocation interceptions per cm^2 of surface, there will be $0.4 \times 10^{11} \ cm/cm^3$ of dislocation lines within the solid (if we assume random distribution and orientation). Also, b is approximately 2.5 A for copper (Table 4-1), since only the shortest Burgers vectors are normally encountered. Therefore,

$$E \cong (0.4 \times 10^{11} \ cm/cm^3)(4 \times 10^{11} \ dynes/cm^2)(2.5 \times 10^{-8} \ cm)^2$$
$$\cong 10^7 \ ergs/cm^3$$
$$\cong 0.25 \ cal/cm^3. \ \blacktriangleleft$$

6-7 BOUNDARIES (Two-dimensional defects)

The *surface* of a liquid was described in terms of surface energy in Section 3-9. Solids also have exterior boundaries, or surfaces. In addition, most crystalline solids have interior boundaries which are often more important for property considerations than the external surfaces.

Grain boundaries. Engineering materials usually contain many crystals.* When more than one crystal is present, an area of mismatch occurs where adjacent crystals come into contact. This is illustrated schematically in Fig. 6-16 by a two-dimensional cut through a crystalline solid. Each separate crystal is called a *grain*, and the region of mismatch a *grain boundary*. Figure 6-17 reveals grain boundaries in a metallic (molybdenum) and a ceramic (MgO) solid. The crystal planes and anisotropic properties are virtually discontinuous across a boundary. This discontinuity is an important factor in mechanical behavior, as we shall see in later chapters. In addition, the atoms along a grain boundary possess more energy than interior atoms. As a result, they are more reactive.

---

* Exceptions include ruby lasers, piezoelectric crystals, transistors, and diamond dressing tools.

6-16 Grain boundaries. Note the area of disorder at the discontinuity (boundary) between the grains (individual crystals).

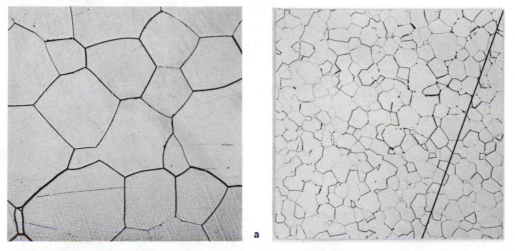

6-17 Grain boundaries. (a) Molybdenum (× 250) (O. K. Riegger). (b) High-density periclase, MgO (× 250). (R. E. Gardner and G. W. Robinson, Jr., "Improved Method for Polishing Ultra-High Density MgO," *Journ. Amer. Ceram. Soc.*, **45**, 1962, p. 46.) The random line is used in Example 6-5.

Grain-boundary area. For reasons just cited, the grain-boundary area per unit volume is important. At first glance this would appear to be difficult to measure, since we can normally see only a two-dimensional section through a three-dimensional solid. Fortunately, however, a very simple statistical procedure can be used. Suppose a random traverse across a microsection intercepts the grain boundaries at P_L points per cm of traverse length. Without giving proof,* we will state that the grain-boundary area per unit volume, S_V in cm²/cm³, is

$$S_V = 2P_L. \qquad (6\text{-}9)$$

* See DeHoff and Rhines in the reference list at the end of the chapter.

Example 6-5

Determine the grain-boundary area per unit volume in the MgO of Fig. 6-17(b).

Answer. A random line intercepts 21 boundaries per 6.2 cm as magnified. Without magnification, P_L is 21/0.025 cm, or 850/cm. Thus

$$S_V = 2(850/\text{cm})$$
$$= 1700 \text{ cm}^2/\text{cm}^3.$$

Note. If we assume that a dislocation is a random line, Eq. (6-9) also provides the basis for the dislocation length calculation of Example 6-4. ◀

Grain-boundary energies. As yet it is impossible to calculate grain-boundary energies as we did dislocation energies (Eq. 6-8). They have been measured experimentally, however; they usually have the order of magnitude of 10^2 to 10^3 ergs/cm², but vary considerably depending on a number of factors such as composition and the orientation of the adjacent grains. Because a boundary has energy, it tends to minimize its area. The driving force for this reduction in interfacial energy is called *surface tension*, and is expressed either as dynes/cm or as energy per unit area, ergs/cm².

Example 6-6

During annealing a thermal groove is formed where a grain boundary intercepts the surface of iron. If this groove has a *dihedral angle*, θ, of 147° and the grain-boundary energy, γ_b, is 850 ergs/cm², what is the surface energy, γ_s, of the iron?

Answer. (See Fig. 6-18.) Since boundary energies appear as tension forces, they may be treated as vectors. Thus,

$$\gamma_b = 2\gamma_s \cos{(\theta/2)}, \tag{6-10}$$
$$\gamma_s = (850 \text{ ergs/cm}^2)/(2)(\cos 73.5°)$$
$$= 1490 \text{ ergs/cm}^2.$$

Note. We measured the dihedral angle, θ, in a plane normal to the line along the "triple point" of Fig. 6-18. As usually observed, a two-dimensional section through a three-dimensional solid does not reveal the true dihedral angle because the boundaries are cut at random angles. The proper value may be obtained, however, through appropriate statistical procedures. ◀

The *thermal grooves* described in Fig. 6-18 can be formed only under special conditions of vacuum at high temperature. In practice, we more commonly observe *etching* grooves in which the atoms are removed from the grain-boundary zone by a solvent. The principle is the same as with thermal grooves; in place of a free surface, however, γ_s is the energy of a metal-solvent phase boundary. Etching grooves reveal the grain boundaries in Fig. 6-17 because they do not give mirror-like reflection of light into the microscope as does the balance of the polished surface.

The energy associated with a boundary may be rationalized in terms of the atomic packing factors and the bonding energies. For example, when three dimensions are considered, the atoms along a boundary of Fig. 6-16 have a lower co-

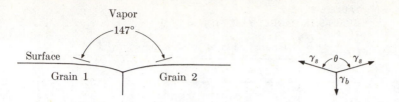

6-18 Dihedral angle. The energies of the three surfaces maintain a vectorial balance (Eq. 6-10).

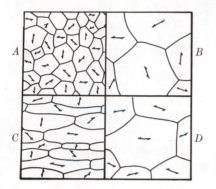

6-19 Microstructural variables of single-phase metals. *A* versus *B*: grain size. *A* versus *C*: grain shape. *B* versus *D*: preferred orientation.

ordination number than the atoms within the crystal proper. Also, we shall find in Chapter 9 that boundaries, with their lower packing, supply effective diffusion channels for atom movements within solids.

6-8 SINGLE-PHASE MICROSTRUCTURES

Our attention in the previous section was focused on grains within a material. Each grain is a single crystal and, since only one phase is present, all grains have the same crystal structure, differing only in size and orientation. This structure of grains is called a *microstructure* because we commonly examine it by microscopic means (Fig. 6-17). Actually, there is no size limit to a microstructure, and in some materials (e.g., devitrified glass) the grain size is below the conventional limit of the electron microscope (×100,000). In other materials (e.g., marble, and castings of brass or stainless steel) the grains may have dimensions of several millimeters or even centimeters.

A single-phase microstructure has three prime features: grain size, grain shape, and grain orientation (Fig. 6-19). Each of these influences the properties of solids. For example, fine-grained materials are stronger than coarse-grained materials, particularly at low temperatures; tungsten wires with elongated grains have enhanced ductility; and transformer sheets in which the grains have a desirable preferred orientation have much lower power losses than sheets with randomly oriented grains.

Grain size. The grain size is often represented as the average "diameter" revealed in a two-dimensional section. This indicates the order of magnitude,

TABLE 6-3

GRAIN-SIZE RANGES (ASTM); $N = 2^{n-1}$

Grain-size number	Grains/in^2 at $\times$ 100 (linear)	
	Mean	Range
$n = 1$	$N = 1$	—
2	2	1.5–3
3	4	3–6
4	8	6–12
5	16	12–24
6	32	24–48
7	64	48–96
8	128	96–192

but leaves much to be desired since (1) the grains are not spherical and (2) the two-dimensional section does not present the full "diameter" of each grain. Corrections must be made for these deficiencies.

A widely used *grain-size index* is one standardized by the American Society for Testing and Materials (ASTM). The index, n, is related to the number of grains/in^2, N, at a linear magnification of $\times 100$ by the equation

$$N = 2^{n-1}. \tag{6-11}$$

This is summarized in Table 6-3. Although called a grain-size index, it could more appropriately be called a grain-count index, since larger numbers indicate more grains and hence more grain-boundary area.

Example 6-7

A steel has an ASTM grain size No. 7. What would be the average area observed for each grain in a polished surface?

Answer

$$N = 2^{7-1} = 64 \text{ grains/in}^2 \qquad \text{at} \times 100,$$

$$\frac{64}{(0.01)(0.01)} = 640,000 \text{ grains/in}^2 \qquad \text{at} \times 1;$$

$$\text{area of 1 grain} = \frac{1}{640,000} \text{ in}^2. \blacktriangleleft$$

Example 6-8

Assume that the grains in the previous calculation are cubic in shape.* What is the grain-boundary area per cubic inch of steel?

---

* Quite obviously, the above calculations give the order of magnitude only. As such, they are useful in determining reaction rates that depend on grain-boundary area. This calculation assumes cubic shape and uniform size. In more accurate calculations, the boundary area would be slightly, but not significantly, smaller.

Answer. From the previous example,

$$640,000 \text{ grains/in}^2 \text{ of surface} = 800 \text{ grains/in. of length.}$$

This equals $(800)^3$, or 5.12×10^8 grains/in^3 of volume. The surface of each grain is $6(1/800)^2 \text{ in}^2$. Since each boundary is composed of two grain surfaces,

$$\text{total boundary} = \frac{6}{2}\left(\frac{1}{800}\right)^2 \text{ in}^2 \ (800/\text{in.})^3$$

$$= 2400 \text{ in}^2 \text{ boundary per cubic inch of steel.} \quad \blacktriangleleft$$

If shape assumptions are made, the number of grains can be calculated from the grain-boundary areas available from Eq. (6-9). By geometric analysis,

$$N_V = (S_V/F)^3, \tag{6-12}$$

where N_V is the number of grains per unit volume. As in Eq. (6-9), S_V is the grain-boundary area per unit volume. The shape factor, F, is 3 for assumed cubic grains, but more nearly 2.7 for the equiaxed, noncubic grains commonly encountered in isotropic microstructures such as those shown in Figs. 6-17 and 6-19(A and B). The latter factor should be used in Eq. (6-12).

Grain shape. Metals and other materials with symmetric crystal structures normally possess *equiaxed* grains (i.e., with no elongation), as shown in Figs. 6-17 and 6-19(A and B). *Elongated* grains may be formed, however, when these materials undergo directional solidification or are subjected to deformation processes.

The amount of elongation is commonly indexed as the length/width ratio observed for the grains. Thus the schematic microstructure of Fig. 6-19(C) would possess a length/width ratio between 3/1 and 4/1, depending on whether maximum or typical widths are used. Here, as with grain size, we also encounter the question of three-dimensional measurements obtained from two-dimensional observations. As a result, it is sometimes desirable to utilize a statistical calculation of shape when the microstructure is elongated. Instead of using the fully random line of Fig. 6-17(b), we use lines randomly selected perpendicular to ($\perp$) and parallel to ($\parallel$) the direction of elongation:

$$\text{grain elongation} = \frac{P_\perp - P_{\parallel}}{P_\perp + (\pi/2 - 1)P_{\parallel}}. \tag{6-13}$$

As with boundary area, P is the number of intercept points per unit length. The grain elongation in Fig. 6-19(C) is about 0.55. This is in contrast to zero elongation for the other sketches of Fig. 6-19, and 1.0 for a long fiber.

In Fig. 6-19(C) it is implied that the dimension of the microstructure in one direction is greater than in the other two, or, in terms of the number of intercepts per unit length, that $P_a = P_b > P_c$. However, microstructures may possess other patterns of elongation. Two alternatives are $P_a = P_b < P_c$ and $P_a \neq P_b \neq P_c$. Because of the existence of such possibilities, it is not safe to draw conclusions about grain shape and microstructural elongation on the basis of *one* section only. As a minimum, it is necessary to examine two perpendicular sections.

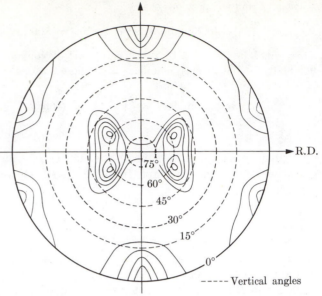

6-20 Preferred orientation (cold-rolled brass sheet). The poles (i.e., normals) to the {111} planes of the various grains are not uniformly distributed. (After B. D. Cullity, *Elements of X-ray Diffraction.* Reading, Mass.: Addison-Wesley, 1956, p. 292; and H. Hu, P. R. Sperry, and P. A. Beck, *Trans. A.I.M.E.,* **194**, 1952, p. 76.)

R.D.

- - - - - Vertical angles

Preferred orientation. In randomly oriented microstructures there are as many planes with normals (i.e., poles) to {hkl} planes pointing in one direction as there are planes with normals pointing in any other direction. This is not true when a preferred orientation develops (Fig. 6-19D). As an example, Fig. 6-20 shows a *pole figure* for cold-rolled brass sheet. To interpret this figure, imagine you are looking down on a hemisphere covering a small section of a sheet surface with the rolling direction, R.D., toward the right. Further, visualize that all of the [hkl] poles are extended to intersect the hemisphere. The orientations of {111} poles, i.e., ⟨111⟩, are concentrated in certain directions, as shown by the heavy contours. The vertical angles in the plot are indicated by the light dotted circles.*

Whether a preferred orientation is to be desired or not depends entirely on the application and the required properties. Previously, the desirability was cited for a preferred [100] orientation in transformer sheet because of favorable magnetic permeability. The extreme case of completely oriented grains is not possible (and may not even be desirable) with present manufacturing methods. Were it obtained, however, our product would be a single crystal, because the mismatch between crystals would have disappeared.

Example 6-9

Align the x-, y-, and z-axes of a cubic crystal with the three prime directions of Fig. 6-20. Locate the directions of the ⟨100⟩, ⟨110⟩, and ⟨111⟩ directional families.

Answer. See Fig. 6-21. ◄

---

* The basis for the vertical angles comes from stereographic projection. See B. D. Cullity, *Elements of X-ray Diffraction,* Reading, Mass.: Addison-Wesley, 1956 p. 60 ff.

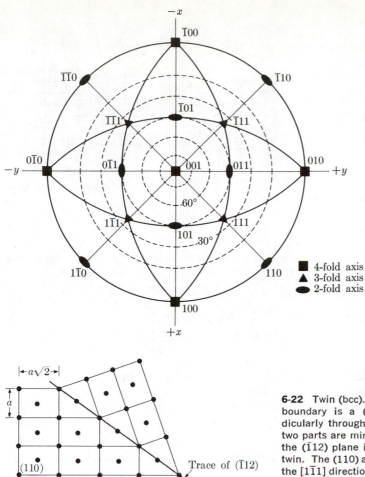

6-21 Principal poles. This is a stereographic projection, with the vertical angles added. (See B. D. Cullity, *Elements of X-ray Diffraction.* Reading, Mass.: Addison-Wesley, 1956, pp. 60ff., or comparable references.)

■ 4-fold axis
▲ 3-fold axis
⬮ 2-fold axis

6-22 Twin (bcc). The (110) plane is shown. The boundary is a ($\bar{1}$12) plane emerging perpendicularly through the plane of the sketch. The two parts are mirror images of one another, and the ($\bar{1}$12) plane is common to both parts of the twin. The (110) and ($\bar{1}$12) planes intersect along the [1$\bar{1}$1] direction.

Trace of ($\bar{1}$12)

6-9 SPECIAL BOUNDARIES

Boundaries are not limited to those which have large angles of mismatch between adjacent grains and possess no special orientational arrangements. There also exist more subtle types, called *twin boundaries* and *subboundaries*, which appear within the confines of a normal grain.

Twin boundaries. Under special conditions a crystal plane may serve as a microstructural discontinuity. This is illustrated in Fig. 6-22, where the ($\bar{1}$12) plane of a bcc lattice emerges vertically from the paper to provide two mirror images, called *twins*. The sketch shows that the atoms adjacent to the twin boundary in the (110) plane have full first-neighbor coordination and the lattice has negligible distortion. Therefore, a twin boundary possesses low energy.*

---

* It does have slight energy because the second-neighbor coordination is altered, as is the coordination with atoms in adjacent (110) planes.

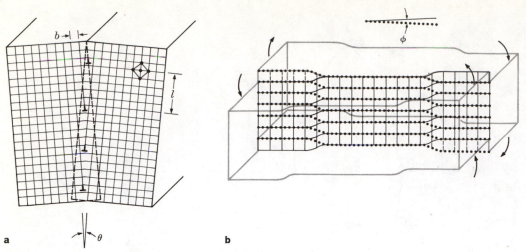

a　　　　　　　　　　　　　　　　　　　**b**

6-23 Small-angle boundaries. (a) Tilt. (b) Twist. Registry across tilt boundaries is through edge disloca-tions. Registry across twist boundaries is through screw dislocations.

Twins are observed in many metals—for example, brass (Fig. 9-19)—and in numerous natural minerals. Deformation twins are important in hexagonal metals such as Zn, Mg, Zr, and Hf, where plastic slip is difficult (Chapter 11).

Tilt and twist boundaries. Two slightly misoriented grains may still retain partial registry, the mismatch being compensated for by edge or screw dislocations. Called tilt and twist boundaries, respectively (Fig. 6-23), these *small-angle* bound-aries, or subboundaries, are widely encountered within almost all crystals and grains, even those which are considered to be perfect at first glance. They have an important effect on properties in a number of materials applications.

Example 6-10

A small-angle tilt boundary which approximately parallels the (110) plane in LiF is revealed in Fig. 6-24(a) by a series of etch pits where each edge dislocation emerges through the polished (001) surface. What is the tilt angle? [*Note.* The LiF structure is the same as NaCl (Fig. 4-1). Its lattice constant is 4.2 A.]

Answer. See Fig. 6-24(b). Since the magnification is ×300, the etch pits are about 0.0006 cm (60,000 A) apart. Thus with

$$\mathbf{b}_{110} = (4.2 \text{ A})/\sqrt{2} = 2.97 \text{ A},$$
$$\theta = b/l$$
$$= (2.97 \text{ A})/60,000 \text{ A} = 0.00005 = 10''. \blacktriangleleft$$

6-10 AMORPHOUS SOLIDS

A solid which is devoid of long-range order, or crystallinity, is termed *amorphous*. Glasses and many plastics provide prime examples. As was discussed in Sec-

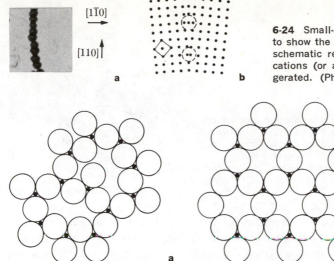

6-24 Small-angle boundary: (a) LiF crystal etched to show the ends of edge dislocations (× 300); (b) schematic representation of the above. Only the cations (or anions) are shown; the angle is exaggerated. (Photomicrograph courtesy of A. S. Keh.)

6-25 Structure of glass. The glass (a) has short-range order only. The crystalline structure (b) has long-range order in addition to short-range order (schematic.)

tion 4-1, each silicon atom of a silica glass is coordinated with four oxygen atoms and each oxygen atom connects two silicon atoms, but there is no long-range order. Figure 6-25 provides a schematic contrast between a B_2O_3 glass and a B_2O_3 crystal. Although each structure of Fig. 6-25 is modified into a two-dimensional presentation, the essential differences are revealed as a difference in the amount of long-range order.

At elevated temperatures, glass structures become viscous liquids. In compliance with free energy considerations (Section 3-8), the liquid has a freezing temperature below which the crystalline structure (Fig. 6-25b) is more stable. This is 460°C for B_2O_3; however, the activation energy necessary for structural rearrangement from the structure shown in Fig. 6-25(a) to that shown in Fig. 6-25(b) is sufficiently high, and the heat of fusion which is released is sufficiently low, so that crystallization proceeds slowly at the freezing temperature. It is even slower at room temperature, where negligibly few of the atoms will have the necessary energy to break existing bonds as required for rearrangement. Thus pure B_2O_3, like other glasses, persists metastably for indefinite periods of time.

Inhibited crystallization has widespread occurrence in (1) covalently bonded materials involving extensive *network structures* or (2) materials which have large, *complex molecules*. Each type of material encounters major obstacles to restructuring. By contrast, metallic atoms, ions, and small molecules like I_2 (Fig. 4-10) crystallize from a liquid in minimal time, with little opportunity for supercooling into a metastable glassy structure.

6-11 GLASS TRANSITION (Fictive) TEMPERATURE

In the absence of crystallization, the thermal contraction of a liquid continues to lower temperatures without volume changes due to solidification. The thermal coefficient for the liquid is greater than for a crystal of the same composition. As

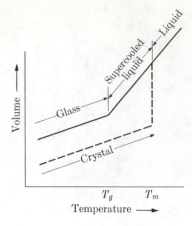

6-26 Glass-transition temperature. Below the melting temperature, T_m, a supercooled liquid has the same expansion coefficient as the stable liquid until the fictive, or glass-transition, temperature, T_g, is reached. Atomic and molecular rearrangements are impossible in the glassy range; hence, the volume changes are limited to changes in the amplitude of thermal vibrations.

was discussed in Section 6-4, the contraction arises from both a reduction in "free volume" and a decrease in vibrational amplitude. There is a temperature, however, below which the atoms, or molecules, cannot be rearranged to give more efficient packing. Further cooling produces only a decrease in vibrational amplitude, so that there is a discontinuity in the thermal expansion or contraction coefficient (Fig. 6-26). This temperature was originally referred to by glass scientists as a *fictive temperature*. It also has major significance in polymer behavior (Chapter 12), and has been labeled by polymer scientists as a *glass-transition temperature*, T_g, or, more simply, the *glass point*. (The glass point is also a function of the cooling rate. With slow cooling, there is an opportunity for a continued decrease of "free volume," which would not be possible with more rapid cooling.)

We use this behavioral discontinuity of an amorphous solid at the glass transition temperature, T_g, to distinguish between a *glass* and a *supercooled liquid*. Below that temperature, the solid is brittle and nondeformable. Just above the transition temperature the supercooled, highly viscous liquid is semirigid, but it progressively reduces its viscosity and becomes a more fluid liquid as the temperature is raised to and above the melting point, T_m, into the true liquid range (Section 12-5).

REFERENCES FOR FURTHER READING

American Society for Metals, *Impurities and Imperfections*. Metals Park, O.: American Society for Metals, 1953. Papers presented at an ASM seminar. Graduate student and professional level.

Azaroff, L. V., *Introduction to Solids*. New York: McGraw-Hill, 1960. Chapter 5 discusses imperfections in the packing and atom movements of solids. Advanced undergraduate level.

Charles, R. J., "The Nature of Glasses," *Scientific American*, **217** [3], September 1967, pp. 126–136. Since glass is the most thoroughly studied amorphous material, this article is recommended reading for amorphous structures. Undergraduate level.

DeHoff, R. T., and F. N. Rhines, *Quantitative Microscopy*. New York: McGraw-Hill, 1968. One of the first quantitative microscopy books. Advanced level.

Hannay, N. B., *Solid State Chemistry*. Englewood Cliffs, N. J.: Prentice-Hall, 1967. Paperback. Chapters 3 and 4 provide good supplementary reading on imperfections and their effect on properties. Sophomore-junior level.

Moffatt, W. G., G. W. Pearsall, and J. Wulff, *The Structure and Properties of Materials: I. Structure*. New York: Wiley, 1964. Paperback. Chapter 4 gives a nonmathematical introduction to imperfections.

Richman, M. H., *Science of Metals*. Waltham, Mass.: Blaisdell, 1967. Chapter 4 provides an alternative presentation of crystalline imperfections.

Rogers, B. A., *The Nature of Metals*, second edition. Metals Park, O.: American Society for Metals, 1964. Chapter 15 provides a simplified presentation of crystal imperfections. Recommended as supplementary reading for the student who has difficulty with the present chapter.

Van Vlack, L. H., *Physical Ceramics for Engineers*. Reading, Mass.: Addison-Wesley, 1964. Chapter 4 introduces glasses. Sophomore level.

Wert, C. A., and R. M. Thomson, *Physics of Solids*. New York: McGraw-Hill, 1964. Chapter 3 summarizes defects in solids at an advanced undergraduate level.

PROBLEMS

6-1 A vertical copper wire has a 5,000-psi tensile stress at 70°F. How much additional stress must be added to restore its original unstressed length after the temperature is dropped to 25°F?

Answer. 1480 psi

6-2 Repeat Problem 6-1 for steel.

6-3 Calculate the radius of the largest atom which could exist interstitially in fcc lead without crowding. [*Hint.* Sketch the (100) face of several adjacent unit cells.]

Answer. 0.725 A

6-4 Determine the radius of the largest atom which can be located in the interstices of bcc iron without crowding. [*Hint.* The largest uncrowded site is at $\frac{1}{2}, \frac{1}{4}, 0$.]

6-5 a) What is the coordination number for the interstitial site of Problem 6-4?
  b) How many of these sites are there per unit cell?

Answer. a) 4   b) 12

6-6 a) What is the coordination number of the interstitial site of Problem 6-3?
  b) What structure would result if every such site were occupied by a smaller atom or ion?

6-7 An activation energy of 2.0 eV is required to form a vacancy in a metal. At 800°C there is one vacancy for every 10^4 atoms. At what temperature will there be one vacancy for every 1000 atoms?

Answer. 930°C

6-8 In copper at 1000°C, one out of every 473 lattice sites is vacant. If these vacancies remain in the copper when it is cooled to 20°C, what will be the density of the copper?

6-9 At 800°C, 1 out of 10^{10} atoms, and at 900°C, 1 out of 10^9 atoms has appropriate energy for movements within a solid.

a) What is the activation energy in cal/mole?

b) At what temperature will 1 out of 10^8 atoms have the required amount of energy?

Answer. a) 58,000 cal/mole   b) 1020°C

6-10 The number of vacancies in crystals increases at higher temperatures. Between 20°C and 1020°C, the lattice constant of a bcc metal increased 0.5% from thermal expansion. In the same temperature range the density decreased 2.0%. Assuming there was one vacancy per 1000 unit cells in this metal at 20°C, estimate how many vacancies there are per 1000 unit cells at 1020°C.

6-11 The lattice constant for lead (fcc) is 4.949 A. If the density of a single crystal of lead is 11.346 gm/cm³, what fraction of the atom sites are vacant? [*Note.* Since a small difference between density values is involved, a slide rule may not give you the required accuracy.]

Answer. 1/1000

6-12 Analysis of a sample of otherwise pure iron showed that 0.02% (by weight) of carbon atoms were present in the interstices among the iron atoms. How many carbon atoms are there per 10,000 unit bcc cells of iron?

6-13 Compare the relative energies of edge dislocations with $\mathbf{b}_{111}$, $\mathbf{b}_{100}$, and $\mathbf{b}_{110}$ as Burgers vectors in bcc tungsten.

Answer. 0.75/1.0/2.0

6-14 Repeat Problem 6-13 for fcc nickel.

6-15 Small-angle (tilt) boundaries are present in some copper because an extra pair of (100) planes of atoms gives a series of aligned edge dislocations. If the tilt boundary accounts for a 1° mismatch between the adjacent crystal areas, how many angstroms are there between succeeding dislocations?

Answer. 206 A

6-16 Repeat Problem 6-15, but with the tilt boundary originating from the extra pair of (110) planes.

6-17 The average grain dimension in a sample of copper is 1.0 mm. How many atoms are there per grain if we assume that the grains are spherical?

Answer. 4.45×10^{19} atoms/grain

6-18 How many grains of austenite per cubic inch exist in a steel with an ASTM grain size (a) No. 2? (b) No. 8? (Assume cubes.)

6-19 Assuming that the grains are cubic in shape, what is the grain-boundary area in a steel with an ASTM grain size (a) No. 2? (b) No. 8?

Answer. a) 425 in²/in³   b) 3400 in²/in³

6-20 a) What is the grain-boundary area per unit volume in the molybdenum in Fig. 6-17(a)?

b) Assign an ASTM grain size to this metal.

6-21 When a rod of iron is swaged (i.e., axially elongated by radial compression) into a wire, the preferred orientation parallel to the length of the wire is [111].

a) Calculate the linear density of atoms in [111].

b) For comparison determine the linear density in [1$\bar{1}$0], which lies radially. (Iron is bcc at 20°C.)

Answer. a) 4×10^7/cm   b) 2.5×10^7/cm

6-22 A pole figure of a cubic crystal possesses a three-fold symmetry. (Cf. Fig. 6-21.)

a) What crystal direction must be vertical?

b) What symmetry is anticipated if a $\langle 110 \rangle$ direction is vertical?

7 □ molecular phases

7-1 MOLECULAR SOLIDS

Many molecular solids have useful combinations of properties for engineering and domestic applications. As molecules, their packing factors and densities are low. The weak intermolecular forces of attraction permit easy processing, particularly at slightly elevated temperatures. If the molecules are large, however, the bulk strength can be sufficient for a multitude of applications. With few exceptions, molecular solids are insulators, they can be either transparent or translucent, and they may be selected to resist a variety of chemical environments.

Micromolecules and macromolecules. The simplest molecules are small—O_2, CH_4, C_2H_4, I_2, C_2H_5Cl, etc. We could call these *micro*molecules because they are several orders of magnitude smaller than the molecules which will engage our interest in this chapter. Our attention will be directed toward *macro*molecules, which contain hundreds to tens-of-thousands of atoms. The bonding principles presented for small molecules in Chapter 3 still hold for macromolecules, of course; in addition, there are various structural features which we will need to consider in order to anticipate the behavior of molecular solids.

Macromolecules are commonly called *polymers* because they may be subdivided into "many units," each unit being a *mer*. These repetitious structural units in polymers are comparable to the unit cells which repeat themselves within crystals. They will serve as our starting point.

7-2 LINEAR POLYMERIZATION

In synthetic polymers the macromolecule is built up from many micromolecules, called *monomers*, which contain *functional* groups that are susceptible to chemical reaction. Some common functional groups that enter into polymerization reactions are hydroxyl,

$$HO\text{—},$$

carboxyl and amino,

and the vinyl radical,

For a monomer to polymerize, it must be at least *bifunctional;* i.e., it must be able to react with at least *two* neighboring monomers so that the molecule may build up repeatedly and produce large molecular sizes. Ethylene glycol,

$$HO—CH_2CH_2—OH,$$

with two hydroxyl groups, and adipic acid,

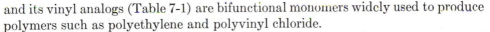

with two carboxyl groups, are examples of bifunctional molecules used to produce polyesters. Ethylene,

and its vinyl analogs (Table 7-1) are bifunctional monomers widely used to produce polymers such as polyethylene and polyvinyl chloride.

The next two subsections will summarize the two most common types of polymerization reactions that are encountered in the manufacture of synthetic macromolecules. From these we will proceed to polymer geometry.

Addition polymerization (chain reaction). Polyethylene is the prototype for addition polymerization:

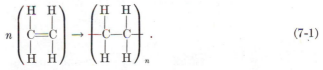

$$(7\text{-}1)$$

Many ethylene (C_2H_4) monomers combine into a single chain of polyethylene, $(C_2H_4)_n$. (Recall Example 3-6.) This is the general type of reaction for the polymerization of all vinyl compounds of

H H
| |
C═C
| |
R H

(Table 7-1).

The addition, or chain polymerization, of Eq. (7-1) actually involves three steps: (1) initiation, (2) propagation, and (3) termination. We will consider each of these steps separately.

TABLE 7-1

ETHYLENE-TYPE MOLECULES (See Appendix D)

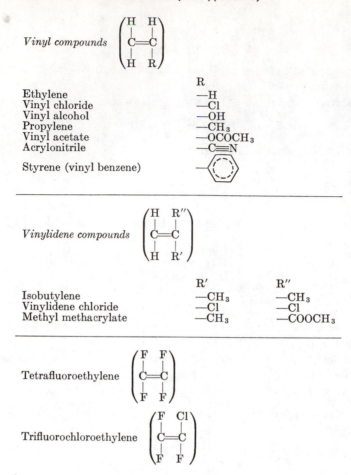

Vinyl compounds	R
Ethylene	—H
Vinyl chloride	—Cl
Vinyl alcohol	—OH
Propylene	—CH$_3$
Vinyl acetate	—OCOCH$_3$
Acrylonitrile	—C≡N
Styrene (vinyl benzene)	

Vinylidene compounds	R′	R″
Isobutylene	—CH$_3$	—CH$_3$
Vinylidene chloride	—Cl	—Cl
Methyl methacrylate	—CH$_3$	—COOCH$_3$

Tetrafluoroethylene

Trifluorochloroethylene

The first step, *initiation*, most commonly utilizes a free radical, X•, from an initiator, I. Peroxides, such as (HO)$_2$ and (C$_6$H$_5$COO)$_2$, serve as specific examples. (See Eq. 3-9.) In general form,

$$I \rightarrow 2X\bullet, \tag{7-2}$$

and

$$X\bullet + \underset{\substack{| \\ R}}{\overset{\substack{H \\ |}}{C}}=\underset{\substack{| \\ H}}{\overset{\substack{H \\ |}}{C}} \rightarrow X-\underset{\substack{| \\ R}}{\overset{\substack{H \\ |}}{C}}-\underset{\substack{| \\ H}}{\overset{\substack{H \\ |}}{C}}\bullet, \tag{7-3a}$$

where the product is the vinyl radical mentioned in the first paragraph of this section. Alternatively, a few monomers, e.g., isobutylene, H$_2$C=C(CH$_3$)$_2$, can form chains from an ionic initiator:

$$H^+ + \overset{\overset{\textstyle H}{\cdot\cdot}}{\underset{\underset{\textstyle H}{\cdot\cdot}}{C}} :: \overset{\overset{\textstyle R}{\cdot\cdot}}{\underset{\underset{\textstyle R}{\cdot\cdot}}{C}} \rightarrow H : \overset{\overset{\textstyle H}{\cdot\cdot}}{\underset{\underset{\textstyle H}{\cdot\cdot}}{C}} : \overset{\overset{\textstyle R}{\cdot\cdot}}{\underset{\underset{\textstyle R}{\cdot\cdot}}{C^+}} . \tag{7-4}$$

Once initiated, the *propagation* of the vinyl group into a chain takes on the sequence

$$X-\underset{\underset{\textstyle R}{|}}{\overset{\overset{\textstyle H}{|}}{C}}-\underset{\underset{\textstyle H}{|}}{\overset{\overset{\textstyle H}{|}}{C}}\bullet + \underset{\underset{\textstyle R}{|}}{\overset{\overset{\textstyle H}{|}}{C}}=\underset{\underset{\textstyle H}{|}}{\overset{\overset{\textstyle H}{|}}{C}} \rightarrow X-\underset{\underset{\textstyle R}{|}}{\overset{\overset{\textstyle H}{|}}{C}}-\underset{\underset{\textstyle H}{|}}{\overset{\overset{\textstyle H}{|}}{C}}-\underset{\underset{\textstyle R}{|}}{\overset{\overset{\textstyle H}{|}}{C}}-\underset{\underset{\textstyle H}{|}}{\overset{\overset{\textstyle H}{|}}{C}}\bullet , \tag{7-3b}$$

or more generally,

$$X\text{-}(\text{mer})_n\text{-}\bullet + \underset{\underset{\textstyle R}{|}}{\overset{\overset{\textstyle H}{|}}{C}}=\underset{\underset{\textstyle H}{|}}{\overset{\overset{\textstyle H}{|}}{C}} \rightarrow X\text{-}(\text{mer})_{n+1}\text{-}\bullet , \tag{7-3c}$$

which continues as long as new monomers are available.

After the supply of monomers is reduced, the active end of the growing free radical has less opportunity to combine with other monomers and a greater probability of encountering another free radical. When it does, *termination* can occur by combination, or *coupling*:

$$X\text{-}(\text{mer})_n\text{-}\bullet + \bullet\text{-}(\text{mer})_m\text{-}X \rightarrow X\text{-}(\text{mer})_{m+n}\text{-}X. \tag{7-3d}$$

In principle, this is just the reverse of the original decomposition of the peroxide initiation (Eq. 7-2). The molecule, however, contains many intervening mers in place of the unstable, central —O—O— peroxide bond.

A second termination mechanism is shown by the following equation and is called *disproportionation*:

$$X\text{-}(\text{mer})_p\text{-}\bullet + \bullet\text{-}(\text{mer})_{q+1}\text{-}X \rightarrow X\text{-}(\text{mer})_p\text{-}H + \underset{\underset{\textstyle H}{|}}{\overset{\overset{\textstyle R}{|}}{C}}=\underset{\underset{\textstyle H}{|}}{\overset{}{C}}\text{-}(\text{mer})_q\text{-}X. \tag{7-5}$$

Polymers such as polystyrene [R of Eq. (7-3) is a benzene ring] terminate primarily by coupling, while other chain polymers have a much higher incidence of disproportionation. Whichever process is prevalent, there is no reason to suspect that termination is controlled by molecular size *per se*. In fact, a spectrum of molecular sizes is found in a normal polymer (Section 7-3).

Condensation polymerization (step reaction). This mechanism of polymerization advances in steps and generally provides a by-product of small molecules. The previously cited bifunctional molecules, ethylene glycol,

$$HO-CH_2CH_2-OH,$$

and adipic acid,

$$HO-\overset{\overset{\displaystyle O}{\|}}{C}-(CH_2)_4-\overset{\overset{\displaystyle O}{\|}}{C}-OH,$$

furnish an example. Abbreviating these formulas to

$$HOM_2OH \qquad \text{and} \qquad HO-CM_4C-OH,$$

where M_2 and M_4 are $(CH_2)_2$ and $(CH_2)_4$, respectively, we can write the reaction as follows:

$$HOM_2O-H + HO-\overset{\overset{\displaystyle O}{\|}}{C}M_4\overset{\overset{\displaystyle O}{\|}}{C}OH \rightarrow HOM_2O\overset{\overset{\displaystyle O}{\|}}{C}M_4\overset{\overset{\displaystyle O}{\|}}{C}OH + H_2O, \qquad (7\text{-}6a)$$

$$HOM_2O\overset{\overset{\displaystyle O}{\|}}{C}M_4\overset{\overset{\displaystyle O}{\|}}{C}-OH + H-OM_2OH \rightarrow HOM_2O\overset{\overset{\displaystyle O}{\|}}{C}M_4\overset{\overset{\displaystyle O}{\|}}{C}OM_2OH + H_2O, \qquad (7\text{-}6b)$$

$$HOM_2O\overset{\overset{\displaystyle O}{\|}}{C}M_4\overset{\overset{\displaystyle O}{\|}}{C}OM_2O-H + HO-\overset{\overset{\displaystyle O}{\|}}{C}M_4\overset{\overset{\displaystyle O}{\|}}{C}OH \rightarrow HOM_2O\overset{\overset{\displaystyle O}{\|}}{C}M_4\overset{\overset{\displaystyle O}{\|}}{C}OM_2O\overset{\overset{\displaystyle O}{\|}}{C}M_4\overset{\overset{\displaystyle O}{\|}}{C}OH + H_2O, \qquad (7\text{-}6c)$$

or, in general,

$$m(H-OM_2O-H) + m(HO-\overset{\overset{\displaystyle O}{\|}}{C}M_4\overset{\overset{\displaystyle O}{\|}}{C}-OH) \rightarrow HO(M_2O\overset{\overset{\displaystyle O}{\|}}{C}M_4\overset{\overset{\displaystyle O}{\|}}{C}O)_mH + (2m-1)M_2O. \qquad (7\text{-}6d)$$

The by-product of condensation polymerization is not always water. Figure 7-1(b) shows the step reaction which forms dacron (a fiber) or mylar (a film) from ethylene glycol and dimethyl terephthalate.* Methyl alcohol, CH_3OH, is the by-product.

The mers of the step reactions just cited contain

linkages which are characteristic of an ester; so the final product is a *polyester*. These linkages are highly polar and provide locations for van der Waals bonding through hydrogen bridges from adjacent molecules (Section 3-6). Several of the common *polar groups* which serve as linkages for condensation reactions are shown in Table 7-2.

A contrast exists between the growth of an addition molecule and that of a step-reaction molecule. Molecular growth proceeds in the former by a sequential

---

* These compounds need not be remembered in detail. The reaction principles should be, however.

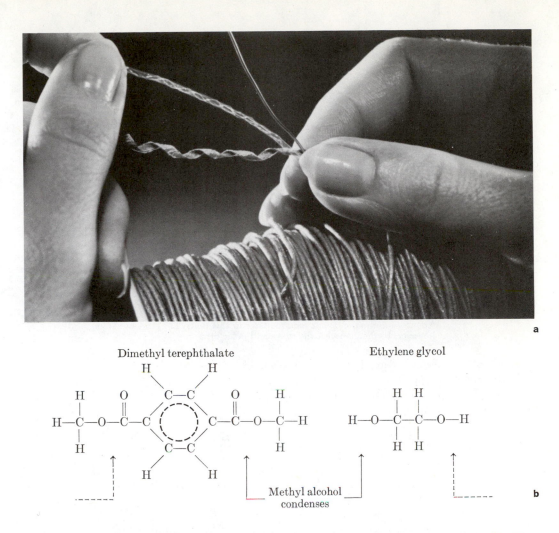

Dimethyl terephthalate

Ethylene glycol

Methyl alcohol
condenses

7–1 Condensation polymerization (dacron, mylar, or terylene). (a) Insulation film of mylar (du Pont). (b) The chemical reaction for polymerization. A by-product is characteristic of condensation reactions.

addition of individual monomers to a few activated molecules. In the latter, any number of many pairs may react, and then pairs may join to continue polymerization without termination.

Example 7-1

Nylon is a condensation polymer of molecules such as adipic acid,

$$HO \cdot CO \cdot (CH_2)_4 \cdot CO \cdot OH,$$

and hexamethylene diamine, $H_2N(CH_2)_6NH_2$.

TABLE 7-2

CONDENSATION POLYMER TYPES

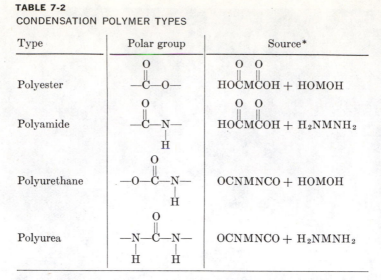

Type	Polar group	Source*
Polyester	$-\overset{\overset{O}{\|}}{C}-O-$	$HO\overset{\overset{O}{\|}}{C}M\overset{\overset{O}{\|}}{C}OH + HOMOH$
Polyamide	$-\overset{\overset{O}{\|}}{C}-\underset{\underset{H}{\|}}{N}-$	$HO\overset{\overset{O}{\|}}{C}M\overset{\overset{O}{\|}}{C}OH + H_2NMNH_2$
Polyurethane	$-O-\overset{\overset{O}{\|}}{C}-\underset{\underset{H}{\|}}{N}-$	$OCNMNCO + HOMOH$
Polyurea	$-\underset{\underset{H}{\|}}{N}-\overset{\overset{O}{\|}}{C}-\underset{\underset{H}{\|}}{N}-$	$OCNMNCO + H_2NMNH_2$

M represents a variety of possible groups; $(CH_2)_x$ is common.

* Source information is not required later in this text. The student should, however, be familiar with the types and polar groups.

a) Sketch the structure of these two molecules.

b) Show how polymerization can occur.

c) Which polar group is formed?

Answer. a) and b)

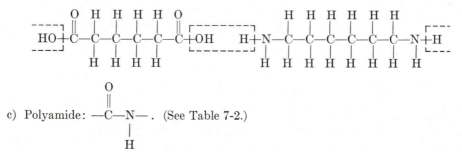

c) Polyamide: $-\overset{\overset{O}{\|}}{C}-\underset{\underset{H}{\|}}{N}-$. (See Table 7-2.)

Note 1. There are various nylon products. Which one is produced depends on the starting materials. This one is commonly called Nylon 66, based on the carbon sequences along the molecular chain.

Note 2. According to Table 3-4 and Example 3-6, 3 kcal are *released* per mole with H_2O as a by-product. If NH_3 had been the by-product, 3 kcal of energy would have been *required* per mole. ◄

7-3 MOLECULAR WEIGHTS

The chief characteristic of polymerized molecules is their high molecular weight. When we are dealing with individual molecules, it is convenient to speak of a degree of polymerization, DP, which is the number of mers per molecule as indi-

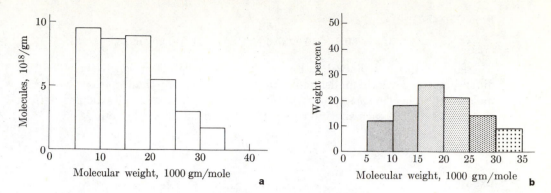

7-2 Molecular weight distribution. (a) Based on numbers (see Example 7-2). (b) Based on weights (see Example 7-3).

cated by n in Eq. (7-1). However, as can be surmised from the previous discussion, the molecules of a plastic are not all identical in size; rather, a range or spectrum of sizes occurs because some are terminated abnormally early and others, by chance, have grown to a very high DP (Fig. 7-2). It is necessary, therefore, to speak of an average molecular weight, $\overline{M}$.

The molecular weight distribution is expressed either by the *number* of molecules in each of several size intervals (Fig. 7-2a) or by the *weight fraction* of the molecules in each size interval (Fig. 7-2b). When it is calculated on the basis of the number of molecules, X_i, in each size interval, where the mean interval size is MW_i, a *"number-average" molecular weight*, $\overline{M}_n$, is obtained:

$$\overline{M}_n = \frac{\sum[(X_i)(MW_i)]}{\sum X_i}. \tag{7-7}$$

Example 7-2

Determine the "number-average" molecular weight for the molecular weight distribution of polyvinyl chloride, $+C_2H_3Cl\frac{}{}_n$, shown in Fig. 7-2(a).

Answer. Based on 1.00 gm.

Size interval, gm/mole	Mean size, MW_i, gm/mole	Number, X_i, molecules/gm	Product, $(X_i)(MW_i)$, molecules/mole
5,000–10,000	7,500	9.6×10^{18}	7.2×10^{22}
10,000–15,000	12,500	8.7	10.8
15,000–20,000	17,500	8.9	15.6
20,000–25,000	22,500	5.6	12.6
25,000–30,000	27,500	3.1	8.5
00,000–25,000	32,500	1.7	5.5
Σ		37.6×10^{18}	60.2×10^{22}

From Eq. (7-7),

$$\overline{M}_n = 60.2 \times 10^{22}/37.6 \times 10^{18} = 16{,}040 \text{ gm/mole.} \quad \blacktriangleleft$$

Figure 7-2(a) is replotted in Fig. 7-2(b) on a weight-fraction basis. The two differ because several smaller molecules weigh no more than one or two larger molecules. The *"weight-average" molecular weight*, $\overline{M}_w$, may be calculated from the weight, W_i, in each size interval, where the mean interval size is MW_i:

$$\overline{M}_w = \frac{\Sigma[(W_i)(MW_i)]}{\Sigma W_i}. \tag{7-8a}*$$

Example 7-3

a) Determine the "weight-average" molecular weight for the molecular weight distribution of polyvinyl chloride, $+\!C_2H_3Cl\!+_n$, shown in Fig. 7-2(b).

b) What is the DP of the average molecule?

Answer. a) Based on 1.00 gm.

Size interval, gm/mole	Mean size, MW_i, gm/mole	Weight, W_i, gm	Product, $(W_i)(MW_i)$, gm^2/mole
5,000–10,000	7,500	0.12	900
10,000–15,000	12,500	0.18	2,250
15,000–20,000	17,500	0.26	4,550
20,000–25,000	22,500	0.21	4,725
25,000–30,000	27,500	0.14	3,850
30,000–35,000	32,500	0.09	2,925
Σ		1.00	19,200

From Eq. (7-8),

$$\overline{M}_w = 19{,}200 \text{ gm/mole.}$$

b) Mer weight $= [(2)(12) + (3)(1) + 35.5] = 62.5 \text{ gm/mer,}$

$$DP = (19{,}200 \text{ gm/mole})/(62.5 \text{ gm/mer})$$

$$= 308 \text{ mers/mole.} \quad \blacktriangleleft$$

The "weight-average" molecular weight has a significance for mechanical properties which will be discussed in Chapter 12. Both averages are used, however, because the ratio $\overline{M}_w/\overline{M}_n$ indicates the width of the size distribution. If only one size is present, the ratio equals one. Normally $\overline{M}_w/\overline{M}_n$ is between 1.5 and 2.5, but may range to 15 or greater. It is often desirable to achieve a low ratio because this indicates more uniformity in the polymer product.

---

* The "weight-average" molecular weight may also be calculated as

$$\overline{M}_w = \frac{\Sigma[(X_i)(MW_i)^2]}{\Sigma[(X_i)(MW_i)]}. \tag{7-8b}$$

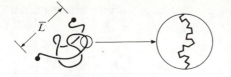

7-3 Molecular lengths. Since linear molecules are naturally coiled and kinked, the root-mean-square length, $\bar{L}$, is appreciably less than the "stretched-out" length (Eq. 7–11).

7-4 MOLECULAR LENGTHS

A single linear polymer molecule may be compared with a cotton or wool fiber. Although the true length/diameter ratios are high, because of kinking and coiling the end-to-end distance is much less than the true length. This is illustrated in Fig. 7-3. If pulled out straight (except for the 109.5° bond angles), an individual polystyrene molecule,

with a DP of 5000 would have an extended length, L, of

$$L = 5000(2)(1.5\ \text{A}) \sin(109.5°/2)$$
$$= 12{,}200\ \text{A}$$

because each C—C bond is about 1.5 A (Table 3-4).

Since all molecules are in continuous thermal agitation, the 109.5° bond angle across each carbon is free to rotate (Fig. 7-4) unless restrained. If the butane, C_4H_{10}, of Fig. 7-4 had the *conformation* of *abcd*, its length, calculated as above, would be 3.7 A. With an *abcd'* conformation, however, its length is only 2.5 A. Various length possibilities for the butane molecule lie between these extremes. In the absence of crystallization, the polystyrene molecule cited in the previous paragraph has a mean length between 1.5 A (essentially closing on itself) and the 12,200 A just calculated. There is very little likelihood, however, that either extreme will be encountered.

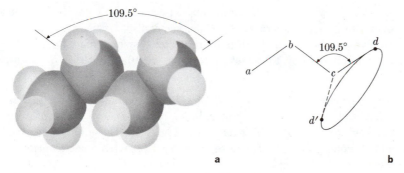

a   **b**

7-4 Bond rotation (butane). (a) Ideal sketch. (b) Cone of rotation. The maximum length *abcd* occurs only a small fraction of the time.

Since a polymer chain can be presented as a series of m vectors, the vectorial length, $\mathbf{L}$, can be shown as follows:

$$\mathbf{L} = \mathbf{l}_1 + \mathbf{l}_2 + \mathbf{l}_3 + \cdots + \mathbf{l}_m. \tag{7-9}$$

The mean square of the end-to-end length, $\overline{L^2}$, is

$$\overline{L^2} = l_1^2 + l_2^2 + l_3^2 + \cdots + l_m^2 + 2l^2 \sum \cos \phi. \tag{7-10a}$$

Assuming that the segments of an unrestrained molecule have equal probabilities for all orientations, ϕ, in space, the last term of Eq. (7-10a) drops out. Further, in most polymers, the large majority of the segments equal the C—C distance, l. Therefore,

$$\overline{L^2} = ml^2, \tag{7-10b}$$

or the *root mean square length*, $\overline{L}$, is related to the degree of polymerization, n, by

$$\overline{L} = l\sqrt{2n}, \tag{7-11}$$

since there are two segments per mer ($m = 2n$). On this basis the *mean length* of the previously cited 5000-mer, unrestrained polystyrene molecule is only 150 A.

Since the mean lengths of noncrystalline polymeric molecules are much less than their extended lengths, a molecular plastic (like a ball of cotton) may undergo considerable elongation when stressed. Unlike a cotton thread, however, the molecules have random thermal motions; consequently, they will again kink and coil when released. In other words, the most stable conformation of an independent polymer molecule is one with the mean length just described.

7-5 CRYSTALLIZATION OF POLYMERS

Our discussion of molecular crystals in Section 4–5 was directed mainly toward crystals of micromolecules. In Fig. 4-11, however, we noted that linear molecules like those of tellurium, +Te+_x, and polyethylene, $\text{+C}_2\text{H}_4\text{+}_n$, also crystallize. The bond angles of adjacent molecules match as shown by the electron-density data of Bunn (Fig. 7-5) to give the orthorhombic unit cell shown in Fig. 7-6 for polyethylene.

The molecule within a crystal lacks the freedom to perform the irregular motion described for noncrystalline polymers in the last section. However, since the molecule may involve a whole distribution of molecular sizes ranging into the hundreds of mers, it is uncommon for crystals to be as regular as those of iodine (Fig. 4-10).

Micelles. It has been proposed that a long polymer chain can encounter crystalline coordination locally and then enter an amorphous region before encountering another crystalline region. Called *fringed micelles* (Fig. 7-7), these small regions (50 to 200 A) could be compared to grains in regular crystals, with one exception—"threads" of molecular continuity exist from one micelle to another. The crystalline volume fraction may range from nil to 0.8 or more, depending on various factors to be discussed later.

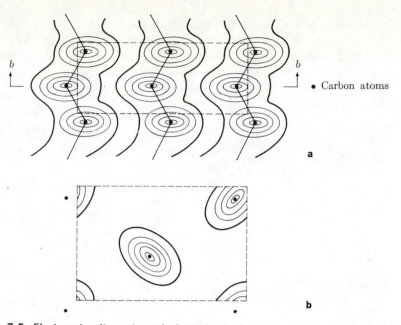

• Carbon atoms

7–5 Electron density contours (polyethylene). Compare with Fig. 4–11(b) and Fig. 7–6. (After C. W. Bunn, *Trans. Faraday Soc.*, **35,** 1939, p. 482.)

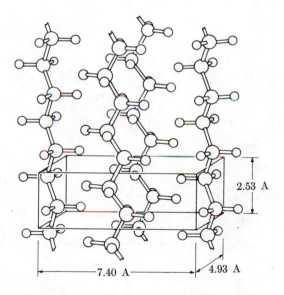

2.53 A

7.40 A   4.93 A

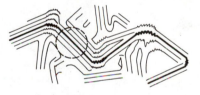

7–7 Fringed micelles. A polymeric molecule may extend through several areas (micelles) of local crystallization. Between micelles, the structure is amorphous.

7–6 Crystalline polyethylene. The chains are aligned to give an orthorhombic unit cell. Compare with Fig. 4–11(b) and Fig. 7–5. (M. Gordon, *High Polymers*, London: Iliffe Books, Ltd., and Reading, Mass.: Addison-Wesley, 1963, p. 90. After C. W. Bunn, *Chemical Crystallography*, London: Oxford University Press, 1945.)

Folded chains. Under certain conditions a polymer chain can be shown to crystallize *with itself*, as indicated in Fig. 7-8! Though they were interpreted at first as a curiosity, the folded chains are proving to have a rather widespread occurrence among crystalline polymers. In fact, they may be shown to be the most stable crystalline conformation for linear molecules, just as coiled or kinked molecules are the most stable for separated molecules.

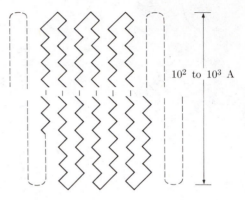

10^2 to 10^3 A

7–8 Folded chains. Polymer crystallization can occur by chain folding. In effect, the polymer molecule crystallizes with itself.

Example 7-4

Calculate the theoretical density of fully crystalline polyethylene.

Answer. From Fig. 7-6, there are two $\left(C_2H_4\right)$ mers per unit cell. Thus

$$
\begin{aligned}
\text{density} &= \frac{\text{mass/unit cell}}{\text{vol/unit cell}} \\
&= \frac{2(28 \text{ gm/mer wt})/(0.602 \times 10^{24}/\text{mer wt})}{(2.53 \times 7.40 \times 4.93)(10^{-24} \text{ cm}^3)} \\
&= 1.005 \text{ gm/cm}^3.
\end{aligned}
$$

Note. Densities of this polymer normally range from 0.92 to 0.95 gm/cm^3, depending on the degree of crystallinity. ◀

7-6 MOLECULAR VARIATIONS

Side radicals. Polyethylene has been useful as a prototype because it contains a simple carbon chain with two hydrogen atoms each. Its regular zig-zag structure permits easy crystallization, at least in local areas. The polymerization process, however, is not always ideal. For example, we sometimes find irregularities like the following:

$$
\text{X}\left(\text{mer}\right)_n\bullet + \underset{\underset{\text{H}}{|}}{\overset{\overset{\text{H}}{|}}{\text{C}}}=\underset{\underset{\text{H}}{|}}{\overset{\overset{\text{H}}{|}}{\text{C}}} \rightarrow \text{X}\left(\text{mer}\right)_n\underset{\underset{\underset{\text{H}}{|}}{\text{H}-\text{C}-\text{H}}}{\overset{\overset{\text{H}}{|}}{\text{C}}}\bullet, \tag{7-12}
$$

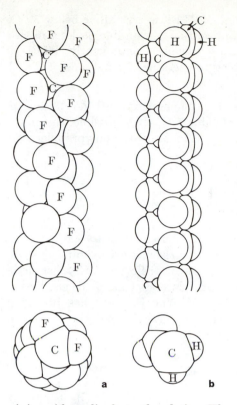

7–9 Steric hindrance. (a) Polytetrafluoroethylene. The larger fluorine atoms introduce a spiral structure to the chain within a crystal. (b) Polyethylene. Although zig-zagged, the chain does not have to spiral within crystals because the hydrogen atoms encounter no steric hindrance from one another. (M. Gordon, *High Polymers*, London: Iliffe Books, Ltd., and Reading, Mass.: Addison-Wesley, 1963, p. 93. After C. W. Bunn and E. R. Howells, *Nature*, **174**, 1954, p. 549.)

a   b

giving side radicals to the chain. The net energy change in Eq. (7-12) is the same as in Eq. (7-3c). In that respect the two reactions are comparable; however, energy must be supplied to relocate a C—H bond as well as to break the double bond in order to activate the reaction step. Therefore, the reaction of Eq. (7-12) is not encountered frequently. When it does occur, complications are introduced into the crystallization process.

Steric hindrance. Side radicals, such as the —CH_3 of Eq. (7-12) and the benzene ring of polystyrene, or even a large atom such as —Cl of polyvinyl chloride, will interfere with crystallization because the adjacent chains cannot readily mesh with one another. These radicals provide a spatial interference called steric hindrance. Whenever crystallization is prevented, there are pronounced effects on mechanical and thermal properties (Section 12-5).

Steric hindrance may also introduce some subtle modifications which give rise to interesting consequences. Consider polytetrafluoroethylene (PTFE), $+C_2F_4+_n$, which is quite similar to polyethylene, $+C_2H_4+_n$, except that PTFE* has larger fluorine atoms. In fact, they are sufficiently large so as to introduce a steric hindrance along the chain which prevents a simple zig-zag alignment. As a result, the chain assumes a helical conformation, as shown in Fig. 7-9, which is markedly

---

* One trade name is teflon.

stiffer than the comparable polyethylene chain. PTFE is crystalline below 30°C. Above this temperature, the stiff chains maintain their crystalline packing in two dimensions but are free to slip in the third dimension. Consequently, PTFE is both soft and has an abnormally high melting temperature for polymers (330°C) —an unusual combination.

Stereoisomers. The ethylene mer has a center of symmetry with no "head" or "tail." This is not true for most vinyl mers (Table 7-1). It is possible, therefore, to have various *configurations* along a molecule. All mers of the polypropylene

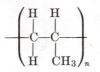

have the *same* orientation in Fig. 7-10(a), whereas there is an *absence* of orientational consistency in Fig. 7-10(b). These two *stereo*isomeric configurations are called *isotactic* and *atactic*, respectively. Because of their regularity, isotactic isomers crystallize while their atactic counterparts do not. One of the significant materials-technology achievements of post–World War II years has been the development of processes for introducing stereoregularity into linear polymers, giving enhanced mechanical properties (Chapter 12).

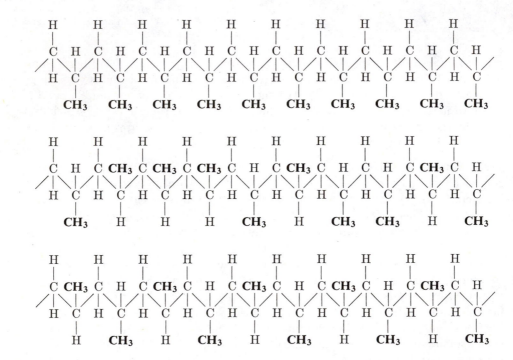

7-10 Stereoisomerism (polypropylene). (a) Isotactic. (b) Atactic. (c) Syndiotactic. The more ordered isomers crystallize more easily.

A third type of stereoisomeric configuration, called *syndiotactic*, is shown in Fig. 7-10(c). Each succeeding *pair* of mers has the same pattern. In practice, configurations may be found which lie between the three prototypes cited in Fig. 7-10.

Branching. Polymer chains can be bifurcated as illustrated in Fig. 7-11. A junction may be found at only one out of a thousand mers. When it occurs, however, it inhibits crystallization.

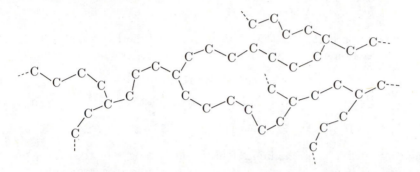

7-11 Branching. Bifurcated chains interfere with crystallization. Therefore, with much branching, the density is lower. (Hydrogen atoms and side radicals are not shown along the carbon chain.)

Plastics of polyethylene that have considerable branching will have densities of <0.925 gm/cm^3 and crystallinity as low as 50%. Varieties with negligible branching have densities of >0.94 gm/cm^3 and about 90% crystallinity. The latter, more crystalline, high-density product has a softening temperature in excess of 110°C, compared with 75°C for the low-density, highly branched variety. This presents a major advantage for use in the home, or wherever hot water is encountered.

Since branching and side radicals are more frequently developed at high polymerization temperatures, the above high-density product had to await the development of appropriate catalyzers so that reaction could be achieved at 100°C rather than the 200°C temperature required in the older, low-density process.

Example 7-5

Show how termination by disproportionation (Eq. 7-5) can lead to branching.

Answer. The second product of Eq. (7-5) has a double bond and may be compared directly to a vinyl compound which may react with other growing molecules:

$$X\!\left(\!\begin{array}{cc} H & H \\ | & | \\ C - C \\ | & | \\ R & H \end{array}\!\right)_{\!m}\!\!+\ \begin{array}{cc} H & H \\ | & | \\ C = C \\ | & | \\ R & \text{(mers)}\!-\!X \end{array}\ +\ n\!\left(\!\begin{array}{cc} H & H \\ | & | \\ C = C \\ | & | \\ R & H \end{array}\!\right)\ \rightarrow\ X\!\left(\text{mers}\right)_{\!m}\!\!\begin{array}{cc} H & H \\ | & | \\ C - C \\ | & | \\ R & \text{(mers)}\!-\!X \\ p \end{array}\!\left(\text{mers}\right)_{\!n}\!\!. \qquad \blacktriangleleft \quad (7\text{-}13)$$

7-7 UNSATURATED POLYMERS

Natural rubber and the majority of our artificial rubbers possess mers with an average of fewer than two hydrogen atoms or side groups for each carbon; thus they are *unsaturated* and possess double bonds along their chain. The polyisoprene mer,

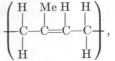

of natural rubber is our prototype. ("Me" stands for the methyl group, CH_3.) The polymerization reaction starts with a monomer containing two double bonds,

$$n \begin{pmatrix} & H & Me & H & H & \\ & | & | & | & | & \\ & C=C-C=C & \\ & | & & | & \\ & H & & H & \end{pmatrix} \rightarrow \begin{pmatrix} & H & Me & H & H & \\ & | & | & | & | & \\ & C-C=C-C & \\ & | & & | & \\ & H & & H & \end{pmatrix}_n, \quad (7\text{-}14)$$

with one being consumed. The remaining double bond changes to the center of the mer to retain four bonds per carbon. Comparable molecules are shown in Table 7-3, in which R may be —H or —Cl in place of —CH_3.

TABLE 7-3
BUTADIENE-TYPE MOLECULES
(See Appendix D)

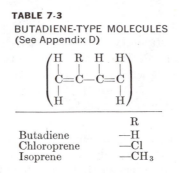

	R
Butadiene	—H
Chloroprene	—Cl
Isoprene	—CH_3

The double bond in these rubbers affects the structure and properties. The C=C—C bond angle is 125°, and the C=C bond *cannot* be rotated as a single bond may. Thus the center of the mer is rigid. (Flexibility still exists between mers.)

Cis- and trans-isomers. Two distinct isomers exist for the polymers which come from Eq. (7-14) and Table 7-3. They are called *cis* or *trans* configurations depending on whether the "vacant" sites are on the *same* side or are *across* the chain (Fig. 7-12). Since the double bond cannot rotate, these two configurations cannot interconvert. Each isomer has its distinct set of properties.

As shown in Fig. 7-12(a), trans-isoprene, called *gutta percha*, has a mer which may take on a zig-zag alignment with adjacent chains in a solid. In contrast, a cis-isoprene, *natural rubber* (Fig. 7-12b), has a kinked mer which complicates alignment, unless stresses are applied to partially unkink the chain. After a stress

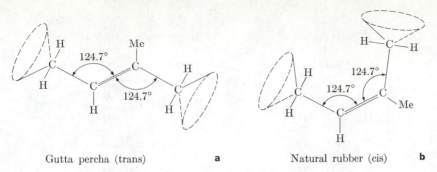

Gutta percha (trans)   **a**   Natural rubber (cis)   **b**

7–12 Cis- and trans-isomers (isoprene: Me = CH₃). Because of the double bond, no rotation can occur between the center carbons. (a) The relatively straight trans-isomer of gutta percha crystallizes more readily than (b) the cis-isomer of natural rubber, which has a significant kink within the mer.

is released, however, the neighboring molecules within the solid are not able to maintain their alignment, and so coiling and kinking recur just as if they were free molecules (Section 7-4). These molecules which recoil within solids are called *elastomers*, and are prevalent in rubbers. They can be stretched several hundred percent and subsequently experience nearly complete recovery. Unlike natural rubber, gutta percha does not have this property at normal temperatures. We shall observe in Chapter 12 that the difference between the behavior of trans-isoprene and cis-isoprene is associated with the glass point (Section 6-11).

7-8 CROSS-LINKING

Heretofore we have limited our discussion to discrete linear molecules, with van der Waals bonds as the only *inter*molecular force of attraction. These forces can be overcome by mechanical stresses, particularly at elevated temperatures, giving us a thermoplastic polymer (Chapter 12).

Under special conditions, molecules can be *cross-linked* through primary bonds. For example, if a fraction of a percent of divinyl benzene (Fig. 7-13a) is added

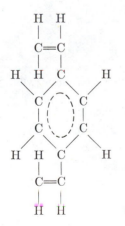

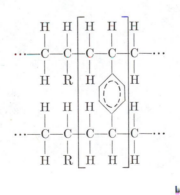

7–13 Divinyl benzene. (a) Monomer. (b) Cross-linking mer. Since this monomer is tetrafunctional, it can be a part of two chains. (See Table 7-1 for a comparison with difunctional styrene, i.e., vinyl benzene.)

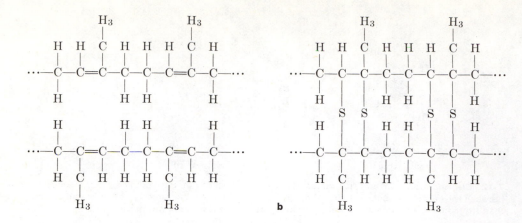

7-14 Vulcanization of natural rubber with sulfur. The sulfur content must be controlled to give the desired number of anchor points. Hard rubber has a high sulfur content.

to styrene, it can polymerize as a member of *two* polystyrene chains, which now are not independent (Fig. 7-13b).

Vulcanization. The commercial process of vulcanization introduces sulfur cross-links into a rubber. Sulfur is added to a linear rubber which, with appropriate heat and activation, reacts with the double-bonded carbon atoms as shown schematically in Fig. 7-14.

The effect of cross-linking is to introduce *anchor points* between molecules. These restrict both elastic and plastic elongation, as can be envisioned schematically in Fig. 7-15. Actual examples for comparison include gum rubber, with no cross-linking, and a hard-rubber comb—highly cross-linked with sulfur.

7-15 Anchor points. Cross-linking interferes with un-coiling because the molecules are tied together at anchor points.

Example 7-6

a) An isoprene rubber is fully cross-linked with sulfur. Assuming no excess sulfur, what is the fraction of sulfur in the final product?

b) What would the sulfur fraction be if the rubber were 50% cross-linked?

Answer. From Fig. 7-14, there is one sulfur atom required per mer of isoprene:

$$\text{mer wt} = (5)(12) + 8(1) = 68 \text{ gm/mer.}$$

a) Fraction sulfur $= \dfrac{32}{32 + 68} = 0.32.$

b) Fraction sulfur $= \dfrac{(32/2)}{(32/2) + 68} = 0.19.$ ◄

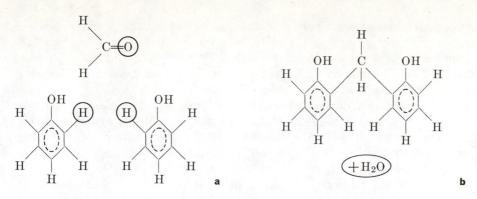

7–16 Phenol-formaldehyde reaction. Each phenol molecule is trifunctional; thus a framework structure forms. This reaction has a by-product.

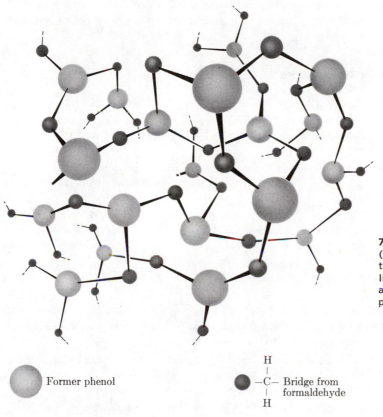

7–17 Framework structures (schematic). Such a structure has insignificant crystallinity, but is relatively strong as a result of its network of primary bonds.

Former phenol

$$-\overset{\displaystyle H}{\underset{\displaystyle H}{C}}-$$ Bridge from formaldehyde

7-9 FRAMEWORK STRUCTURES

Cross-linking changes a linear structure into a three-dimensional structure. As more and more cross links are introduced, there is less and less opportunity for a polymer to be plastic, because the individual molecules are no longer independent.

Polyfunctional monomers. For cross-linking to occur, the divinyl benzene (Fig. 7-13) and isoprene (Fig. 7-14) had to be tetrafunctional. Each had four sites for reaction. Actually, the third and fourth sites of isoprene and the other mers of Table 7-3 are less reactive than the end sites. As a result, a linear molecule is preferentially formed, and cross-linking occurs as a second step.

On the other hand, trifunctional molecules, as exemplified by phenol, C_6H_5OH, can react with formaldehyde, CH_2O, to give the first of the synthetic resins, phenol-formaldehyde (Fig. 7-16). The reaction of Fig. 7-16 may occur at three places around the benzene ring of phenol* to build a three-dimensional structure of covalent bonds (Fig. 7-17).

The possibility of long-range order in a phenol-formaldehyde plastic is remote, since many permutations of bond angles and rotations are possible before each successive CH_2 bridge is introduced. Thus the framework structure of phenol-formaldehyde and other similar plastics is directly comparable with that of the glasses of Section 6-11. Organic glasses of this nature are widespread and have many applications, particularly where heat might soften a linear polymer which relies on weak intermolecular bonding.

REFERENCES FOR FURTHER READING

Alfrey, T., and E. F. Gurnee, *Organic Polymers*. Englewood Cliffs, N. J.: Prentice-Hall, 1967. This short text (127 pp.) is excellent as supplementary reading on the structure and properties of polymers. Organic chemistry is not required in the first two chapters but would be helpful in Chapter 3.

Billmeyer, F. W., *Textbook of Polymer Science*. New York: Interscience, 1962. Recommended text for the student who is wishing to study polymers in more detail. Background in organic chemistry is helpful.

DiBenedetto, A. T., *The Structure and Properties of Materials*. New York: McGraw-Hill, 1967. The structure of polymers is discussed in Chapters 8 and 9 on an undergraduate level. Organic chemistry is presented as needed.

"Giant Molecules," *Scientific American*, **197,** September 1957. This is a special issue devoted to polymeric materials. The student will find many of the articles interesting as supplementary reading.

Gordon, M., *High Polymers*. London: Iliffe, and Reading, Mass.: Addison-Wesley, 1963. Supplementary reading for the relationship of structure and physical properties. Except for Chapter 1, a chemistry background is not needed.

Mark, H. F., "The Nature of Polymeric Materials," *Scientific American*, **217** [3], September 1967, pp. 148–154. An introduction to the structure of polymers and the effect on properties. Introductory level.

Modern Plastics Encyclopedia. New York: McGraw-Hill (Annual). Published yearly. Contains considerable property data for all types of plastics.

O'Driscoll, K. F., *The Nature and Chemistry of High Polymers*. New York: Reinhold, 1964. Paperback. Recommended background reading for the student who has had general college chemistry.

---

* It does not occur at adjacent carbons because of steric hindrance.

Van Vlack, L. H., *Elements of Materials Science*, second edition. Reading, Mass.: Addison-Wesley, 1964. Polymers are presented in a more introductory manner in the first part of Chapter 3 and in Chapter 7. Freshman-sophomore level.

PROBLEMS

7-1 a) What is the "number-average" molecular weight of a $75CO_2$-$25N_2$ gas?

  b) The "weight-average" molecular weight? [*Note.* Gas analyses are reported on mole or volume percent *unless stated otherwise.* Condensed phases are reported on weight percent *unless stated otherwise.*]

Answer. a) 40 gm/mole   b) 41.3 gm/mole

7-2 Air has 5% water vapor added. How much is the "number-average" molecular weight changed?

7-3 What are (a) the "weight-average molecular weight" and (b) the degree of polymerization of a polyethylene, $\left(C_2H_4\right)_n$, plastic with the following analysis?

(Molecular weight)$_i$, (gm/mole)	Weight fraction
5,000	0.15
10,000	0.22
15,000	0.31
20,000	0.18
25,000	0.09
30,000	0.05

Answer. a) 14,950 gm/mole   b) 533

7-4 Assume you have a polyvinyl chloride containing molecules of only three sizes: 6000 gm/mole, 10,000 gm/mole, and 12,000 gm/mole. There are 1000 grams of each size. What fraction of the total number of molecules are of the smallest size (6000 gm/mole)?

7-5 A polyvinyl chloride has the following molecular size distribution:

Range	Weight percent
0– 5,000 gm/mol wt	15
5–10,000	25
10–15,000	40
15–20,000	20

  a) What is the average degree of polymerization?

  b) What is the ratio of $\overline{M}_w$ to $\overline{M}_n$?

Answer. a) 172   b) 1.49

7-6 Fifty polyethylene molecules are selected so that 10 of them have 10 mers, 10 have 20 mers, 10 have 30 mers, 10 have 40 mers, and 10 have 50 mers.

a) What is the weight-average molecular weight?

b) What is the number-average molecular weight?

7-7 If HCl is to be used as an ionic initiator for isobutylene polymerization, how much would have to be added to produce an average molecular weight of 8100 gm/mole? Assume 30% efficiency of the HCl. [*Note.* Both H^+ and Cl^- serve as initiators *and* terminators.]

Answer. 1.5 w/o

7-8 Two-tenths of one percent by weight of H_2O_2 was added to ethylene prior to polymerization. What would the average DP be if all the H_2O_2 were used as terminals for the molecules? (Mer = C_2H_4)

7-9 A polymer crystallizes in the hexagonal system with the following unit-cell dimensions: $a_1 = a_2 = 74.7$ A; $c = 30.6$ A. The density is 1.296 gm/cm^3 and the molecular weight is 5733. Calculate the number of polymer molecules per unit cell. Choose either the full hexagonal unit cell or the rhombic unit cell, but state clearly which you are using.

Answer. 20 molecules per rhombic unit cell (60 per hexagonal unit cell)

7-10 Vinyl chloride molecules have a DP of 900. What would their length be if they were stretched from their normal kinked conformation to a straight molecule (except for the 109.5° bond angles)?

7-11 What percent sulfur would be present if it were used as a cross-link at every possible point (a) in polyisoprene? (b) in polychloroprene?

Answer. a) 32% S   b) 26.5% S

7-12 A rubber contains 91% polymerized chloroprene and 9% sulfur. What fraction of the possible cross-links are joined in vulcanization? (Assume that all the sulfur is used for cross-links.)

7-13 A rubber contains 54% butadiene, 34% isoprene, 9% sulfur, and 3% carbon black. What fraction of the possible cross-links are joined by vulcanization, if you assume that all the sulfur is used in cross-linking?

Answer. 0.188

7-14 A rubber containing 47 w/o isoprene, 38 w/o butadiene, and 15 w/o of an inert filler is exposed to ozone, gaining 0.6 gm per 100 gm of the original product. Assume that the gain in weight is a consequence of cross-linking. What fraction of the possible cross-links are completed?

7-15 One pound of divinyl benzene (Fig. 7-13) is added to 50 lb of styrene. What is the maximum possible number of cross-links per pound of product?

Answer. 4×10^{22}/lb of product

7-16 One hundred grams of isoprene rubber are partially cross-linked with 4 gm of sulfur and then permitted to receive further cross-linking by oxygen. Assume that all the sulfur and all the oxygen enter the rubber as cross-links. How much will the rubber weigh when the number of oxygen cross-links equals the number of sulfur cross-links?

7-17 An isoprene rubber gains 5 w/o by oxidation. Assume that this oxygen produced cross-linkage. What fraction of the possible anchor points (fix points) contain oxygen atoms?

Answer. 0.21

8 □ solid solutions

8-1 SOLUTIONS

A *phase*, defined in Section 4-1, may have more than one component. For example, an aqueous solution may contain NaCl in addition to water; and carbon tetrachloride can dissolve C_8H_{18} within its liquid structure. Multiple-component phases existing over a range of compositions are called *solutions*. A characteristic of a primary solution is that the basic structure of the pure solvent accommodates solute additions. This occurs rather readily in a liquid because its structure is not rigorously fixed. Still, there are limits; for example, only very limited amounts of C_8H_{18} can be dissolved in water because the molecules are not fully compatible.

Solid structures can also accommodate a second component in an analogous way. To illustrate, the fcc structure of copper can contain nearly 40% zinc without destroying the fcc structure. This product, which has been known for centuries as *brass*, has copper as the solvent and zinc as the solute. In general, a *solid solution* demands a more special solvent-solute relationship than a liquid solution, because the structure is less adaptable. Zinc, for instance, can be incorporated into the structure of copper because the atomic radii of zinc and copper are reasonably comparable (1.332 A versus 1.278 A) and the two have similar electronic characteristics. Since copper and nickel have even closer atomic radii, a complete solid solution series is possible from 100% copper to 100% nickel.

Solubility limits. At 20°C, no more than 28 w/o NaCl can be dissolved in water under equilibrium conditions. Any excess NaCl remains undissolved and maintains its own crystal structure as a second phase. The solubility limit is sensitive to temperature. The change in solute concentration with temperature at saturation (dX/dT) may be either positive or negative (Fig. 8-1).

8-2 INTERSTITIAL SOLID SOLUTIONS

A very small atom may be dissolved interstitially among larger atoms. As an example, fcc iron can dissolve carbon atoms into the interstices of its structure, as shown in Fig. 8-2, where a {100} plane is illustrated in approximate scale. Although the small carbon atom fits into the structure better than a self-interstitial (Fig. 6-6b), the lattice distortion is still sufficient to make it impossible to locate a solute atom at every equivalent interstitial site in the fcc lattice.

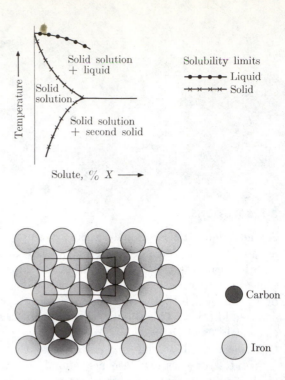

8–1 Solubility limits. The solubility may either increase or decrease with temperature. If the solubility limit is exceeded, a second phase forms.

8–2 Interstitial solution (C in fcc Fe). A limited number of atoms may be dissolved into a crystal structure at interstitial sites. (A {100} plane is shown.)

Only a few atoms are small enough to be interstitial solutes in close-packed structures. These include hydrogen, boron, carbon, nitrogen, and sometimes oxygen. Certain covalent crystals, however, have large interstitial sites which allow them to dissolve large atoms and even molecules. For example, one ceramic product, called a molecular sieve, has interstices large enough to contain small hydrocarbons such as propane, C_3H_8.

Example 8-1

a) How much displacement will there be between the $0, 0, 0$ and $0, 1, 0$ sites in a unit cell of fcc iron if a carbon atom is placed at $0, \frac{1}{2}, 0$ at 1350°F?

b) Repeat for bcc iron. [*Note*. At 1350°F, $R_{\text{fcc Fe}} = 1.292$ A; $R_{\text{bcc Fe}} = 1.258$ A; $R_C = 0.75$ A.]

Answer

a) $a_{\text{fcc}} = (4)(1.292 \text{ A})/\sqrt{2} = 3.654$ A,

$2(1.292 \text{ A}) + 2(0.75 \text{ A}) - 3.654 \text{ A} = 0.43$ A.

b) $a_{\text{bcc}} = (4)(1.258 \text{ A})/\sqrt{3} = 2.905$ A,

$2(1.258 \text{ A}) + 2(0.75 \text{ A}) - 2.905 \text{ A} = 1.11$ A.

Note. Since atoms are not actually rigid spheres, each nearby atom is distorted to some extent. ◀

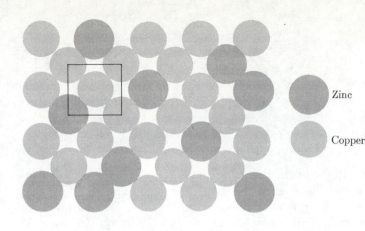

Zinc

Copper

8-3 Substitutional solid solution (zinc in copper, i.e., *brass*). (A {100} plane is shown.)

Example 8-2

If there are two carbon atoms per 25 unit cells of fcc iron, what is the weight percentage (w/o) carbon?

Answer. Each fcc unit cell has 4 iron atoms (Fig. 4-5).

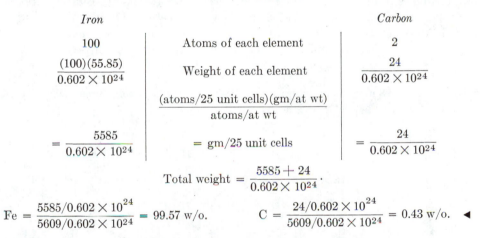

	Iron		*Carbon*
	100	Atoms of each element	2
	$\dfrac{(100)(55.85)}{0.602 \times 10^{24}}$	Weight of each element	$\dfrac{24}{0.602 \times 10^{24}}$
		$\dfrac{\text{(atoms/25 unit cells)(gm/at wt)}}{\text{atoms/at wt}}$	
	$= \dfrac{5585}{0.602 \times 10^{24}}$	$= \text{gm/25 unit cells}$	$= \dfrac{24}{0.602 \times 10^{24}}$

$$\text{Total weight} = \frac{5585 + 24}{0.602 \times 10^{24}}.$$

$$\text{Fe} = \frac{5585/0.602 \times 10^{24}}{5609/0.602 \times 10^{24}} = 99.57 \text{ w/o.} \qquad \text{C} = \frac{24/0.602 \times 10^{24}}{5609/0.602 \times 10^{24}} = 0.43 \text{ w/o.} \blacktriangleleft$$

8-3 SUBSTITUTIONAL SOLID SOLUTIONS

In fcc brass, copper atoms are replaced by zinc atoms. The substitutional alloy is *random* if the probability that any atom site is occupied by a copper atom is equal to the fraction of copper atoms.

Extended substitutional solid solutions can occur in close-packed structures only when the atoms have similar size and electrical characteristics. If sufficiently similar—that is, if the size discrepancy is less than about 10 or 15% and if the ionization characteristics are related—a complete series of solid solutions is possible. A metallic example is shown in Fig. 8-3; other examples include Cu-Ni (fcc), Cr-Fe (bcc), and Cd-Mg (hcp) alloys. Ceramic examples include (Mg, Fe)O

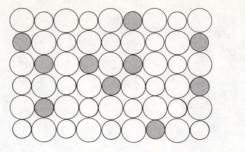

8-4 Substitutional solid solution. Fe^{2+} ions are substituted for Mg^{2+} ions in MgO to give (Mg, Fe)O.

(Fig. 8-4) and Mn(S, Se). In the oxide, there is a complete series of cation substitutions, because Mg^{2+} and Fe^{2+} ions carry the same charge and have radii close to 0.8 A. Likewise, S^{2-} and Se^{2-} produce anionic substitutions. We shall observe later that hardening generally accompanies solid-solution formation (Chapters 11 and 20); thus alloys of this type are often useful.

Size has less, and often negligible, importance in substitutional solid solutions of covalent structures. Often these are open structures, and consequently an odd-sized atom does not introduce excessive lattice strain.

Copolymers may be viewed as substitutional polymeric solid solutions (Fig. 8-5). Two examples from among the many commercial copolymers are Buna-S rubbers, a copolymer of butadiene (Table 7-3) and styrene (Table 7-1), and ABS plastics, a triple polymer of acrylonitrile-butadiene-styrene. The components of these copolymers must possess similar polymerization reactions (Section 7-2).

Example 8-3

A copolymer of 88 m/o* polyvinyl acetate (PVA) and 12 m/o polystyrene (PS) is used for textile fibers. What is the weight percent PVA?

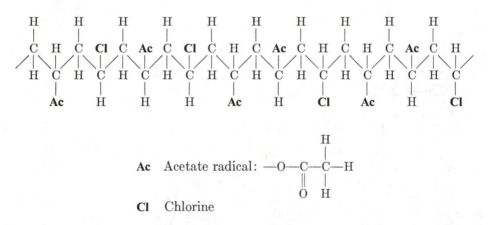

Ac   Acetate radical: $-O-C-C-H$

Cl   Chlorine

8-5 Copolymers (vinyl chloride in vinyl acetate). Copolymers generally have lower melting points and lower glass points than single-component polymers.

---

*   The symbol m/o stands for mer percent.

Answer. Basis: 100 mers. From Table 7-1, or Appendix D,

$$\text{Wt PVA} = 88[4(12) + 2(16) + 6] = 7568 \text{ amu}$$
$$\text{Wt PS} = 12[8(12) + 8] \qquad\quad = \underline{1248 \text{ amu}}$$
$$\qquad\qquad\qquad\qquad\qquad\qquad\quad 8816 \text{ amu.}$$

$$\text{PVA} = 7568/8816 = 85.9 \text{ w/o,}^*$$
$$\text{PS} = 1248/8816 = 14.1 \text{ w/o.} \quad \blacktriangleleft$$

8-4 ORDER-DISORDER TRANSITIONS

Not all substitutional solid solutions are random like the copper-rich fcc brass of Fig. 8-3 (called *α-brass*). In some cases, *ordering* occurs in which the atoms arrange themselves in a preferred manner within the solid solution. As an example, let us consider bcc brass (called *β-brass*) containing equal numbers of copper and zinc atoms. A strong tendency exists for the copper atoms to surround themselves with zinc atoms, and for the zinc atoms to surround themselves with copper atoms. This preference for the formation of *unlike* pairs of atoms, rather than *like* pairs, is expressed by the inequality

$$E_{AB} < \tfrac{1}{2}(E_{AA} + E_{BB}), \qquad\qquad\qquad (8\text{-}1)$$

where E_{AB} is the energy of an unlike bond, and E_{AA} and E_{BB} are the energies of like bonds.

If the above inequality is relatively large, the ordering can become *long-range* in nature. This condition is readily understood by imagining that a bcc lattice is composed of two sets of sites: *α*-sites at the corners of the unit cells, and *β*-sites at the body centers. When the atomic fractions of the *A*- and *B*-components are the same, respectively, as the fractions of the *α*- and *β*-sites, we have a *stoichiometric composition*, and perfect ordering can result (Fig. 8-6a). Here, all the *α*-sites are

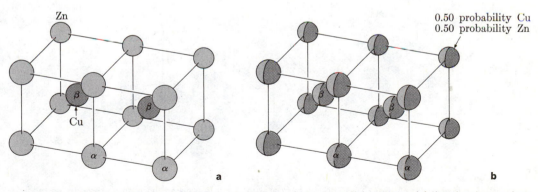

8-6 Order and disorder (*β*-brass). With equal numbers of copper and zinc atoms, (a) the structure orders below 460°C; (b) above that temperature it disorders.

---

* The symbol w/o stands for weight percent.

occupied by zinc atoms, and all the β-sites by copper atoms.* In this structure, each zinc atom is then surrounded by eight copper atoms, and each copper atom by eight zinc atoms.

Given sufficient time for equilibrium to be achieved, the degree of ordering decreases with increasing temperature because of the disordering effect of thermal agitation. In fact, the long-range order of β-brass is lost with heating above 460°C; such heating gives a random bcc solid solution (Fig. 8-6b).

Order parameters. The degree of *long-range order* can be expressed as

$$\text{LRO} = r_\alpha - w_\beta, \tag{8-2a}$$

where r_α is the fraction of α-sites "rightly" occupied (i.e., by A-atoms) and w_β is the fraction of β-sites "wrongly" occupied (i.e., by A-atoms). This relationship is equivalent to the more commonly used long-range order parameter:

$$\text{LRO} = \frac{r_\alpha - X_A}{1 - X_A}, \tag{8-2b}$$

where X_A is the atomic fraction of component A. It will be seen that LRO is unity for perfect order ($r_\alpha = 1$ and $w_\beta = 0$ as in Fig. 8-6a), and is zero when the α- and β-sites are randomly occupied ($r_\alpha = X_A$ and $w_\beta = X_A$ as in Fig. 8-6b). Of course, the long-range order parameter may have intermediate values as well.

Figure 8-7 illustrates the long-range order in an fcc gold-copper solid solution of composition $X_{Au} = 0.25$, $X_{Cu} = 0.75$, or AuCu₃. With the corner sites designated as α and the face-centered sites as β, perfect long-range order corresponds to all the α-sites being occupied by gold atoms, and all the β-sites by copper atoms. When there is no long-range order, however, one-fourth of both the α- and β-sites are occupied by Au atoms, and three-fourths by copper atoms. Example 8-4 shows how the occupancy of the α- and β-sites varies with the degree of long-range order.

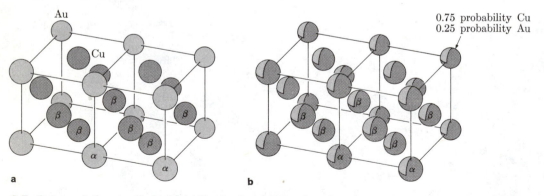

8-7 Order and disorder (AuCu₃). (a) The low-temperature form is ordered, with gold atoms in α-sites and copper atoms in β-sites. (b) Above 380°C, the fcc structure is disordered.

---

* As such, the ordered lattice is no longer bcc, but simple cubic (Fig. 4-7 and Example 5-2b).

Example 8-4

The alloy composed of 25 a/o Mn and 75 a/o Ni forms an ordered solution at low temperature of the AuCu$_3$ type. It becomes disordered at higher temperatures. Indicate the "rightly" and "wrongly" occupied sites as a function of the long-range order parameter, LRO.

Answer. From Eq. (8-2b),

$$r_\alpha = \text{LRO}(1 - X_{\text{Mn}}) + X_{\text{Mn}}$$

and

$$r_\beta = \text{LRO}(1 - X_{\text{Ni}}) + X_{\text{Ni}}.$$

At LRO = 0.5,

$$r_\alpha = 0.625 \quad \text{and} \quad w_\beta = 0.125.$$

Also,

$$r_\beta = 0.875 \quad \text{and} \quad w_\alpha = 0.375.$$

For other values:

LRO	r_α	r_β	w_α	w_β	Fig.
1.00	1.00	1.00	0	0	8-7(a)
0.75	0.81	0.94	0.19	0.06	—
0.50	0.62	0.88	0.38	0.12	—
0.25	0.44	0.81	0.56	0.19	—
0	0.25	0.75	0.75	0.25	8-7(b) ◀

Example 8-5

The long-range order parameter for a 55Cu-45Zn β-brass (atom percent) is 0.67. What fractions of the α- and β-sites are "rightly" and "wrongly" occupied?

Answer. From Eq. (8-2b),

$$r_\alpha = (0.67)(1 - 0.45) + 0.45 = 0.82,$$
$$r_\beta = (0.67)(1 - 0.55) + 0.55 = 0.85;$$

therefore,

$$w_\alpha = 0.18, \quad w_\beta = 0.15. \quad ◀$$

When long-range order is absent, a different type of order (called *local order*) may still be present. Local order does not depend on lattice-site occupancy over long distances, and may even exist in a liquid or amorphous phase. Local order constitutes a departure from randomness in the sense that each *A*-atom may have other than a random number of *A*- or *B*-atoms surrounding it, and similarly for the *B*-atoms. More specifically, the local-order parameter is

$$\alpha = 1 - \frac{P_{A(B)}}{X_A}, \tag{8-3}$$

where $P_{A(B)}$ is the probability that an *A*-atom is located next to a *B*-atom. If this probability happens to equal X_A, it means that an average *B*-atom will have a random number of A-neighbors, and so $P_{A(B)} = X_A$. The local-order parameter is then zero.

On the other hand, if B-atoms prefer to surround themselves with A-atoms, $P_{A(B)} > X_A$ and α is negative. This case corresponds to a tendency for the formation of unlike bonds rather than like bonds [in line with Eq. (8-1)], and is described as *short-range order*. Alternatively, the inequality in Eq. (8-1) may be reversed in some systems, and then there is a tendency for the formation of like bonds rather than unlike bonds. Under such circumstances, $P_{A(B)} < X_A$ and α is positive. This situation is called *clustering*, and the solid solution will have both A-rich and B-rich clusters. An example of the technical significance of clustering will be presented in Section 20-4.

In solid solutions containing local order (whether short-range or clustering), the lattice is continuous across the nonrandom regions, and hence comprises a single phase. Both short-range order and clustering tend toward randomness ($\pm\alpha \to 0$) with increasing temperature.

Compounds. The structure of the ordered β-brass of Fig. 8-6(a) is the same as that of CsCl shown in Fig. 4-7. When the order is essentially perfect and the composition is essentially fixed, as it is in CsCl, we apply the term "compound," particularly if elevated temperatures introduce negligible disorder.

Most ionic structures are compounds because their positive and negative charges demand near perfect neighbor-to-neighbor order and exact proportions of the component atoms. Metallic structures can have more variation; small solute additions of copper to aluminum, for example, undergo virtually no ordering within the fcc aluminum structure. At 33 a/o copper, however, the ordering is nearly perfect; so we identify $CuAl_2$ as an *intermetallic compound*. A gradation can exist between these extremes.

Molecular order. Order-disorder transitions occur readily in metals even when the energy difference of Eq. (8-1) is not large, because then the atoms can move individually. In a polymer, however, mer rearrangements within the chain require fracturing of strong directional, covalent bonds and relocation of groups of atoms. Therefore, ordering rearrangements of this type rarely occur after polymerization. Any ordering of component mers within a polymer chain must arise during the polymerization process.*

Near-perfect order is realized in condensation polymers where a step reaction occurs between the two components. To illustrate, the nylon polymerization of Example 7-1 can be written

$$m[\text{HO} \cdot \text{CO} \cdot (\text{CH}_2)_4 \cdot \text{CO} \cdot \text{OH}] + m[\text{H}_2\text{N}(\text{CH}_2)_6\text{NH}_2] \rightarrow$$
adipic acid   hexamethylene diamine

$$\underbrace{\text{CO} \cdot (\text{CH}_2)_4 \cdot \text{CO}}_{\text{adipate}} \cdot \underbrace{\text{HN}(\text{CH}_2)_6\text{NH}}_{\text{amide}} \Big)_m;$$

in this reaction there is a one-to-one alternation (Fig. 8-8a).

---

* Subtle and somewhat complex order-disorder reactions can occur by conformation changes; e.g., the helix of Fig. 7-9(a) tightens at 20°C to produce a shorter "pitch."

$$\cdots -A-B-A-B-A-B-A-B-A-B-A-B-A-B-A-B- \cdots \qquad \text{a}$$

$$\cdots -C-C-C-C-C-C-D-D-D-D-C-C-C-C-C-D- \cdots \qquad \text{b}$$

8-8 Order in polymers. (a) Linear step-reaction polymer, e.g., $\{$adipate—amide$\}$ in nylon. (b) Block copolymer of addition components.

Block copolymers, as shown in Fig. 8-8(b), possess domains of segregation along the chain. Copolymers of this type have proved to have certain enhanced mechanical properties.

In nature, the large number of proteins arises from the ordered copolymeric permutations of amino acids. One major achievement of structural analysis has been the deciphering of insulin into an ordered sequence of 102 mers of 16 types of amino acids.

8-5 DEFECT STRUCTURES

An ionic compound may contain vacancies to maintain the charge balance. To illustrate, a fraction of the ferrous ions in FeO are oxidized to ferric ions:

$$Fe^{2+} \rightleftarrows Fe^{3+} + e^{-}. \qquad (8\text{-}4)$$

Two Fe^{3+} ions are equivalent to 3 Fe^{2+} in balancing the O^{2-} charge in this iron oxide. Thus, for every two Fe^{3+} ions, there must be one vacancy in the host FeO structure (Fig. 8-9). Situations like this commonly occur in those compounds containing ambivalent ions, such as iron (Fe^{2+} and Fe^{3+}), copper (Cu^{+} and Cu^{2+}), and uranium (U^{3+} and U^{4+}). The vacancies enhance diffusion (Chapter 9), and usually modify electronic properties to produce semiconduction (Chapter 15).

O^{2-}	Fe^{2+}	O^{2-}	Fe^{2+}	O^{2-}	Fe^{2+}	O^{2-}	Fe^{2+}
Fe^{2+}	O^{2-}	Fe^{2+}	O^{2-}	Fe^{2+}	O^{2-}	Fe^{2+}	O^{2-}
O^{2-}	Fe^{3+}	O^{2-}	Fe^{2+}	O^{2-}	$\square$	O^{2-}	Fe^{2+}
Fe^{2+}	O^{2-}	$\square$	O^{2-}	Fe^{3+}	O^{2-}	Fe^{3+}	O^{2-}
O^{2-}	Fe^{3+}	O^{2-}	Fe^{2+}	O^{2-}	Fe^{2+}	O^{2-}	Fe^{2+}
Fe^{2+}	O^{2-}	Fe^{2+}	O^{2-}	Fe^{2+}	O^{2-}	Fe^{2+}	O^{2-}

8-9 Nonstoichiometric compound ($Fe_{1-x}O$). One vacancy ($\square$) must be present for each two Fe^{3+} ions. Thus the number of vacancies is increased under more highly oxidizing conditions.

Nonstoichiometric compounds which have a range of chemical compositions depending on the oxidation level present have defect structures. Thus, the above iron oxide usually exists as $Fe_{1-x}O$. At 1200°C, $Fe_{1-x}O$ may range from 48.5 a/o Fe to only 46.0 a/o Fe.

Example 8-6

In our example of wüstite (which is the name for the iron oxide structure of Fig. 8-9), the ratio of Fe^{3+} to Fe^{2+} may be as high as 0.5.

a) With this ratio, what fraction of the normal cation sites are vacant?

b) What is the weight fraction of oxygen in this composition?

Answer. Basis: 100 Fe^{2+} ions.

a) 50 Fe^{3+} ions and 25 $\square$.

    Total cation sites = $100 + 50 + 25 = 175$.

    Vacancy fraction = $25/175 = 0.143$. (This is also the value of x in $Fe_{1-x}O$.)

b)
$$
\begin{array}{rl}
100\ Fe^{2+}\ \text{ions} = & 100\ O^{2-}\ \text{ions} \\
\underline{50\ Fe^{3+}\ \text{ions} =} & \underline{75\ O^{2-}\ \text{ions}} \\
150\ Fe\ \ \ \text{ions} & 175\ O^{2-}\ \text{ions}
\end{array}
$$

$$\text{Weight of oxygen} = (175)(16.0/AN) = 2800/AN.*$$
$$\text{Weight of iron} = (150)(55.8/AN) = 8370/AN.$$

$$\text{Therefore, weight fraction of oxygen} = \frac{2800/AN}{(2800 + 8370)/AN}$$
$$= 0.251. \ \blacktriangleleft$$

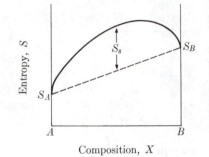

8-10 Solution entropy. Equation (8-7) applies for mixtures of two components.

Composition, X

8-6 FREE ENERGY OF SOLUTIONS†

In solids and liquids, where the PV energy is relatively minor and essentially constant,

$$E \cong H = F + TS. \tag{8-5}$$

The first term, E, is the internal energy of the material, while H, F, and S are, respectively, the enthalpy, free energy, and entropy, as discussed in Section 1-4 (Eq. 1-7). The free energy, F, is of interest to us because it governs reactions and phase stability (Section 3-8).

    The free energy of solutions may be analyzed by looking at Eq. (1-7b):

$$F = H - TS. \tag{8-6}$$

Two components, A and B, have entropy values of their own, S_A and S_B (Fig. 8-10).

---

*   AN = Avogadro's number.

†   This section may be omitted if spinodal reactions (Section 18–7) are to be omitted.

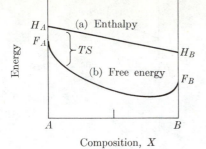

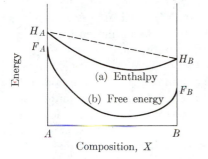

8-11 Ideal solution: $F = H - TS$. The entropy, S, for a solution is shown in Fig. 8-10.

8-12 Exothermic solutions (negative deviation from ideality). Free-energy considerations strongly favor solubility for all solution compositions.

Intermediate solutions have higher entropy than the interpolated mean because a solution, by its very nature, has a high probability of randomness. The entropy increase, S_s, due to solution can be shown to be

$$S_s = -R[X_A \ln X_A + X_B \ln X_B]. \qquad (8\text{-}7)$$

The values X_A and X_B are the mole fractions of the components A and B respectively, and R is the gas constant, 1.98 cal/mole · °K. Therefore, a 50–50 solution has an entropy value which is $R \ln 2$, or 1.38 cal/mole · °K, greater than the linear value of Fig. 8-10.

The enthalpy, H, of Eq. (8-6) may fall into any of three categories. In an *ideal solution* where there is no preferential ordering, $E_{AB} = \frac{1}{2}(E_{AA} + E_{BB})$, and H may be interpolated between H_A and H_B to give the free energy, F, for the solution (Fig. 8-11).

If $E_{AB} < \frac{1}{2}(E_{AA} + E_{BB})$, as expressed in Eq. (8-1), the enthalpy has a *negative* deviation from ideality; hence, the free energy of the solution is markedly less than that of a corresponding mixture of the pure components (Fig. 8-12). The two components therefore readily dissolve in one another.

The third category involves a solution in which $E_{AB} > \frac{1}{2}(E_{AA} + E_{BB})$; this gives a *positive* deviation from ideality. The free-energy-versus-composition plot which results is shown in Fig. 8-13.

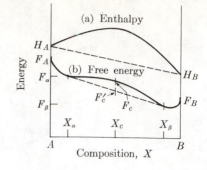

(a) Enthalpy

(b) Free energy

Composition, X

8-13 Endothermic solutions (positive deviation from ideality). Immiscibility develops because the free energy of intermediate solutions, X_c, is higher than that of mixtures of X_α and X_β. Minimum free energy favors phase separation.

Immiscibility. As a solution, the composition X_c near the middle of the range of Fig. 8-13 will have the energy F_c. However, this can be lowered to F_c' if the solution separates into a mixture of two phases, X_α and X_β, with free energies (per mole) of F_α and F_β, respectively. Since the system with the minimum free energy is more stable, such a separation is a move toward equilibrium.

The solubility limits occur at the points of contact of a common tangent drawn to the free-energy curve. Between these solubility limits the solution is said to be *immiscible*. We may interpret this as the preferential segregation of like components into an A-rich α-phase and a B-rich β-phase. (Compare clustering, Section 8-4.)

Since TS_s increases with temperature, an analysis of Fig. 8-14 reveals why the width of the miscibility gap generally decreases with higher temperatures.

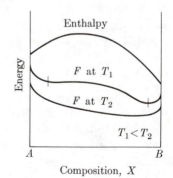

Enthalpy

F at T_1

F at T_2

$T_1 < T_2$

Composition, X

8-14 Effect of temperature on free energy. Since $F = H - TS$, the free energy decreases at higher temperatures and the miscibility gap disappears.

REFERENCES FOR FURTHER READING

Brophy, J. H., R. M. Rose, and J. Wulff, *The Structure and Properties of Materials: II. Thermodynamics of Structure.* New York: Wiley, 1964. Paperback. Section 2.3 is recommended as alternate reading for the study of composition as a variable in solid solutions. Introductory level.

Massalski, T. B., "Structure of Solid Solutions," *Physical Metallurgy.* Amsterdam, North-Holland Publishing Co., 1965, pp. 149–208. Advanced-level reading on solid solutions.

Rogers, B. A., *The Nature of Metals*, second edition. Metals Park, O.: American Society for Metals, 1964. Chapter 4 is recommended as background for this chapter. Introductory level.

Van Vlack, L. H., *Elements of Materials Science*, second edition. Reading, Mass.: Addison-Wesley, 1964. Solid solutions are presented in Chapter 4. Introductory level.

PROBLEMS

8-1 A solid solution contains 5 w/o Cu and 95 w/o Al. What is the atomic percent copper?

Answer. 2.2 a/o

8-2 The maximum solubility of tin in fcc copper is 9 a/o. What is this solubility limit in weight percent?

8-3 What is the atom percent lead in a 70Cu-29Zn-1Pb alloy? [*Note.* Analyses are presented as weight percent in condensed phases, *unless stated otherwise.* In gases, they are expressed as volume or mole percent, *unless stated otherwise.*]

Answer. 0.3 a/o

8-4 A sample of 18-8 stainless steel contains 18 w/o Cr, 8 w/o Ni, and 0.10 w/o C, with the balance (about 74 w/o) being Fe. The steel has a crystal structure with an fcc unit cell. Calculate the average number of carbon atoms per unit cell. For this steel, the density is 7.9 gm/cm^3 and the lattice parameter is 3.6 A.

8-5 A 70Cu-30Zn brass has a lattice constant, a, of 3.65 A.

a) How many zinc atoms are there per cm^3?
b) How many zinc atoms are there per cm^2 on the (100) plane?

Answer. a) 0.025×10^{24}/cm^3 b) 0.045×10^{16}/cm^2

8-6 An alloy contains 80 w/o Ni and 20 w/o Cu in substitutional fcc solid solution with $a = 3.54$ A. Calculate the density of this alloy.

8-7 A copolymer of vinyl chloride and vinyl acetate has a mer ratio of 10/1 respectively.

a) What are the weight percents of the two?
b) What are the weight percents of C, O, H, and Cl?
c) What is the atom fraction of each?

Answer. a) 87.9 w/o VC; 12.1 w/o VAc   b) 40.5 w/o C; 4.5 w/o O; 5.1 w/o H; 49.9 w/o Cl   c) 33.3 a/o C; 2.8 a/o O; 50.0 a/o H; 13.9 a/o Cl

8-8 A copolymer of vinyl chloride (H_2C=$CHCl$) and vinylidene chloride (H_2C=CCl_2) contains equal numbers of the two types of monomers. The molecular weight of one specific molecule is 500,000 gm/mole.

a) How many mers are in this molecule?
b) Making use of the bond lengths of Table 3-4, calculate the maximum length of the above molecule, assuming that the bond angles across all carbons are stretched to 120°.

8-9 Calculate the radius of the largest atom which can exist interstitially in fcc iron without crowding.

Answer. 0.53 A

8-10 An analysis of a copolymer of vinyl acetate and vinyl chloride shows 28.0 w/o chlorine. What is the mer percent vinyl acetate?

8-11 Analysis of a sample of otherwise pure iron showed that 0.1 w/o of hydrogen was present in the interstices among the iron atoms. How many hydrogen atoms are there per 1000 unit cells?

Answer. 112

8-12 If all the iron ions of Fig. 8-4 were changed to Ni ions, what would be the w/o MgO?

8-13 A sample of $Fe_{1-x}O$ contains 23.7 w/o oxygen.

   a) What is the Fe/O ion ratio?
   b) Determine the Fe^{2+}/Fe^{3+} ion ratio.

Answer. a) 48/52   b) 5.35

8-14 A ceramic material containing 8 w/o Cr_2O_3 and 92 w/o MgO forms a solid solution with an NaCl structure. All anion sites are occupied by O^{2-} ions.

   a) What is the ratio of Mg^{2+} to Cr^{3+} ions?
   b) What fraction of the cation sites are vacant?

8-15 Three solid solutions of gold and copper have respective atom ratios of (a) 20/80, (b) 25/75, and (c) 30/70. Compare the maximum long-range order parameter for each alloy as based on Eq. (8-2a) and Fig. 8-7.

Answer. a) 0.80   b) 1.00   c) 0.93

8-16 Eighty percent of the α-sites contain Ti atoms in a 52Ti-48Ni solid solution (atom percent) of the β-brass type. What fraction of the β-sites contain Ni atoms?

part III □ atomic processes in solids

9 □ atom movements

9-1 SELF-DIFFUSION

Atomic displacements (Section 6-5) and molecular coiling (Section 7-4) illustrate the fact that the interior of a solid is not static, but rather a combination of energetic, dynamic atomic movements. In this chapter we shall examine the mechanisms of atom movements and see how their rates are influenced.

Radioactive tracers. Atom movements can be followed by radioactive isotopes. For example, radioactive nickel, Ni^{63}, can be plated onto the surface of regular nickel (primarily Ni^{58} and Ni^{60}). Since Ni^{63} decays with a β-ray emission, and hence can be detected by photographic or other means, it can serve as a tracer for nickel diffusion. A *microradiographic* procedure, sketched in Fig. 9-1, is one of the more common tracer methods used.

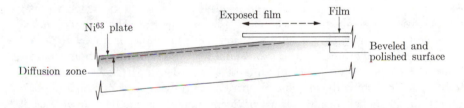

9-1 Microradiography. The photographic density of the exposed film reveals the profile of the radioactive nickel.

Since Ni^{63} and regular nickel form an ideal solution (Section 8-6), Fig. 8-11 presents the free energy of intermediate isotopic solutions and shows that the free energy is reduced if homogenization occurs; homogenization is illustrated in Fig. 9-2.

The homogenization process of Fig. 9-2 can also be interpreted as follows. Although there is an equal probability that an individual atom will move in each direction, the concentration gradient of this figure favors a net movement of the tracer atoms to the right. At point A in Fig. 9-2(b), there are more tagged atoms than there are at point B. Thus, even with the same probability *per atom* for the tagged atoms at A to move to the right as for the tagged atoms at B to move to the

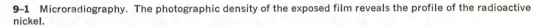

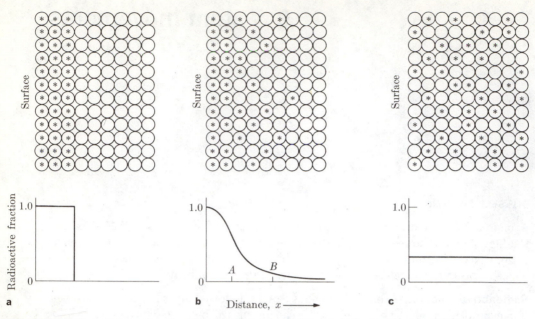

9-2 Self-diffusion. Here radioactive nickel (Ni^{63}) has been plated onto the surface of nonradioactive nickel. (a) Time, $t = t_0$. (b) Diffusion gradient, $t_0 < t < t_\infty$. (c) Homogenized, $t = t_\infty$.

left, the difference in numbers produces a differential in movements and increases the uniformity and randomness in the structure. (A converse description could have been written if our attention had been focused on the nonradioactive atoms.)

The description of *self-diffusion*, shown in Fig. 9-2, can also be applied to the presence of solute atoms within a solid solution. For example, if nickel is plated onto the surface of copper, atomic diffusion should bring about nickel homogenization within the copper after sufficient time at elevated temperatures. However, the rate of nickel diffusion down the concentration gradient in copper happens to be greater than in nickel, because the activation energy needed to move nickel atoms among copper atoms is less than that required to move nickel atoms among nickel atoms. This difference can be anticipated because the lower melting temperature of copper (1083°C versus 1455°C for nickel) indicates that the Cu-Cu bonds are weaker than the Ni-Ni bonds.

Diffusion mechanisms. Vacancies and interstitials play an important role in atomic diffusion. Neighboring atoms can move into a vacancy. After one such move the new vacancy can receive another atom, and the process is continued (Fig. 9-3). The net effect is homogenization, for reasons cited in the previous subsection. Atoms can also diffuse through crystals via interstitial sites (Fig. 9-4). Small solute atoms move readily in this way. Diffusion of solvent and larger solute atoms most commonly involves cooperative movements within the lattice in the manner shown in Fig. 9-5(a, b, and c). Additional diffusion mechanisms are possible (Fig. 9-5d and e), but probably are important only in special situations.

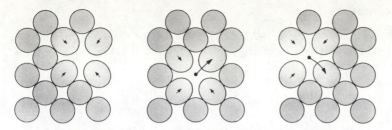

9–3 Diffusion through vacancies. The vacancy moves in the opposite direction to the atoms. [*Note.* In Chapter 15, we will observe an analogy for movements of electron holes.]

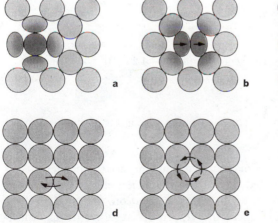

9–4 Interstitial diffusion of small solute atoms. Although the atom is small, an activation energy is required to move it from one position to another, equivalent interstitial location.

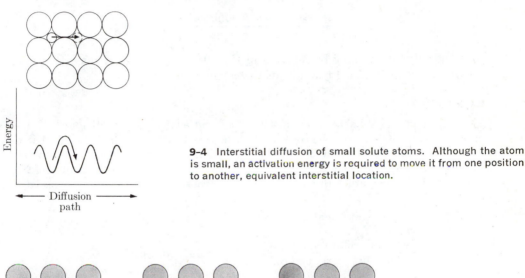

9–5 (a, b, c) Diffusion by interstitialcies. Because compressive stresses are already present with large interstitial atoms, a cooperative movement of two atoms limits further distortion of the lattice. This type of movement is called the *interstitialcy* mechanism. (d) Interchange. (e) Ring diffusion.

9-2 DIFFUSION COEFFICIENTS

We observed in the discussion of Fig. 9-2 that the number of diffusing atoms was proportional to the concentration. Similarly, the net flux, J, is taken to be proportional to the concentration gradient, dC/dx:

$$J = -D \frac{dC}{dx}. \tag{9-1a}$$

The proportionality constant, D, is called the *diffusion coefficient* or *diffusivity*, and Eq. (9-1a) is known as *Fick's first law*. The negative sign in Eq. (9-1a) indicates that a flux occurs in the down-gradient direction as anticipated. The units for the above equation are

$$\frac{\text{atoms}}{\text{cm}^2 \cdot \text{sec}} = \frac{(\text{cm}^2/\text{sec})(\text{atoms}/\text{cm}^3)}{\text{cm}}. \tag{9-1b}$$

Alternatively, atoms may be replaced in Eq. (9-1b) by grams or mass fraction if consistency is maintained throughout the calculation.

Example 9-1

The surface of a nickel sheet contains 50% Ni^{63} and 50% nonradioactive nickel. Four microns below that surface, the concentration ratio is 48:52. We shall assume that the gradient is linear between the two points. Experiments indicate that the self-diffusion coefficient of nickel in nickel is 1.6×10^{-9} cm²/sec at 1000°C. What is the flux of Ni^{63} atoms at that temperature through a plane 2 microns below the surface? The lattice constant of nickel is 3.6 A at 1000°C.

Answer. Basis: One unit cell at each position.

$$C_2 = \frac{(4 \text{ Ni/unit cell})(0.48 \text{ Ni}^{63}/\text{Ni})}{(3.6 \times 10^{-8} \text{ cm})^3/\text{unit cell}} = 4.12 \times 10^{22} \text{ Ni}^{63}/\text{cm}^3.$$

$$C_1 = \frac{(4 \text{ Ni/unit cell})(0.50 \text{ Ni}^{63}/\text{Ni})}{(3.6 \times 10^{-8} \text{ cm})^3/\text{unit cell}} = 4.29 \times 10^{22} \text{ Ni}^{63}/\text{cm}^3.$$

$$\text{Flux} = \frac{-D(C_2 - C_1)}{x_2 - x_1} = \frac{-(1.6 \times 10^{-9} \text{ cm}^2/\text{sec})(4.12 - 4.29)(10^{22} \text{ Ni}^{63}/\text{cm}^3)}{4 \times 10^{-4} \text{ cm}}$$

$$= 0.7 \times 10^{16} \text{ Ni}^{63}/\text{cm}^2 \cdot \text{sec}.$$

Note. This flux is equal to approximately 9 atoms passing through each unit cell every second! ◄

Activation energy for diffusion. For atoms to move from one site to another, bonds must be broken and the lattice distorted; hence, an activation energy is required. The number of atoms which have the necessary activation energy increases exponentially with the reciprocal of temperature (Eq. 6-3). As a result, the diffusion coefficient, D, has a similar temperature relationship:

$$D = D_0 e^{-E^*/kT}, \tag{9-2a}$$

where k is Boltzmann's constant again, T is the absolute temperature, and D_0 is an experimentally determined factor for each diffusion couple and may vary with concentration. The *activation energy*, E^*, may be visualized (in a simplified way) as the necessary energy to raise an atom over a barrier (Figs. 6-7 and 9-4).

Equation (9-2a) is more commonly expressed as

$$D = D_0 e^{-Q/RT}, \tag{9-2b}$$

where R is 1.987 cal/mole $\cdot$ °K, the "gas constant," and Q is the activation energy expressed as calories per *mole* of atoms. Thus,

$$\log_{10} D = \log_{10} D_0 - Q/(2.3RT), \tag{9-2c}$$

the factor of 2.3 being required to change to logarithms of base ten.

Example 9-2

The diffusion data at 500°C and 1000°C in Table 9-1 have come from experiments. Verify the activation energy which has been calculated for the self-diffusion of copper.

Answer. At 1000°C, Eq. (9-2c) gives

$$-8.8 = \log D_0 - Q/(4.575 \text{ cal/mole} \cdot °\text{K})(1273°\text{K}).$$

TABLE 9-1
DIFFUSION DATA

Solute	Solvent	D_0, cm²/sec	Q, cal/mole	$D_{500°C}$, cm²/sec	$D_{1000°C}$, cm²/sec	Reference*
1 C	fcc Fe	0.25	34,500	$[10^{-10.4}]$	$10^{-6.5}$	(1)
2 C	bcc Fe	0.2	20,100	$10^{-6.4}$	$[10^{-4.1}]$	(1)
3 N	bcc Fe	0.14	17,700	$10^{-5.8}$	$[10^{-3.9}]$	(2)
4 Fe	fcc Fe	0.58	67,900	$[10^{-20.0}]$	$10^{-12.2}$	(2)
5 Fe	bcc Fe	118	67,200	$10^{-16.9}$	$[10^{-9.5}]$	(1)
6 Ni	fcc Fe	0.5	66,000	$[10^{-18.9}]$	$10^{-11.7}$	(3)
7 Mn	fcc Fe	0.35	67,500	$[10^{-19.6}]$	$10^{-12.1}$	(3)
8 C	hcp Ti	5.06	43,500	$10^{-11.6}$	$[10^{-6.8}]$	(2)
9 Cu	Al	0.08	32,600	$10^{-10.3}$	$10^{-6.7}$M	(2)
10 Al	Cu	0.38	43,000	$10^{-12.6}$	$10^{-7.8}$	(2)
11 Cu	Cu	0.2	47,100	$10^{-14.0}$	$10^{-8.8}$	(1),(2)
12 Zn	Cu	0.34	45,600	$10^{-13.4}$	$10^{-8.3}$	(2)
13 Ag	Cu	0.012	35,600	$10^{-12.0}$	$10^{-8.0}$	(2)
14 Au	Cu	0.01	44,900	$10^{-14.6}$	$10^{-9.7}$	(2)
15 Ag	Au	0.024	37,000	$10^{-12.1}$	$10^{-8.0}$	(2)
16 Au	Ag	0.26	45,500	$10^{-13.4}$	$10^{-8.4}$M	(2)
17 Cu	Ag	1.23	46,100	$10^{-13.0}$	$10^{-7.8}$M	(2)
18 Zn	Ag	0.54	41,700	$10^{-12.1}$	$10^{-7.4}$M	(2)
19 Ag	Ag	0.89	45,900	$10^{-13.0}$	$10^{-7.9}$M	(1),(2)
20 Th	W	1.0	120,000	$10^{-33.9}$	$10^{-20.6}$	(2)

[ ] Metastable at indicated temperature.
M Calculated, although temperature is above the melting point.
* References: (1) P. G. Shewmon, *Diffusion in Solids.* New York: McGraw-Hill, 1963. (2) C. J. Smithells, *Metals Reference Book,* third edition. Washington, D. C.: Butterworths, 1962, pp. 583 ff. (3) A. G. Guy, *Physical Metallurgy for Engineers.* Reading, Mass.: Addison-Wesley, 1962, p. 251.

At 500°C,

$$-14.0 = \log D_0 - Q/(4.575 \text{ cal/mole} \cdot °\text{K})(773°\text{K}).$$

Solving simultaneously, we obtain

$$Q = 47{,}000 \text{ cal/mole},$$
$$D_0 = 0.2 \text{ cm}^2/\text{sec}.$$

Note. The activation energy is positive, since energy is required for the atoms to pass the energy barrier. ◀

Factors affecting diffusion coefficients. The data of Table 9-1 are plotted in Fig. 9-6. The coefficients are highly dependent on temperature. It is also apparent that there are major differences (as much as 10^{10}) in the diffusion coefficients between various solute-solvent couples. These differences can be rationalized. *With other*

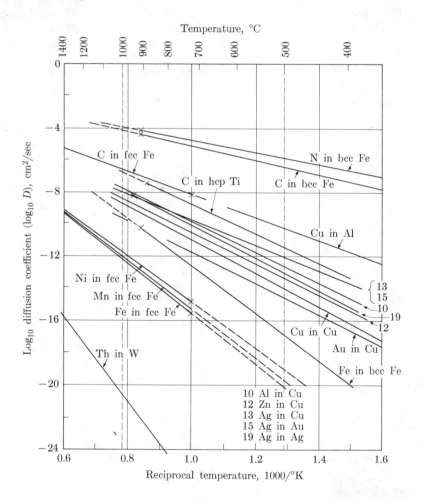

9–6 Diffusion coefficients versus temperature.

factors equal:

1) The activation energy for diffusion is lower with a small interstitial solute atom than with a larger substitutional solute atom. [Compare C ($R \cong 0.75$ A) in fcc Fe with Ni ($R \cong 1.24$ A) in fcc Fe, or with Fe ($R \cong 1.27$ A) in fcc Fe.]

2) The activation energy for diffusion is lower in a solvent with a low melting point than in a solvent with a higher melting point. [Compare Cu in Al (melting point = 660°C) with Cu in Cu (melting point = 1083°C).]

3) The activation energy for diffusion is lower through a structure with a low atomic packing factor than through a structure with more dense packing. [Compare C in bcc Fe (APF = 0.68) with C in fcc Fe (APF = 0.74). In addition, compare D for Fe in bcc Fe with D for Fe in fcc Fe.]

In addition to the above factors, the diffusion coefficient is dependent on concentration when the solutions are not dilute. This may be anticipated on the basis of the data in Fig. 9-6. Copper diffusion in an aluminum matrix has the coefficient shown. In higher concentrations, however, some copper atoms would have to move between or past other copper atoms where higher activation energies are required, thus reducing the overall diffusion coefficient.

9-3 DIFFUSION PROFILES

Except in special cases, the diffusion process in solids is not steady-state; i.e., the concentration as a function of position changes with time. This is illustrated schematically in Fig. 9-2. At any given time and position, the time rate of concentration change, $\partial C/\partial t$,* is related to the second derivative of the concentration gradient through the diffusion coefficient,

$$\partial C/\partial t = D(\partial^2 C/\partial x^2). \tag{9-3}$$

This is Fick's second law†; as presented here it assumes that the diffusion coefficient,

---

* Partial derivatives are used since concentration depends on both time and location.
† Fick's second law as presented in Eq. (9-3) may be obtained as follows. In Fig. 9-7, an incremental volume of thickness dx and unit cross section is taken at position x in the diffusion zone. The rate of accumulation in this zone is the product of the time-rate increase in concentration, $\partial C/\partial t$, and the incremental volume, dx:

$$\text{accumulation} = (\partial C/\partial t)\,dx. \tag{9–4}$$

This rate of accumulation is also equal to the difference between the entering flux from the left, J_x, and the exit flux to the right, J_{x+dx}, or

$$\text{accumulation} = J_x - J_{x+dx} = -(\partial J_x/\partial x)\,dx. \tag{9–5}$$

Equating these two rates of accumulation, we get

$$(\partial C/\partial t)\,dx = -(\partial J_x/\partial x)\,dx.$$

From Eq. (9-1),

$$\partial C/\partial t = -\partial(-D\,\partial C/\partial x)/\partial x, \tag{9–6}$$

or

$$\partial C/\partial t = D\,\partial^2 C/\partial x^2.$$

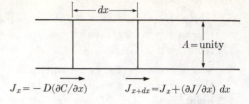

$$J_x = -D(\partial C/\partial x) \qquad \overrightarrow{J_{x+dx}} = J_x + (\partial J/\partial x)\, dx$$

9-7 Fick's second law. [See Eqs. (9-3) through (9-6).]

D, is independent of concentration. Further, it applies to a one-dimensional gradient, which is sufficient for our purposes. Appropriate modifications are available for more complex situations. (See the References for Further Reading at the end of this chapter.)

Semi-infinite slabs. We shall not solve the above differential equation for C as a function of x, D, and t, but will present a solution for a special case of an infinite source (or sink) in a semi-infinite slab. This is illustrated in Fig. 9-8, where the supplying concentration at the surface remains fixed (i.e., an infinite source). The atoms are diffusing into the slab, which extends indefinitely to the right (i.e., it is infinite in one direction only, or semi-infinite).

The solution* of Eq. (9-3) under these conditions is

$$\frac{C_s - C_x}{C_s - C_o} = \mathrm{erf}\left(\frac{x}{2\sqrt{Dt}}\right), \tag{9-7}$$

where C_s is the concentration at the surface, C_o is the original or base concentration, and C_x is the concentration at the distance, x, into the slab. The value $\mathrm{erf}(x/2\sqrt{Dt})$ is the *Gaussian error function* as presented in Table 9-2. This

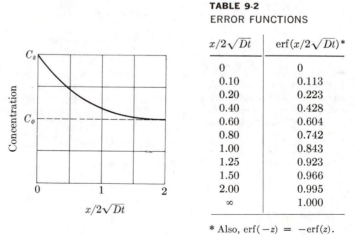

9-8 Diffusion profile (infinite source).

TABLE 9-2

ERROR FUNCTIONS

$x/2\sqrt{Dt}$	$\mathrm{erf}(x/2\sqrt{Dt})$*
0	0
0.10	0.113
0.20	0.223
0.40	0.428
0.60	0.604
0.80	0.742
1.00	0.843
1.25	0.923
1.50	0.966
2.00	0.995
∞	1.000

* Also, $\mathrm{erf}(-z) = -\mathrm{erf}(z)$.

function is widely encountered in statistical studies and is applicable here because diffusion behaves in a "random walk" or statistical manner.

---

* See References for Further Reading (Darken and Gurry).

Example 9-3

The end of a 0.1C-99.9Fe (weight percent) slab is exposed at 1000°C to a carburizing gas that maintains the surface constantly at 0.93 w/o carbon. Plot the carbon profile behind the surface after (a) 10 min, (b) 100 min. (Assume that the diffusion coefficient is constant with concentration below 1.0 w/o carbon.)

Answer. From Table 9-1, $D = 0.31 \times 10^{-6}$ cm^2/sec.

a) 10 min = 600 sec; $\sqrt{Dt} = 0.0137$ cm

Let $x =$	$x/2\sqrt{Dt}$	erf$(x/2\sqrt{Dt})$ (from Table 9-2)	$C_x =$ $C_s - (C_s - C_o)$erf $\dfrac{x}{2\sqrt{Dt}}$
0.0548 cm = $4\sqrt{Dt}$	2.0	0.995	0.104 w/o
0.0274 = $2\sqrt{Dt}$	1.0	0.843	0.23
0.0137 = $\sqrt{Dt}$	0.5	0.521	0.50
0 = 0	0.0	0.000	0.93

b) 100 min = 6000 sec; $\sqrt{Dt} = 0.0431$ cm

Let $x =$	$x/2\sqrt{Dt}$	erf$(x/2\sqrt{Dt})$	C_x
0.20 cm	2.32	1.00	0.10 w/o
0.10	1.16	0.894	0.19
0.05	0.58	0.582	0.45
0	0.0	0.00	0.93

Note. These data are plotted in Fig. 9-9(a). ◄

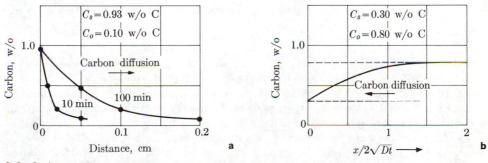

9-9 Carbon profiles. (a) During carburization (see Example 9–3). (b) During decarburization.

The solution of Eq. (9-3) for an infinite sink is identical to that for an infinite source; thus Eq. (9-7) still applies. The constant surface concentration, however, is lower than the original composition. An example of an infinite sink is presented in Fig. 9-9(b), where the surface of a 0.8C-99.2Fe (weight percent) piece of steel is exposed to an atmosphere which has a carbon potential equivalent to 0.3 w/o C in 00.7 w/o Fe.

Thermal diffusivity. At this point it is desirable to digress to consider thermal diffusivity, h, which is a combination of thermal conductivity, k, heat capacity, c, and density, ρ:

$$h = k/c\rho. \tag{9-8a}$$

The dimensions are

$$cm^2/sec = \frac{(cal \cdot cm)/(cm^2 \cdot sec \cdot {}^{\circ}K)}{[cal/(gm \cdot {}^{\circ}K)](gm/cm^3)}. \tag{9-8b}$$

In the context of transport properties, thermal diffusivity is the "diffusion coefficient" of thermal energy, and can be treated as such in heat-flow calculations. This is permissible because each atom acts as a random vibrator in transferring thermal energy. The example which is given will be pertinent to quenching calculations in Chapter 20.

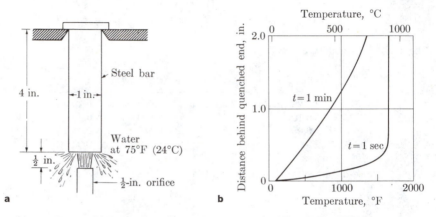

9–10 End quenching. (a) Procedure. (b) Thermal profile during quenching. (See Example 9–4.)

Example 9-4

A round bar of steel (diameter = 1 in.) is heated to 915°C (1680°F) and end-quenched by 75°F (24°C) water, as illustrated in Fig. 9-10(a). Plot the temperature profile after (a) 1 sec, (b) 1 min. [*Note.* We will make several assumptions which are valid for normal calculations. They would need to be reconsidered for stainless steels or if extreme accuracy were necessary in unusual situations. These assumptions are: (1) the values of k, c, and ρ are constant between the two temperatures and equal to 0.11 (cal · cm)/ (cm^2 · °K · sec), 0.111 cal/(gm · °K), and 7.8 gm/cm^3, respectively; (2) heat is removed only through the end of the round bar, the end being immediately quenched to 75°F (24°C); and (3) the heat of transformation from fcc iron at 1680°F to bcc iron at 75°F is negligible.]

Answer. For thermal diffusivity, Eq. (9-7) may be rewritten for quenching as

$$\frac{T_x - T_s}{T_o - T_s} = \mathrm{erf}(x/2\sqrt{ht}). \tag{9-9}$$

a) See Fig. 9-10(b) for results.

b) $h = k/c\rho = 0.127$ cm^2/sec; $t = 60$ sec

x, cm	$x/2\sqrt{ht}$	erf$(x/2\sqrt{ht})$	T_x, °F	T_x, °C
5.08	0.92	0.80	1360	737
2.54	0.46	0.48	845	452
1.27	0.23	0.25	475	246
0.635	0.115	0.13	285	141
0.32	0.06	0.07	185	85
0	0	0	75	24

Note. Since each side of Eq. (9-9) is dimensionless, we can use either °F or °C on the left and cgs units on the right. ◄

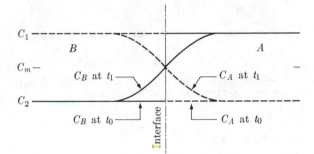

9–11 Interdiffusion. Component A diffuses into B, and B into A. If the diffusion couple is symmetric, the interface remains fixed.

Interdiffusion. The concurrent diffusion of A into B and B into A is called interdiffusion (Fig. 9-11). For this type of diffusion couple, Eq. (9-7) is modified to

$$\frac{C_x - C_m}{C_1 - C_m} = \text{erf}\,\frac{x}{2\sqrt{Dt}}, \qquad (9\text{-}10)$$

where C_m equals the mean between C_1 and C_2. You should note from Table 9-2 that erf$(-z) = -$erf(z). The two curves are symmetrical about C_m in Fig. 9-11 for the special case of $D_A = D_B$.

Kirkendall effect. The symmetry just mentioned is not present when $D_A > D_B$. Under such conditions A-atoms move more readily in one direction than B-atoms in the other. This effect was first described by Kirkendall, who studied diffusion at elevated temperatures by placing inert molybdenum wire markers at an original copper–brass interface (Fig. 9-12a). Since $D_{Zn} > D_{Cu}$ in this case, there are more zinc atoms moving to the right into the copper than copper atoms moving to the left into the brass! This produces a net flow of atoms to the right past the markers. These excess atoms at the right produce a corresponding *bulk movement* of material, including the wires, to the left. Accordingly the final location of the wire markers is on the brass side of the original interface (Fig. 9-12b). The differ-

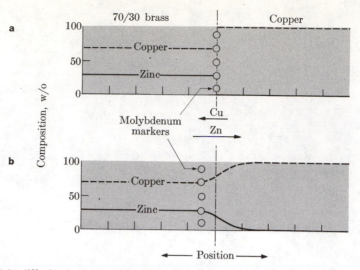

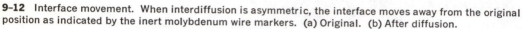

9–12 Interface movement. When interdiffusion is asymmetric, the interface moves away from the original position as indicated by the inert molybdenum wire markers. (a) Original. (b) After diffusion.

ence in D_{Zn} and D_{Cu} may be determined by analyzing the movement of the markers. If, for one reason or another, the bulk movement cannot occur, pores will be introduced as a consequence of the nonequal diffusion rates.

9-4 DIFFUSION IN COMPOUNDS

In each of the examples discussed so far, no ordering was present; thus any adjacent atom could move into an available lattice site. As discussed in the previous chapter, however, ordering is nearly perfect within a compound, as is illustrated by NiO in Fig. 9-13. The two vacancies of the Schottky defect in NiO cannot accept *nearest* neighbors because they have the "wrong" charge. Therefore, diffusion requires ion movements from second-neighbor positions across a relatively high-energy barrier.

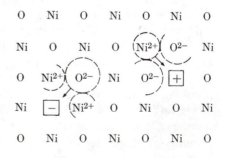

9–13 Diffusion in ionic compounds (NiO). Unlike diffusion in metals (Fig. 9-5), diffusion in compounds involves second-neighbor movements. Since the activation energies are high, the diffusion coefficients are low unless vacancies are present from nonstoichiometric ratios (Fig. 8–9).

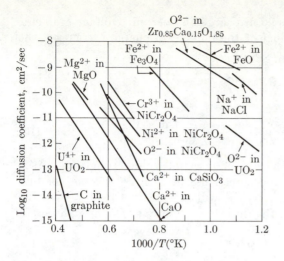

O^{2-} in Zr$_{0.85}$Ca$_{0.15}$O$_{1.85}$

Fe^{2+} in Fe$_3$O$_4$

Fe^{2+} in FeO

Mg^{2+} in MgO

Na$^+$ in NaCl

Cr^{3+} in NiCr$_2$O$_4$

Ni^{2+} in NiCr$_2$O$_4$

O^{2-} in NiCr$_2$O$_4$

O^{2-} in UO$_2$

U^{4+} in UO$_2$

Ca^{2+} in CaSiO$_3$

C in graphite

Ca^{2+} in CaO

$1000/T(°K)$

Log$_{10}$ diffusion coefficient, cm^2/sec

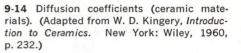

9-14 Diffusion coefficients (ceramic materials). (Adapted from W. D. Kingery, *Introduction to Ceramics*. New York: Wiley, 1960, p. 232.)

The diffusion coefficients of the two or more components of a compound are seldom comparable because of size, charge, or structural differences. At 1000°K, for example, the diffusion coefficient of an Na$^+$ ion in NaCl is approximately five times greater than that of a Cl$^-$ ion. At 825°K, the factor is approximately 50. This difference is primarily a result of size (the radii are 1 A and 1.8 A, respectively). Another example is indicated in Fig. 9–14. If the curves for U^{4+} and O^{2-} ions are extrapolated to 1000°C [i.e., $1000/T(°K) \cong 0.8$], their diffusion coefficients differ by approximately 10^7. This is partially a consequence of charge, since it takes more energy to activate a tetravalent uranium ion than a divalent oxygen ion. In addition, uranium oxide is not stoichiometric, having a few U^{3+} ions to result in UO$_{2-x}$. The anion vacancies contribute much to the increased O^{2-} ion mobility.

Solid solutions within compounds lead to a variety of interesting diffusion situations. As shown in Fig. 9–14, the diffusion coefficient of Mg^{2+} in MgO is low except at very high temperatures. If Fe$_{1-x}$O is dissolved in MgO under reducing conditions, the above coefficient is only slightly altered. Under more oxidizing conditions, however, the Mg^{2+} diffusion coefficient increases because one vacancy appears for every two Fe^{3+} ions formed. (Cf. Fig. 8-9.) Likewise, if MgF$_2$ is dissolved in LiF, the diffusion coefficient of Li$^+$ ions increases because each Mg^{2+} ion requires the omission of two Li$^+$ ions, but replaces only one of these in the structure, leaving a vacancy.

Ionic conductivity. We are aware that both molten salts and aqueous electrolytes can conduct charges when they are placed in an electric field, the positive and negative ions moving in opposite directions. The same occurs, although at a much slower rate, in solids. Each ion possesses a charge which is the product of the valence, Z, and the electron charge, q (1.6×10^{-19} amp · sec). Therefore, ion movements introduce charge movements and ionic conductivity results. In

Eq. (1-3), conductivity, σ, was expressed as charge flux per unit voltage gradient. It may also be expressed as the product of the number of charge carriers, n, the charge carried by each, Zq, and their mobility, μ:

$$\sigma = nZq\mu. \qquad (9\text{-}11a)$$

The units for this equation are

$$\frac{1}{\text{ohm} \cdot \text{cm}} = \left(\frac{\text{carriers}}{\text{cm}^3}\right)\left(\frac{\text{amp} \cdot \text{sec}}{\text{carrier}}\right)\left(\frac{\text{cm}^2}{\text{sec} \cdot \text{volt}}\right), \qquad (9\text{-}11b)$$

the mobility being the velocity of the carrier per unit voltage gradient. As a transport equation, Eq. (9-11a) has widespread use, not only for ionic conductivity, but also for electronic conductivity, as will be seen in Chapter 15. The *mobility* term, μ, is closely related to the diffusion coefficient, D, through one of Einstein's equations,

$$\mu = ZqD/kT. \qquad (9\text{-}12)$$

Example 9-5

What conductivity will result from Ca^{2+} ion movements in CaO at 1727°C? CaO has the NaCl structure (Fig. 4-1) with a lattice constant of 4.81 A.

Answer. From Fig. 9-14, $D \cong 10^{-10}$ cm²/sec at 2000°K. Also, from Eqs. (9-11) and (9-12),

$$\sigma = nZ^2q^2D/kT. \qquad (9\text{-}13)$$

Thus

$$n_{Ca^{2+}} = (4/\text{unit cell})/(4.81 \times 10^{-8} \text{ cm})^3/\text{unit cell}$$

$$= 3.6 \times 10^{22}/\text{cm}^3,$$

$$\sigma = \frac{(3.6 \times 10^{22}/\text{cm}^3)(2 \times 1.6 \times 10^{-19} \text{ amp} \cdot \text{sec})^2(10^{-10} \text{ cm}^2/\text{sec})}{(1.38 \times 10^{-16} \text{ erg}/°\text{K})(2000°\text{K})(10^{-7} \text{ volt} \cdot \text{amp} \cdot \text{sec}/\text{erg})}$$

$$\cong 1.3 \times 10^{-5} \text{ ohm}^{-1} \cdot \text{cm}^{-1}. \blacktriangleleft$$

Since the diffusion coefficient itself is a function of temperature (Eq. 9-2), ionic conductivity, σ_i, can be expressed as follows:

$$\sigma_i = \frac{nZ^2q^2D_0}{kT} e^{-Q/RT}, \qquad (9\text{-}14)$$

or, if we recognize that the preexponential factors of Eq. (9-14) have only minor variations as compared to the exponential factor,[*]

$$\log_{10} \sigma_i \cong \text{constant} - Q/(2.3RT). \qquad (9\text{-}15)$$

A plot of ionic conductivity versus temperature is given in Fig. 9-15 for NiO. The slope of this curve, if corrected for dimensional units, is the activation energy for self-diffusion, Q, divided by $-2.3R$. In practice, electrical conductivity provides a feasible method of determining diffusion data for ionic solids.

---

[*] Log $(1/T)$ varies much less than $(1/T)$.

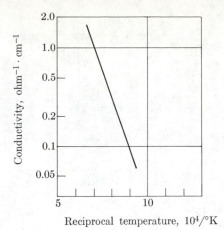

Conductivity, $ohm^{-1} \cdot cm^{-1}$

2.0
1.0
0.5
0.2
0.1
0.05

5   10

Reciprocal temperature, $10^4/°K$

9-15 Ionic conductivity versus temperature (NiO). Since conductivity occurs by ionic diffusion, the same laws apply in both cases. Equations (9-15), and (9-2c) are comparable. (Data are from S. P. Mitoff, *Journ. Chem. Phys.*, **31**, 1961, p. 1261.)

Transport numbers. Since positive and negative ions may have markedly different diffusion coefficients, one or the other becomes the majority charge carrier in an ionic material. We use the term *transport number*, $t_{\oplus}$ or $t_{\ominus}$, to indicate the fraction of conductivity arising from positive and negative ions, respectively:

$$t_{\oplus} = \sigma_{\oplus}/\sigma \qquad (9\text{-}16a)$$

and

$$t_{\ominus} = \sigma_{\ominus}/\sigma. \qquad (9\text{-}16b)$$

Sample values of $t_{\oplus}$ are 0.98 and 0.80 for Na^+ conductivity with NaCl at 550°C and 727°C, respectively.

Impurities in ionic conductors. Figure 9-15 shows the *intrinsic* conductivity, σ_{in}, of NiO as a function of temperature, i.e., the conductivity which arises from the movements of the Ni^{2+} and O^{2-} ions. Invariably, some solid-solution impurities are present and add an *extrinsic* conductivity, σ_{ex}, to the total conductivity:

$$\sigma = \sigma_{in} + \sigma_{ex}. \qquad (9\text{-}17)$$

Since the solvent and the solute have a different set of values in Eqs. (9-14) and (9-15), each will have its own $\log \sigma$-versus-$1/T$ plots, as shown for NaCl in Fig. 9-16(a). The total conductivity is the arithmetic sum of the two and is shown as the heavy curve. The amount of solute expressed as carriers/cm^3, n, is contained within the constant of Eq. (9-15) and is represented by the series of curves for different degrees of purity (Fig. 9-16b).

Example 9-6

From the data of Fig. 9-16(a) calculate the activation energy of Na^+ diffusion in NaCl. (As indicated above, the transport numbers for Na^+ are 0.98 and 0.80 at 550°C and 727°C, respectively.)

Answer. From Fig. 9-16(a) and above data.

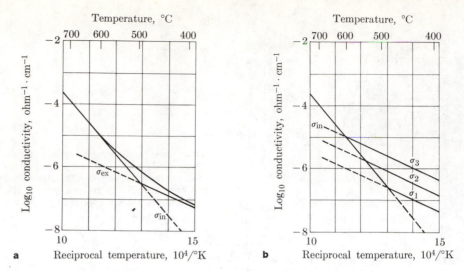

9-16 Intrinsic and extrinsic conductivity (NaCl). (a) $\sigma = \sigma_{in} + \sigma_{ex}$. (b) The values σ_1, σ_2, and σ_3 are for progressively higher impurity levels.

1) At 550°C (823°K),

$$\sigma = (2 \times 10^{-6}\ \text{ohm}^{-1} \cdot \text{cm}^{-1})(0.98)$$
$$= 1.96 \times 10^{-6}\ \text{ohm}^{-1} \cdot \text{cm}^{-1} = 10^{-5.708}\ \text{ohm}^{-1} \cdot \text{cm}^{-1}.$$

2) At 727°C (1000°K),

$$\sigma = (2.5 \times 10^{-4}\ \text{ohm}^{-1} \cdot \text{cm}^{-1})(0.80)$$
$$= 2 \times 10^{-4}\ \text{ohm}^{-1} \cdot \text{cm}^{-1} \doteq 10^{-3.699}\ \text{ohm}^{-1} \cdot \text{cm}^{-1}.$$

$$\text{Slope of curve} = -Q/2.3R = \frac{\log \sigma_1 - \log \sigma_2}{1/T_1 - 1/T_2},$$

$$Q_{\text{Na}^+} = -\left(4.575\ \frac{\text{cal}}{\text{mole} \cdot \text{°K}}\right)\left(\frac{-5.708 + 3.699}{0.001214 - 0.0010000}\right)$$
$$= 43,000\ \text{cal/mole.} \quad \blacktriangleleft$$

9-5 DIFFUSION IN POLYMERS

Diffusion of polymeric molecules differs from that of atoms or ions in at least two ways. First, the macromolecules are large and cumbersome. Because they may coil and kink (Section 7-4), their entanglements inhibit significant movements past other molecules. Second, the use of polymeric solids is confined to relatively low temperatures (10^2 to 10^{3}°K); hence, the thermal energies encountered in many of the earlier diffusion discussions are not operative here.

Small solute molecules, however, can diffuse within and through polymeric solids because the activation energy for molecular distortion is lower, particularly in linear molecules with weak intermolecular bonds. This has major practical importance, because many polymers are used as film or fiber products where one or two of the dimensions are very small. A few examples will emphasize this point. (1) Food-packaging films are only a fraction of a millimeter thick; so even a low diffusion coefficient may permit excessive quantities of O_2 to pass inward or CO_2 to pass outward. (2) A gradual diffusion of O_2 and N_2 molecules through an inner tube of a tire will lower the internal pressures. (3) A textile fiber utilizes diffusion as a means of dye penetration. (4) Selective diffusion through membranes is a biological function; it is also receiving emphasis in commercial practices, e.g., water desalinization.

Example 9-7

If an inner tube (1000 in^2 surface and 2000 in^3 volume) loses pressure from 30 psi to 29 psi in 100 days, what is the diffusion flux?

Answer. Assume a temperature of 68°F (20°C). Switch to cm · atm units.

$$\text{Volume} = 33{,}000 \text{ cm}^3; \quad \text{area} = 6400 \text{ cm}^2.$$

From Eq. (2-10),

$$\Delta n = n_2 - n_1 = (P_2 - P_1)V/RT$$

$$= \left[\left(\frac{14.7 + 29.0}{14.7}\right) - \left(\frac{14.7 + 30.0}{14.7}\right) \text{atm}\right] \frac{33{,}000 \text{ cm}^3}{(82.06 \text{ cm}^3 \cdot \text{atm/mole} \cdot \text{ K})(293 \text{ K})}$$

$$= -0.093 \text{ mole.}$$

$$\text{Flux} = \frac{(-0.093 \text{ mole})(0.602 \times 10^{24} \text{ molecules/mole})}{(6400 \text{ cm}^2)(8.64 \times 10^6 \text{ sec})}$$

$$= -10^{12} \text{ molecules/cm}^2 \cdot \text{sec.}$$

Note. It is negative because the molecules are leaving the tube. ◄

Factors affecting diffusion in polymers. As might be expected, atoms and small molecules diffuse more rapidly through polymers than large molecules do. Likewise, amorphous polymers with lower densities permit more rapid diffusion than highly crystalline polymers; thus branching and atactic structures facilitate diffusion.

Diffusion through an amorphous polymer varies sharply, depending on whether the temperature is below or above the glass point. When it is above, there is an opportunity for diffusion paths to open up for solute movements; when the temperature is below the glass point, the structure is rigid, requiring high activation energies for diffusion. Unfortunately, systematic diffusion data are meager for polymeric materials because it is difficult to characterize the structure.

Swelling. An obvious consequence of diffusion is the swelling or *absorption* which can occur in noncrystalline linear polymers. Added volume is required when micromolecules enter into a polymer structure. In turn, these additions *plasticize* a polymeric solid to lower its glass transition temperature (glass point) and alter its mechanical properties. Large additions of micromolecules can eventually dissolve the polymer. Commercial implications are apparent when one considers materials such as rubber hoses for gasoline or other organic liquids. The structural

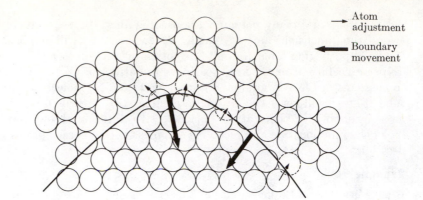

9-17 Diffusion across boundaries. The average atom is more stable on the concave grain, where it may have more neighbors than would normally be expected on the convex grain. As a result, the boundary moves toward the center of curvature.

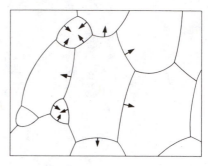

9-18 Grain-growth mechanism. The boundaries move toward the center of curvature (arrows). As a result, the small grains eventually disappear.

factors which favor diffusion also accentuate swelling. For this reason, crystalline or cross-linked polymers are specified where volume stability is necessary.

9-6 GRAIN GROWTH

Atoms move across grain boundaries as well as within crystals. Movements are balanced when the boundary is a plane, with as many atoms crossing in one direction as the other. If the boundary is curved, an atom has a slightly higher probability of having more neighbors, and therefore less energy, when it is part of a concave surface than when it is part of a convex surface (Fig. 9-17). Therefore, the movements in the two directions across the boundary are not equal; consequently, the boundary itself moves toward the center of curvature.* The driving force for such movement is the minimization of the grain-boundary area and its energy (Section 6-7).

The grain-growth mechanism of a single-phase microstructure may be idealized as in the sketch of Fig. 9-18. For grains of a single-phase material to fill three-dimensional space with a minimum of grain-boundary area and 120° grain junctions,

---

* In a three-dimensional solid, the boundary moves so as to minimize $1/\overline{R} = 1/R_1 + 1/R_2$, where R_1 and R_2 are the two principal radii of curvature.

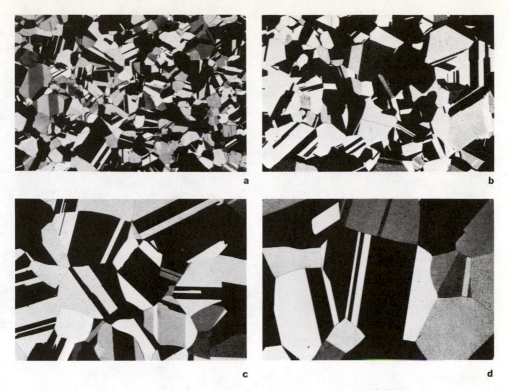

9-19 Grain growth (brass at ×40). Each grain contains several twins (black and white bands). (Courtesy of J. E. Burke, General Electric Co., Schenectady, N.Y.)

the boundaries *cannot* be planar, but must be curved. Topologically, this gives convex surfaces to small grains and concave surfaces to large grains. Consequently, as the grain boundaries move toward their center of curvature, large grains grow at the expense of the small ones, and the latter eventually disappear. Fewer, but larger, grains result (Fig. 9-19).

Since the rate of grain growth depends on the curvature of the boundary, it also is inversely related to a power, n, of the dimensions of the grains, δ*:

$$d\delta/dt = k'/\delta^n. \qquad (9\text{-}18)$$

The proportionality constant, k', includes the surface energy and a diffusion constant for the movement of atoms across the boundaries. From Eq. (9-18),

$$\delta^{n+1} - \delta_i^{n+1} = kt; \qquad (9\text{-}19)$$

or if the initial dimension, δ_i, is much smaller than the subsequent grain size at time t,

$$\delta = (kt)^{1/(n+1)}. \qquad (9\text{-}20)$$

---

* Often called the "diameter," even though the grain is not spherical.

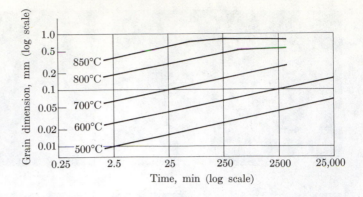

9-20 Isothermal grain growth (brass). (J. E. Burke, "Some Factors Affecting Rate of Grain Growth in Metals," *Metals Technology*, **T.P. 2472**, October 1948.)

The value $n + 1$ is commonly between three and six in commercial materials. The larger values may result from minor impurity phases which inhibit grain growth. Furthermore, external surfaces eventually arrest growth as δ approaches the cross-sectional dimensions of the material because the grain boundaries then tend to become planar where they intersect the surface.

Diffusion movements increase exponentially with temperature (Eq. 9-2). Thus we find a pronounced increase in the *rate* of grain growth (and the grain size reached) at higher temperatures (Fig. 9-20). Note, however, that growth, on the average, is an irreversible process. The grains do *not* contract at low temperatures.

Example 9-8

Determine the growth exponent, $n + 1$, for the brass of Fig. 9-20.

Answer. From Eq. (9-20),

$$1/(n + 1) = (\ln \delta_2 - \ln \delta_1)/(\ln t_2 - \ln t_1).$$

At 500°C,

$$n + 1 = \ln (2500/2.5)/[\ln (0.04/0.01)]$$
$$= 5. \quad \blacktriangleleft$$

Example 9-9

A sample of the brass used in Fig. 9-20 was annealed at 600°C until the mean dimension of the grains was 0.02 mm. It was cooled immediately to 500°C. What will be the mean dimension after 30 min at that temperature?

Answer

1) Graphical solution: This is the dimension which would have developed in 80 min at 500°C. An extra 30 min will give the dimension shown in Fig. 9-20 for 110 min, or $\delta = 0.022$ mm.

2) Analytical solution: Use Eq. (9-19) and $n + 1 = 5$ from Example 9-8, and also Fig. 9-20 at 500°C. Solving for k, we obtain

$$(0.07)^5 - (0.01)^5 = k(25,000),$$
$$k = 10^{-10.2}.$$

Solving for δ at 30 min, we obtain

$$\delta^5 - (0.02)^5 = 10^{-10.2}(30),$$

$$\delta = 0.022 \text{ mm}. \quad \blacktriangleleft$$

REFERENCES FOR FURTHER READING

Birchenall, E., *Physical Metallurgy*. New York: McGraw-Hill, 1959. Diffusion in solid metals and oxides is presented in Chapter 9. Advanced undergraduate level.

Darken, L. S., and R. W. Gurry, *Physical Chemistry of Metals*. New York: McGraw-Hill, 1953, p. 442. A formal derivation of the error function is presented for advanced undergraduate and graduate students.

Girifalco, L. A., *Atomic Migration in Crystals*. New York: Blaisdell, 1964. A helpful small book for use as supplementary reading on diffusion. Same mathematical level as this book.

Guy, A. G., *Elements of Physical Metallurgy*, second edition. Reading, Mass.: Addison-Wesley, 1959. Chapter 11 provides a good summary of diffusion in metals. Undergraduate level.

Hannay, N. B., *Solid State Chemistry*. Englewood Cliffs, N. J.: Prentice-Hall, 1967. Paperback. Chapter 6 introduces diffusion on the sophomore-junior level. A good supplement for this chapter.

Kingery, W. D., *Introduction to Ceramics*. New York: Wiley, 1964. Electrical conductivity of ionic solids is discussed in Chapter 19. Undergraduate level.

Richman, M. H., *Science of Metals*. Waltham, Mass.: Blaisdell, 1967. Diffusion is discussed in an introductory manner in Chapter 11.

Shewmon, P., *Diffusion in Solids*. New York: McGraw-Hill, 1963. The preeminent textbook on diffusion for advanced undergraduate and graduate students.

PROBLEMS

9-1 A solid solution of copper in aluminum has 10^{20} atoms of copper per cm^3 at point X, and 10^{18} atoms of copper per cm^3 at point Y. Points X and Y are one micron apart. What will be the diffusion flux of copper atoms between these two points at 100°C?

Answer. 6000 atoms/$cm^2 \cdot$ sec

9-2 Assume that a copper gradient in an aluminum bar decreases from 0.4 a/o Cu at the surface to 0.2 a/o Cu one millimeter below the surface. Thus there will be a net movement of atoms toward the interior.

    a) If the above gradient is maintained, what is the net movement of atoms across a plane 0.5 mm below the surface if the bar is at 500°C?

    b) If the bar is at 100°C?

[*Hint.* Convert a/o to atoms/cm^3.]

9-3 How much should the concentration gradient be for nickel in iron if a flux of 10^6 nickel atoms/$cm^2 \cdot$ sec is to be realized (a) at 800°C? (b) at 1400°C?

Answer. a) 10^{20} (atoms/cm^3)/cm   b) 10^{15} (atoms/cm^3)/cm

9-4 The following data were obtained for the diffusion of aluminum in silicon:

Temperature, °C	D, cm^2/sec
1380	3.11×10^{-10}
1300	7.1×10^{-11}
1250	4.1×10^{-11}
1200	1.74×10^{-11}

a) Determine the constants D_0 and Q.
b) Calculate the diffusion constant for this system at 800°C.

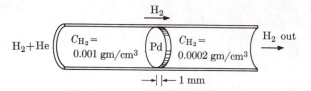

9-21 See Problem 9-5.

9-5 Hydrogen gas diffuses through hot palladium metal quite readily, whereas other common gases do not. Hot Pd membranes can therefore be used to separate H_2 from other gases. If the type of apparatus shown in Fig. 9-21 is to be used to separate H_2 from helium, calculate the area of the Pd membrane that would be required to pass 100 cm^3 (STP) of H_2 per hour. The diffusion constant for H_2 in Pd at the operating temperature is 10^{-4} cm^2/sec.

Answer: 3.1 cm^2

9-6 Aluminum is to be diffused into a silicon single crystal. At what temperature will the diffusion coefficient be 10^{-10} cm^2/sec? ($Q = 78,000$ cal/mol and $D_0 = 5.55$ cm^2/sec.)

9-7 A large piece of iron (fcc) with 0.1 w/o carbon is placed in a carburizing furnace which is at 1000°C. The surface immediately attains and maintains 1.0 w/o carbon.
a) Plot the carbon content versus depth below the surface after 10 seconds.
b) After one hour.

Answer. a) 1.0 w/o at 0 cm; 0.7 w/o at 0.001 cm; 0.1 w/o at 0.01 cm   b) 1.0 w/o at 0 cm; 0.98 w/o at 0.001 cm; 0.85 w/o at 0.01 cm; 0.13 w/o at 0.1 cm

9-8 At 800°C, $D = 10^{-13}$ cm^2/sec for the self-diffusion of germanium in its own structure. The diffusion coefficient of copper in germanium is 3×10^{-9} cm^2/sec. Account for the difference of 30,000/1.

9-9 Nitrogen was diffused into a bcc iron slab at 475°C for 9 hours. The resulting nitrogen profile developed:

Surface	0.040 w/o
1.0 mm below	0.031
2.0 mm below	0.023
5.0 mm below	0.010
10.0 mm below	0.010

Determine the diffusion coefficient.

Answer. 10^{-6} cm^2/sec

9-10 A polyethylene balloon which contains hydrogen loses the gas by diffusion. The rate of loss is 10 ft^3/hr at 27°C and 22 ft^3/hr at 42°C. (Volumes are reported at STP.) What is the activation energy for diffusion of hydrogen through polyethylene?

9-11 As shown in Appendix C-1, the thermal conductivity of stainless steel is about one-third that of 1080 steel. (Their densities and heat capacities are comparable.) The latter steel was used in Example 9-4(b). Determine the temperature profile for stainless steel and similar quenching conditions.

Answer. Surface, 24°C; 0.32 cm, 125°C; 0.64 cm, 230°C; 1.27 cm, 407°C; 2.54 cm, 690°C; 5.08 cm, 890°C

9-12. a) Estimate the approximate thermal diffusivity of copper from data in Appendix C and Section 6-2.
   b) Repeat Example 9-4(b), but for copper.

9-13 Electrical conductivity in cubic $(Zr_{0.85}Ca_{0.15})O_{1.85}$ occurs almost entirely ($t_\ominus = 0.99$) by anion movements through anion vacancies at 750°C. The lattice constant, a, is 5.45 A. Although the 4 cation sites per unit cell are all filled, the 8 anion sites contain some vacancies. Determine the electrical conductivity at 750°C if the diffusion coefficient for the vacancies is 1.25×10^{-5} cm^2/sec.

Answer. 0.34 ohm$^{-1} \cdot$ cm^{-1}

9-14 Assume that the bulk of the ionic electrical conductivity in Fig. 9-15 is from cation movement (Fig. 9-13). Calculate the diffusion coefficient for Ni^{2+} in NiO at 1000°C. [*Hints.* NiO has an NaCl structure, and the Ni^{2+} concentration may be calculated from the ionic radii (Appendix B).]

9-15 It has been proposed that the equation

$$\delta = \delta_0 e^{-E/kT}$$

could apply for grain size (for a given time) as a function of temperature. Check this proposition with the data in Fig. 9-20.

Answer. A plot of log δ versus $1/T$ does not provide a constant slope for a unique activation energy, as would be required for the Arrhenius equation to apply. It can, however, be used over limited temperature ranges of $\Delta T \leq 200$°C.

10 □ elastic behavior of solids

10-1 ELASTIC DEFORMATION

Elastic deformation is a *reversible deformation;* i.e., the original dimensions are recovered after externally applied stresses are removed. This is in contrast to plastic deformation (Chapter 11) and viscous flow (Chapter 12).

Elastic strain is linear with stress in most materials, particularly at low stresses. Exceptions, which will be described later, have a rational basis with respect to structures.

We usually assume that elastic deformation occurs instantaneously with the application of the external stress. In a practical sense this is true, since the time lag for the resulting elastic strain is usually less than our measuring interval. There are exceptions, however, in which the time lag cannot be ignored. The basis for the slower responses will be considered in the section on anelasticity which follows later in the chapter.

Elastic deformation is anisotropic within single crystals or in materials with preferential orientations (Section 6-8). We will consider these variations in Section 10-3 but will not include rigorous mathematical treatments, which require tensors for analysis.

Elastic behavior is a function of temperature because thermal disorder affects the bonding between atoms and the rearrangement of atoms under directional stresses. In this case, molecular and atomic solids have significantly different responses (Section 10-5).

Young's modulus. The first of three elastic moduli to receive consideration is Young's modulus, Y, which relates strain, ϵ, to uniaxial stress, σ:

$$Y = \sigma/\epsilon. \tag{10-1}$$

This equation, which is often called Hooke's law, shows a linear relationship between strain and stress. Such an assumption is reasonable for strains of less than one or two percent, and therefore can be applied to all solids except those molecular solids which undergo chain straightening.

The elastic modulus is a measure of stiffness. Since force can be related to stress, and changes in distance can be related to strain within any given material, the elastic modulus originates from the slope of the net force-versus-distance curve

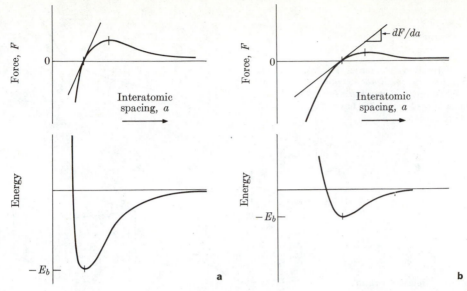

10-1 Young's modulus versus bonding energy. (a) Strongly bonded materials. (b) Weakly bonded materials. Since Young's modulus is proportional to dF/da and to d^2E/da^2, a material with a high bonding energy, E_b, also has a high elastic modulus.

in Fig. 3-2. The slope of this curve is a continuous function and does not vary significantly within the first one or two percent strain on either side of the equilibrium distance. Thus, we can use the same modulus for both tensile (+) and compressive (−) stresses.

The values of Young's modulus are directly related to the interatomic bonding energies. Materials with deep and narrow energy troughs have a high elastic

TABLE 10-1
YOUNG'S MODULI VERSUS MELTING TEMPERATURES

Compound	Average modulus of elasticity (10^6 psi)	Melting temperature, °C
Titanium carbide	45	3180
Tungsten	50	3410
Silicon carbide	50	>2800
Periclase (MgO)	35	2800
Corundum (Al_2O_3)	53	2050
Iron	30	1539
Copper	16	1083
Halite (NaCl)	5	801
Aluminum	10	660
Magnesium	6	650
Lead	2.2	327
Polystyrene	0.4	<300
Nylon	0.4	<300
Rubber	0.01	<300

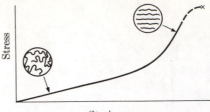

10-2 Elastic deformation (elastomers). Initially, relatively low stresses are required to unkink polymeric molecules. The strain is reversible after the stress is removed because kinked molecules have less free energy than stretched molecules.

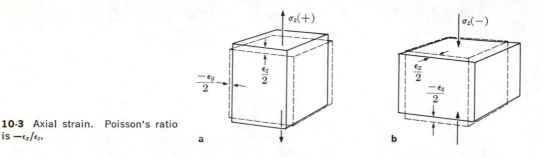

10-3 Axial strain. Poisson's ratio is $-\epsilon_x/\epsilon_z$.

modulus, since dF/da is equal to the second derivative of energy with respect to distance. This correlation is shown qualitatively in Fig. 10-1 and quantitatively in Table 10-1.

Nonlinear stress-strain relationships are important in rubbers where, because of the unkinking of the molecules, it is possible to have elastic strains exceeding 100%. The initial stresses produce considerable strain (Fig. 10-2), but as the chains are unkinked there is greater resistance to further strain because the molecules then have to be stretched out of their natural conformation (Section 7-4).

Poisson's ratio. Lateral dimensional changes accompany axial strains. As illustrated in Fig. 10-3(a), a tensile stress, σ_z, produces an axial strain, $+\epsilon_z$, and lateral contractions of $-\epsilon_x$ and $-\epsilon_y$. If the material is isotropic, ϵ_x and ϵ_y are equal. The ratio

$$\nu = -\epsilon_x/\epsilon_z \tag{10-2}$$

is called Poisson's ratio. It is shown in Example 10-1 that this ratio is 0.5 for an "ideal" material which maintains constant volume when stressed elastically. In practice, few if any materials are ideal in this sense, as is shown by the values of the Poisson ratio in Table 10-2. This is not surprising inasmuch as the attractive and repulsive forces within a material are different functions of interatomic distances [Eq. (3-4) and Fig. 3-2].

Example 10-1

Show that Poisson's ratio for a constant-volume material is 0.5 under low uniaxial stresses. (For actual Poisson's ratios, see Table 10-2.)

TABLE 10-2
POISSON RATIOS

Material	$-\epsilon_x/\epsilon_z$
Lead	0.4
Aluminum	0.34
Copper	0.35
Iron	0.28
Tungsten	0.27
Glass (lime-soda)	0.25
Polyethylene	0.4
Polyisoprene	0.49

Answer. Use a cube, size l.

$$V_0 = V_f,$$
$$l^3 = (l + \Delta l_z)(l + \Delta l_x)^2,$$
$$1 = (1 + \Delta l_z/l)(1 + \Delta l_x/l)^2,$$
$$0 = \epsilon_z + 2\epsilon_x + \cdots,$$
$$\nu = -\epsilon_x/\epsilon_z = 0.5.$$

Note. The above calculation applies equally for tension or compression. ◀

Bulk modulus. Hydrostatic compression introduces a volume contraction, $\Delta V/V$, which is initially proportional to the pressure, σ_{hyd}:

$$\Delta V/V = \beta\sigma_{\text{hyd}}, \tag{10-3a}$$

or

$$K = 1/\beta = \sigma_{\text{hyd}}/(\Delta V/V), \tag{10-3b}$$

where β is *compressibility* and K is the bulk modulus that equals the ratio of pressure to the resulting deformation.

The Young's modulus, Y, and the bulk modulus, K, can be related through Poisson's ratio. Refer to Fig. 10-3(b), and consider the step-wise process of first applying σ_z, then adding σ_x and σ_y, so all three stresses are equal to σ_{hyd}, the hydrostatic pressure. First,

$$\epsilon_z = \sigma_z/Y.$$

Then, as σ_x and σ_y are added, ϵ_z is modified by $-\nu\epsilon_x$ and $-\nu\epsilon_y$, respectively, so that

$$\epsilon_z = \frac{\sigma_z}{Y} + (-\nu\epsilon_x) + (-\nu\epsilon_y) = \frac{\sigma_z}{Y} - \frac{\nu\sigma_x}{Y} - \frac{\nu\sigma_y}{Y},$$

or

$$\epsilon_z = [\sigma_z - \nu(\sigma_x + \sigma_y)]/Y, \tag{10-4}$$

which is the general form of Hooke's law. Since, for hydrostatic pressure,

$$\sigma_{\text{hyd}} = \sigma_x = \sigma_y = \sigma_z$$

and

$$\Delta V/V \cong \epsilon_z + \epsilon_x + \epsilon_y = 3\epsilon_z,$$

then

$$\Delta V/V = 3\sigma_{\mathrm{hyd}}(1 - 2\nu)/Y.$$

So

$$Y = 3K(1 - 2\nu). \tag{10-5}$$

Example 10-2

a) What hydrostatic pressure would be required if iron were to be transformed from fcc to bcc with no dimensional change?

b) What stress is required to provide constraint in one dimension only?

Answer

a) From Example 4-4, $\Delta V/V = 0.016$. From Eq. (10-5) and Table 10-2,

$$K = (30{,}000{,}000 \text{ psi})/3(1 - 0.56) = 22{,}700{,}000 \text{ psi},$$
$$\sigma_{\mathrm{hyd}} = (K)(\Delta V/V)$$
$$= (22{,}700{,}000 \text{ psi})(0.016)$$
$$= 360{,}000 \text{ psi}.$$

b) If $\Delta V/V = 0.016$, then

$$\Delta l/l \cong 0.0053.$$

Let the transformation occur, then compress to the original dimension:

$$\sigma = (30{,}000{,}000 \text{ psi})(0.0053)$$
$$= 160{,}000 \text{ psi}.$$

Note. Less stress is necessary in one-dimensional restraint because there can be lateral relief. ◀

Shear modulus. Elastic shear strains, γ, arise from shear stresses, τ. The ratio of these two is called the shear modulus, G:

$$G = \tau/\gamma. \tag{10-6}$$

In this case, the shear strain is defined as the tangent of the shear angle, α, as sketched in Fig. 10-4:

$$\gamma = \tan \alpha = x/y. \tag{10-7}$$

Like other moduli, the shear modulus is a constant for the levels of strain normally encountered before plastic deformation occurs.

If we were to go through a moderately simple proof, we could show that the shear modulus is related to the Young's modulus through Poisson's ratio:

$$G = Y/2(1 + \nu). \tag{10-8a}$$

Using Eqs. (10-5) and (10-8), we obtain

$$G = \frac{3K(1 - 2\nu)}{2(1 + \nu)}. \tag{10-8b}$$

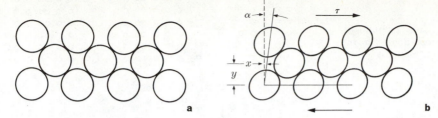

10-4 Elastic shear strain. (a) No strain. (b) Shear strain. The shear is elastic so long as the atoms maintain their original neighbors.

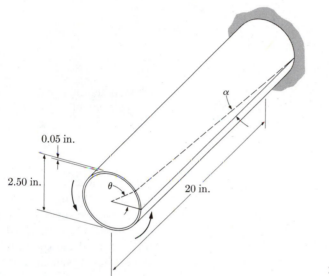

10-5 Shear strain. (See Example 10-3.)

Since Poisson's ratio, ν, is never more than 0.5 and is typically about 0.3, the shear modulus, G, is usually 35 to 40% of Young's modulus, Y.

Example 10-3

A hollow copper tube, 20 in. long, has a radius of 1.25 in. and a wall thickness of 0.05 in. A radial torque of 2000 in-lb is applied between the tube ends. What is the elastic displacement in degrees of rotation around its axis? (See Fig. 10-5.)

Answer. The force at the tube wall is

$$F = 2000 \text{ in-lb}/1.25 \text{ in.} = 1600 \text{ lb.}$$

Thus

$$\tau = 1600 \text{ lb}/[(0.05 \text{ in.})(2\pi)(1.25 \text{ in.})]$$
$$= 4080 \text{ psi.}$$

From Table 10-2, Appendix C, and Eq. (10-8a),

$$G = 6.0 \times 10^6 \text{ psi} = 4080 \text{ psi}/\gamma,$$

10-6 Longitudinal wave (simplified). The velocity is proportional to the square root of Young's modulus.

10-7 Transverse wave (simplified). The velocity is proportional to the square root of the shear modulus.

Therefore,

$$\gamma = 0.00068 = x/20 \text{ in.} \ (= \tan \alpha),$$
$$x = 0.0136 \text{ in.},$$
$$\theta = (0.0136 \text{ in.}/1.25 \text{ in.}) \text{ rad} = 0.6°.$$

Note. The above method, in reverse, is commonly employed to determine the shear modulus, G. When it is combined with experimental data for Young's modulus (Fig. 1-3) and Eq. (10-8a), it is possible to calculate Poisson's ratio. ◄

10-2 ELASTIC WAVES

On the basis of the spring model of interatomic forces (Fig. 3-4 and Section 3-2), it is easy to see how *sound waves* and other elastic waves travel through solids. Consider Fig. 10-6(a), where the initial plane of atoms is compressed against the next plane. In order to balance forces the second plane is displaced, then the third plane, and so the pulse moves along (Fig. 10-6b and c). In the meantime the first plane recoils. If the pulse is driven by a sound of frequency f/sec, a second displacement is introduced $1/f$ sec later. Thus, a series of "pulses," i.e., a wave, moves through the material.*

The amount of displacement in Fig. 10-6 is related to the elastic modulus, and the time lag is related to the mass of the atoms. As a result, the wave travels at a velocity, v, which is simply expressed as

$$v_l = \sqrt{Y/\rho}, \qquad (10\text{-}9a)$$

where ρ is the density. The subscript l indicates the velocity of a longitudinal or compressional wave, as illustrated in Fig. 10-6. In this case Young's modulus, Y, must be used. Alternatively, a shear or transverse wave has a velocity, v_t, of

$$v_t = \sqrt{G/\rho}. \qquad (10\text{-}9b)$$

The shear modulus, G, must be used in this case for reasons that are obvious from Fig. 10-7.

---

* We will limit our discussion to a long rod, in which the "wavefront" remains planar.

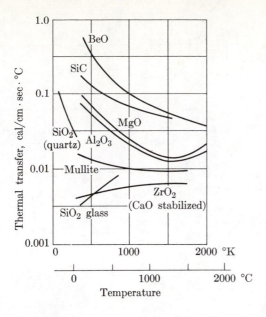

Thermal transfer, cal/cm·sec·°C

- BeO
- SiC
- MgO
- SiO₂ (quartz)
- Al₂O₃
- Mullite
- ZrO₂ (CaO stabilized)
- SiO₂ glass

Temperature

10-8 Thermal transfer. As the temperature is raised, there is more scattering of elastic waves in crystals because of thermal disorder. Therefore, the conductivity decreases until the temperature is reached where radiation transfer becomes significant. At very low temperatures, the conductivity falls off sharply to zero because the specific heat approaches zero (Fig. 6-2). (Adapted from C. Kittel, *Introduction to Solid State Physics.* New York: Wiley, 1953; and W. D. Kingery, "Thermal Conductivity: XII. Temperature Dependence of Conductivity for Single-Phase Ceramics," *Amer. Ceram. Soc.*, **38,** 1955, pp. 251–255.)

Dynamic versus static moduli. Sound waves and other related dynamic elastic behaviors provide means whereby elastic moduli can be measured. Unlike the "static" stress-strain curves of Figs. 1-3 and 10-2, the elastic responses in a "dynamic" test procedure occur in very short time intervals. We shall see later (Section 10-4) that certain elastic displacements have measurable delay, or relaxation, periods. Thus, a dynamically measured strain may be less, and the modulus higher, than the corresponding statically determined data. [Roughly speaking, however, the two are comparable for metallic and ceramic materials at normal temperatures, and for polymers below their glass-transition temperature (Section 6-11 and Chapter 12).]

Thermal conductivity. Except for thermal energy transferred by electrons (Section 14-4), heat is conducted through opaque solids by elastic waves. If we were to pay detailed attention to this type of conductivity, we would have to utilize quantum mechanics and speak of phonons. Short of that, we may make some qualitative comparisons of the data shown in Fig. 10-8.

1) Close-packed compounds of light elements, e.g., BeO, have high thermal conductivity because (a) they have low densities as a result of their small atomic weight and (b) they have high elastic moduli because of their smaller radii. Both situations lead to a high velocity for the elastic waves (Eq. 10-9).

2) Glass and other amorphous solids have lower thermal conductivities than their crystalline counterparts because the elastic waves move less readily through amorphous structures than through simple, highly ordered structures (cf. quartz and SiO₂ glass).

3) Higher temperatures reduce thermal conductivity in crystalline solids because the added, random thermal vibrations deflect and diffract the elastic waves.

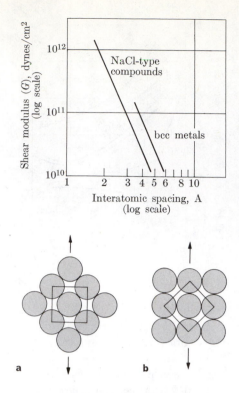

10-9 Shear modulus versus interatomic spacing. For a given structure, the modulus varies inversely with the fourth power of the interatomic spacing.

10-10 Stress versus crystal orientation. Because the atom spacings vary with crystal orientation, elastic moduli also vary.

4) Total thermal transfer increases in *transparent* materials at still higher temperatures as a result of radiation and the fourth-power Stefan-Boltzmann relationship.*

10-3 VARIATIONS IN ELASTIC MODULI

Elastic moduli vary widely among materials (Table 10-1); furthermore, they change for different structures and temperatures of any one material.

Lattice constants. For a given structural type, the elastic moduli have an inverse fourth-power variation with the lattice constant. This is illustrated in Fig. 10-9 in plots of the shear modulus for both bcc and NaCl-type structures. The experimental slope of approximately −4 (on a log scale) could have been predicted, since (1) the attractive forces of Eq. (3-4) vary inversely with the square of the distance and (2) the stress is a force per unit area, another exponent of −2.

Anisotropy. If we were to reorient the schematic crystal structures shown in Fig. 10-6(a) so that the compression wave paralleled a different crystal plane, we could anticipate a different wave velocity. Since the density, ρ, of Eq. (10-9) is constant, this indicates that the elastic constant varies with orientation. Likewise, Fig. 10-10(a and b) suggests different strains per unit stress for different orientations.

---

* See a physics text for information about the Stefan-Boltzmann relationship.

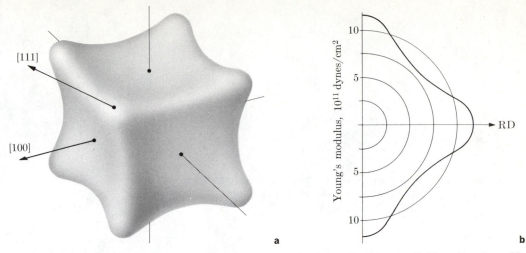

10-11 Elastic anisotropy (copper). (a) Single crystal. (b) Cold-rolled sheet. Rolling direction—RD. (Adapted from J. Weerts, *Zeit. Metallunde*, **25,** 1933, p. 101.)

TABLE 10-3
ANISTROPY OF YOUNG'S MODULI*

Metal	Maximum [111]	Minimum [100]	Random
Lead	4×10^6 psi	1×10^6 psi	2×10^6 psi
Aluminum	11	9	10
Gold	16	6	12
Copper	28	10	16
Iron (bcc)	41	19	30
Tungsten	50	50	50

* Adapted from E. Schmid and W. Boas, *Plasticity in Crystals*, English translation, London: Hughes and Co.

Crystals are anisotropic, having different elastic constants for different orientations. This is modeled three-dimensionally in Fig. 10-11(a) for copper, and is represented by data in Table 10-3. Elastic anisotropy is significant in two respects. First, if a material has a preferred orientation, as does the cold-rolled copper sheet of Fig. 10-11(b), we should not use the mean values of Table 10-3. This fact is important in the case of cold-drawn steel wire, e.g., a coat hanger, which develops a preferred orientation and a Young's modulus of close to 35,000,000 psi along the wire's length. Without the preferred orientation, the mean modulus is only about 30,000,000 psi for steel. (The mean values of Table 10-3 continue to have widespread use in engineering design because there is a nearly random orientation of the grains in most polycrystalline solids. The cold-drawn wire example is not typical.)

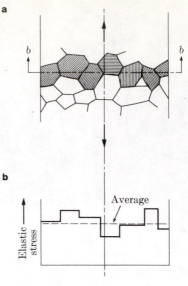

10-12 Stress heterogeneities (schematic). To a first approximation, the grains of a polycrystalline material are equally strained. Therefore, they are stressed differently as a result of anisotropic moduli.

The second consequence of elastic anisotropy is shown in Fig. 10-12. Depending on their orientation, adjacent grains of copper may have a Young's modulus as high as 28,000,000 psi in the [111] direction, or as low as 10,000,000 psi in the [100] direction. Thus, if Fig. 10-12 represents a cross section of a wire in which the *average* stress is 32,000 psi and if each grain deforms equally, the average strain will be 0.002 in./in. since the mean modulus is 16,000,000 psi. However, the stress in some grains will be as high as 56,000 psi, and only 20,000 psi in others. This quick calculation illustrates one source of stress heterogeneity which can occur within a material. Other sources arise from anisotropies of thermal expansion (Section 6-3) or local phase changes (Example 10-2a).

Solid solutions. If a solid solution is ideal, i.e., if Fig. 8-11 applies, the modulus of elasticity varies linearly with atom fraction. If ordering occurs (Eq. 8-1), the modulus is increased because under such conditions additional energy is required to separate the favored pairs of atoms. For this reason, intermetallic compounds, which may be considered highly ordered solid solutions, usually have higher elastic moduli than their component metals; e.g., in the [100] direction, ordered Cu_3Au has a Young's modulus of 11,000,000 psi, while copper and gold have values of 10,000,000 psi and 6,000,000 psi, respectively, in the same direction.

Temperature. Elastic moduli usually decrease at higher temperatures (Fig. 10-13a) because thermal energy helps overcome the interatomic forces. An exception occurs with rubbers and other elastomers (Fig. 10-13b) because of their natural kinked conformation (Section 7-4). At higher temperatures, the factors favoring the kinked conformation are increased; so additional stress is required to produce elastic strains. (These factors can also produce the contraction or negative thermal expansion coefficient cited in Section 10-5.)

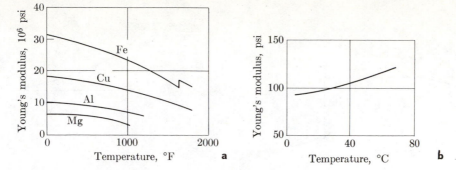

10-13 Young's modulus versus temperature. (a) Metals. (b) Isoprene rubber. (Adapted from A. G. Guy, *Elements of Physical Metallurgy*. Reading, Mass.: Addison-Wesley, 1959; and R. L. Anthony, R. H. Caston, and E. Guth, *Jour. Phys. Chem.*, **46,** 1942, p. 826.)

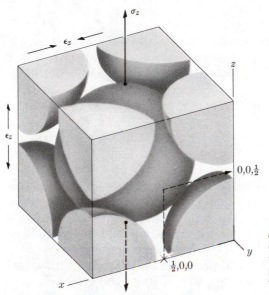

10-14 Stress-induced ordering (C in bcc Fe). The axial stress, σ_z, alters the interstitial sites so that the carbon preferentially moves to the 0, 0, $\frac{1}{2}$-type sites from the $\frac{1}{2}$, 0, 0 and 0, $\frac{1}{2}$, 0 sites. This movement permits additional strain because the lattice distortion is reduced.

10-4 ANELASTICITY

Time-dependent, or anelastic, deformation arises because some atoms or molecules can relocate themselves within the total structure. As is the case with diffusion (Chapter 9), time is required for these movements. (It is desirable to interrupt our discussion at this point to emphasize that elastic deformation, including anelasticity, is recoverable after the stress is removed. Thus the atomic and molecular movements to which we are giving attention in this section are temporary. In the next two chapters we will consider irreversible or plastic deformation.)

To illustrate anelastic behavior, consider Fig. 10-14. As was discussed in Example 8-1, a few carbon atoms may reside interstitially along the edge of a unit cell of bcc iron at $\frac{1}{2}$, 0, 0 or related positions. Their size, however, distorts the surrounding structure. If an axial stress is now applied as indicated, the elastic strain

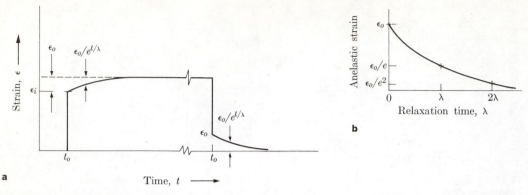

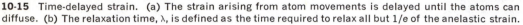

10-15 Time-delayed strain. (a) The strain arising from atom movements is delayed until the atoms can diffuse. (b) The relaxation time, λ, is defined as the time required to relax all but 1/e of the anelastic strain.

distorts the unit cell so that it is no longer ideally cubic. As a result, the interstitial positions equivalent to 0, 0, $\frac{1}{2}$ (as drawn) open up, and those equivalent to $\frac{1}{2}$, 0, 0 and 0, $\frac{1}{2}$, 0 become slightly smaller. Therefore, the carbon atoms preferentially move to the 0, 0, $\frac{1}{2}$ positions,* which have a lower strain energy. This is a stress-induced *ordering* process, because atoms achieve a preference among the interstitial sites. When such ordering occurs, the bcc iron undergoes a reduction in total strain energy because of the better fitting of interstitial atoms into the slightly enlarged sites.

Now let us look at the strain-time curve for the above process (Fig. 10-15). The time-independent strain, ϵ_i, occurs immediately as the stress is applied. The additional strain accompanying the ordering is time-dependent, or anelastic, because atom movements are required. The total strain, $\epsilon_i + \epsilon_o$, is approached exponentially. When the stress is removed, the time-independent strain is recovered immediately, while the anelastic strain relaxes exponentially.

Anelastic behavior, or time-dependent elasticity, is also associated with rubber deformation (Example 10-4), thermoelasticity (Section 10-5), and grain-boundary gliding. An electrically induced "anelasticity" will be discussed under dielectric losses (Chapter 13); viscoelasticity (Chapter 12) and creep (Chapter 21) also possess anelastic components.

Example 10-4

Stretch a 3-in. rubber band over a 12-in. ruler and place it in a home freezer.

a) After chilling to below 0°C, remove it from the freezer and "snap" it off the ruler.
b) Repeat these steps with a warm band. Explain the difference in observed behavior.

Answer. During the stretching, the rubber molecules are straightened from their normal kinked and coiled conformations (Fig. 7-3 and Section 7-4). This is a stress-induced ordering process. When the rubber band is released at 0°C, the rekinking and recoiling of

---

* Close scrutiny will reveal that the $\frac{1}{2}$, $\frac{1}{2}$, 1 positions of body-centered unit cells are also equivalent to the 0, 0, $\frac{1}{2}$ positions through an $(\mathbf{a}/2 + \mathbf{b}/2 + \mathbf{c}/2)$-translation (Section 5–2).

the molecules occurs slowly enough so that one can see the relaxation delay. At room temperature the disordering occurs appreciably more rapidly, so that strain relaxation is almost immediate.

Note. There are a variety of rubber compositions. Rubbers may also have different degrees of cross-linking (Section 7-8). Therefore, the effect just described will be observed more dramatically with some rubber bands than with others. ◄

Relaxation time. The time-dependent strain is logarithmic with respect to time, because the rate of atom movements, dn/dt, from those sites with a higher strain energy is proportional to the number of atoms, n, *remaining* in those less stable sites:

$$\lambda(-dn/dt) = n. \tag{10-10a}$$

Therefore,

$$dn/n = -dt/\lambda$$

and

$$n = n_0 e^{-t/\lambda}. \tag{10-10b}$$

In this equation, n_0 is the total number of atoms at $t = 0$ which may be relocated, and λ is the *relaxation time*. At $t = \lambda$, $n/n_0 = 1/e$. The number of atoms, n_a, which have been relocated at any time t is

$$n_a = n_0 - n$$
$$= n_0(1 - e^{-t/\lambda}). \tag{10-11}$$

Inasmuch as the time-dependent strain, ϵ_a, is directly proportional to these atom movements,

$$\epsilon_a = \epsilon_0(1 - e^{-t/\lambda}). \tag{10-12a}$$

At $t = 0$, ϵ_0 is the potential anelastic strain which may develop through atom relaxation. As shown in Fig. 10-15(a), the anelastic strain remaining, ϵ_r, is

$$\epsilon_r = \epsilon_0 e^{-t/\lambda}. \tag{10-12b}$$

Equation 10-12(b) also applies after stress removal, except that ϵ_0 must be reassigned to a new t_0. Furthermore, the directions of both ϵ_0 and ϵ_r must be reversed.

The relaxation time, λ, is *inversely* proportional to the diffusion coefficients. Thus, from Eq. (9-2),

$$\log_{10} \lambda = \text{constant} + Q/2.3RT, \tag{10-13}$$

where R and T have their previous connotations. The activation energy, Q, must, of course, be the energy which pertains to the specific atomic (or molecular) movements that produce the anelastic strain.

Example 10-5

When a long wire was stressed 3000 psi, the immediate elastic elongation was 0.1121 in. Five minutes later the elongation was 0.1171 in., and several days later it was 0.1174 in. Estimate the remaining elongation which is still present two minutes after the stress is removed. Assume that the relaxation times following stress application and following stress removal are the same.

Answer $\epsilon_o = (0.1174 - 0.1121)/l = 0.0053/l.$

Using the 300 sec data and Eq. (10-12b) to calculate the relaxation time, we find that

$$\epsilon_r = (0.0053/l)e^{-300/\lambda}$$
$$= (0.1174 - 0.1171)/l,$$
$$e^{300/\lambda} = (0.0053)/(0.0003),$$
$$\lambda = 105 \text{ sec}.$$

After 120 sec,

$$-\epsilon_r = -(0.0053/l)e^{-120/105},$$
$$\epsilon_r = 0.0017/l.$$

Remaining elongation = 0.0017 in. ◀

10-5 THERMOELASTICITY

Under certain conditions, elastic strain energy can be converted into thermal energy. The effect is called thermoelasticity. We shall consider two examples and then look at the energy losses which may arise.

If a nonpolymeric material is adiabatically* stressed in tension, it will be strained as shown in Fig. 10-16 ($O \rightarrow A$). A slight, but finite, volume increase accompanies this tension because Poisson's ratio is less than 0.5 (cf. Example 10-1). This requires a small amount of PV energy which must come from thermal sources. As a result, there is a small temperature drop of the specimen. Now, if heat is supplied at constant stress and the temperature is raised to the original value, the temperature increase permits more expansion ($A \rightarrow B$). Of course, if the material had been stressed isothermally, the (σ/ϵ)-curve would have moved directly from O to B. Adiabatic stress removal reverses the sequence to A', with final thermal contraction from A' to O completing the loop. With slow stressing, the modulus

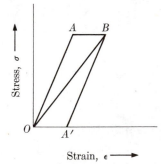

10-16 Thermoelastic deformation. *OA*—adiabatic deformation; *OB*—isothermal deformation. (See text.)

---

* Without a change in the heat content of the specimen.

of elasticity is σ/ϵ_B; with extremely rapid stressing, which can occur in a vibrating plate or rotating shaft, the modulus of elasticity is σ/ϵ_A. At intermediate rates, an elastic hysteresis loop develops. This will be discussed in Section 10-6.

Rubbers and other elastomers have two characteristics which require another interpretation of their thermoelastic behavior. First, elastic elongation aligns the molecules and permits greater crystallinity. This ordering gives off a heat of fusion (Eq. 3-16) which more than counteracts the PV energy produced by volume changes. Thus the temperature is *increased* during rapid loading. Second, *stretched* rubber has a negative thermal expansion in the direction of stressing (Section 10-3), so that the eventual temperature equalization permits additional elongation. Therefore, the sequence of Fig. 10-16 is applicable for rubbers as well as for other materials, although for somewhat different reasons.

Thermoelastic effects are anelastic because time is required to diffuse the heat necessary to equalize the temperature. As the heat diffuses, there are corresponding dimension changes.

Example 10-6

Detect evidence of crystallization of a rubber band during deformation.

Answer. This is a simple experiment using a heavy but easily deformable rubber band. Your lip can serve as a sensitive detector of temperature changes. Place the band in contact with your lower lip. Stretch it rapidly, noting that heat is transferred to your lip. After the band cools to your lip temperature, return it quickly (without snapping it) to its original length, noting that it feels cool to your lip.

Adiabatic crystallization during deformation, like all crystallization, evolves a heat of fusion (Eq. 3-16). This heat appears as a temperature rise, but is eventually dissipated to the surroundings. During contraction, a heat of fusion is required to increase the entropy (disorder) within the rubber band. This energy change decreases the temperature of the rubber band.

10-6 ANELASTIC ENERGY LOSSES (Damping)

Elastic deformation requires energy because a force is applied over a distance (even though it is a short distance). This energy may be represented graphically as the area under the elastic stress-strain curve. When Hooke's law applies, i.e., when the (σ/ϵ)-curve is linear, the elastic energy per unit volume, U, may be expressed as

$$U = \tfrac{1}{2}\sigma\epsilon, \qquad (10\text{-}14a)$$

or as

$$U = \sigma^2/2Y, \qquad (10\text{-}14b)$$

where Y is Young's modulus.

If time is not a factor, all of the elastic energy is recoverable when the stress is removed. Under anelastic conditions, however, there may be an energy loss. This was illustrated for the extreme case in Fig. 10-16, where the stress-strain cycle formed a hysteresis loop. The area within the loop represents an energy dissipation:

$$\Delta U = \int \sigma \, d\epsilon. \qquad (10\text{-}15)$$

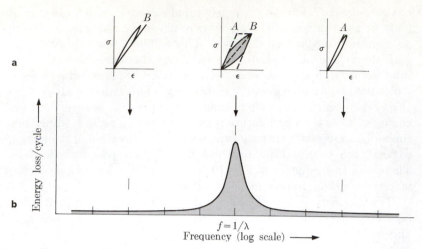

10-17 Energy losses versus frequency. Losses are greatest when the frequency, *f*, corresponds to the reciprocal of the relaxation time, λ. At lower frequencies, deformation is isothermal; at higher frequencies, deformation is adiabatic.

In practice, the ideal "Carnot-like" cycle of Fig. 10-16 is seldom encountered. Figure 10-17(a) illustrates three possibilities. With extremely fast cycling, the time is so short that there is no time-dependent strain and the adiabatic line is approximated in both directions, $O \to A \to O$. With very slow stressing, the isothermal curve is closely followed, $O \to B \to O$. An intermediate cycle period is shown in the center sketch of Fig. 10-17(a), where energy consumption is high. Since there are gradations, the energy consumption versus frequency can be plotted as in Fig. 10-17(b).

To understand this better, let us consider a cyclic stress application where

$$\sigma = \sigma_m \sin \omega t. \tag{10-16}$$

Here, σ is the stress at any particular time, t; σ_m is the maximum elastic stress, and ω is the angular velocity of stress cycling. With no anelastic strain, the strain, ϵ, is in phase with the stress:

$$\epsilon = \epsilon_m \sin \omega t, \tag{10-17a}$$

and Fig. 10-18(a) applies. However, with a lag for time-dependent strain,

$$\epsilon = \epsilon_m \sin (\omega t - \phi). \tag{10-17b}$$

The term ϕ is called the *phase angle*, and the hysteresis loop of Fig. 10-18(b) is applicable. The maximum stress and the maximum strain are not coincident. Also, a "coercive" stress is required to return the strain to zero.

The energy loss per cycle, ΔU, is the area of the hysteresis loop; it is obtained by substituting Eqs. (10-16) and (10-17b) in Eq. (10-15):

$$\Delta U = \int_0^{2\pi} \sigma \, d\epsilon = \int_0^{2\pi} \sigma_m \epsilon_m \sin \omega t \, d(\sin (\omega t - \phi))$$

$$= \pi \sigma_m \epsilon_m \sin \phi. \tag{10-18}$$

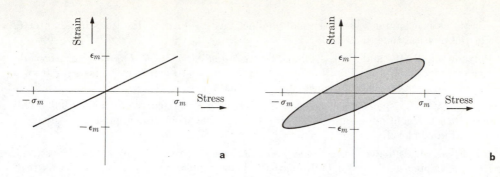

10-18 Stress-strain cycle. (a) Loss-free. (b) High loss. Equation (10-17b) applies. [*Note.* The stress-strain axes have been interchanged to be comparable with $\mathcal{E}$-$\mathcal{P}$ losses in Chapter 13 and *B-H* losses in Chapter 16.]

Since the maximum elastic strain energy is $\sigma_m \epsilon_m/2$ from Eq. (10-14), each cycle dissipates

$$\Delta U/U = 2\pi \sin \phi \qquad (10\text{-}19)$$

as the fractional energy loss.

The importance of Eqs. (10-18) and (10-19) becomes apparent in several engineering situations. In a rubber tire, for example, the strain lags the stress sufficiently so that there is a large hysteresis loop as pictured in Fig. 10-18(b). The total energy dissipated is large at high speeds (many stress cycles per second) or with underinflation (high strain energy). This energy loss shows up as heat in the tire, thus weakening the tire.

As a second example, vibrations can be damped through the ΔU energy dissipation. In this respect, lead is a good damping medium because it possesses a large phase angle, ϕ, at sound frequencies. As a final example, these energy losses and their activation energies (Eq. 10-13) are used for the experimental determination of atom movements within solids; the analysis of carbon-atom movements in Fig. 10-14 was obtained by *internal-friction* or *damping-capacity* measurements.

REFERENCES FOR FURTHER READING

Biggs, W. D., *The Mechanical Behavior of Engineering Materials.* Oxford: Pergamon, 1965. Paperback. Chapters 4 and 5 provide alternate reading on elastic behavior. Introductory level.

Cottrell, A. H., *The Mechanical Properties of Matter.* New York: Wiley, 1964. Chapter 4 presents a good discussion of elasticity. For the advanced undergraduate and instructor.

Dieter, G. E., *Mechanical Metallurgy.* New York: McGraw-Hill, 1961. Stress-strain relationships for elastic behavior are presented in Chapter 2. Advanced level.

Hayden, H. W., W. G. Moffatt, and J. Wulff, *The Structure and Properties of Materials: III. Mechanical Behavior.* New York: Wiley, 1965. Paperback. Chapter 2 on elastic properties and Chapter 3 on anelasticity give introductory presentations.

Kittel, C., *Introduction to Solid State Physics,* third edition. New York: Wiley, 1966. Elastic constants and elastic waves are presented in Chapter 4, phonons and lattice vibrations in Chapter 5, and thermal conductivity by lattice vibrations in Chapter 6. For the physics student.

Polakowski, N. H., and E. J. Ripling, *Strength and Structure of Engineering Materials*. Englewood Cliffs, N. J.: Prentice-Hall, 1966. Chapter 10 emphasizes elasticity from a continuum point of view. Advanced undergraduate level.

Richards, C. W., *Engineering Materials Science*. San Francisco: Wadsworth, 1961. This book ties the continuum approach with the internal-structural approach. Advanced undergraduate level.

Rosenthal, D., *Introduction to Properties of Materials*. Princeton, N. J.: Van Nostrand, 1964. Chapters 6, 7, and 8 of Rosenthal's book supplement Chapter 10 of this text. Undergraduate level.

Teggart, W. J. M., *Elements of Mechanical Metallurgy*. New York: Macmillan, 1966. Elastic properties are presented in Chapter 4. Advanced undergraduate and graduate level.

Ziman, J., "The Thermal Properties of Materials," *Scientific American*, **217** [3], September 1967, pp. 179–188. Thermal conductivity by lattice vibrations. Undergraduate level.

PROBLEMS

10-1 A steel block which has the same elastic moduli as iron is compressed 300,000 psi (a) hydrostatically, (b) uniaxially. What are the changes in volume?

Answer. a) -1.3 v/o   b) -0.44 v/o

10-2 When iron is compressed hydrostatically with 22,500 psi, its volume is changed by 0.10%. On the basis of this information and Poisson's ratio of Table 10-2, (a) what is Young's modulus and (b) how much will the volume of the iron change when it is stressed axially with 90,000 psi?

10-3 An alloy has a compressibility of 10^{-7} in^2/lb and a Young's modulus of 1.5×10^7 psi. What is the shear modulus?

Answer. 6,000,000 psi

10-4 Assume that copper has a modulus of elasticity of 16,000,000 psi, a Poisson's ratio of 0.3, and is under a tensile stress of 12,000 psi. What are the dimensions of the unit cell? (Stress is parallel to the axes.)

10-5 A test bar 0.5051 in. in diameter, with 2-in. gage length, is loaded elastically with 35,000 lb and is elongated 0.014 in. Its diameter is 0.5040 in. under load.
   a) What is the bulk modulus of the bar?
   b) What is the shear modulus?

Answer. a) 22,000,000 psi   b) 9,500,000 psi

10-6 A 1.0021-cm^3 cube of iron is compressed hydrostatically with a 225,000-psi pressure. Its volume is reduced 0.0099 cm^3 to 0.9922 cm^3. Determine (a) the compressibility and (b) Poisson's ratio.

10-7 Compare the maximum velocity of sound in copper and aluminum.

Answer. $v_{Cu} = 4.65 \times 10^5$ cm/sec; $v_{Al} = 5.3 \times 10^5$ cm/sec. (*Note.* These values are for longitudinal waves, since $Y > G$, and for the [111] direction.)

10-8 Compare the minimum velocity of longitudinal waves in copper and aluminum.

10-9 Fifty percent of the anelastic strain in a rubber is recovered in 73 sec after the stress is removed.
  a) What is the relaxation time?
  b) How much anelastic strain remains after an additional 18 sec?

Answer. a) 105 sec   b) 42%

10-10 After a 7000-psi stress was applied to a plastic, the immediate strain was 0.0191 in./in. The total elastic strain rose to 0.0252 after 2 min, and to 0.0270 in./in. after a second 2-min period. At that time the stress was removed. How long would it take for this strain to drop to 0.0010 in./in.? (Assume that the relaxation time is the same for both elongation and contraction.)

10-11 A plastic has a Young's modulus of 200,000 psi and a Poisson's ratio of 0.4 above its glass-transition temperature. It is compressed uniaxially by 10,000 psi within a rigid steel cylinder; consequently, it cannot undergo lateral expansion. What is the axial strain after sufficient time is available to remove the frictional wall stresses and homogenize the stress distribution?

Answer. 3 l/o

10-12 A 20,000-ft steel wire is lowered vertically into the ocean.
  a) How strong must the wire be to support itself without yielding?
  b) How much longer will it be when it is fully extended?

10-13 An aluminum rod is stress-cycled 30 cps from $+2000$ psi to -2000 psi according to Eq. (10-16). The strain phase angle is 0.1°. If no heat is lost, how much will the temperature rise in 10 min?

Answer. 0.1°C

10-14 The temperature of an iron rod rises 0.5°C under the conditions of Problem 10-13. What is the strain phase angle?

11 □ plastic deformation

11-1 PERMANENT DEFORMATION IN CRYSTALS

Irreversible deformation in crystals is called *plastic deformation*. On a macroscale, plastic deformation is revealed by a permanent set, or strain. On an atomic scale, plastic deformation commonly occurs by slip, or shearing of the crystal lattice on crystal planes. In general, however, the atoms retain the same geometric coordination as before deformation, because the slip displacements occur over an integral number of interatomic distances.

In this chapter we shall examine the deformation mechanism, both on the basis of crystal planes and on the basis of atom movements. Our considerations must account for the strength of materials as a function of (1) interatomic forces, (2) alloying elements, and (3) the amount of deformation. Likewise, we will examine the softening which is caused by annealing after plastic deformation.

11-2 PLASTIC SLIP IN ELEMENTAL CRYSTALS

Plastic deformation can be studied most simply by considering single crystals of only one kind of element. The fcc, bcc, and hcp metals provide good examples. Figure 11-1 shows a rod of a single crystal of an hcp metal which was pulled in tension. X-ray analysis verified that slip occurred by displacement on the (0001) planes, as shown in the (b) and (c) parts of the figure.

From the above example and comparable experiments on single crystals of other metals, metallurgists have concluded that slip occurs most readily on the planes and in the directions listed in Table 11-1. The two facts which stand out

TABLE 11-1
SLIP SYSTEMS IN METALS

Structure	Examples	Slip plane	Slip direction	Number of combinations
bcc	α-Fe, Mo, Na, W	$\{101\}$	$\langle \bar{1}11 \rangle$	12
bcc	α-Fe, Mo, Na, W	$\{211\}$	$\langle \bar{1}11 \rangle$	12
fcc	Ag, Al, Au, Cu, γ-Fe, Pb	$\{111\}$	$\langle \bar{1}10 \rangle$	12
hcp	Cd, Mg, α-Ti, Zn	$\{0001\}$	$\langle 11\bar{2}0 \rangle$	3
hcp	α-Ti	$\{10\bar{1}0\}$	$\langle 11\bar{2}0 \rangle$	6

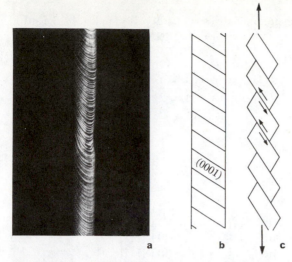

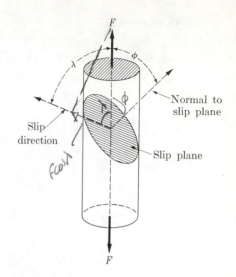

11-1 Slip in a single crystal (hcp). Slip paralleled the (0001) plane, which contains the shortest Burgers vector. (Constance Elam, *The Distortion of Metal Crystals,* Oxford: Clarendon Press.)

11-2 Resolved shear stresses. Schmid's law (Eq. 11-1) defines the shear stress, τ, in terms of the tensile stress, σ_t, or F/A.

in that table are the following: (1) the slip direction for each crystal is the direction with the highest linear density, or shortest Burgers vector (Section 6-6); and (2) the slip planes are planes which have a wide interplanar spacing.

Resolved shear stresses. The easy slip planes can operate only if the stresses exceed a specific value, the *critical shear stress.* Obviously, a shear stress does not exist if either the slip plane or slip direction is perpendicular to, or parallel to, the uniaxial-stress axis. The greatest shear stress occurs when both the slip plane and slip direction are at 45° to this stress axis. This is shown in Fig. 11-2, where ϕ is the angle between the uniaxial-stress axis and the normal to the slip plane. Likewise, λ is the angle between the stress axis and the slip direction. If A is the cross-sectional area, $A/\cos\phi$ is the area of the slip plane; and if F is the applied force, $F\cos\lambda$ is the resolved shear force. Thus the resolved shear stress, τ, may be expressed in terms of an axial stress, σ:

$$\tau = F\cos\lambda/(A/\cos\phi) = \sigma\cos\lambda\cos\phi. \qquad (11\text{-}1a)^*$$

Slip occurs when the critical shear stress, τ_c, is exceeded, or when

$$\pm\tau_c = \sigma\cos\lambda\cos\phi. \qquad (11\text{-}1b)$$

This is known as *Schmid's law.*

---

* The sum $\lambda + \phi$ must be equal to or greater than 90°; the product of their cosines is a maximum when each equals 45°.

Example 11-1

Slip occurs according to Table 11-1 when an axial stress of 100 psi is applied in the [010] direction of an aluminum crystal. What is the critical shear stress for $(111)[\bar{1}10]$ slip?

Answer. Use Schmid's law (Eq. 11-1):

$$\phi = [010] \not\perp [111]; \quad \cos\phi = 1/\sqrt{3}$$
$$\lambda = [010] \not\perp [\bar{1}10]; \quad \cos\lambda = 1/\sqrt{2}$$
$$\tau_c = (100 \text{ psi})/\sqrt{6} = 40.8 \text{ psi.} \quad \blacktriangleleft$$

Example 11-2

A stress is applied in the [110] direction of a bcc iron crystal. Which of the 12 $\{101\}\langle\bar{1}11\rangle$ sets referred to in Table 11-1 will develop a shear stress?

Answer. First we should list the 12 combinations:

$(110)[\bar{1}11]^*$	$(101)[11\bar{1}]$	$(011)[1\bar{1}1]^*$
$(110)[1\bar{1}1]^*$	$(101)[\bar{1}11]^*$	$(011)[11\bar{1}]$
$(1\bar{1}0)[111]\dagger$	$(10\bar{1})[111]$	$(0\bar{1}1)[111]$
$(1\bar{1}0)[11\bar{1}]\dagger$	$(10\bar{1})[\bar{1}1\bar{1}]^*$	$(0\bar{1}1)[1\bar{1}\bar{1}]^*$

The resolved shear stresses in those sets marked (*) will be zero because the slip directions are perpendicular to the stress axis. The sets marked (†) have their plane normals perpendicular to the stress axis. The rest of the sets will possess a resolved shear stress, τ, of $\sigma/\sqrt{6}$.

Note. The dot product of each of the above sets should equal zero if the direction lies within the plane (Eq. 5-10). The reader is asked to verify that other possible permutations, such as $(\bar{1}0\bar{1})[\bar{1}11]$, are parallel to one of the above sets. ◄

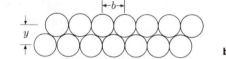

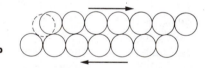

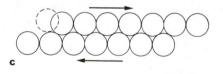

11-3 An assumed mechanism of slip (schematic). Metals actually deform with less shear stress than this mechanism requires.

Theoretical shear strengths. It is possible to estimate the shear stress required to move one plane of atoms over another as shown in Fig. 11-3. First, we note that a shear stress is required to move the planes from Fig. 11-3(a) to 11-3(b). Beyond that point, slip is spontaneous to 11-3(c). Thus the stress sequence presented in Fig. 11-4 is plausible. The period is **b**, the Burgers vector; so, if we assume a sinusoidal relationship, we can calculate the maximum stress at **b**/4.

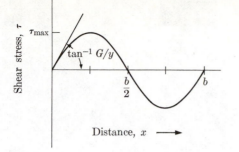

11-4 Stress for slip in a perfect crystal. (See text.)

The slope of the curve near $x = 0$ obeys Hooke's law for shear (Eqs. 10-6 and 10-7):

$$\tau/\gamma = G = \tau y/x,$$

where y is the interplanar spacing. Also, if the relationship is sinusoidal,

$$\tau = \tau_{max} \sin (2\pi x/b). \tag{11-2}$$

For very small strains, Eq. (11-2) becomes

$$\tau = \tau_{max} 2\pi x/b. \tag{11-3}$$

From Fig. 11-3, $y \cong b$, and combining Eqs. (10-7) and (11-3), we obtain

$$\tau_{max} \cong G/2\pi. \tag{11-4}$$

Although the assumptions chosen here can be refined, the results still indicate that the shear stress for slip, as described in Fig. 11-3, should be approximately $(0.1)G!$ If so, the shear strength of steel would exceed 10^6 psi, and that of copper would be approximately 600,000 psi. In reality, the initial shear stresses for single crystals of pure iron and copper are a few hundred psi at the most. Furthermore, according to Fig. 11-4, an elastic strain of about 25% (where $\tau = \tau_{max}$) would be required before a permanent displacement would occur. Thus, while experiments reveal that crystal planes serve as slip planes, *slip does not occur* simply as one crystal "block" moving over another part of the crystal.

Slip by dislocation movements. The extreme weakness of single crystals as compared to their theoretical strength led several independent investigators of the mid-1930's to propose that dislocations are involved in crystal deformation.* Figure 11-5 revises Fig. 11-3 to show a mechanism for slip that involves the displacement of only a few atoms at any one time. The stress required under these conditions is only a fraction of that calculated in Eq. (11-4).

---

* The proof of this hypothesis had to await the development of the transmission electron microscope some 15 to 20 years later (Fig. 6-14a).

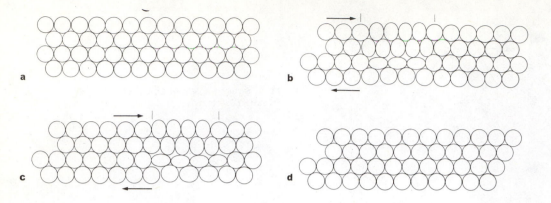

a   b

c   d

11-5 Slip by dislocation movements. In this model only a few atoms are moved from their low-energy positions at any one time. Less stress is therefore required to produce slip.

11-6 Wrinkle analogy to slip. A carpet can be moved across the room with less effort if a wrinkle, or ruck, is worked along the carpet. As with dislocations, another ruck must be formed if the total displacement distance is to be greater than AA'.

11-7 Whiskers (sapphire; ×3500). These are slender single crystals of Al_2O_3 without edge dislocations. Thus they cannot slip in the manner presented in Fig. 11-5, and as a result have strengths approaching the theoretical (Eq. 11-3). (Courtesy of E. Scala.)

    The mechanism of slip by dislocation movements may be compared to moving a carpet across a room by working a wrinkle (or ruck) from one end of the carpet to the other (Fig. 11-6). In this way, the force necessary is much less than if the whole carpet had to be moved at once. However, the total distance through which the force works is much greater.

    Further evidence that the "weakness" of metal crystals is a consequence of dislocation movements is available from studies of *whiskers* (Fig. 11-7). These single crystal filaments of materials can be grown without any edge dislocations. As such, they can undergo elastic elongation of several percent without yielding plastically, compared to much less than one percent normally. Furthermore, dislocation-free whiskers can resist a shear stress of as much as five percent of the shear modulus, $G/20$ (which is several orders of magnitude greater than is observed

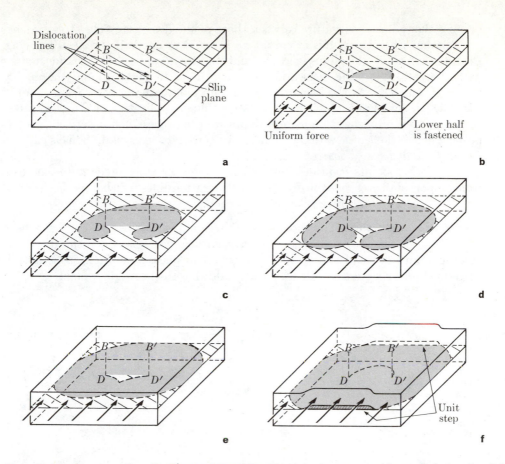

11-8 Dislocation generation. $D-D'$ = anchor points. The stress for movement is greatest when the dislocation loop is semicircular because its radius of curvature is then shortest. As the loop closes on itself (d and e), it forms a second dislocation loop (f). [A. G. Guy, *Elements of Physical Metallurgy* (2nd ed.), Reading, Mass.: Addison-Wesley, 1959, p. 114.]

for regular crystals). There is no doubt that plastic deformation requires dislocation movements.*

Dislocation generation. If the carpet of Fig. 11-6, or the crystal slip of Fig. 11-5, is to glide by more than one unit distance, a series of dislocations must be generated. The following discussion will describe one such mechanism.

Consider Fig. 11-8(a), where a dislocation line lies between two points, D and D', which are fixed. If a shear stress is directed as indicated, the dislocation will bow forward, as shown in Fig. 11-8(b). The shear stress must be increased

* We will observe in Section 11-5 that dislocations interfere with the movements of other dislocations; therefore, plastic deformation is retarded when the dislocation density becomes large.

because the radius of curvature between those points decreases. The stress reaches a maximum when the radius of curvature is equal to one-half the D–D' spacing. Beyond that point, the radius increases and the stress required for shear decreases, with the result that the dislocation loop continues to pivot around the anchor points, D and D'. As the two arms of the loop join [between parts (d) and (e) of Fig. 11-8], a new dislocation is formed. A continued application of the shear stress causes the second dislocation loop to duplicate the process just described, and repeating generations of new dislocations move through the crystal. This is called a *Frank-Read dislocation source.*

The anchor points mentioned above may originate from precipitate particles, from impurity phases, or even from other dislocation lines which have been introduced into the crystal by previous deformation.

Climb and cross slip. The dislocation movements described so far have been limited to one plane, the slip plane. Under suitable conditions, a dislocation loop can move into other planes. This is important in plastic deformation because it permits slip to progress on other planes in addition to the original plane.

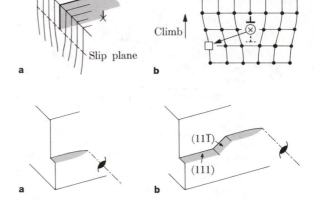

11-9 Dislocation climb. (a) Jog at the edge of the half-plane. If an atom at a jog diffuses to a vacancy, a grain boundary, or an interstitial site, the dislocation can climb (b).

11-10 Dislocation cross slip. A screw dislocation can shift from one plane to another of the same {*hkl*} form. The Burgers vector remains the same.

Dislocation *climb* from one plane to another requires diffusion of atoms (or vacancies). Figure 11-9 shows a jog in the extra half-plane of an edge dislocation. If atoms diffuse away from the edge of the half-plane (or vacancies diffuse to it), there is *positive climb.* Conversely, *negative climb* occurs if extra atoms diffuse onto the edge of the half-plane. Because climb requires diffusion, it becomes increasingly favored at higher temperatures, where thermal energy is available to activate atom movements. The mechanism of dislocation climb is important in creep at high temperatures (Section 21-7).

Cross slip involves the movement of screw dislocations from one plane to another. Figure 11-10(a) shows a screw dislocation in a conventional sketch. Because it lies parallel to the Burgers vector, it can cross-slip or expand on any plane which contains that same Burgers vector. Thus, if the slip plane of Fig. 11-10(a) is a (111) plane of an fcc metal (and the slip direction is $[\bar{1}10]$), the screw dislocation can expand on a $(11\bar{1})$ plane with no extra energy (Fig. 11-10b). This screw-dislocation expansion can subsequently continue onto another (111) plane, which may serve as a new slip plane parallel to the original one. Cross slip does not require diffusion. Therefore, it is not temperature-sensitive, as is dislocation climb. Cross slip is more prevalent than climb at low temperatures.

11-3 PLASTIC SLIP IN COMPOUNDS

An atom in a single-component metal always has like atoms for neighbors, even during deformation. A compound, however, has two or more types of atoms with a preference for unlike neighbors. [Equation (8-1) applies.] On many potential slip planes, deformation brings together like atoms and separates a fraction of the unlike atoms. Within a compound this means higher energy, and it shows up as a resistance to shear. Consequently, the critical shear stresses on some planes are sufficiently high so that slip is essentially impossible. In effect, the number of possible slip systems is reduced, and ductility decreases. This is revealed in Table 11-2, which cites the slip systems for several relatively simple compounds. For example, only six sets are operative in MgO. In more complex compounds such as $Ni_8Fe_{16}O_{24}$, used in magnet ceramics, and $PbZrO_3$, used in piezoelectric transducers, the possibility of slip is nil at normal temperatures; hence, the materials behave in a brittle manner.

TABLE 11-2
SLIP SYSTEMS IN SIMPLE COMPOUNDS

Structure	Examples	Slip plane	Slip direction	Number of combinations
NaCl	LiF, MgO, MnS, TiC	$\{1\bar{1}0\}$	$\langle 110 \rangle$	6
NaCl	PbS	$\{001\}$	$\langle 110 \rangle$	6
NaCl	MnSe	$\{\bar{1}11\}$	$\langle 110 \rangle$	12
CsCl	CsCl	$\{001\}$	$\langle 100 \rangle$	6
Al_2O_3	Al_2O_3	$\{0001\}$	$\langle 11\bar{2}0 \rangle$	3

As revealed in Fig. 4-1, the shortest repeating distance (i.e., Burgers vector) in an NaCl-type crystal is along the several $\langle 110 \rangle$ directions. Quite expectedly, therefore, the $\langle 110 \rangle$ directions have been found by experiment to be the slip directions. The most common slip plane for NaCl-type crystals is one of those in the $\{110\}$ form (Fig. 11-11). This is particularly true in those compounds, such as LiF and MgO, which have small, "nondeformable" ions. Other NaCl-type compounds, such as PbS and MnSe, possess other slip planes, but still have the $\langle 110 \rangle$ slip directions (Table 11-2).

The preference for the $\{1\bar{1}0\}\langle 110 \rangle$ slip system in the oxides and fluorides can be understood from Fig. 11-11. Anion A is able to move past B or C with less net coulombic repulsion than is possible for shear in the $\langle 110 \rangle$ directions of $\{001\}$ or

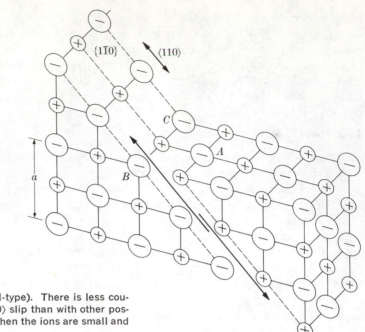

11-11 Slip in compounds (NaCl-type). There is less coulombic repulsion with $\{1\bar{1}0\}\langle 110\rangle$ slip than with other possible slip systems, particularly when the ions are small and "hard," i.e., nonpolarizable.

$\{\bar{1}11\}$ planes. Sulfur and selenium atoms are sufficiently large, however, so that they can be distorted (or polarized—see Chapter 13) as they move past other ions to give less coulombic repulsion than prevails with "hard" ions like O^{2-} or F^-.

We conclude that both intermetallic and ceramic compounds are inherently less deformable than metals, because critical shear stresses are high and the number of slip planes is small. Consequently, fracture stresses are commonly exceeded before plastic deformation is initiated (Chapter 21).

11-4 DEFORMATION BY TWINNING

Twinning, as presented in Section 6-9, may occur as a stacking fault where the sequence of stacking had been reversed in the growth process. Such twins (Fig. 9-19) are called *annealing* twins.

Of interest to us in this section are *mechanical* twins. Although identical in geometry to annealing twins, they arise from deformation rather than from crystal growth. They are illustrated in Fig. 11-12(a), where twinning in an fcc crystal occurs by shear in the $[\bar{1}1\bar{2}]$ direction on a $(\bar{1}11)$ plane. Twinning deformation combinations for bcc and hcp structures are listed in Table 11-3, together with the amount of shear strain which is possible. Four contrasts may be made between deformation by slip and by twinning:

1) Twinning shear is uniform; slip is heterogeneous and occurs on relatively few planes out of many.

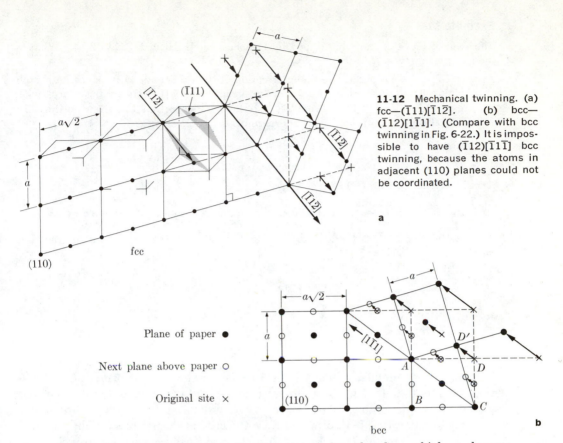

11-12 Mechanical twinning. (a) fcc—$(\bar{1}11)[\bar{1}1\bar{2}]$. (b) bcc—$(\bar{1}12)[1\bar{1}\bar{1}]$. (Compare with bcc twinning in Fig. 6-22.) It is impossible to have $(\bar{1}12)[\bar{1}1\bar{1}]$ bcc twinning, because the atoms in adjacent (110) planes could not be coordinated.

Plane of paper •

Next plane above paper ○

Original site ×

2) Twinning shear can occur only in the direction and on the plane which produces a mirror image (Fig. 11-12b); slip may be in either a positive or negative direction within the slip plane.

3) Twinning shear strain is limited to the values shown in Table 11-3; slip may continue until failure.

4) Twinning causes an abrupt reorientation of the lattice; slip does not.

TABLE 11-3
TWINNING SYSTEMS IN METALS

	Twin direction	Twin plane	Normal plane	Shear strain	
bcc	$[1\bar{1}1]$	$(\bar{1}12)$	(110)	0.707	
fcc	$[11\bar{2}]$	$(\bar{1}11)$	(110)	0.707	
hcp	$[10\bar{1}1]$	$(10\bar{1}2)$	$(1\bar{2}10)$	0.14 Zn*	Compression
				0.17 Cd*	Compression
hcp	$[\bar{1}011]$	$(10\bar{1}2)$	$(1\bar{2}10)$	0.13 Mg*	Tension
				0.19 Ti*	Tension

* (c/a)-values: Zn = 1.86, Cd = 1.89, Mg = 1.624, Ti = 1.59.

Example 11-3

The twin of Fig. 6-22 may be formed by the shear sketched in Fig. 11-12(b). What is the twinning direction?

Answer. The shear is parallel to the intersection of the (110) and ($\bar{1}$12) planes. A sketch of the unit cell will reveal that these planes intercept along a [1$\bar{1}$1] direction. [Check with the cross product (Eq. 5-9).] From Fig. 11-12(b), observe that the atom positions in adjacent (110) planes preclude [$\bar{1}$1$\bar{1}$] and permit only [1$\bar{1}$1] deformation. ◄

Example 11-4

Show that the twinning strain in a bcc crystal is 0.707.

Answer. (Refer to Fig. 11-12b.)

$$\overline{AD'} = \overline{CD} = a, \qquad \overline{AD} = \overline{CD'} = a\sqrt{2}, \qquad \text{and} \qquad \overline{AC} = a\sqrt{3}.$$

Place E and E' on $\overline{AC}$ so that $\overline{ED} \perp \overline{AC}$ and $\overline{E'D'} \perp \overline{AC}$. Solving by similar triangles, we obtain

$$\overline{D'E'} = \overline{DE} = a\sqrt{2/3},$$

$$\overline{AE} = \overline{CE} = a/\sqrt{3} = \overline{AC}/3,$$

$$\overline{EE'} = a/\sqrt{3} = \overline{DD'},$$

$$\gamma = \frac{\overline{DD'}}{DE} = \frac{a/\sqrt{3}}{a\sqrt{2/3}} = 0.707. \quad ◄$$

Twinning in hcp metals. Usually the critical shear stress for deformation by twinning is higher than for slip. Therefore, we do not find twinning except in situations where slip is unfavorable. This may occur with deformation at low temperatures or at high strain rates, both of which raise the critical shear stress for slip. It may also occur in hcp metals (e.g., Be, Cd, Mg, Ti, and Zn) because there are relatively few possible slip systems. None of the hexagonal slip systems listed in Table 11-1 will be activated by tensile or compressive stresses along the c-axis, or [0001] direction. The most important twinning mode in hcp metals is $\{10\bar{1}2\}\langle10\bar{1}\bar{1}\rangle$. The

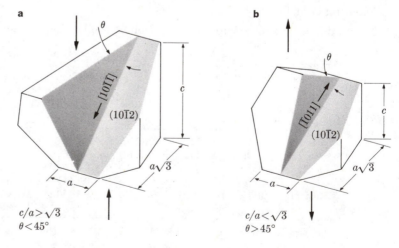

11-13 Twinning in hcp structures. (a) $c/a > \sqrt{3}$. (b) $c/a < \sqrt{3}$. The former can twin only by compression, and the latter only by tension in the ⟨0001⟩ directions. (Alternatively, the former can twin only by tension, and the latter only by compression in the ⟨10$\bar{1}$0⟩ directions.)

$c/a > \sqrt{3}$
$\theta < 45°$

$c/a < \sqrt{3}$
$\theta > 45°$

response of hcp metals to this form of twinning depends on whether the c/a ratio for the unit cell is greater than, or less than, $\sqrt{3}$.

Cadmium and zinc have (c/a)-values of 1.89 and 1.86, respectively. Therefore, as shown in Fig. 11-13(a), the mirror image in these metals must be formed by compression along the c-axis. The (c/a)-value of magnesium is 1.624; thus the only way the twin can form is by tension along the c-axis (or $[10\bar{1}0]$ compression), because the twinning direction ($\bar{1}011$) is more than 45° from the c-axis of the crystal $[0001]$ (Fig. 11-13b).

Example 11-5

A grain of magnesium within a polycrystalline wire is oriented so that the direction of tensile stress, σ_t, is in the $(1\bar{2}10)$ plane but 7.5° from the $[0001]$ direction. (In this case, the tension direction is 86° from the $[\bar{2}110]$ slip direction.)

a) Compare the resolved shear stresses for twinning and for slip.

b) Assuming that the critical stresses for slip and twinning are 80 psi (0.55×10^7 dynes/ cm²) and 500 psi (3.45×10^7 dynes/cm²), respectively, which mechanism will initiate plastic deformation?

c) How much tensile stress is required?

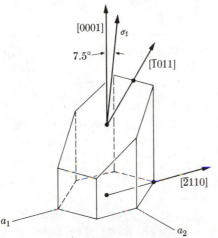

11-14   See Example 11-5.

Answer. Refer to Fig. 11-14 and to the data for magnesium in Table 11-3.

a)
$$\lambda_{\text{twin}} = \tan^{-1}\left(\frac{a\sqrt{3}}{c}\right) - 7.5° = 39.5°,$$

$$\phi_{\text{twin}} = \tan^{-1}\left(\frac{c}{a\sqrt{3}}\right) + 7.5° = 50.5°,$$

$$\tau_{\text{twin}} = \sigma_t \cos 39.5° \cos 50.5° = 0.49\sigma_t;$$

$$\lambda_{\text{slip}} = 86°, \qquad \phi_{\text{slip}} = 7.5°,$$

$$\tau_{\text{slip}} = \sigma_t \cos 86° \cos 7.5° = 0.07\sigma_t.$$

b) and c) $\qquad \sigma_t$ for twinning $= 500 \text{ psi}/0.49$

$$= 1020 \text{ psi} \quad (0.7 \times 10^8 \text{ dynes/cm}^2),$$

$\qquad\qquad \sigma_t$ for slip $\qquad = 80 \text{ psi}/0.07$

$$= 1150 \text{ psi} \quad (0.8 \times 10^8 \text{ dynes/cm}^2).$$

Therefore, the first plastic deformation will occur by twinning. Had the stresses been compressive, however, [$10\bar{1}2$] twinning could *not* have occurred (Fig. 11-13b). Hence, the crystal would have deformed by slip. ◄

11-5 STRAIN HARDENING

When the stress-strain curve was presented in Fig. 1-3, we noted that the resistance to deformation increases as the amount of plastic strain is increased. In other words, the first plastic deformation occurs relatively easily; then higher stresses are required for further deformation. We call this strain hardening.

Strain hardening follows the empirical relationship,

$$\sigma_{\text{tr}} = A(B + \epsilon_{\text{tr}})^n, \tag{11-5a}*$$

or

$$\ln \sigma_{\text{tr}} = \ln A + n \ln (B + \epsilon_{\text{tr}}), \tag{11-5b}$$

where the subscript, tr, refers to true stress and true strain. The constants, A, B, and n must be evaluated experimentally for each material. Of these, the *strain-hardening exponent*, n, is most significant because it gives the clearest indication of the variation of strength with strain. It is always less than one, so that the strength does not increase in direct proportion to the amount of strain.

Cold work. For convenience, we define the amount of strain which is incorporated into a material during mechanical processing at ambient temperatures as cold work, CW:

$$\text{CW} = \frac{A_o - A_f}{A_o}. \tag{11-6}$$

The subscripts refer to the original and final cross-sectional areas, before and after deformation. As shown in Fig. 11-15, cold work increases strength and hardness, but decreases ductility. The latter correlation is a logical consequence, because if the material has been deformed in the cold-working process, the sample which is selected for elongation or reduction-of-area tests (Table 1-1) will need to undergo less additional deformation before fracture.

Strain hardening is important in engineering practice: (1) it provides a practical method for strengthening metals (Chapter 20); (2) the properties of the metal change during deformation, and care must be taken to avoid cracking as a result of the loss in ductility; and (3) the cold-working process requires a greater input of power in the later stages of deformation.

---

* Sometimes simplified to $\sigma_{\text{tr}} = A \epsilon_{\text{tr}}^n$.

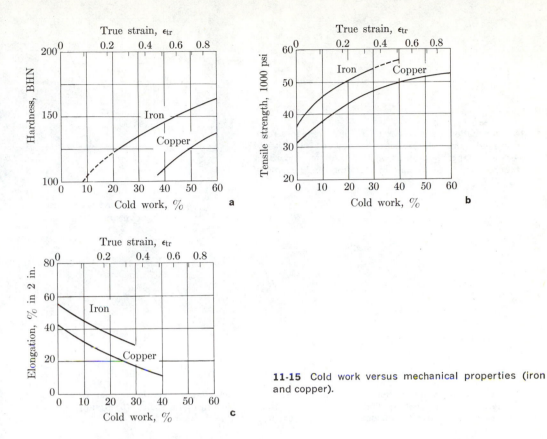

11-15 Cold work versus mechanical properties (iron and copper).

Hardening mechanisms. Figure 11-16 shows that there is a direct relationship between the number of dislocations per cm^2 (or length of dislocations per cm^3) and the strength of copper. Similar data are available for other metals. Thus, although dislocations are required for plastic deformation and contribute to the easy yielding of materials (Section 11-2), dislocations can also contribute to strength if their numbers are great.

Several mechanisms for strain hardening have been proposed. Since each predicts that the resolved shear stress required for deformation (i.e., strength) is proportional to G, **b**, and the square root of the dislocation density, $\rho_d^{1/2}$, it has not been possible to formulate critical experiments to ascertain the exact mechanism of strain hardening. Two of the several mechanisms proposed will illustrate the possibilities. One of them suggests that back stresses in a *dislocation pile-up* (Fig. 11-17a) tend to prevent further generation of dislocations. This arises because dislocations have a stress field around them (Fig. 6-13). Thus the first dislocations to be generated interfere with the formation and movements of subsequent dislocations. Though climb and glide divert dislocations to other planes, as the numbers of dislocations multiply their movements become restricted and the stress necessary to cause further deformation increases noticeably. A second

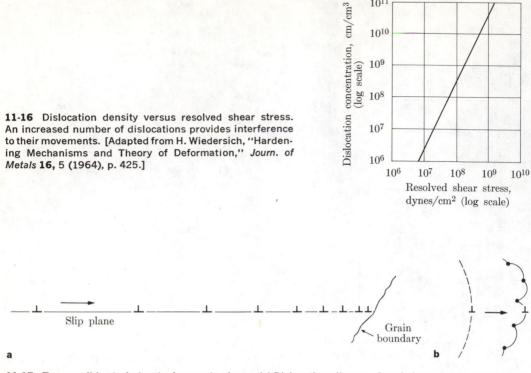

11-16 Dislocation density versus resolved shear stress. An increased number of dislocations provides interference to their movements. [Adapted from H. Wiedersich, "Hardening Mechanisms and Theory of Deformation," *Journ. of Metals* **16,** 5 (1964), p. 425.]

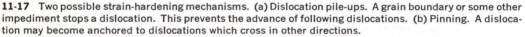

11-17 Two possible strain-hardening mechanisms. (a) Dislocation pile-ups. A grain boundary or some other impediment stops a dislocation. This prevents the advance of following dislocations. (b) Pinning. A dislocation may become anchored to dislocations which cross in other directions.

explanation envisions dislocations on one plane as forming obstacles or "forests" which inhibit the movements of other dislocations, which must cut through them. These points of intersection can be regarded as anchor points which interfere with the generation of new dislocations and their movements (Fig. 11-17b).

11-6 ANNEALING, RECOVERY, AND RECRYSTALLIZATION

It is well known that strain hardening can be removed by heating or annealing. The metallurgist uses *annealing* to soften metals when more ductility is required. Figure 11-18 shows the changes in hardness as a function of annealing temperatures. The pronounced softening at about half the absolute melting temperature is associated with the *recrystallization temperature* because new crystals, or grains, are formed (Fig. 11-19). Even before recrystallization has started, however, internal changes have begun to occur, and the early stages of a *recovery* of the strain-free properties may be detected. For example, the additional electrical resistivity which accompanies strain hardening (Section 14-3) starts to disappear.

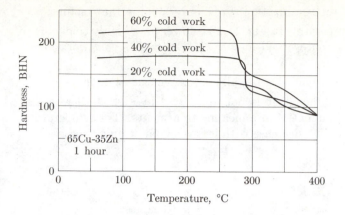

11-18 Softening during annealing (65Cu-35Zn brass). The more highly strain-hardened brass softens at a lower temperature and with less thermal energy during the one-hour heating. (ASM data.)

11-19 Recrystallization of strain-hardened brass (×40). (Courtesy of J. E. Burke, General Electric Co., Schenectady, N.Y.)

Studies of recovery and recrystallization reveal several important factors:

1) The recrystallization temperature in Fig. 11-18 is lower for a highly cold-worked metal than for a lightly cold-worked metal (275°C for 60% cold work versus 290°C for 40% cold work).

2) Above the recrystallization temperature, all curves eventually merge.

3) Time is also important, although it is not illustrated in Fig. 11-18 where all the data were obtained from one-hour annealing treatments. The temperatures necessary for recrystallization decrease with increasing annealing times.

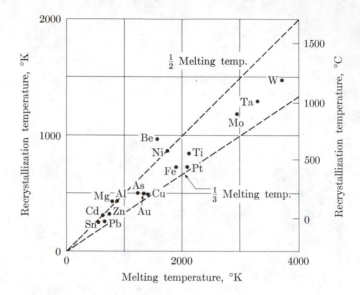

11-20 Recrystallization temperature versus melting temperature. The average recrystallization temperature is roughly one-half the absolute melting temperature.

The atomic mechanism of recovery and recrystallization can be readily pictured. Since atoms can move more easily at higher temperatures, dislocations can climb out of slip planes, or can be removed by recombination where positive (⊥) and negative (⊤) dislocations annihilate each other.* Thus the amount of disorder within a crystal can be reduced. New grains can also be nucleated along slip planes (Fig. 11-19b). These new grains are essentially free of dislocations and therefore soft and ductile. Inasmuch as both dislocation elimination and recrystallization require an activation energy for atom movements, it is not surprising to find a correlation between the recrystallization temperature and the melting temperature (Fig. 11-20).

---

* A positive edge dislocation (⊥) has an extra half-plane above the slip plane, while a negative dislocation (⊤) has an extra half-plane below.

Hot-working versus cold-working of metals. In production operations, the distinction between *hot-working* and *cold-working* does not rest on temperature alone, but on the relationship of the processing temperature to the recrystallization temperature. Hot-working is performed above the recrystallization temperature; cold-working is performed below it. Thus the temperature for cold-working copper may be higher than for hot-working lead.

The choice of the recrystallization temperature for distinguishing between hot- and cold-working is quite logical from the production point of view. Below the recrystallization temperature, the metal becomes stronger, harder, and less ductile with additional mechanical working. More power is required for deformation, and there is a greater chance of cracking during the process. Above the recrystallization temperature, the metal anneals itself during, or immediately after, the mechanical working. It remains soft and relatively ductile.

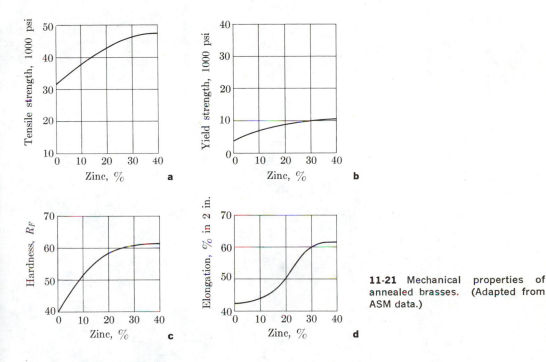

11-21 Mechanical properties of annealed brasses. (Adapted from ASM data.)

11-7 SOLUTION HARDENING

Solid solutions have higher strengths than their solute-free counterparts, as shown in Fig. 11-21 for brass and Fig. 11-22 for Cu-Ni alloys. This fact has engineering importance, not only for alloy strengthening (Chapter 20) but also because it may make it possible to use a solute which is cheaper than the pure element. For example, since zinc costs less than half as much as copper, a brass is not only stronger than pure copper but also is normally less expensive. (Brass also has lower electrical

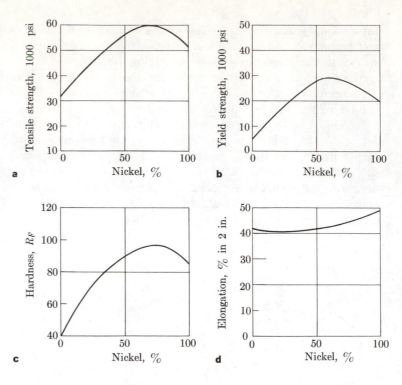

11-22 Mechanical properties of annealed copper-nickel alloys. (Adapted from ASM data.)

a

b

c

d

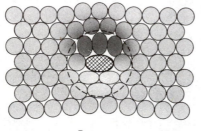

Larger atom

Smaller atom

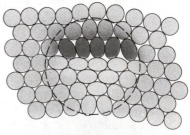

Same size atom

11-23 Solute atoms and dislocations. An odd-sized atom decreases the stress around a dislocation. As a result, the dislocation requires more stress for continued movement.

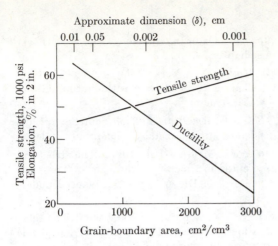

Approximate dimension (δ), cm

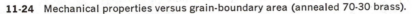

11-24 Mechanical properties versus grain-boundary area (annealed 70-30 brass).

conductivity, for reasons to be discussed in Chapter 14; therefore, it cannot replace copper in all applications.)

Dislocations become "locked" onto solute atoms because the latter can reduce the strain energy of the dislocation. This is shown schematically in Fig. 11-23. Smaller atoms become stabilized at the edge of the extra half-plane; likewise, larger atoms can reduce the strain energy when they are present in the tension zone below the dislocation. In either situation, the strain energy is reduced, and a greater shear force is required to pull the dislocations from the solute and to continue their movements.

Other aspects of solution hardening will be presented in Chapter 20 on strengthening mechanisms.

11-8 PLASTIC DEFORMATION OF POLYCRYSTALLINE MATERIALS

The grain boundary presents a discontinuity which interferes with slip in a polycrystalline material, particularly if the temperature is low enough so that the dislocations cannot climb out of their initial slip planes. Thus fine-grained materials with a large boundary area per unit volume are stronger than coarse-grained materials (Fig. 11-24). N. J. Petch has expressed this in terms of grain dimensions, δ:

$$\sigma_y = \sigma_i + k/\sqrt{\delta}, \tag{11-7}$$

where σ_y is the yield stress and σ_i is the "inherent strength" of the material in single-crystal form. Both σ_i and k are experimentally determined constants.

At high temperatures, a grain boundary may weaken a material rather than strengthen it by producing *creep*, which will be considered in Chapter 21.

REFERENCES FOR FURTHER READING

Biggs, W. D., *The Mechanical Behavior of Materials*. Oxford: Pergamon, 1965. Paperback. Introductory supplementary reading for this chapter.

Cottrell, A. H., *The Mechanical Properties of Matter*. New York: Wiley, 1964. Chapter 9 discusses the crystallography of plasticity. Chapter 10 ties this in with continuum mechanics. For the advanced undergraduate and instructor.

Cottrell, A. H., "The Nature of Metals," *Scientific American*, **217** [3], September 1967, pp. 80–100. Special attention is given to the deformation of metals. Undergraduate level.

Dieter, G. E., *Mechanical Metallurgy*. New York: McGraw-Hill, 1961. The structural fundamentals for plastic deformation are presented in Chapters 4 through 6. These are tied in with the continuum approach of Chapter 3 in Dieter's book. Advanced undergraduate level.

Edelglass, S. M., *Engineering Materials Science*. New York: Ronald, 1966. The morphology of plastic deformation and of dislocations is presented in Chapter 9, and plastic flow in Chapter 10. These two chapters are at a sophomore-junior level.

Friedel, J., *Dislocations*. Reading, Mass.: Addison-Wesley, 1964. Advanced-level presentation of dislocation theory.

Gilman, J. J., "Deformation and Fracture of Ionic Crystals," *Progress in Ceramic Science*, Vol. 1 (J. E. Burke, Ed.). Oxford: Pergamon, 1961. An excellent review of deformation mechanisms in simple compounds. Advanced undergraduate and graduate level.

Hayden, H. W., W. G. Moffatt, and J. Wulff, *The Structure and Properties of Materials: III. Mechanical Behavior*. New York: Wiley, 1965. Paperback. An excellent introductory presentation of dislocations, microplasticity, and plastic deformation.

Hull, D., *Introduction to Dislocations*. Oxford: Pergamon, 1965. Paperback. A thorough discussion of dislocations and their movements. In general, it is nonmathematical, but the reader needs the ability to visualize crystals in three dimensions. For the advanced undergraduate and instructor.

McClintock, F. A., and A. S. Argon, *Mechanical Behavior of Materials*. Reading, Mass.: Addison-Wesley, 1966. Chapter 4 discusses dislocation mechanics and Chapter 5 discusses plastic deformation in crystalline materials. Advanced undergraduate level. Excellent reference list.

Rosenthal, D., *Introduction to Properties of Materials*. Princeton, N. J.: Van Nostrand, 1964. Plasticity is presented in Chapter 9. Undergraduate level.

Teggart, W. J. M., *Elements of Mechanical Metallurgy*. New York: Macmillan, 1966. Paperback. Plastic properties of single crystals and polycrystalline aggregates are presented in Chapters 5 and 6. Advanced undergraduate level.

Weertman, J., and J. R. Weertman, *Elementary Dislocation Theory*. New York: Macmillan, 1964. Paperback. The physics and mechanics of dislocations are presented. Advanced undergraduate level.

PROBLEMS

11-1 A tensile stress of 1000 psi is applied in the [112] direction of an iron (bcc) crystal. What is the shear stress in the [010] direction of the (001) plane?

Answer. 333 psi

11-2 A shear stress of 600 psi will produce $(1\bar{1}1)[011]$ slip in a certain silver crystal. What compressive stress in the [111] direction will produce slip of the above orientation?

11-3 Copper is an fcc metal and deforms by slip on {111} planes and in $\langle 1\bar{1}0 \rangle$ directions. A single crystal is stressed in tension along its [110] direction until slip starts.
 a) On which {111} planes and in which $\langle 110 \rangle$ directions will slip occur?
 b) If the critical shear stress for slip is 120 psi, what is the applied tensile stress?

Answer. a) $[101](11\bar{1})$; $[10\bar{1}](111)$; $[011](11\bar{1})$; $[01\bar{1}](111)$ b) 294 psi

11-4 The critical shear stress, τ_c, for $\{110\}\langle 1\bar{1}0 \rangle$ slip in MgO is 1300 psi. No other slip system operates.
 a) In which, if any, of the $\langle 111 \rangle$ directions will compressive stresses produce slip?
 b) What compressive stress is required?

11-5 Aluminum deforms by a $\{111\}\langle 1\bar{1}0 \rangle$ slip mechanism. Which of the 12 slip combinations could operate from a stress parallel to [111]?

Answer. $(11\bar{1})[101]$, $(11\bar{1})[011]$, $(1\bar{1}1)[110]$, $(1\bar{1}1)[011]$, $(\bar{1}11)[110]$, $(\bar{1}11)[101]$, and their negative counterparts

11-6 If a single crystal of aluminum were tested in tension along the [111] direction, on which of the following slip systems would you expect slip to occur first?
 a) $(111)[1\bar{1}0]$ b) $(11\bar{1})[101]$
 c) $(1\bar{1}1)[110]$ d) $(1\bar{1}1)[10\bar{1}]$

11-7 One slip system of NaCl is $(110)[1\bar{1}0]$.
 a) If the critical shear stress is 100 psi, what is the lowest tensile stress necessary to produce slip on this system?
 b) In what direction should this force be applied?

Answer. a) 200 psi   b) [100] or [010]

11-8 An fcc crystal of aluminum is stressed in the [101] direction with 615 psi tension. Consider possible slip in the following systems: $(111)[\bar{1}01]$, $(101)[\bar{1}11]$, $(111)[1\bar{1}0]$.
 a) Sketch each of these three slip systems, together with the [101] direction.
 b) Calculate the shear stresses for each slip system.

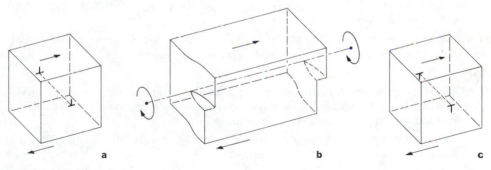

11-25 See Problem 11-9.

11-9 Consider the crystals sketched in Fig. 11-25. Indicate which way each of the dislocations would move under the shear stress indicated.

11-10 A shear stress applied in the direction shown in Fig. 11-26 has caused slip to occur within a circular region on a slip plane. There has been no slip outside this circle. The

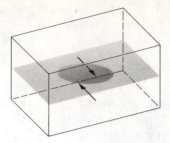

11-26 See Problem 11-10.

circle is therefore a dislocation. Indicate which part of the loop is an edge dislocation and which part is a screw dislocation.

11-11 Verify the twinning strain of 0.707 as listed for an fcc metal in Table 11-3. [*Hint.* draw the (110) plane of Fig. 11-12(a) to scale.]

11-12 In which of the following crystal directions may compressive forces introduce twinning in Cd? In hcp Ti?

    a) [0001]   b) [$10\bar{1}\bar{1}$]   c) [$\bar{1}011$]   d) [$10\bar{1}0$]

11-13 Can (111)[$11\bar{2}$] twinning occur in NaCl?

Answer. Sketch the atoms and their displacements on the ($1\bar{1}0$) plane; [$11\bar{2}$] cannot occur, but [$\bar{1}\bar{1}2$] can.

11-14 A pure iron sheet 0.10 in. thick is annealed before being cold-rolled to 0.08 in. (negligible change in width).

    a) What would be the ductility of the iron after cold-rolling?

    b) Estimate the approximate temperature of recrystallization for this iron.

    c) Give two reasons why the recrystallization temperature of any metal is not fixed.

11-15 A copper wire 0.10 in. in diameter was annealed before being cold-drawn through a die 0.08 in. in diameter. What tensile strength does the wire have after this cold-drawing?

Answer. 48,000 psi

11-16 Iron is to have a BHN of at least 125 and an elongation of at least 32%. How much cold work should the iron receive?

11-17 Copper is to be used in a form with at least 45,000 psi tensile strength and at least 18% elongation. How much cold work should the copper receive?

Answer. 26% cold work

11-18 A company makes 500 units of a product per day. Each requires 6 copper-nickel alloy rods 20 in. long and 0.2 in^2 in cross-sectional area. Each rod must support 5000 lb in tension without yielding. Suggest the best alloy.

11-19 A severely cold-worked metal was found to be 50% recrystallized at the following combinations of heating times and temperatures:

$$
\begin{array}{ll}
1 \text{ min} & 162°\text{C} \\
100 \text{ min} & 97°\text{C}
\end{array}
$$

Consider that the recrystallization is a self-diffusion process. Estimate the temperature required for 50% recrystallization in 10^4 min (i.e., ~1 week).

Answer. 49°C

12 □ viscoelastic deformation

12-1 VISCOUS FLOW

Viscosity is a property of fluids, including amorphous solids. In fact, the *viscosity*, η, and *fluidity*, f, are reciprocals of each other:

$$\eta = 1/f. \tag{12-1}$$

The definition of viscosity may be formalized as the ratio of shear stress, τ, to the velocity gradient of flow, dv/dy, that is encountered in fluid flow (Fig. 12-1):

$$\eta = \frac{\tau}{dv/dy}; \tag{12-2a}$$

or, since $x = vt$ and $\gamma = x/y$, we can see from Fig. 10-4 that

$$\eta = t\tau/\gamma. \tag{12-2b}$$

The unit of viscosity is the *poise*, which, from Eq. (12-2), has the dimensions of dyne $\cdot$ sec/cm^2 in the cgs system. Various viscosities are given in Table 12-1 for purposes of comparison.

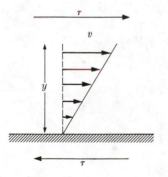

12-1 Viscosity. More viscous materials require a higher shear stress per unit flow rate. Since $x = vt$ and $\gamma = x/y$, either Eq. (12-2a) or Eq. (12-2b) may be used.

Viscosity values are temperature sensitive, decreasing exponentially with the reciprocal of absolute temperature, T:

$$\eta = \eta_0 e^{Q/RT}, \tag{12-3a}$$

or

$$\ln \eta = \ln \eta_0 + Q/RT. \tag{12-3b}$$

TABLE 12-1
VISCOSITIES OF NONCRYSTALLINE MATERIALS

Material	Viscosity (20°C), poises*
Air	0.00018
Pentane, C_5H_{12}	0.0025
Water	0.01
Phenol, C_6H_5OH	0.1
Syrup (60% sugar)	0.56
Oil, machine	1 to 6
Glycerin	9
Sulfur (120°C)	10^3
Window glass (515°C)	10^{13}
Window glass (800°C)	10^5
Polymers, T_g†	$\sim 10^{13}$
Polymers, $T_g + 10°C$	$\sim 10^{10}$
Polymers, $T_g + 20°C$	$\sim 10^8$

* Poise = sec · dyne/cm².
† T_g = glass-transition temperature.

Combined with Eq. (12-1), this shows that

$$\log_{10} f = \log_{10} f_0 - Q/2.3RT. \tag{12-4}$$

Both η_0 and f_0 are constants for a given fluid, and Q is an activation energy for the viscous shear of the atoms as they pass one another. On this basis, fluidity, f, is a diffusion coefficient per unit shear force, $(cm^2/sec)/dyne$, and may be compared to other displacement-type coefficients.*

Viscosity of supercooled liquids. Equation (12-4) remains valid below the melting point if the liquid is supercooled and does not crystallize (Fig. 6-26). Therefore, the viscosity continues to increase, and the fluidity to decrease, as cooling proceeds toward the transition temperature, T_g (Fig. 12-2). Below that temperature, how-

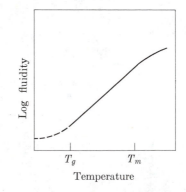

12-2 Fluidity versus temperature. Between T_m and T_g, the temperature dependence is a consequence of the activation energy for atom rearrangements to produce viscous flow. Below T_g, flow is precluded and only bond straightening occurs.

* Diffusion coefficient: $D = cm^2/sec$ (see Eq. 9-1b). Electrical mobility: $\mu = (cm^2/sec)/volt$ (see Eq. 9-11b). Fluidity: $f = (cm^2/sec)/dyne$.

ever, the atoms no longer have the ability to rearrange themselves in response to shear stresses. Any further deformation must depend on bond straightening, which is elastic in nature and relatively insensitive to temperature. This means that the viscosity curves of Fig. 12-3 do not extend indefinitely to higher values as cooling proceeds below the glass temperature; rather the viscosity approaches a maximum of about 10^{20} poises.

12-2 VISCOSITY AND STRUCTURE

Other than temperature, the chief factor affecting the viscosity of amorphous materials is their structure. As with diffusion, small ions and molecules can move past their neighbors with comparative ease and provide low viscosities. Other factors, such as molecular shape, polarity, and cross-linking, are also important in the viscosity and deformation of amorphous engineering solids.

The relationship between viscosity and structure is illustrated in Fig. 12-3 for two glasses. Fused silica has a framework structure. Since every SiO_4 unit is coordinated with four neighbors, the effect is one of cross-linking saturation. The shear stress for flow is very high because primary bonds must be broken. In contrast, a soda-glass (Fig. 12-3b) has a much less rigid structure with fewer tetrafunctional units (Section 7-9), with the result that lower shear stresses are required for flow.

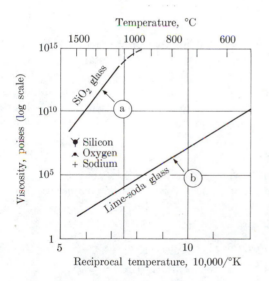

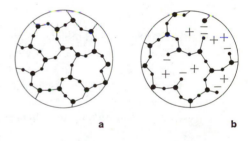

12-3 Viscosity versus structure. A silica glass (fused silica) has a rigid framework structure of covalently bonded silicon and oxygen atoms. This network is more open and the bonds less specific in a lime-soda glass. [*Note.* In three dimensions, the silicon is also bonded to a fourth oxygen to give an "SiO_4" unit.]

Figure 12-4 demonstrates that molecular materials with only weak *inter*molecular bonds in contrast to ionic bonds in Fig. 12-3(b) and covalent bonds [in Fig. 12-3(a)] have low viscosities. Among the examples presented, pentane (C_5H_{12}) is lowest because it is not polar; i.e., it has a symmetrical molecule and no dipole moment for intermolecular attractions (Section 3-6). The viscosities for other paraffin compounds (C_nH_{2n+2}) are higher than pentane inasmuch as the molecular

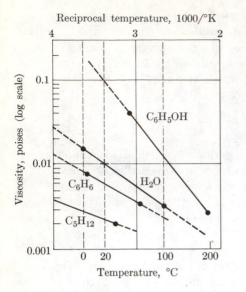

Reciprocal temperature, 1000/°K

12-4 Viscosities for molecules. Viscosities depend on molecular size and the presence of hydrogen bridges. The latter is significant in water.

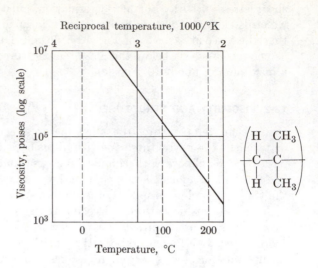

Reciprocal temperature, 1000/°K

12-5 Viscosity of polyisobutylene. (See Example 12-1.)

weights are greater and molecular entanglements are possible. The extreme example in the paraffin series is polyethylene, which, because of its molecular size, has a viscosity high enough to give it solid-like characteristics.

The viscosity of water is worthy of note. Its molecular weight is much smaller than that of pentane, but its relatively strong hydrogen bridge between molecules requires a measurably higher shear stress for viscous flow.

Example 12-1

As shown in Fig. 12-5, the viscosity of amorphous polyisobutylene with $\overline{M}_w$ of 100,000 gm/mole is $10^{3.5}$ and $10^{7.0}$ poises at 217°C and 30°C, respectively. At what temperature is the viscosity 10^{10} poises?

Answer. From Eq. (12-3) (changed to base 10),

$$\log_{10} \eta = \log_{10} \eta_0 + Q/2.3(1.987 \text{ cal/mole} \cdot °K)T,$$
$$3.5 = \log \eta_0 + Q/(4.575 \text{ cal/mole} \cdot °K)(490°K),$$
$$7.0 = \log \eta_0 + Q/(4.575 \text{ cal/mole} \cdot °K)(303°K).$$

Solving simultaneously, we obtain

$$Q \doteq 12,700 \text{ cal/mole},$$
$$\log \eta_0 = -2.16,$$
$$10.0 = -2.16 + (12,700 \text{ cal/mole})/(4.575 \text{ cal/mole} \cdot °K)(T),$$
$$T = 228°K \quad (= -45°C). \blacktriangleleft$$

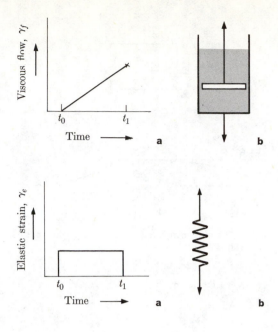

12-6 Ideal (Newtonian) fluid. (a) Displacement. (b) Dashpot model. The flow rate is proportional to time. The flow is not reversible when the shear stress is removed at *t*.

12-7 Ideal (Hookian) solid. (a) Displacement. (b) Spring model. The strain is constant but reversible when the stress is removed at *t*.

12-3 DISPLACEMENT MODELS

Since an *ideal (Newtonian) fluid* obeys the viscosity relationship (Eq. 12-2), Fig. 12-6(a) may be used schematically to show flow displacement, γ_f, at constant shear stress as a function of time. For convenience, we use a "dashpot" to represent the behavior of the fluid (Fig. 12-6b). A dashpot is a loosely fitting piston which is gradually displaced as fluid flows around its edges when a force is applied. Displacement is more rapid when the liquid is less viscous. If the force is removed, the flow is not reversed. For reversal to occur with a dashpot, the force must be reversed.

An *ideal (Hookian) solid* exhibits the elastic behavior discussed earlier (Eqs. 10-1 and 10-6). The elastic displacement, γ_e, is not time dependent, but occurs immediately as shown schematically in Fig. 12-7(a). For convenience, we use a spring to represent the elastic behavior (Fig. 12-7b). Displacement is greater and the curve of Fig. 12-7(a) lies higher when the solid has a lower elastic modulus. If the force is removed, the displacement is recovered.

Most materials behave ideally only in a limited number of situations. More commonly, they have characteristics which are *combinations* of the two ideal extremes. For example, an asphalt road surface responds elastically when loaded by a passing car. Under a parked car, however, the asphalt gradually flows. Glass and most plastics behave similarly, but, of course, with different time and temperature parameters.

Example 12-2

On a simplified basis, the behavior of asphalt may be represented as a spring and dashpot in *series*. That is, it has both an elastic component and a viscous component. Formulate an expression to show the amount of displacement that will occur as a function of time if the material is exposed to a constant shear force.

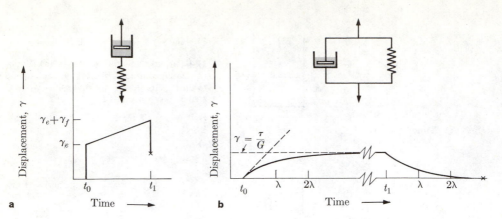

12-8 Viscous and elastic combinations. (a) Maxwell model. The elastic and viscous elements are in series (Eq. 12-5). (b) Voigt model. The two elements are parallel (Eq. 12-6).

Answer. Since

$$\gamma = \gamma_e + \gamma_f,$$

then, from Eqs. (10-6) and (12-2b),

$$\gamma = \tau/G + t\tau/\eta,$$
$$\gamma = \tau(1/G + t/\eta). \tag{12-5}$$

This is shown schematically in Fig. 12-8(a). At time t_1, when the force is removed, the elastic part of the displacement is recovered. ◄

The time-dependent strain of Section 10-4 may be represented as a spring and dashpot in *parallel* loading under constant stress. In this case we shall look at shear properties (γ, τ, and G) rather than the Young's-modulus relationships that we used in Chapter 10. With a parallel loading, γ_e must always equal γ_f. Initially, however, all the resistance to deformation comes from the viscous component; consequently, $d\gamma/dt$ is equal to τ/η (from Eq. 12-2b). Eventually the elastic component carries the full load and the displacement ceases at an elastic strain of τ/G (from Eq. 10-6).

In terms of the shear properties which are applicable in viscous flow, Eq. (10-12a) may now be rewritten:

$$\gamma = \frac{\tau}{G}(1 - e^{-t/\lambda_v}), \tag{12-6a}$$

because τ/G corresponds to γ_0. The relaxation time, λ_v, for viscous flow may be related to viscosity and the shear modulus:

$$\lambda = \eta/G. \tag{12-7}$$

Thus the anelastic shear strain (in Fig. 12-8) may be presented as

$$\gamma = \frac{\tau}{G}(1 - e^{-tG/\eta}). \tag{12-6b}$$

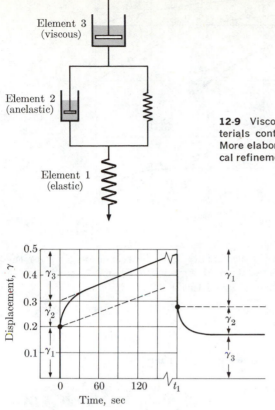

Element 3
(viscous)

Element 2
(anelastic)

Element 1
(elastic)

12-9 Viscoelastic model. The behavior of amorphous materials contains elastic, anelastic, and viscous components. More elaborate models could be formulated to provide empirical refinements.

12-10 Viscoelastic displacement. Equation (12-8) applies to the model shown in Fig. 12-9. The viscous displacement, γ_3, is not recoverable.

12-4 VISCOELASTICITY

Amorphous solids (and polycrystalline solids under certain conditions) are *viscoelastic;* that is, they combine elastic and viscous behaviors. Again it is possible to analyze their deformation with appropriate combinations of springs and dashpots. The deformation of a polymeric material, for example, can be explained for a wide range of temperatures and strain rates by the schematic representation in Fig. 12-9. The total displacement, γ, of such a combination is equal to the sums of the displacements in each element of the series:

$$\gamma = \gamma_1 + \gamma_2 + \gamma_3$$
$$= \tau/G_1 + (\tau/G_2)(1 - e^{-tG_2/\eta_2}) + t\tau/\eta_3. \qquad (12\text{-}8)$$

Equation (12-8) gives the time-strain relationship shown in Fig. 12-10, and will be the basis for our discussion in Section 12-5. Since each component of Fig. 12-9 is separate, each has its own values of G and η. After t_1, when the stress is released, the elastic and anelastic parts of the deformation are recovered, but the viscous part is not.

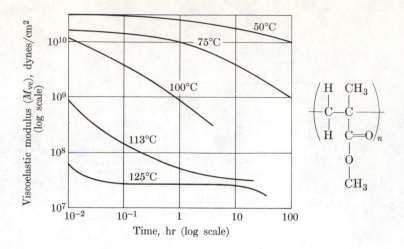

12-11 Viscoelastic modulus versus time and temperature (polymethylmethacrylate).

Example 12-3

Determine the shear moduli, the viscosity coefficients, and the relaxation time for the polymer whose deformation follows the pattern of Fig. 12-10 when a shear stress of 7×10^6 dynes/cm^2 is applied. ($\gamma = 0.263$ at $t = 10$ sec.)

Answer. Element 1: At $t = 0$, $\gamma_1 = \tau/G_1 = 0.20$. Therefore,

$$G_1 = 3.5 \times 10^7 \text{ dynes/cm}^2 \ (= 500 \text{ psi}).$$

Element 3: At $t > 80$ sec, $d\gamma/dt = 0.08/80$ sec $= \tau/\eta_3$. Therefore,

$$\eta_3 = 7 \times 10^9 \text{ dyne} \cdot \text{sec/cm}^2.$$

Element 2: At $t > 80$ sec, $\gamma_2 = 0.10 = \tau/G_2$. Therefore,

$$G_2 = 7 \times 10^7 \text{ dynes/cm}^2 \ (= 1000 \text{ psi}).$$

At $t = 10$ sec,

$$\gamma_2 = 0.063 = \tau/G_2(1 - e^{-10G_2/\eta_2}),$$
$$0.63 = 1 - e^{-10G_2/\eta_2},$$
$$\ln 0.37 = -10G_2/\eta_2 = -1.0,$$
$$\eta_2 = 7 \times 10^8 \text{ dyne} \cdot \text{sec/cm}^2.$$

Therefore,

$$\text{(relaxation time)}_2 = \eta_2/G_2$$
$$= 10 \text{ sec.} \ \blacktriangleleft$$

Time-temperature relationships. A *viscoelastic modulus*, M_{ve}, may be defined as the ratio of the shear stress, τ, to the total shear displacement, γ, of Eq. (12-8):

$$M_{ve} = \tau/\gamma. \tag{12-9}$$

Since the denominator is a function of both time and viscosity, which in turn is a function of temperature (Eq. 12-3), we can plot M_{ve} versus t for different tempera-

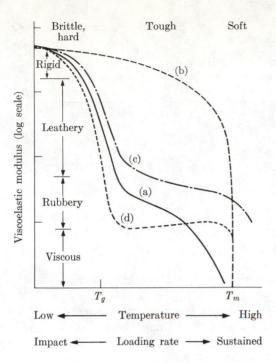

12-12 Viscoelastic modulus versus structure. (a) Amorphous linear polymer. (b) Crystalline polymer. (c) Cross-linked polymer. (d) Elastomer.

tures. This is done in Fig. 12-11 for polymethylmethacrylate at a series of temperatures. It may be shown that longer times correspond to higher temperatures. Therefore, presentations like Fig. 12-12 are more commonly used.

12-5 VISCOELASTIC BEHAVIOR OF POLYMERS

Before analyzing the behavior of polymers as presented in Fig. 12-12, several pertinent facts should be cited:

a) Polymers range from amorphous to moderately crystalline (Section 7-5).

b) Amorphous polymers possess a glass-transition temperature which is the lower limit of molecular rearrangement (Section 6-11 and Table 12-2).

c) Cross-linked structures (Section 7-8) and framework structures (Section 7-9) cannot deform as readily as linear structures.

d) Linear molecules of amorphous polymers are most stable in a kinked conformation (Section 7-4).

In order to simplify our considerations we shall categorize polymers into four arbitrary types: (1) amorphous, (2) crystalline, (3) cross-linked, and (4) elastomeric.

Amorphous polymers (Fig. 12-12a). Our example is a polymer such as polyvinyl chloride or polystyrene. The viscosity is very high below the glass point, T_g. At these temperatures, γ_1 is the major term of Eq. (12-8). Since there is no oppor-

TABLE 12-2
TRANSITION TEMPERATURES FOR POLYMERS

Material	R (Table 7-1)	Melting temperature, °C	Glass-transition temperature, °C
Polyethylene	—H	137	—120
Polyvinylchloride	—Cl	175	87
Polypropylene	—CH_3	175	—15
Polyacrylonitride	—CN	320	110
Polystyrene (isotactic)	—C_6H_5	240	100
Polybutadiene (isotactic)	—	120	—90
Polychloroprene	—	80	—50
Polyisoprene (cis)	—	30	—73
Polyisoprene (trans)	—	75	—
Polyformaldehyde	—	—	—85

tunity for stress adjustments, the plastic is hard and brittle, and behaves in a glass-like manner.

At high temperatures $(T > T_m)$, the polymer is a true liquid; hence, the viscosity is very low and elastic stresses cannot build up. Under these conditions, γ_3 is the predominant term in Eq. (12-8). Deformation occurs rapidly by viscous flow.

At intermediate temperatures $(T_g < T < T_m)$, γ_2 is the major term of Eq. (12-8). Within this important temperature range, the polymer is anelastic in behavior. In general, the structure is too rigid for viscous flow (η_3 is high); however, η_2 is sufficiently low so that the molecules can rearrange themselves as stress is applied. Deformation occurs by an unkinking of chains; thus, when the stress is relieved, the deformation is recovered. The plastic is rubbery if G_2 of Eq. (12-8) is low, or leather-like when the spring component gets stiffer. Many polymer products are developed with these characteristics, e.g., squeeze bottles, fibers for cloth, and plastic film. As T_m is approached, γ_3 becomes more significant so that polymeric products lose their utility; however, many plastic products are made by being heated in this soft range, then injected under pressure into a mold where they are chilled and retain the shape of the mold.

Crystalline polymers (Fig. 12-12b). A fully crystalline polymer has no leathery or rubbery ranges but only a gradual decrease in its elastic modulus until an abrupt drop is encountered at the melting temperature. Such a gradual decrease is typical of metals and crystalline materials in general. The elastic component, γ_1, is the major term of Eq. (12-8) and controls deformation up to the melting point.

Since a polymer is seldom, if ever, fully crystalline (Section 7-5), those polymers which are called "crystalline" usually have a performance curve which is intermediate between curves (a) and (b) of Fig. 12-12.

Crystalline polymers are highly desired if the plastic product is to be employed for stress-carrying applications at temperatures approaching the melting point. Thus we find considerable technical interest in producing materials like polyethylene with maximum crystallinity.

Cross-linked polymers (Fig. 12-12c). Cross-linking restricts the amount of viscous flow which is possible. Thus a cross-linked polymer never obtains high fluidity (low viscosity) even in the liquid range. Like other polymers, a cross-linked one is brittle and hard below the glass-transition temperature, but develops molecular freedom in the range from T_g to T_m, to give recoverable anelastic displacement. The freedom, however, is never as great as it is in a polymer without cross-linking, and so the performance curve lies higher than for an amorphous polymer.

A framework polymer of polyfunctional mers (Section 7-9) presents the extreme example of cross-linking. Since the viscosities for both γ_2 and γ_3 of Eq. (12-8) are high, τ and G of γ_1 present the major feature of the deformation. Examples of framework polymers include phenol-formaldehyde (Fig. 7-16), and rubber which has been fully vulcanized, i.e., cross-linked with sulfur.

Elastomers (Fig. 12-12d). Rubbers are polymers with very low moduli of elasticity (10^2 to 10^4 psi). The term *elastomer* is used technically to classify these materials. The low modulus of elasticity is a result of the straightening of molecules which have abnormally severe coiling and kinking. [See the discussion of cis-isoprene (Section 7-7).]

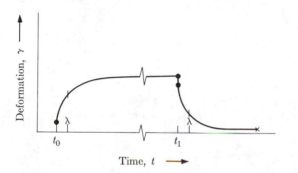

12-13 Deformation of rubber. In a rubber tire, the relaxation time must be much less than a second.

In terms of Eq. (12-8), a rubber has low values of G_2 and η_2, and hence the γ_2-term far exceeds the γ_1- and γ_3-terms. In fact, we characterize the M_{ve} "plateau" of Fig. 12-12 as the rubbery range. Above the glass point, there is minimal resistance to stresses as the molecules are unkinked, either elastically from the bond straightening or viscously from the internal friction caused by the displacement of adjacent molecules. Thus Fig. 12-13 can represent the viscoelastic deformation of a rubber.

Immediately above the glass point, the elastomer curve of Fig. 12-12 drops as much as or more than it does for any other type of polymer. At still higher temperatures, where the thermal energy available for kinking and coiling is greater, it then takes more stress to produce a given strain. Thus, the elastic modulus, G_2, of Eq. (12-8) is increased slightly (Fig. 10-13b). In elastomers, unlike the other polymers, the viscoelastic modulus also has a slight increase, until eventually the melting temperature is approached and viscous flow, γ_3, becomes the major factor.

12-6 STRESS RELAXATION

The discussion in the previous three sections has been in terms of constant loads or stresses. However, there are numerous engineering applications where a *constant strain* is encountered. Under the latter condition, a viscous component of deformation permits stress relaxation with no change in dimensions. A simple model for stress relaxation under constant strain (as opposed to viscoelastic strain under constant stress) is that of a dashpot and spring in series,* with

$$\gamma = \gamma_e + \gamma_f = \text{constant},$$

where γ_e is the elastic displacement and γ_f is the flow displacement. Thus

$$\frac{d\gamma_e}{dt} = -\frac{d\gamma_f}{dt},$$

or, from Eqs. (10-6) and (12-2b),

$$\frac{d\tau/G}{dt} = -\tau/\eta.$$

Consequently,

$$d\tau/\tau = -(G/\eta)\,dt; \tag{12-10a}$$

so

$$\tau = \tau_0 e^{-t/\lambda_v}, \tag{12-10b}$$

where τ_0 is the initial shear stress and η/G, or λ_v, is the *relaxation time* for viscoelastic flow. Since a resolved shear stress, τ, is proportional to an axial stress, σ, it follows that

$$\sigma = \sigma_0 e^{-t/\lambda_v}. \tag{12-10c}$$

The viscosity, η, is a logarithmic function of temperature (Eq. 12-3); thus we find that the relaxation time for viscous flow (Eq. 12-7) also decreases logarithmically with temperature:

$$\lambda_v = \lambda_0 e^{Q/RT}. \tag{12-11}$$

The exponents have the usual connotations, and λ_0 is the constant of integration.

Example 12-4

A stress of 1200 psi is required to stretch a 4-in. rubber band to 5.6 in. After 42 days in the same stretched position, the band exerts a stress of only 600 psi.

a) What is the relaxation time?

b) What stress would be exerted by the band in the same stretched position after 90 days?

---

* The reader will note that we must use a series loading to represent relaxation at constant strain, whereas we use a parallel loading to depict relaxation at constant stress (Fig. 10-15b). In actual viscoelastic materials, it is necessary to use more complex models to give precise mathematical analogies.

Answer. From Eq. (12-10c),

a)
$$\ln (600 \text{ psi}/1200 \text{ psi}) = -42/\lambda_v,$$
$$\lambda_v = 61 \text{ days};$$

b)
$$\sigma_{90} = (1200 \text{ psi})e^{-90/61} = 274 \text{ psi}.$$

Alternative solution for (b), with 48 *additional* days:

$$\sigma_{+48} = (600 \text{ psi})e^{-48/61} = 274 \text{ psi}. \blacktriangleleft$$

Example 12-5

The temperature used in Example 12-4 was 20°C. A similar test at 25°C gave a value of only 400 psi after 42 days. What will the remaining stress be after 90 days at 15°C if the initial stress is 1000 psi?

Answer. At 20°C (from Example 12-4),

$$\lambda_{20}° = 61 \text{ days}.$$

At 25°C,
$$\ln (400 \text{ psi}/1200 \text{ psi}) = -42/\lambda_{25}°,$$
$$\lambda_{25}° = 38 \text{ days}.$$

Using Eq. (12-11), and changing to base 10, we find that, at 20°C (293°K),

$$\log_{10} 61 = \log_{10} \lambda_0 + Q/2.3(1.987)(293) = 1.787;$$

and at 25°C (298°K),

$$\log_{10} 38 = \log_{10} \lambda_0 + Q/2.3(1.987)(298) = 1.580.$$

Solving simultaneously, we obtain

$$Q = 16{,}350 \text{ cal/mole},$$
$$\log_{10} \lambda_0 = -10.53.$$

At 15°C (288°K),

$$\log_{10} \lambda = -10.54 + 16{,}530/(4.575)(288) = 2.01.$$

Thus,

$$\lambda_{15}° = 102 \text{ days},$$
$$\sigma_{(90 \text{ days},15°C)} = (1000 \text{ psi})e^{-90/102} = 413 \text{ psi}. \blacktriangleleft$$

Glass processing. The engineer in glass manufacturing pays close attention to the viscosity of the product during production. First, the glass must be very fluid in the *melting range* so that gas bubbles can escape and homogenization can occur. Experience has shown that in this range the viscosity coefficient must not exceed 300 poises. The *working range* (Fig. 12-14) covers several orders of magnitude. Fast operations, such as the production of light bulbs, require a rather fluid glass ($\sim 10^{9.1}$ poises); the drawing operation on heavy glass sheet requires a viscous

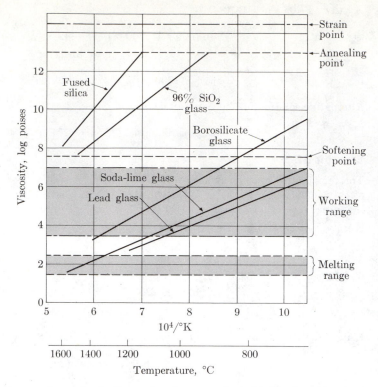

12-14 Viscosity versus temperature (glass). Processing steps are adapted to the viscosity values.

product ($\sim 10^7$ poises) so that only a limited amount of deformation will occur during handling before the glass cools.

Two other viscosity points are important. The next is the *annealing point*, defined as the temperature at which the viscosity is 10^{13} poises. At that temperature, the atoms can move sufficiently so that residual thermal stresses may be relieved within a 15-minute period, an interval that is compatible with production schedules. Below the *strain point*, the temperature at which the viscosity is $10^{14.5}$ poises, the glass is sufficiently rigid so that it may be handled without the generation of additional residual stresses.

12-7 THERMOPLASTIC AND THERMOSETTING RESINS

As linear polymers are heated toward their melting points, they become sufficiently soft so that they may be injected into molds. If the molds are water-cooled, the resin, i.e., plastic material, retains the shape of the mold. Although oversimplified, this is a description of the process used in making many plastic products. Polymers which are amenable to this process are called "thermoplastics." The chief structural requirement is that there be no cross-linking, and that the relaxation time be sufficiently short for stress relief to occur during forming. For obvious reasons, these products cannot be used at elevated temperatures.

Polymers with polyfunctional mers, and those which are heavily cross-linked, cannot soften as they are heated. In fact, heating often causes the polymerization and vulcanization reactions to continue, with the result that curve (c) of Fig. 12-12 is raised.* These materials take on a "set" and are called "thermosetting" materials. Quite commonly, the raw material for a thermosetting resin will, by intent, be only partially polymerized, so that it may be deformed into the heated mold. Then there must be a cure period for the final reactions. Obviously, production is slower than for thermoplastic products; however, the thermoset products are desired for higher-temperature applications.

REFERENCES FOR FURTHER READING

Alfrey, T., and E. F. Gurnee, *Organic Polymers*. Englewood Cliffs, N. J.: Prentice-Hall, 1967. The viscoelastic behavior of amorphous polymers is discussed in Chapter 4. More mathematical detail is given than in this text. For the advanced undergraduate student.

DiBenedetto, A. T., *The Structure and Properties of Materials*. New York: McGraw-Hill, 1967. Mechanical properties of polymers are presented on a thermodynamic basis in Chapter 12. For the graduate student and instructor.

Nielsen, L. E., *Mechanical Properties of Polymers*. New York: Reinhold, 1962. Written for the mechanical engineer who has become involved in the use of polymers. Advanced undergraduate and graduate level.

Orowan, E., "Deformation in Polymers: Viscoelasticity," in *Mechanical Behavior of Materials* (edited by F. A. McClintock and A. S. Argon). Reading, Mass.: Addison-Wesley, 1966. Chapter 6 relates the structure of polymers to their deformation. Excellent reference list. Advanced undergraduate level.

Tobolsky, A. V., "The Mechanical Properties of Polymers," *Scientific American*, **197** [3], September 1957, pp. 120–134. The glassy, leathery, and rubbery qualities of solid polymer materials are explained in terms of structure. Undergraduate level.

Tobolsky, A. V., *Properties and Structure of Polymers*. New York: Wiley, 1960. One of the definitive references on viscoelastic behavior of materials. Advanced level.

PROBLEMS

12-1 From Fig. 12-4, determine the activation energy for the viscous flow (a) of phenol (C_6H_5OH), (b) of pentane (C_5H_{12}).

Answer. a) 5500 cal/mole   b) 1550 cal/mole

12-2 What are (a) the activation energy and (b) the preexponential constant for the *fluidity* of water?

' In Fig. 12-12, as drawn, it is assumed that no chemical reactions occur. We noted earlier, however, that the location of curve (c) depends on the *amount of cross-linking.*

12-3 What are the viscosity and shear modulus for a polymer which behaves as a dashpot and spring in parallel if the shear strains are:

1 hr	0.0060
2 hr	0.0084
10 hr	0.010
20 hr	0.010

when the shear stress is 10^7 dynes/cm^2?

Answer. $\eta = 4 \times 10^{12}$ dyne $\cdot$ sec/cm^2, $G = 10^9$ dynes/cm^2

12-4 What is the relaxation time for the polymer in Problem 12-3?

12-5 After the stress is removed in Problem 12-3, how long will it take for the strain to return from 0.010 to 0.0005?

Answer. 200 min

12-6 The data for Problem 12-3 were obtained at 20°C. Suppose that the relaxation time is reduced from 4000 to 3000 sec at 30°C.
   a) Determine the activation energy for relaxation.
   b) Determine the new strain values for 1 hr and 2 hr, and 10^7 dynes/cm^2.
(Assume that G is constant over this 10°C temperature interval.)

12-7 A stress relaxes from 70 psi to 50 psi in 123 days.
   a) What is the relaxation time?
   b) How long would it take to relax to 30 psi?

Answer. a) 365 days   b) an additional 186 days (total of 309 days)

12-8 An initial stress of 1500 psi is required to strain a piece of rubber 0.5 in./in. After the strain has been maintained constant for 40 days, the stress required is only 750 psi. What would be the stress required to maintain the strain after 80 days?

12-9 The relaxation time for a plastic is known to be 45 days and the modulus of elasticity is 10^4 psi (both at 100°C). The plastic is compressed 0.05 in./in. and held at 100°C. What is the stress (a) initially? (b) After 1 day? (c) After 1 month? (d) After 1 year?

Answer. a) 500 psi   b) 490 psi   c) 256 psi   d) 0.15 psi

12-10 Scrap phenol-formaldehyde is worthless. Scrap polyvinyl chloride can be reused. Why?

part IV □ electrical processes in solids

13 □ dielectric behavior of materials

13-1 POLARIZATION

If an *electric field*, $\mathcal{E}$, is established between two parallel electrodes, a charge will develop on them. The resulting *charge density*, $\mathfrak{D}$, is directly proportional to the applied field:

$$\mathfrak{D} = \epsilon_0 \mathcal{E}. \tag{13-1a}$$

When a vacuum is between the plates, the proportionality constant, ϵ_0, is called the *electric permittivity of a vacuum*. Since we express the electric field in volts/m* and charge density in coul/m², the units for electric permittivity are (coul/volt)/m, or farads/m:

$$\text{coul/m}^2 = (\text{farads/m})(\text{volts/m}). \tag{13-1b}$$

With *these* units, the value of ϵ_0 is 8.854×10^{-12} farad/m. All of our discussions will be based on these units.

When a material is introduced between the electrodes, the ratio of the charge density to the electric field is always increased. We can list permittivity values, ϵ, for each material in units of farads/m. Alternatively, we can give relative values for materials as a multiplying factor for ϵ_0. Following the latter procedure, we use the equation

$$\mathfrak{D} = \epsilon_0 \kappa \mathcal{E}. \tag{13-2}$$

The term κ, which equals ϵ/ϵ_0, is the relative dielectric constant, or more simply, the *dielectric constant*.† It has no dimensions because the units are carried in the basic permittivity value.

Figure 13-1 divides the charge density into two parts: (a) the charge density contributed by the material and (b) the charge density existing in the absence of a

---

* Mks units will be used in Chapters 13 through 16 because they permit easier cross correlation. Admittedly, cgs units are more widely used in many industries, particularly for magnetic materials.
† This procedure is comparable to listing specific gravities rather than densities. It lets us use the same table of values for dielectric calculations, regardless of which system of units we use, just as specific-gravity values can be applied in calculations with both metric and English units

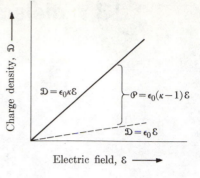

13-1 Charge density versus electric field. Polarization, $\mathcal{P}$, is that part of the charge density contributed by the material.

material. In the study of materials, we are naturally more interested in the former part of the charge density, which is called *polarization*, $\mathcal{P}$, and of course has the same units as charge density, coul/m². Since

$$\mathcal{P} = \mathfrak{D} - \epsilon_0\mathcal{E}, \tag{13-3}$$

and since $\mathfrak{D}$ is defined in Eq. (13-2) as equal to $\epsilon_0\kappa\mathcal{E}$,

$$\mathcal{P} = (\kappa - 1)\epsilon_0\mathcal{E}. \tag{13-4}$$

The *electric susceptibility*,

$$\chi = \kappa - 1, \tag{13-5}$$

is the part of the total dielectric constant which is a consequence of the material's presence. For simplicity, we shall use $\kappa - 1$ rather than χ.

Dipole moments. We wish to understand the source of the polarization which is contributed by the material. This can be facilitated by changing the units of polarization to represent a *dipole moment per unit volume:*

$$\frac{\text{coul}}{\text{m}^2} = \frac{\text{coul} \cdot \text{m}}{\text{m}^3}. \tag{13-6}$$

(Cf. Section 1-3 and Example 2-5.)

Polarization can be divided into four categories, each having a distinct mechanism based on the type of dipole moment which is established:

$$\mathcal{P} = \mathcal{P}_e + \mathcal{P}_i + \mathcal{P}_o + \mathcal{P}_s. \tag{13-7}$$

Electronic polarization, $\mathcal{P}_e$, is shown schematically in Fig. 13-2(a) as an electron displacement within the atom. *Ionic polarization*, $\mathcal{P}_i$, arises from comparable displacements of ions and atoms (Fig. 13-2b). Both electronic and ionic polarization must be *induced* by an electric field. They disappear when the field is removed. These two types of polarization will be discussed in the next section. *Orientation polarization*, $\mathcal{P}_o$, arises from *permanent dipoles* which are normally randomly

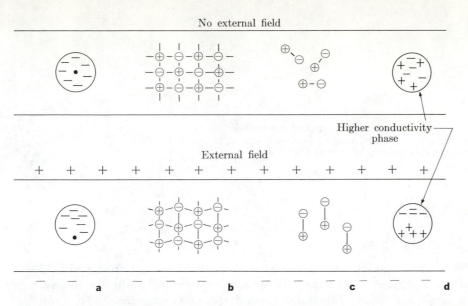

No external field

External field

Higher conductivity phase

a   b   c   d

13-2 Polarization mechanisms (schematic): (a) electronic displacements around atoms or ions; (b) ionic displacements within molecules and crystal structures; (c) dipole orientation; (d) space charge. In (d), the higher the conductivity of a phase, the higher the effective potential gradient within the remainder of the dielectric.

oriented but become preferentially oriented by an electric field (Fig. 13-2c). There is a fourth category, *space-charge polarization*, $\mathcal{P}_s$, which is the result of the presence of lower-resistivity materials within the dielectric (Fig. 13-2d). Discussion of this latter type of polarization will be deferred until Chapter 19, where we study the properties of multiphase materials.

As discussed in Section 1-3, the dipole moment, p, is the product of the charge, Q, and the length of the moment arm, d, which separates the positive and negative charges:

$$p = Qd. \tag{13-8a}$$

If we are discussing ions or other charges which are integer multiples of the electron charge, q, then

$$p = Zqd. \tag{13-8b}$$

In reality, the polarization is an integrated sum of all the individual dipole moments per unit volume:

$$\mathcal{P} = [\textstyle\sum p_e + \sum p_i + \sum p_o + \sum p_s]/V. \tag{13-9}$$

It will often be convenient, however, to envision the polarization as a large dipole in which the "centers of gravity" for the positive and negative charges within a volume of material are displaced with respect to each other (Fig. 13-3).

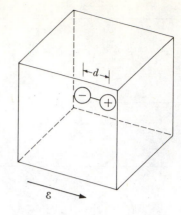

13–3 Polarization. It is convenient to represent polarization as the net dipole moment per unit volume. The moment arm, d, is the distance between the centers of positive and negative charges.

13-2 INDUCED DIPOLES

Electronic polarizability. Although at any small instant of time ($\sim 10^{-16}$ sec) individual atoms are not electrically symmetrical (Section 3-6), on the average electrons are uniformly distributed around the atomic nucleus. Thus, on the average, atoms have no net dipole moment and therefore no polarization. If an electric field is applied, however, the electrons and the nuclei are shifted in opposite directions, as sketched in Fig. 13-2(a), so that a dipole moment develops. The length of the moment arm, d, is proportional to the applied field, as is shown in the following paragraph.

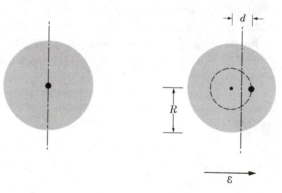

13-4 Electronic polarizability. The nucleus, ●, and the electrons are displaced in opposite directions (see text).

If an atom of atomic number Z and "hard-ball" radius R is placed in an electric field $\mathcal{E}$, and if we assume uniform distribution of electron charge through the atom sphere, the displacement distance, d, will be as shown in Fig. 13-4. Two forces act on the nucleus: (1) the force, F_f, of the electric field is $Zq\mathcal{E}$; (2) the coulombic force, F_c, is in the opposite direction toward the center of electron charge, and may be calculated on the basis of the amount of electron charge which is within the dashed circle.*

---

* The charge outside the dashed circle will have zero net force on the nucleus, according to the Gauss theorem.

From Coulomb's law,

$$F_c = \frac{-1}{4\pi\epsilon_0} \frac{Zq(Zqd^3/R^3)}{d^2} .$$

Since

$$F_f = Zq\mathcal{E} = -F_c,$$

then

$$d = \left[\frac{4\pi\epsilon_0 R^3}{Zq}\right]\mathcal{E}. \tag{13-10}$$

The terms within the brackets are all constants for a given type of atom. We call $4\pi\epsilon_0 R^3$ the polarizability, α.

Since

$$p = Zqd,$$

Eq. (13-10) gives

$$p_e = (4\pi\epsilon_0 R^3)\mathcal{E} \tag{13-11a}$$

$$= \alpha_e\mathcal{E} \tag{13-11b}$$

and

$$\mathcal{P}_e = N\alpha_e\mathcal{E}, \tag{13-11c}$$

where N is the number of atoms per unit volume. *Electronic polarizability*, α_e, for a given type of atom is constant; hence, the induced electronic dipole moment is proportional to the electric field affecting the atom.

TABLE 13-1
TYPICAL POLARIZABILITIES

Material	α,* 10^{-40} farad · m^2
Ne	0.35
Ar	1.4
Kr	2.2
Xe	3.5
CH$_4$	2.7
CCl$_4$	13.0
LiF	6.8
NaCl	8.9
KI	15.3

* Electronic plus ionic.

We might note several consequences of Eq. (13-11). (1) Larger atoms are more polarizable. This may be verified from the data in Table 13-1. (2) On the basis of electronic polarizability, we can estimate the radii of atoms as shown in Example 13-1 below. (3) There is no temperature factor present.

Example 13-1

The dielectric constant of argon gas as a result of electronic polarization was measured to be 1.00044 at 760 mm Hg and 0°C.
a) Estimate the radius of the argon atom.
b) How much are the electrons displaced with respect to the nucleus when an argon atom is in a field of 10^4 volts/m?

Answer. a) Since $\mathcal{P}_e = \epsilon_0(\kappa - 1)\mathcal{E} = N\alpha_e\mathcal{E}$, $\alpha_e = \epsilon_0(\kappa - 1)/N$. At 760 mm Hg and 0°C (cf. Example 2-2),

$$N = (0.602 \times 10^{24} \text{ atoms/mole})(P/RT)$$

$$= \frac{(0.602 \times 10^{24} \text{ atoms/mole})(1 \text{ atm})}{(0.000082 \text{ atm} \cdot \text{m}^3/\text{mole} \cdot \text{°K})(273\text{°K})} = 2.68 \times 10^{25} \text{ atoms/m}^3,$$

$$\alpha_e = (8.854 \times 10^{-12} \text{ farad/m})(0.00044)/(2.68 \times 10^{25}/\text{m}^3)$$

$$= 1.45 \times 10^{-40} \text{ farad} \cdot \text{m}^2,$$

$$R^3 = \alpha_e/4\pi\epsilon_0 = 1.45 \times 10^{-40} \text{ farad} \cdot \text{m}^2/[4\pi(8.854 \times 10^{-12} \text{ farad/m})],$$

$$R = 1.09 \times 10^{-10} \text{ m} \quad (= 1.09 \text{ A}).$$

Note. This is less than the value of 1.92 A for the radius of argon in the fcc crystal. A larger interatomic distance is maintained in solids because of electronic repulsions (Section 3-2).]

b) From Eq. (13-10),

$$d = \alpha_e\mathcal{E}/Zq = \frac{(1.45 \times 10^{-40} \text{ farad} \cdot \text{m}^2)(10^4 \text{ volts/m})}{(18)(1.6 \times 10^{-19} \text{ coul})}$$

$$= 0.5 \times 10^{-18} \text{ m} \quad (= 5 \times 10^{-9} \text{ A}).$$

Note. The actual displacement is a very small fraction of the radius. The resulting dipole moment is sufficient, however, to condense argon at 87°K (see Section 3-6). ◄

Ionic polarizability. The polarizability of methane, CH_4, is higher than that of neon, even though each has 10 electrons (Table 13-1). The difference lies in the displacement of the atoms as well as of the electrons. This conclusion is warranted because the polarizability of methane drops to the polarizability of neon at frequencies which are greater than the natural vibrational frequencies of the atoms ($> 10^{13}$ sec^{-1}). This will be discussed in Section 13-5.

Example 13-2

A methane molecule, CH_4, is in a field of 10^4 volts/m. What is the distance between the centers of positive and negative charges?

Answer. From Table 13-1, $\alpha_{e+i} = 2.7 \times 10^{-40}$ farad $\cdot$ m^2. From Eqs. (13-8b) and (13-11b),

$$Zqd = \alpha_{e+i}\mathcal{E},$$

$$d = \frac{(2.7 \times 10^{-40} \text{ farad} \cdot \text{m}^2)(10^4 \text{ volts/m})}{(10)(1.6 \times 10^{-19} \text{ coul})}$$

$$= 1.7 \times 10^{-18} \text{ m} \quad (= 1.7 \times 10^{-8} \text{ A}).$$

Note. We have combined electronic and atomic displacements in this case. This procedure gives a larger separation than in argon (Example 13-1), even though argon has more electrons than methane. ◀

The atomic displacements in CCl_4 within an electric field are identical in principle to the ionic displacements in Fig. 13-2(b). Although covalently bonded, we must consider the Cl atoms as more negative than the central carbon atoms, since they have a greater affinity for the electrons (Table 2-4). Thus their displacement to form the negative end of a dipole satisfactorily explains the polarizability of CCl_4. This displacement is called *ionic polarizability*. Ionic polarizability, like electronic polarizability, is insensitive to temperature. Therefore, the dielectric constant and electric susceptibility do not vary with temperature (Fig. 13-5).

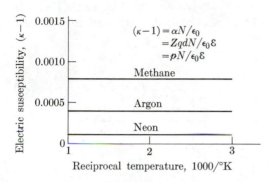

13-5 Electric susceptibilities from induced polarization (gas at 22.4 l/mole). The polarizability, α, is the dipole moment, p, per unit field, $\mathcal{E}$.

Example 13-3

Magnesium oxide has a relative dielectric constant of 9.6. Since it contains no permanent dipoles, the dielectric constant is a result of the electronic and ionic polarization. Separate determinations (Example 13-7) indicate that the electronic polarization is $2.0\epsilon_0\mathcal{E}$. How much are the Mg^{2+} ions displaced from their normal positions (with respect to the O^{2-} ions) when a field of 10^4 volts/m is applied? [MgO has the NaCl structure (Fig. 4-1) and a lattice constant of 4.2 A.]

Answer

$$\mathcal{P} = \mathcal{P}_e + \mathcal{P}_i = \epsilon_0(\kappa - 1)\mathcal{E},$$

$$\mathcal{P}_i = [(9.6 - 1) - 2.0](8.854 \times 10^{-12} \text{ farad/m})(10^4 \text{ volts/m})$$

$$= 58.5 \times 10^{-8} \text{ coul/m}^2 = ZqdN_{Mg^{2+}},$$

$$N_{Mg^{2+}} = 4 \text{ Mg}^{2+}/(4.2 \times 10^{-10} \text{ m})^3 = 5.4 \times 10^{28}/\text{m}^3,$$

$$d = \frac{(58.5 \times 10^{-8} \text{ coul/m}^2)}{(2)(1.6 \times 10^{-19} \text{ coul})(5.4 \times 10^{28}/\text{m}^3)}$$

$$= 3.4 \times 10^{-17} \text{ m} \quad (= 3.4 \times 10^{-7} \text{ A}).$$

Note. We used $Z = 2$, since the displaced ion has a charge of 2. In Examples 13-1 and 13-2, Z was larger because we paid attention to the displacement of the nucleus. ◀

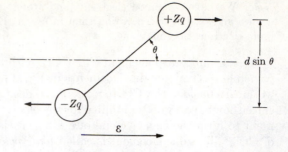

13-6 Torque on permanent dipoles. In the absence of thermal agitation, the dipole would align with the electric field (Eq. 13-13).

13-3 PERMANENT DIPOLES

In Section 3-6 we observed that some molecules are polar; i.e., they have permanent dipoles. Selected values of molecular dipole moments are given in Table 13-2. In gases and liquids with no net electric fields present, the dipoles are randomly aligned.

TABLE 13-2
DIPOLE MOMENTS

Molecule	p_p, debyes*	coul · m	Molecule	p_p, debyes*	coul · m
CO	0.1	0.3×10^{-30}	CH_4	0	0×10^{-30}
CO_2	0	0	CH_3Cl	1.87	6.2
H_2O	1.85	6.2	CH_2Cl_2	1.54	5
H_2S	0.92	3.1	$CHCl_3$	1.02	3
HF	1.91	6.4	CCl_4	0	0
HCl	1.05	3.5			

* 1 debye = 3.33×10^{-28} coul · m.

An applied field, however, tends to align the molecules in the direction of the field. The torque, T, on a dipole may be calculated as indicated in Fig. 13-6. Consider the dipole lying at an angle θ to the direction of the field $\mathcal{E}$:

$$T = (Zqd)\mathcal{E} \sin \theta \qquad (13\text{-}12\text{a})$$

$$= p_p \mathcal{E} \sin \theta, \qquad (13\text{-}12\text{b})$$

where p_p is the permanent dipole moment. The potential energy, PE, of the dipole in an electric field is the integrated value of the torque:

$$PE = \int_0^\theta p_p \mathcal{E} \sin \theta \, d\theta$$

or

$$PE = p_p \mathcal{E}(1 - \cos \theta). \qquad (13\text{-}13)$$

Thus the dipole has lowest potential energy when aligned with the field, and highest when at 180°. Were it not for thermal agitation, all the dipoles would order themselves in alignment with the field. However, like atom vibrations, thermal energy

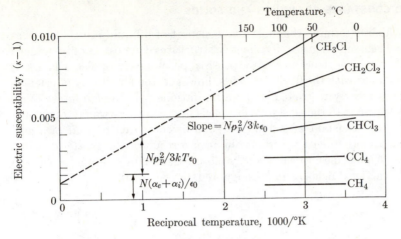

Temperature, °C

Slope $= Np_p^2/3k\epsilon_0$

$Np_p^2/3kT\epsilon_0$

$N(\alpha_e + \alpha_i)/\epsilon_0$

Electric susceptibility, $(\kappa - 1)$

Reciprocal temperature, $1000/°K$

CH₃Cl
CH₂Cl₂
CHCl₃
CCl₄
CH₄

13-7 Electric susceptibilities from permanent dipoles (gas at 22.4 l/mole). (Cf. Fig. 13-5.) (Adapted from data by R. Sanger, *Physik Z.*, **26**, 1926, pp. 556ff.)

disorders the dipoles. Thus the polarization from permanent dipoles is not a simple sum of all the dipoles per unit volume. Rather it is expressed as an orientation polarization, $\mathcal{P}_o$, which is a function of temperature as well as of the electric field and the permanent dipole moment:

$$\mathcal{P}_o = N\left[\frac{(p_p)^2}{3k}\right]\frac{\mathcal{E}}{T}. \qquad (13\text{-}14)^*$$

Figure 13-7 shows the electric susceptibility $(\kappa - 1)$ of several gases versus temperature. Since only the orientation polarization of the dipoles is temperature sensitive, it is possible to distinguish between the susceptibility which arises from induced polarization and that which arises from permanent dipoles.

Example 13-4

What fraction of the polarization of CH_3Cl arises from induced dipoles at 25°C? At 100°C? *Answer.* Refer to Fig. 13-7. At 298°K $(1/T = 0.00336)$,

$$\kappa - 1 = 0.0090.$$

At 373°K $(1/T = 0.00268)$,

$$\kappa - 1 = 0.0075.$$

Thus,

$$\frac{\Delta(\kappa - 1)}{\Delta(1/T)} = -2.2°K,$$

$$N(\alpha_e + \alpha_i)/\epsilon_0 = 0.0090 - 2.2/298 = 0.0016,$$

$$\left(\frac{\mathcal{P}_i}{\mathcal{P}_o + \mathcal{P}_i}\right)_{25°C} = \frac{0.0016\epsilon_0\mathcal{E}}{0.0090\epsilon_0\mathcal{E}} = 17.8\%,$$

$$\left(\frac{\mathcal{P}_i}{\mathcal{P}_o + \mathcal{P}_i}\right)_{100°C} = \frac{0.0016\epsilon_0\mathcal{E}}{0.0075\epsilon_0\mathcal{E}} = 21.3\%. \quad \blacktriangleleft$$

---

* This is a special form of the Langevin equation and is derived through hyperbolic functions, as shown in many solid-state physics texts.

13-4 DIELECTRIC CONSTANTS OF LIQUIDS AND SOLIDS

The principles which determine the dielectric constant of a liquid are the same as those previously discussed for gases. Symmetric structures can possess only induced polarization; asymmetric structures also possess permanent dipoles which, in liquids, can be oriented with the field. In solids, however, the possibility of orientation polarization is greatly restricted. These limitations will be of interest to us because most dielectric materials are solids. Let us first consider two factors which increase the dielectric constant of a liquid or solid as compared to that of a gas. (1) There are many more dipoles per unit volume as a result of the greater density. (2) The atoms and molecules are sufficiently close together in a liquid or solid so that the effect of the local fields cannot be ignored.

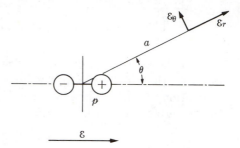

13-8 Local field. A dipole has its own field in addition to any externally applied field. The average local field thus exceeds the external field (Eq. 13-16).

Local fields. Since each dipole has a positive and a negative end, it establishes a field in the surrounding area that is a function of distance, a, and angle, θ. An analysis on the basis of field theory gives the following values for the vector components, $\mathcal{E}_r$ and $\mathcal{E}_\theta$, of the field (Fig. 13-8):

$$\mathcal{E}_r = \frac{1}{4\pi\epsilon_0}\left(\frac{2p\cos\theta}{a^3}\right), \tag{13-15a}$$

$$\mathcal{E}_\theta = \frac{1}{4\pi\epsilon_0}\left(\frac{p\sin\theta}{a^3}\right). \tag{13-15b}$$

Thus when a dipole with a moment p is oriented with an external field $\mathcal{E}$, a point immediately ahead (or behind) is exposed to a field $\mathcal{E} + p/2\pi\epsilon_0 a^3$, where a is the distance from the dipole. Atoms or molecules in these positions "see" a local field which is greater than the external field, and exhibit greater polarization than would be derived from the external field alone. If all dipoles were in one straight line, the local field on each dipole of the chain would be

$$\mathcal{E}_{\text{loc}} = \mathcal{E} + \left(\frac{1.2}{\pi}\right)\frac{\mathcal{P}}{\epsilon_0}. \tag{13-15c}$$

In a more realistic, three-dimensional array with an 8-fold coordination, e.g., bcc, the local field, called the *Lorentz field*, is

$$\mathcal{E}_{\text{loc}} = \mathcal{E} + (1/3)(\mathcal{P}/\epsilon_0). \tag{13-16}$$

Since this latter relationship is sufficient for all condensed phases, we will use it as a general relationship.*

The polarization in solids and liquids must be rewritten to account for the local field:

$$\mathcal{P} = N\alpha\mathcal{E}_{loc}, \tag{13-17}$$

where the symbols have the earlier connotations of polarization, number per unit volume, polarizability, and local field, respectively. The combination of Eqs. (13-16) and (13-17) gives

$$\mathcal{P} = \frac{N\alpha\mathcal{E}}{1 - N\alpha/3\epsilon_0} = \epsilon_0(\kappa - 1)\mathcal{E}. \tag{13-18a}$$

Thus the dielectric constant is a function of the density through N and the polarizability:

$$\kappa = 1 + N\alpha/(\epsilon_0 - N\alpha/3). \tag{13-18b}$$

Example 13-5

The dielectric constant of silica glass is 3.8 at atmospheric pressure. What is the dielectric constant at a pressure of 10,000 atm? Its bulk modulus is 5,000,000 psi.

Answer

$$10{,}000 \text{ atm} = 147{,}000 \text{ psi.}$$

From Eq. (10-3b),

$$\Delta V/V = 147{,}000 \text{ psi}/5{,}000{,}000 \text{ psi}$$
$$= 0.03.$$

Therefore,

$$N_{10{,}000} = 1.03 N_1.$$

From Eq. (13-18b),

$$N\alpha = 3\epsilon_0(\kappa - 1)/(\kappa + 2),$$
$$N_1\alpha = 12.82 \times 10^{-12} \text{ farad/m},$$
$$N_{10{,}000}\alpha = 13.21 \times 10^{-12} \text{ farad/m},$$
$$\kappa = 1 + \frac{13.21 \times 10^{-12}}{8.854 \times 10^{-12} - 13.21 \times 10^{-12}/3}$$
$$= 4.0. \quad \blacktriangleleft$$

Dielectric constants versus temperature. A general plot of electric susceptibility, $\kappa - 1$, and therefore of the dielectric constant, κ, versus temperature is given in Fig. 13-9(a). In the liquid range the dielectric constant increases as the temperature is decreased until the freezing temperature, T_m, is reached, where orientation polarization becomes impossible. Major exceptions occur, however, as illustrated by HCl and water (Fig. 13-9b). Although they are in a solid, the

---

* No attempt has been made to derive Eq. (13-15) or Eq. (13-16) here. The interested student is referred to texts on solid-state physics.

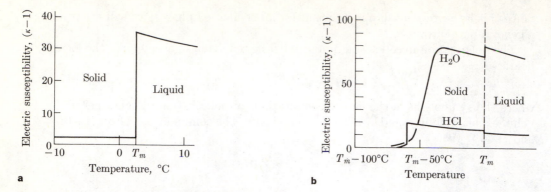

13-9 Electric susceptibilities versus temperature. (a) Nitrobenzene. The permanent dipole of this large molecule cannot respond to the applied field within a solid as it can within a liquid. (b) Water and HCl. These small molecules can reorient themselves within a solid below the melting temperature to give dipole alignment. At still lower temperatures, however, such reorientation becomes impossible. (Adapted from data by C. P. Smyth and C. S. Hitchcock, *Journ. Amer. Chem. Soc.*, **54**, 1932, p. 4631; and **55**, 1933, p. 1830.)

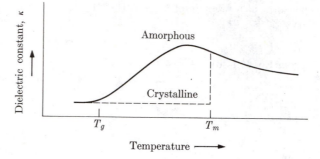

13-10 Dielectric constants of polymers. Since most plastics are at least partially amorphous below their melting point, they exhibit an increase in their dielectric constant with increased temperature. In effect, the permanent dipoles are more readily polarized.

hydrogen atoms of the molecules, which are simply protons, are able to align with the field. Thus, after a discontinuity caused by a volume change at freezing, the dielectric constant of these solids continues to rise until the temperature is low enough to preclude even the hydrogen displacement.

Amorphous solids. These materials present a special case which is of interest to us because glasses and many insulating plastics have amorphous structures. Above the glass-transition temperature, the atoms and molecules have limited freedom for movement, so that it is possible for orientation polarization to occur (Fig. 13-10).

The actual mechanism of polarization involves displacements of the polar sites of a molecule within a plastic (Fig. 13-11a), or anelastic jumps of ions within a glass (Fig. 13-11b). Since charge sites are never identical in noncrystalline materials, the dielectric constant changes over a temperature range rather than abruptly at a single temperature.

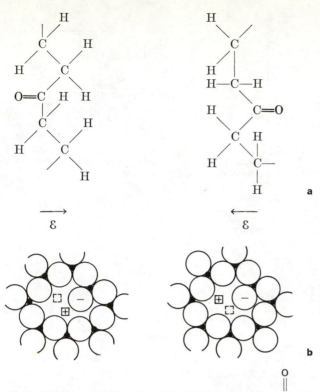

13-11 Dipole orientations in solids. (a) Polymer. The —C— group is polar because oxygen is more electro-
negative than carbon. Likewise the hydrogen atoms (a proton at the end of a bond) are displaced with the
field. (b) Glasses. A positive ion, e.g., Na^+, within a semirigid Si—O network is displaced with the field.

13-5 ALTERNATING FIELDS

Our discussions of dielectric constants so far have centered around a static field.
Since the majority of dielectric applications involve a fluctuating field, it is necessary
to examine the effect of a changing field on polarization.

Dielectric constants versus frequency. Figure 13-12(a) provides a plot of dielectric constants of MgO versus frequency. The obvious feature is the steplike
changes which occur. Above about 10^{13} Hz, the natural frequency of atomic
vibrations, only electronic polarization is possible. Finally, above about 10^{16} Hz,
the natural frequency of electronic vibrations about an atom, the polarization
drops to zero. Figure 13-12(b) presents similar data for polyvinyl chloride and
polytetrafluoroethylene. The difference between the two curves at lower frequencies is due to the permanent dipole of the C_2H_3Cl mer, which introduces
orientational polarization.

The combined effect of temperature and frequency on orientation polarization
(Fig. 13-13) may be explained simply by noting that the polarization requires more
time at lower temperatures because of restrictions on atom and molecular movements for charge displacements. As a result, the reorientation cannot keep abreast
of higher frequency fluctuations.

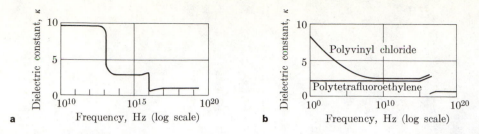

13-12 Dielectric constant versus frequency. (a) MgO. Ionic polarization does not occur above about 10^{13} Hz. (b) Polyvinyl chloride (pvc) and polytetrafluoroethylene (ptfe). Pvc has orientation polarization as well as electronic polarization. Electronic polarization does not occur above about 10^{16} Hz.

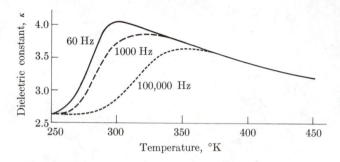

13-13 Combined effects of temperature and frequency on the dielectric constant (rubber with 12 w/o S). (After A. H. Scott, A. T. McPherson, and H. L. Curtis, *J. Res. N.B.S.*, **11**, 1933, p. 173.)

Dielectric losses. In general, polarization is a reversible behavior of a material within an electric field. Its reversibility, however, depends on whether there is sufficient time available for the necessary electronic, atomic, and dipole movements. Moreover, polarization requires finite times, and consequently energy losses do occur when the period of cycling approximates these time delays.

To understand this better, let us consider a cyclic field application where

$$\mathcal{E} = \mathcal{E}_m \sin \omega t. \tag{13-19}$$

Here, $\mathcal{E}$ is the electric field at any particular time, t; $\mathcal{E}_m$ is the maximum field; and ω is the angular velocity of field reversal. If there were no time delays, the charge density, $\mathfrak{D}$, would be in phase with the field:

$$\mathfrak{D} = \mathfrak{D}_m \sin \omega t, \tag{13-20a}$$

and Fig. 13-14(a) would apply. However, with a lag in polarization, the charge density changes as follows:

$$\mathfrak{D} = \mathfrak{D}_m \sin (\omega t - \delta). \tag{13-20b}$$

The term δ is called the *loss angle*, and the hysteresis loop of Fig. 13-14(b) is applicable. The maximum field and the maximum polarization are not coincident. Also, a coercive field is required to return the polarization to zero.

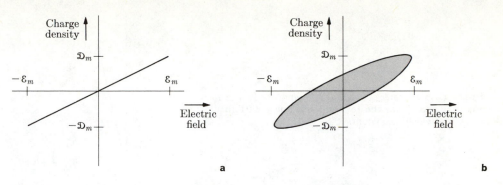

13-14 $\mathcal{E}$-$\mathfrak{D}$ cycle. (a) Loss free. (b) High loss. Equation (13-20b) applies. (Compare with Fig. 10-18.)

The energy loss per cycle, ΔU, is the area of the hysteresis loop; it is obtained by using Eq. (13-20b) and the differential of Eq. (13-19):

$$\Delta U = \int_0^{2\pi} \mathfrak{D}\, d\mathcal{E}$$

$$= \int_0^{2\pi} \mathfrak{D}_m \mathcal{E}_m \cos(\omega t) \sin(\omega t - \delta)\, d(\omega t)$$

$$= \pi \mathcal{E}_m \mathfrak{D}_m \sin \delta. \tag{13-21a}$$

From Eq. (13-2),

$$\Delta U = \pi \mathcal{E}_m^2 \epsilon_0 \kappa \sin \delta. \tag{13-21b}$$

The last two paragraphs intentionally parallel those in Section 10-6 where anelastic energy losses were described, because dielectric and anelastic losses have much in common. The electrical engineer usually considers *power losses*, PL, in electrical applications, instead of working with fractional energy losses as for damping (Eq. 10-19). This power loss is the rate of energy loss per unit volume and is a product of Eq. (13-21b) and frequency, $\omega/2\pi$:

$$PL = \frac{\omega \mathcal{E}^2 \epsilon_0}{2} (\kappa \tan \delta), \tag{13-22}$$

where $\tan \delta$ is called the *loss tangent* (or *dissipation factor*), and $\kappa \tan \delta$ the *loss factor*.

The replacement of $\sin \delta$ with $\tan \delta$ in Eq. (13-22) is permissible because any practical electrical material for high-frequency applications must have a low value of δ. By using $\tan \delta$, it is possible to separate the dielectric constant into the in-phase component, κ', and the out-of-phase component, κ'', as shown in Fig. 13-15:

$$\tan \delta = \kappa''/\kappa'. \tag{13-23}$$

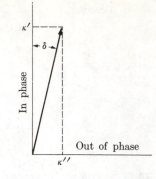

13-15 Loss tangent: $\tan \delta = \kappa''/\kappa'$, where δ is the loss angle; κ' and κ'' are the real and imaginary parts of the complex dielectric constant.

The out-of-phase component permits us to analyze the power loss as a function of frequency. To do this, we must start with the relationship

$$\alpha = \frac{\alpha_0}{1 + i\omega\tau},$$

(13-24)[*]

where α_0 is the static polarizability, α is the complex polarizability for fluctuating fields, ω is the angular velocity, and τ is a dielectric relaxation time much as λ was a mechanical relaxation time in Eq. (10-10). At the natural or resonant frequency, ω_0, we have $\tau = 1/\omega_0$. Finally, i has the usual connotation of $\sqrt{-1}$. Since

$$\mathcal{P} = N\alpha\mathcal{E}_{loc} = \epsilon_0(\kappa - 1)\mathcal{E}_{loc},$$

then

$$\kappa - 1 = \left(\frac{N\alpha_0/\epsilon_0}{1 + i\omega\tau}\right)\left(\frac{1 - i\omega\tau}{1 - i\omega\tau}\right),$$

$$\kappa = 1 + \underbrace{\frac{N\alpha_0/\epsilon_0}{1 + \omega^2\tau^2}}_{\text{real} = \kappa'} - \underbrace{i\frac{N\alpha_0\omega\tau/\epsilon_0}{1 + \omega^2\tau^2}}_{\text{imaginary} = i\kappa''},$$

(13-25)

$$\kappa = \kappa' - i\kappa''.$$

(13-26)

Let us examine three special cases.

1) At *low frequencies*, $\omega \ll \omega_0$ and $\omega\tau \to 0$. Therefore, $\kappa'' \to 0$, and

$$\kappa' = (1 + N\alpha/\epsilon_0) = \kappa.$$

Since the out-of-phase component drops to zero, so do $\tan \delta$ and the loss factor, $\kappa \tan \delta$. There are negligible energy losses.

2) At *high frequencies*, $\omega \gg \omega_0$ and $\omega\tau \to \infty$. Therefore, $\kappa' \to 1$, and $\kappa'' \to 0$. Again κ''/κ' and the loss factor, $\kappa \tan \delta$, are almost nil.

---

[*] This equation can be derived in various ways. One analysis may be observed in C. Kittel, *Elementary Solid State Physics: A Short Course*, New York: Wiley, 1962, p. 87.

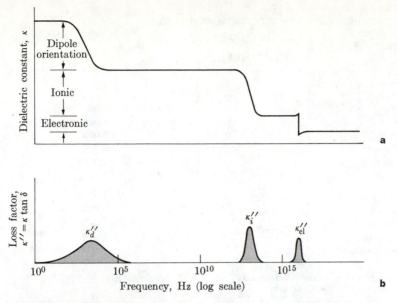

13-16 Dielectric constant and loss factor versus frequency. The dielectric constant, κ, approximates κ' in any usable dielectric insulator.

3) Near *resonant frequencies*, $\omega \sim \omega_0$ and $\omega\tau \rightarrow 1$. By differentiating the imaginary part of Eq. (13-25), we find that κ'' is at a maximum when $\omega\tau = 1$, so that the power losses are at a maximum.

A plot of κ' and κ'' versus frequency is given in Fig. 13-16. A frequency limit for each type of polarization occurs when the frequency matches the natural frequency. At these frequencies, there is a large value of κ'' and therefore of energy absorption.

Electrical insulators. The dielectric constant, κ, and the loss tangent, tan δ, are usually reported for dielectric materials rather than κ'', the imaginary part of the complex dielectric constant. Sample data are given in Table 13-3, where it becomes apparent that there may be variations with frequency.

Electrical insulators fall conveniently into two categories: plastic (polymer) and ceramic. The former category contains many materials which can be made into film and fiber products, but are restricted to relatively low-temperature applications, particularly when dimensional stability is critical. Although the dielectric losses of some of these materials, such as polyethylene, polystyrene, and nylon, are low, the dielectric constants are also relatively low, since they arise only from induced polarization or the restricted movements of polar radicals along the sides of polymer chains as shown below from Fig. 13-11,

$$\begin{matrix} & O \\ & \| \\ -\!\!\!\!&C\!\!\!\!&- \end{matrix} .$$

TABLE 13-3

DIELECTRIC PROPERTIES OF ELECTRICAL INSULATORS*

Material	κ		tan δ	
	60 Hz	10^6 Hz	60 Hz	10^6 Hz
Ceramics				
Porcelain	6	—	0.010	—
Steatite	6	6	0.005	0.003
Zircon	9	8	0.035	0.001
Alumina	—	9	—	<0.0005
Soda-lime glass	7	7	0.1	0.01
E-glass	—	4	—	0.0006
Fused silica	4	3.8	0.001	0.0001
Polymers				
Nylon 6/6	4	3.5	0.02	0.03
Polyethylene	2.3	—	<0.0005	—
Polystyrene	2.5	2.5	0.0002	0.0003
Polyvinyl chloride	7.0	—	0.1	—

* These are typical values. Values may vary for different commercial products as a result of fillers and other compositional variables.

Ceramic dielectrics, including glass, are generally stable at higher temperatures, but are commonly more difficult to fabricate than polymeric insulators. They can have relatively high dielectric constants, particularly if ions have choices of more than one site, or if the ions are polyvalent. Of course, this can also lead to power losses if the period for atom movements is approximately the time cycle of an alternating circuit. This is the situation for the Na^+ ion movements in soft soda-lime glass at 10^6 Hz (Table 13-3). As a result, electrical glasses are made with very low sodium content.

Example 13-6

Assuming no heat loss, how much will the temperature rise in a polyethylene insulator 1 cm $\times$ 1 cm $\times$ 0.01 cm thick if it is used in a 1.3-volt circuit at 10^9 Hz for 6 min? (The specific heat and density of polyethylene are 0.54 cal/gm $\cdot$ °C and 0.94 gm/cm³, respectively; κ tan δ = 0.001.)

Answer. From Eqs. (13-21) and (13-22),

$$\text{PL} = \pi(10^9/\text{sec})(1.3 \text{ volts}/0.0001 \text{ m})^2(8.854 \times 10^{-12} \text{ amp} \cdot \text{sec/volt} \cdot \text{m})(0.001)$$
$$= 4.7 \times 10^3 \text{ watts/m}^3.$$

Therefore,

$$\Delta U = (4700 \text{ watts/m}^3)(10^{-8} \text{ m}^3)(860 \text{ cal/watt} \cdot \text{hr})(0.1 \text{ hr})$$
$$= 0.004 \text{ cal}.$$
$$\Delta T = (0.004 \text{ cal})/(0.54 \text{ cal/gm} \cdot \text{°C})(0.94 \times 10^6 \text{ gm/m}^3)(10^{-8} \text{ m}^3)$$
$$= 0.8\text{°C}. \blacktriangleleft$$

13-6 OPTICAL PROPERTIES

Optical behavior is related to dielectric properties because photons of light provide an electromagnetic field within the material. We shall consider only two properties, *index of refraction* and *absorption*. The first is related to electronic polarization; the second, to dielectric losses.

Index of refraction. The electronic part of the dielectric constant, κ_e, is equal to the square of the index of refraction, n:

$$\kappa_e = n^2. \tag{13-27}$$

The index of refraction is easily determined in transparent solids by Snell's law:

$$\frac{n_2}{n_1} = \frac{\sin \phi_1}{\sin \phi_2} = \frac{v_1}{v_2}, \tag{13-28}$$

where the angles ϕ_2 and ϕ_1 are measured from the interface normal, and v indicates the respective light velocities. Therefore, one may readily separate electronic and ionic polarizability of a solid because only the former is effective at visible frequencies, while both are effective at lower frequencies.

Example 13-7

Calculate the average electronic and ionic polarizabilities of MgO, assuming that its dielectric constant at 10^{10} Hz is 9.6 and its index of refraction is 1.74. MgO has an NaCl structure (Fig. 4-1) and a lattice constant of 4.2 A.

Answer

$$\mathcal{P}_e = [(1.74)^2 - 1]\epsilon_0 \mathcal{E} = 2.0\epsilon_0 \mathcal{E},$$

$$\mathcal{P}_i = (9.6 - 1)\epsilon_0 \mathcal{E} - 2.0\epsilon_0 \mathcal{E} = 6.6\epsilon_0 \mathcal{E},$$

$$\alpha_e = \mathcal{P}_e/N\mathcal{E}$$

$$= \frac{(2.0)(8.854 \times 10^{-12} \text{ farad/m})}{8/(4.2 \times 10^{-10} \text{ m})^3} = 1.64 \times 10^{-40} \text{ farad} \cdot \text{m}^2,$$

$$\alpha_i = \frac{(6.6)(8.854 \times 10^{-12} \text{ farad/m})}{8/(4.2 \times 10^{-10} \text{ m})^3} = 5.4 \times 10^{-40} \text{ farad} \cdot \text{m}^2. \blacktriangleleft$$

Local fields are not symmetric in noncubic crystals. Therefore, the electronic polarizability varies with the direction of light vibration. This means that the index of refraction is also anisotropic, or directional. For example, calcite ($CaCO_3$) has its CO_3^{2-} ions oriented so that the four atoms are coplanar and in the (0001) plane of the hexagonal crystal. The electronic polarizability is 24% greater perpendicular to the [0001] direction than parallel to it. Thus light vibrating perpendicular to the c-axis has higher values of κ_e and n than does light vibrating parallel to the c-axis. The two indices are 1.66 and 1.49, respectively. This difference is called *birefringence* and appears as double refraction (Fig. 13-17).

Cubic and amorphous phases do not possess birefringence because the three axial directions are electrically symmetric. Tetragonal, hexagonal, and rhombic

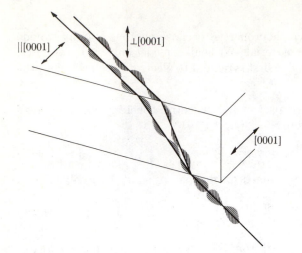

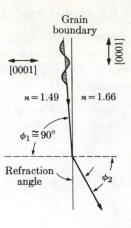

13-17 Double refraction. The index of refraction depends on the vibration plane of the light and the relative polarizability of the electrons in the planes.

13-18 Grain-boundary refraction. (See Example 13-8.)

crystals have two indices of refraction, while the remaining crystal systems (Table 5-1) have three indices.

A transparent material becomes *translucent* if light is scattered by internal reflections or refracted at grain and phase boundaries (Example 13-8). As a rule, only glasses, amorphous plastics, and single crystals lack internal heterogeneities which produce scattering. Exceptions of current technological interest are the pore-free, polycrystalline, single-phase ceramics with cubic structures. These include LiF and MgO, which are transparent and rigid up to their melting temperatures.

Example 13-8

Two grains of calcite, $CaCO_3$, join so that their [0001] directions are perpendicular (Fig. 13-18). What is the maximum angle of refraction which can occur as a ray of light crosses the grain boundary?

Answer. The greatest refraction will occur when the incident angle in the lower-index orientation is 90° to the boundary normal.

From Eq. (13-28) and the index data in previous paragraphs,

$$\sin \phi_2 = \frac{1.49}{1.66} \sin 90°,$$

$$\phi_2 = 64°,$$

$$\text{refraction} = 90° - 64° = 26°.$$

Note. The above example shows why light cannot pass straight through polycrystalline calcite, i.e., marble, without scattering. ◄

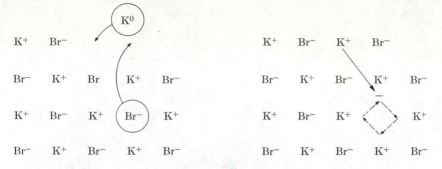

13-19 Color center. An ionized vacancy is a low-energy trap for an electron. When the vibrational frequency of the electron within an ion vacancy lies in the visible light range, the electron absorbs the light.

Absorption. The visible part of the electromagnetic spectrum lies near 5×10^{14} Hz, between two of the peaks shown in Fig. 13-16(b). Therefore, simple ionic and covalent materials are transparent because the value of κ'' is essentially zero for both ionic and electronic polarizability. The exceptions to this general rule are of interest to us because they include the opaque and colored materials.

Metals are opaque because valence electrons are not fixed to individual atoms. These electrons can oscillate in the range of visible frequencies, thus absorbing energy and precluding light transmission, except in the case of submicron thicknesses.

Colored materials are the consequence of partial absorption of white light and the resulting selective transmission of the balance of the spectrum. We will consider one of several mechanisms of light absorption, that of *color centers*. If KBr, for example, is heated in the presence of potassium vapors, excess potassium atoms can deposit on the surface as shown in Fig. 13-19(a). Recall from Fig. 6-6(d) and Eq. (6-5) that there is a tendency for vacancies to form within a crystal. This tendency is accentuated for Br^- ions in the presence of excess potassium atoms, leaving a local excess of positive charge around each Br^- vacancy. An electron readily moves from the surface potassium atom to *ionize* the vacancy, where it is free to hop from the vicinity of one to another of the six* adjacent K^+ ions. Its vibrational period, τ, is 2.06×10^{-15} sec; hence, its resonant frequency is 4.85×10^{14} Hz. This frequency corresponds to an optical wavelength of 6200 A. The structural defect just described is called an *F-center* in deference to the original German description as a Farbe-point, or color-point. Other defects provide other absorptions.

Optical measurements on KBr present the absorption curve of Fig. 13-20, which removes most of the orange and part of the red colors and transmits more of the blue and green. This bell-shaped curve corresponds directly with the absorption curve shown in Fig. 13-16(b), which is the κ''-component of the dielectric constant.

---

*   In three dimensions.

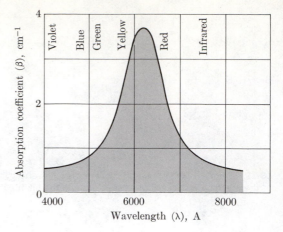

13-20 Optical absoprtion versus wavelength (KBr with F-centers).

The decrease of light intensity during transmission, $-dI/dx$, is proportional to the light intensity, I:

$$-dI/dx = \beta I, \tag{13-29a}$$

where β is called the *absorption coefficient;* or

$$\int_{I_0}^{I} \frac{dI}{I} = -\int_{0}^{x} \beta \, dx, \tag{13-29b}$$

and so

$$I = I_0 e^{-\beta x}. \tag{13-29c}$$

Example 13-9

Assume that equal intensities of yellow-green (5600 A) and orange (6200 A) light enter a KBr crystal which is 3 mm thick.
a) Compare the transmitted intensities of the above two colors when there are sufficient color centers to give the curve of Fig. 13-20.
b) Compare their transmitted intensities when there are twice as many color centers.

Answer. From Fig. 13-20, $\beta_{6200} = 3.7$/cm and $\beta_{5600} = 1.8$/cm.

a)
$$\frac{I_{5600}}{I_{6200}} = \frac{I_0 e^{(-1.8)(0.3)}}{I_0 e^{(-3.7)(0.3)}} = 1.8.$$

b)
$$\frac{I_{5600}}{I_{6200}} = \frac{I_0 e^{(-3.6)(0.3)}}{I_0 e^{(-7.4)(0.3)}} = 3.1.$$

Note. The change in ratios gives a change in apparent color. ◄

13-7 PIEZOELECTRICITY

Not all crystalline materials have a center of symmetry. Quartz, SiO_2, for example, does not; another example is $BaTiO_3$. When *below* 120°C, $BaTiO_3$ is tetragonal, with a Ba^{2+} ion at each corner of the unit cell, an O^{2-} ion *near* the center of each

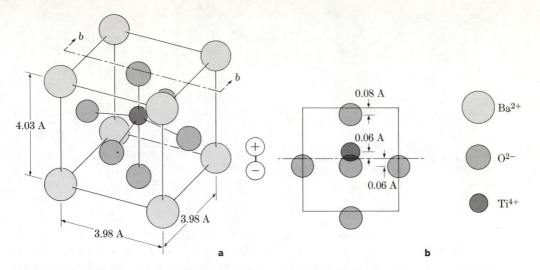

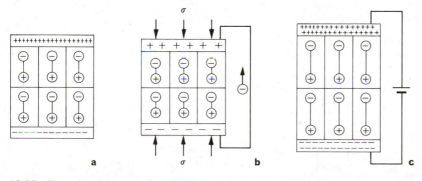

13-21 Tetragonal BaTiO$_3$. Each unit cell has a dipole moment. (Not drawn to scale.)

13-22 Piezoelectricity. (a) The dipoles within the material produce a charge difference between the two ends. (b) Pressure changes the dipole moment and changes the charge density (or introduces a voltage difference if the two ends are not shorted). (c) An external field changes the polarization and alters the dipole moment arm (introducing dimensional changes).

face, and a Ti^{4+} ion *near* the center of the cell (Fig. 13-21).* Thus, each unit cell contains an identifiable dipole.

As a result of *spontaneous polarization* arising from strong local fields, all cells tend to align in the same electrical direction. This gives a positive and a negative end to a local *domain* within a crystal. Electrodes placed on the ends carry opposite charges (Fig. 13-22a). Now consider two alternatives. (1) Pull (or compress) the crystal. Its dipole moment is changed; therefore, a voltage difference is realized (Fig. 13-22b). (2) Place a voltage across the crystal (Fig. 13-22c). The charges respond to the field to alter the moment arm and change the dimensions. We have

---

* Above 120°C, BaTiO$_3$ is cubic, with the O^{2-} ions *at* the center of each face and a Ti^{4+} ion *at* the center of the cube.

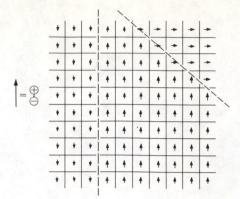

13-23 Ferroelectric domains. Domain boundaries (----)
are interfaces of higher energy. Domains, like grains,
contain many unit cells; however, adjacent domains are
crystallographically related.

a voltage-strain, or electromechanical, *transducer*. Materials of this type are called *piezoelectric*, i.e., pressure-electric. Materials such as these are used for pressure gages, for phonograph cartridges to transpose mechanical movements into electronic signals, for high-frequency sound generators, etc.

Example 13-10

The polarization of a piezoelectric ceramic is 0.04 coul $\cdot$ m/m^3. A pressure is applied in the polarity direction of a 1.0 cm $\times$ 1.0 cm $\times$ 1.0 cm cube of the material to give a strain of 0.004. How many electrons will flow between the ends if they are shorted?

Answer

$$\Delta\mathcal{P} = \frac{\Delta d}{d}\frac{p}{V} = (0.004)(0.04) \text{ coul} \cdot \text{m/m}^3$$

$$= 1.6 \times 10^{-4} \text{ coul} \cdot \text{m/m}^3.$$

$$\text{Number of electrons} = \frac{(1.6 \times 10^{-4} \text{ coul} \cdot \text{m/m}^3)(10^{-2} \text{ m})^2}{1.6 \times 10^{-19} \text{ coul/electron}}$$

$$= 10^{11} \text{ electrons.} \quad \blacktriangleleft$$

13-8 FERROELECTRICITY

A limited number of piezoelectric materials also possess *ferroelectricity*. Ferroelectric materials polarize spontaneously; i.e., many dipoles assume the same orientation. Our prototype material is again $BaTiO_3$. As previously noted, $BaTiO_3$ is cubic *above* 120°C, with Ba^{2+} ions at the corners, O^{2-} ions at face centers, and a Ti^{4+} ion at the cube center. Since it is symmetric, there is no permanent dipole moment. Below 120°C, there is less thermal agitation and greater density. Both factors permit a spontaneous distortion of the structure to that shown in Fig. 13-21. *Domains* are established as shown in Fig. 13-23. Although each domain involves many ($\sim 10^{18}$) unit cells, each crystal commonly contains a number of domains oriented in all six of the coordinate directions. Therefore, the net polarization is normally near zero.

The net polarization can be increased if an external electric field is applied, because those domains which are favorably oriented grow at the expense of those

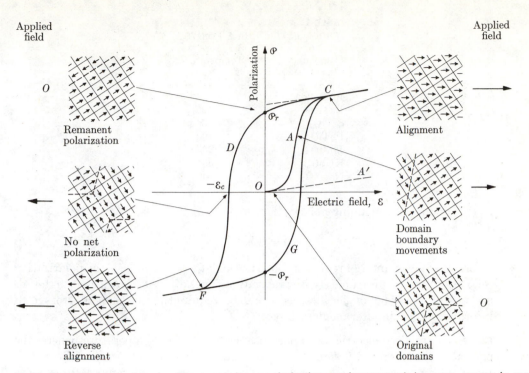

Applied field (top left) — O — Remanent polarization

Applied field (top right) — Alignment

No net polarization

Domain boundary movements

Reverse alignment

Original domains

Polarization, $\mathcal{P}$

$\mathcal{P}_r$

C

A

A'

D

$-\mathcal{E}_c$

O

Electric field, $\mathcal{E}$

G

$-\mathcal{P}_r$

F

13-24 Ferroelectric hysteresis. The spontaneous polarization can be reversed; however, energy is consumed with each cycle (see text).

which are not oriented with the field. This growth occurs by a shift of the titanium ions in the direction of the field and the oxygen ions in the opposite direction, so that the unit cell of Fig. 13-21(b) has its dipole oriented with the field. With domain growth, the net polarization increases rapidly (curve OAC in Fig. 13-24) until most of the domains are aligned in the same direction. This is called *saturation polarization*. Additional field strength can only add small amounts of electronic and ionic polarization, or rotate dipoles more strongly into the direction of the field.

The polarization does not disappear immediately when the field is removed; rather, the electronic and ionic polarizations are relaxed and a residual, or *remanent polarization*, $\mathcal{P}_r$, persists. An opposing *coercive field*, $-\mathcal{E}_c$, is required to balance the domains and reduce the net polarization to zero. Cyclic fields produce a *hysteresis loop*, as indicated by the completed $CDFGC$ path of the $\mathcal{P}$-$\mathcal{E}$ curves. The energy consumed per cycle is equal to the $\mathcal{P}\mathcal{E}$ area within the loop.

Ferroelectric materials, like ferromagnetic materials (Chapter 16), may be used for information storage. An applied signal may be $+\mathcal{E}$ or $-\mathcal{E}$ to store $+\mathcal{P}_r$ or $-\mathcal{P}_r$; the value of $\mathcal{P}_r$ may later be sensed by appropriate circuitry.

Relaxation. The remanent polarization decays if domains of other orientations grow within the material. Depolarization occurs with time because it permits a slight reduction in energy. Since the rate of depolarization, $-d\mathcal{P}/dt$, is proportional

TABLE 13-4

FERROELECTRIC CURIE POINTS

Compound	Temperature, °C
Rochelle salt	24
$BaTiO_3$	120
$SrTiO_3$	−200
$PbTiO_3$	490
$PbZrO_3$	233
$KNbO_3$	434
$NaNbO_3$	640

to the amount of remaining polarization, $\mathcal{P}$,

$$\mathcal{P} = \mathcal{P}_z e^{-t/\tau_p}, \tag{13-30}$$

where $\mathcal{P}_z$ is the amount of remanent polarization at $t = 0$, and τ_p is the polarization relaxation time. Unfortunately, the relaxation times of most ferroelectric materials are not of the order of decades, as would be required for permanent storage of information in computer memory tapes.

Curie point. Ferroelectric behavior disappears at the temperature where the crystal assumes greater symmetry. Consistent with magnetic terminology, this temperature is called a Curie point. The phase-transition temperatures for a number of ferroelectric materials are listed in Table 13-4.*

Above the Curie point, these materials have a linear $\mathcal{P}$-$\mathcal{E}$ behavior which is sometimes called *paraelectric*. It simply involves electronic and (large) ionic dielectric polarizabilities.

Example 13-11

On the basis of Fig. 13-21, calculate the saturation polarization of $BaTiO_3$ when all of the dipoles of the unit cells are oriented in the same direction.

Answer. Use one unit cell containing one-half of each face atom and one-eighth of each corner atom. With reference to a plane at $c/2$,

$$\left.\begin{array}{l} p_{Ba2+} = (4/8 \text{ ions})(2q/\text{ion})(+c/2) + (4/8 \text{ ions})(2q/\text{ion})(-c/2), \\ p_{Ti4+} = (1 \text{ ion})(4q/\text{ion})(+0.06), \end{array}\right\} +0.24q$$

$$\left.\begin{array}{l} p_{O2-} = (4/2 \text{ ions})(-2q/\text{ion})(-0.06), \\ p_{O2-} = (1/2 \text{ ion})(-2q/\text{ion})(c/2 - 0.08), \\ p_{O2-} = (1/2 \text{ ion})(-2q/\text{ion})(-c/2 - 0.08), \end{array}\right\} \quad +0.40q$$

$$\overline{ +0.64q}$$

$$\mathcal{P} = \frac{(1.6 \times 10^{-19})(0.64) \text{ coul} \cdot \text{A}}{\sim 64 \text{ A}^3} = 1.6 \times 10^{-21} \text{ coul/A}^2 \quad (= 0.16 \text{ coul} \cdot \text{m/m}^3). \blacktriangleleft$$

---

* Many readers will recognize the similarity in appearance between the ferroelectric hysteresis loop (Fig. 13-24) and a ferromagnetic hysteresis loop (Chapter 16). Because they are analogous, comparable terms are used for both. While this is advantageous the prefix *ferro-* is misleading because ferroelectric materials seldom, if ever, contain iron, as do most ferromagnetic materials.

REFERENCES FOR FURTHER READING

Azaroff, L. V., and J. J. Brophy, *Electronic Processes in Materials*. New York: McGraw-Hill, 1963. Dielectric processes are discussed in Chapter 12. The theoretical bases are developed more fully than in this text. Otherwise it is on the same level as this text. Recommended as supplementary reading.

Dekker, A. J., *Electrical Engineering Materials*. Englewood Cliffs, N. J.: Prentice-Hall, 1959. Dielectric properties of insulators in static fields are discussed in Chapter 2. Chapter 3 covers dielectrics in alternating fields. Excellent followup reading for this chapter. Advanced undergraduate level.

Hutchinson, T. S., and D. C. Baird, *The Physics of Engineering Solids*. New York: Wiley, 1963. Hutchinson and Baird's chapter on the properties of dielectric crystals is at about the same level as this chapter. Recommended as alternate reading.

Javan, A., "The Optical Properties of Materials," *Scientific American*, **217** [3], September 1967, pp. 238–248. A simply presented quantum-mechanical interpretation of optical properties. Advanced undergraduate level.

Kittel, C., *Elementary Solid State Physics: A Short Course*. New York: Wiley, 1962. Dielectric properties are discussed in Chapter 4. Advanced undergraduate level.

Moore, W. J., *Seven Solid States*. New York: Benjamin, 1967. Paperback. Salt (NaCl) is used as an example of an insulator. From an introductory base, the author builds a detailed explanation of crystal behavior, including color centers. Undergraduate level.

Nussbaum, A., *Electronic and Magnetic Properties of Materials*. Englewood Cliffs, N. J.: Prentice-Hall, 1967. Paperback. A very concise summary of dielectrics is given in Chapter 3. For the advanced student. A more thorough coverage is given in Nussbaum's book, *Electromagnetic and Quantum Properties of Materials*.

Rose, R. M., L. A. Shepard, and J. Wulff, *The Structure and Properties of Materials: IV. Electronic Properties*. New York: Wiley, 1966. Paperback. Chapters 12 and 13 may be used as background reading for this chapter.

Rosenthal, D., *Introduction to Properties of Materials*. Princeton, N. J.: Van Nostrand, 1964. The chapter on dielectrics can serve as supplementary reading for this chapter. Advanced undergraduate level.

Wert, C. A., and R. M. Thomson, *Physics of Solids*. New York: McGraw-Hill, 1964. Static dielectric behavior is discussed in Chapter 16 at the same level as this text. Dielectric losses are discussed in Chapter 17 at the advanced undergraduate level.

PROBLEMS

13-1 Determine the dielectric constant for neon gas at 100 psi and 20°C.

Answer. 1.00067

13-2 To what pressure should methane be compressed to have the same dielectric constant as neon has at 200 psi?

13-3 What polarization is realized in Problem 13-1 when a voltage gradient is 1900 volts/cm?

Answer. 1.13×10^{-8} coul · m/m^3

13-4 The polarization of xenon at 27°C is 10^{-10} coul · m/m^3 in a voltage gradient of 10^4 volts/m. What is the pressure?

13-5 Determine the dipole moment of a CH_3Cl molecule on the basis of the data in Fig. 13-7. [*Hint.* $k = 8.63 \times 10^{-5}$ eV/°K, or 1.38×10^{-23} coul · volt/°K.]

Answer. 5.5×10^{-30} coul · m

13-6 What is the electric susceptibility at 0°C and 1 atm pressure which arises from the permanent dipole in CH_2Cl_2?

13-7 Estimate the dielectric constant of liquid CCl_4. Its density is 1.6 gm/cm³.

Answer. 2.3

13-8 A liquid has a relative dielectric constant, κ, of 3.0 at 25°C and 2.7 at 160°C. At each of these temperatures, what is the fraction of polarizability, expressed as $\mathcal{P}/\mathcal{E}$, which arises from induced polarization. [*Note.* It is probably simpler to solve this partially by graphical methods. Plot $\mathcal{P}/\mathcal{E}$ versus $1/T$.]

13-9 The loss factor for a borosilicate glass is 0.02 at 20°C. What is the maximum frequency which may be used in an electric field of 20 volts/cm if the power loss is to be less than 0.001 watt per cm³?

Answer. 4.5×10^9 Hz

13-10 What is the power loss when steatite is used in an electric field of 300 volts/cm and 10^6 Hz?

13-11 The index of refraction of Al_2O_3 is 1.76 and its density is 4 gm/cm³. What are (a) its ionic polarization at 10^6 Hz and 100 volts/cm and (b) its ionic polarizability?

Answer. a) 4.3×10^{-7} coul · m/m³   b) 3.7×10^{-40} farad · m²

13-12 The two indices of calcite ($CaCO_3$) are 1.49 and 1.66. Compare the electronic polarizabilities in the two vibration directions.

13-13 Quartz and tridymite are two polymorphs of SiO_2, with indices of refraction of 1.55 and 1.485, respectively. Quartz has a density of 2.65. Estimate the density of tridymite. [*Hint.* The density changes the number of atoms per cm³ but not the electronic polarizability.]

Answer. 2.3 gm/cm³

13-14 The absorption coefficients in a plastic of blue and green light are 20/cm and 23/cm, respectively. Assuming equal initial intensities, compare their ratios upon emergence from a 2-mm thickness.

13-15 A leaded glass has 98% transmission through a 1.0-cm thickness. How much light would be transmitted through a 10-in. glass window in a radioactive laboratory?

Answer. 60%

13-16 Lead zirconate is cubic in one of its polymorphs. The unit cell may be chosen so that each corner has a Zr^{4+} ion; the center of each edge, an O^{2-} ion; and the center of the cell, a Pb^{2+} ion. (a) Give the Pb/Zr/O ion ratio. (b) Relocate and sketch the cell so that the Zr^{4+} ion is in the center of the cell. (c) Show that this cell is simple cubic and not fcc or bcc.

13-17 If a piezoelectric crystal has a Young's modulus of 3×10^7 psi, what stress must be applied to change its polarization from 0.1085 to 0.1070 coul · m/m³?

Answer. 450,000 psi compression

13-18 Two materials have the same piezoelectric coefficient, $d\mathcal{E}/d\epsilon$, but have Young's moduli of 10^7 and 5×10^7 psi, respectively. Which will develop the larger spark if a stress of 10,000 psi is utilized in a spark pump? Why?

14 □ electronic properties of solids

14-1 ELECTRONIC CONDUCTIVITY (Diffusion model)

Ionic conductivity was considered in Section 9-4 as the diffusion of ions in an electrical field. Negative ions move against the field; positive ions move with the field. Ionic mobility was viewed as the diffusion coefficient per unit voltage [$(m^2/sec)/volt$], or as the drift velocity per unit field [$(m/sec)/(volt/m)$].

We may consider electronic conductivity in comparable terms, and speak of the drift velocity of *electrons* against the field and *electron holes* in the direction of the field.* This diffusion model can be amplified further by assigning high activation energies to the mobilities of the charge carriers (electrons or electron holes) in insulators, intermediate values to activation energies in semiconductors, and very low values to activation energies of the charge carriers in metals. Actually, we shall use a *band model* which involves an energy gap. Whether we use a diffusion model or a band model, we can calculate the conductivity, σ, in terms of the number of carriers, n, their charge, q, and their mobility, μ:

$$\sigma = nq\mu. \tag{14-1}$$

This equation differs from Eq. (9-11) by having only a single electron charge for each carrier; i.e., $Z = 1$.

Variations in conductivity with temperature, composition, and structure depend on how these factors affect the number of available carriers and their mobility. To understand this, it is necessary to develop a band model.

14-2 BAND MODEL

We observed in Fig. 2-6 that electrons of an isolated atom can occupy only specific energy states, or levels. In effect, they establish standing waves around the atom, and, as is the case with all standing waves, it is possible to have only specific frequencies and harmonics. The Pauli exclusion principle (Section 2-2) states that

---

* The concept of electron holes could have been applied to positive ions, because these atoms are short one or more electrons. It will be a very useful concept to us in our discussion of electronic conduction, because a missing electron serves the same function for conduction as a lattice vacancy does for atomic diffusion.

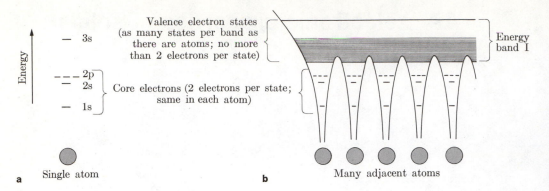

14-1 Energy bands (sodium). When atoms are close together, the valence electrons interact. No more than two may have the same state, or energy level. Therefore many discrete levels exist in the valence energy band.

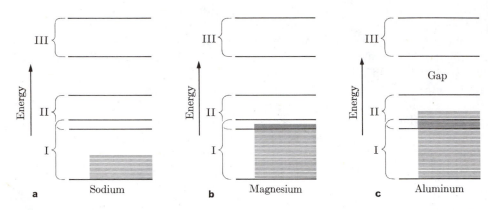

14-2 Energy bands of metals. There are N levels, or states, in each band. Since each band will hold $2N$ electrons, there are vacant energy states at the top of the bands. In Chapter 15 we will observe that the energy states of bands I and II of silicon are completely filled.

no more than two electrons may have the same wave characteristics. In fact, the wave behavior of the two electrons is different because of their opposite magnetic orientations.

In solids, the less tightly bonded valence electrons are not independent.* The movements of *all* of the valence electrons are interrelated. Since the Pauli exclusion principle still holds, as it must for wavelike behavior, the number of discrete energy states becomes enormous, and the differences in their energy values become minute.

An energy band exists corresponding to each valence energy level, or state, in the isolated atom (Fig. 14-1). Each valence band has as many individual states as

---

* For our purposes we may consider that the subvalence electrons are independent and have the same wave behavior they would have in independent atoms. An exception will occur in the case of magnetic materials (Chapter 16).

there are interacting atoms in the system! Therefore, with n atoms, each band can accommodate $2n$ electrons. This means that alkali metals with only one valence electron per atom have the first valence band only half-filled, and aluminum, with three valence electrons, has its second band half-filled (Fig. 14-2). The lower halves are filled first because they represent the lower energy states.

Without giving a detailed explanation, we will simply note that the first and second bands overlap, while the second and third bands do not. This means that magnesium, with two electrons per atom, has its electron states filled into the overlap zone.

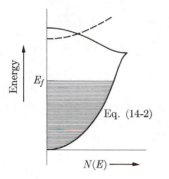

14-3 Density of states (schematic of sodium). The number of states per unit energy, $N(E)$, increases parabolically (Eq. 14-2) at the lower end of the energy band. The second band overlaps the first band.

Density of states. The simplification of Fig. 14-2 overlooked the fact that the states are not uniformly distributed across the energy band. On the basis of introductory quantum mechanics, it may be shown that the density of states, $N(E)$, is greatest in the center of the energy band (Fig. 14-3). Further, the density of states in the lower part of the band increases parabolically with energy according to the formula

$$N(E) = \frac{2\pi}{h^3}(2m)^{3/2}E^{1/2}, \qquad (14\text{-}2)$$

where h is Planck's constant and m is the effective electron mass. It is not necessary for us to derive the relationships that lead to Fig. 14-3, but the student who plans to give closer attention to electrical materials in more advanced courses will need to understand them. We will, however, utilize this relationship whenever necessary.

Fermi energy. Electrons, like atoms, occupy the lowest energy states except when heated or otherwise activated. Thus we will find that at 0°K all the lowest states of the band are filled, and higher states are empty. This is shown in Fig. 14-4(a) for sodium. At higher temperatures some of the electrons can be raised to higher states by thermal activation, with the result that the density profile of occupied states is altered. Although the profile changes with temperature, the energy level at which 50% of the states are occupied remains constant. This is called the Fermi energy level, E_f.

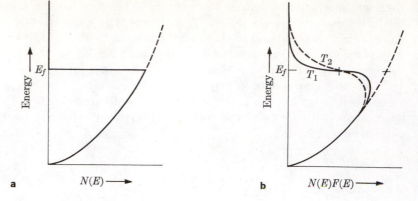

14-4 Density-of-states profile (sodium). (a) $0°K$. (b) $T_2 > T_1 > 0°K$. There is a 50% probability that the states will be occupied at the Fermi energy, E_f.

Fermi distribution. In Chapter 2 we observed that the kinetic-energy distribution of molecules within a gas was an exponential function of temperature; we called this a Boltzmann distribution. It obeys the following relationship:

$$B(E) = \frac{1}{e^{(E-\bar{E})/kT}}. \tag{14-3}$$

With integration, this led to Eq. (2-14) when $E \gg \bar{E}$.

In a Boltzmann distribution, there is nothing to prevent more than one molecule from possessing the same energy. However, in the case of electron energies, only two electrons may occupy a specific energy state; i.e., the Pauli exclusion principle applies. Thus the energy distribution is slightly altered, particularly at lower energy levels where a large fraction of the energy states are occupied. Equation (14-4) is applicable, and the distribution is called a Fermi distribution, $F(E)$:

$$F(E) = \frac{1}{1 + e^{(E-E_f)/kT}}, \tag{14-4}$$

where E_f is again the Fermi energy and k is Boltzmann's constant. This function, as plotted in Fig. 14-5, indicates the probability that a state is occupied, and has wide application in this and the next chapter. The density profile of Fig. 14-4(b) is a product of the number of states (Eq. 14-2) and the Fermi distribution (Eq. 14-4).

Example 14-1

Determine the Fermi distribution for $E - E_f$, assigning different multiples of kT as values for $E - E_f$.

Answer. At $E - E_f = kT$,

$$F(E) = \frac{1}{1 + e^1} = 0.269.$$

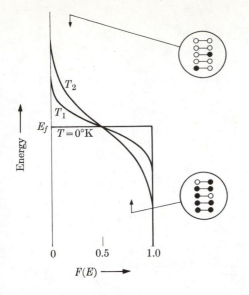

14-5 Fermi distribution, *F(E)*. This function of energy and temperature (Eq. 14-4) gives the probability that an energy state is occupied.

At other energies, $F(E)$ has the following values:

$E - E_f$	$F(E)$
$+4kT$	0.018
$+3kT$	0.048
$+2kT$	0.118
$+1kT$	0.269
0	0.500
$-1kT$	0.731
$-2kT$	0.882
$-3kT$	0.952
$-4kT$	0.982

Note 1. At room temperature, $kT = 0.025$ eV.* Thus few electrons exceed the Fermi energy by more than a fraction of an electron volt.

Note 2. The Fermi distribution is symmetrical. In other words, occupied states above the Fermi energy are balanced by vacated states below the Fermi energy level. ◄

Example 14-2

a) Compare the number of electrons *at* $E = E_f + 0.5$ eV when the temperature is 20°C and 40°C.

b) Repeat for the number of electrons *above* $E = E_f + 0.5$ eV.

Answer

a) $$\frac{n_{20°C}}{n_{40°C}} = \frac{1/[1 + e^{0.5\,eV/(8.63\times10^{-5}\,eV/°K)(293°K)}]}{1/[1 + e^{0.5\,eV/(8.63\times10^{-5}\,eV/°K)(313°K)}]} = \frac{1 + e^{0.5/0.0270}}{1 + e^{0.5/0.0253}} \cong \frac{e^{18.5}}{e^{19.75}} = \frac{1}{3.5}.$$

---

* In studying electron behavior it is convenient to use electron volts, eV, as a unit of energy; $k = 8.63 \times 10^{-5}$ eV/°K.

b) At $E \gg kT$,

$$F(E) \cong B(E) \cong \frac{1}{e^{(E-E_f)/kT}} .$$

Integrating from 0.5 to ∞ for *all* the states occupied, we obtain

$$\int_{E_f+0.5}^{\infty} \frac{1}{e^{(E-E_f)/kT}} \, dE = \left[\frac{-kT}{e^{(E-E_f)/kT}} \right]_{E_f+0.5}^{\infty} ,$$

$$n = kT/e^{0.5/kT},$$

$$\frac{n_{20°C}}{n_{40°C}} = \frac{(293°K)e^{0.5/0.0270}}{(313°K)e^{0.5/0.0253}} \cong \frac{1}{3.7} . \quad \blacktriangleleft$$

Work function and contact potential. Electrons must be given energy to lift them from the Fermi energy level and remove them from the material. This is shown schematically in Fig. 14-6 as a product of the electron charge, e,* and a voltage, ϕ. The product, $e\phi$, is called the *work function*, and has energy dimensions.

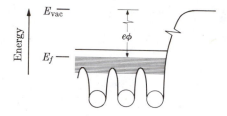

14-6 Work function. An energy of $e\phi$ would be required to remove an electron from the metal into a vacuum.

The work function is directly analogous to the ionization potential of Table 2-3, except for the fact that we are now talking about a material with many atoms. Also, there is only an infinitesimal difference between the energy required to remove succeeding electrons, because their levels are closely spaced.

Since each material has its own Fermi-energy level, the work functions differ between any pair of materials, $e\phi_1 - e\phi_2$, as shown in Fig. 14-7. This energy difference shows up as a *contact potential*, $V_c \, (= \phi_1 - \phi_2)$, which permits electrons to move from the material with the higher Fermi energy to the one with the lower Fermi energy. The contact potential will be important to us in the next chapter when we discuss semiconducting devices.

[The origin of the contact potential may also be described as follows. The Fermi energy is the level at which 50% of the states are filled. When two metals are brought into contact, it is impossible for 50% of the states at one energy level to be filled in one metal and not in the other. Therefore, electrons move from the metal with the higher initial Fermi energy to the one with the lower initial Fermi energy, as shown in Fig. 14-7(b). This equalizes the Fermi energies in the two

---

* Although we have used q to indicate the electron charge elsewhere in this text, we will use e when we speak of electron energies. This matches the convention of using eV for electron volts, i.e., (coul) (volts), to give joules. Here $e\phi$ has the same energy units.

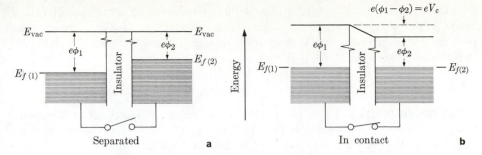

14-7 Contact potential. After contact, electrons move so that the Fermi levels of the two metals are equal. Therefore, a potential difference ($\phi_1 - \phi_2$) is established between the two metals.

metals; i.e., it gives 50% occupancy to the same absolute energy level in each metal. Of course, the atoms do not move across the boundary; consequently, an excess positive charge develops in the right metal and an excess negative charge in the left metal. This difference provides the contact potential shown in Fig. 14-7(b).]

14-3 ELECTRICAL CONDUCTIVITY (And resistivity)

All conductors have vacant energy states above the Fermi-energy level. In contrast, the valence bands of insulators are completely filled. Thus it is not the number of electrons *per se* that determines the conductivity, but rather it is the number of electrons which are able to be moved into a higher energy state.

For an electron to have a net movement in an electric field, it must gain velocity and momentum as it moves toward the positive electrode (and lose velocity and momentum as it moves toward the negative electrode). The increased velocity and momentum require more energy, and therefore an unoccupied energy state must be present at a higher level within the valence band to accept the electron.

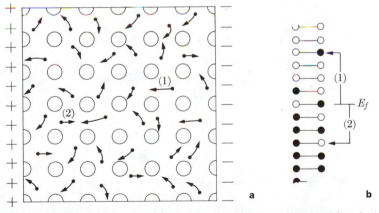

14-8 Electronic conduction. The net movement is toward the positive electrode. Electrons moving toward this electrode are accelerated if they can occupy a higher energy level (1). Those moving away from this electrode drop to a lower energy level when decelerating (2).

This is possible in the unfilled band of a metallic conductor. Further, an energized electron leaves a vacant state below the Fermi level (Fig. 14-8b), providing an opportunity for some other electron which is moving toward the negative electrode to drop to a lower energy level and thereby lose momentum.

Since neither of the above energy changes is possible when the energy band of the valence electrons is filled, conduction is impossible in an insulator.

Conductivity in metals. From the previous discussion, it is apparent that only the uppermost electrons near the Fermi-energy level are active charge carriers. Those deep in the energy band require too much energy to enter a vacant state. Even so, there are still many electrons in the upper states available for conduction, and the limiting factor for conductivity (in metals) is not the n of Eq. (14-1), but rather the mobility, μ, of the electrons.

The mobility of an electron is determined by the *mean free path*. To define this, we need to look briefly at the mechanism of electron movements in metals. Electrons do not move independently of the atoms that are present. If they did, as they moved toward the positive electrode they would continue to gain velocity until they approached the speed of light. In reality, they are diffracted and scattered by irregularities in the crystal structure. When an accelerated electron is scattered (elastically) so that it moves in the reverse direction, it undergoes deceleration, in accord with the above discussion. The distance between deflections is important because larger distances represent longer times for acceleration (and deceleration) and a greater *drift velocity*. The average distance between scattering points is called the mean free path, Λ (Table 14-1). We will want to examine structural factors which affect the mean free path and hence conductivity.

Example 14-3

A 0.56-mm copper wire carries 0.67 amp. If the charge is carried by 1% of the 4s-electrons, what is their drift velocity? [The lattice constant, a, is 3.61 A.]

Answer

$$\text{Velocity} = \frac{\text{carriers/sec}}{\text{carriers/m of wire length}}.$$

$$\text{Electrons/sec} = 0.67 \text{ amp}/1.6 \times 10^{-19} \text{ coul/electron}$$

$$= 4.2 \times 10^{18} \text{ electrons/sec.}$$

$$\frac{\text{Electrons}}{\text{(m of wire length)}} = \frac{(0.01 \times 4 \text{ electrons/unit cell})}{(3.61 \times 10^{-10} \text{ m})^3/\text{unit cell}} \frac{(\pi/4)(0.56 \times 10^{-3} \text{ m})^2(1 \text{ m})}{\text{(m of wire length)}}$$

$$= 2.1 \times 10^{20} \text{ electrons/m.}$$

$$\text{Drift velocity} = (4.2 \times 10^{18} \text{ electrons/sec})/(2.1 \times 10^{20} \text{ electrons/m})$$

$$= 0.02 \text{ m/sec.}$$

Note. Since there is considerable scattering, the actual velocity is much greater. ◄

Effect of temperature on resistivity. Higher temperatures introduce more thermal agitation and more structural disorder to diffract and scatter the electrons, consequently producing lower conductivity (higher resistivities). The resistivities of

TABLE 14-1

ELECTRON MEAN FREE PATHS (0°C)*

Metal	Mean free path, Λ, A	Valence electrons, N, m^{-3}	Conductivity, σ, ohm$^{-1}\cdot$m^{-1}
Na	350	2.5×10^{28}	0.23×10^{10}
Ag	570	5.8	0.69
Cu	420	8.5	0.65
Brass (70-30)	110	8.0	0.16

* Adapted from data by C. Kittel, *Elementary Solid State Physics: A Short Course*, New York: Wiley, 1962, p. 112.

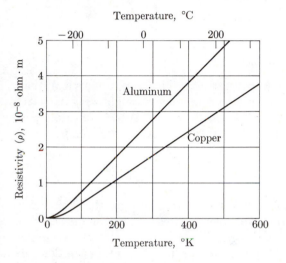

14-9 Resistivity versus temperature (aluminum and copper). Except at very low temperatures, electrical resistivity varies linearly with temperature (Eq. 14-5).

pure copper and aluminum are shown in Fig. 14-9. Except for very low temperatures, electrical resistivities vary linearly with temperature:

$$\rho_T = \rho_{273}[1 + y_T(T - 273)], \tag{14-5}$$

where ρ_{273} is the resistivity at 273°K, or 0°C. The *temperature resistivity coefficient*, y_T, has a value of about 0.005/°K in pure metals; this suggests that the mean free path is reduced by a factor of two between 0°C and 200°C.

Effect of solid solution on resistivity. The local field around a solute atom is different from the field in the remainder of the material. These local-field irregularities scatter the electrons and reduce the conductivity, as shown for Cu-Zn and Cu-Ni alloys in Fig. 14-10. The resistivity due to solid solution may be expressed as

$$\rho_x = y_x X(1 - X), \tag{14-6a}$$

where X is the mole fraction of solute. The *solution resistivity coefficient*, y_x, is larger for greater differences in the valence and atom size between solute and solvent. Table 14-2 shows values of the solution resistivity coefficient in copper.

TABLE 14-2

SOLUTION RESISTIVITY COEFFICIENTS
(In copper at 20°C)

Solute	Solution resistivity coefficient, ohm · m
Ag	0.2×10^{-6}
Al	0.8
Fe	9.5
Ni	1.2
Pb	2.3
Si	2.0
Sn	2.9
Zn	0.2

In dilute solutions,

$$\rho_x \cong y_x X. \tag{14-6b}$$

An *ordered solid solution* has a lower resistivity (higher conductivity) than random solid solutions because the electrons are moving through a regularly repeating lattice. This contrast is shown in Fig. 14-11 for a 25 Au-75 Cu (atom percent) alloy which has an order-disorder reaction at 380°C. (Above this temperature, the fcc structure has a random distribution of gold and copper atoms. Below 380°C, all the gold atoms are at cube corners and all the copper atoms are at face centers, to give $AuCu_3$.)

Effect of strain hardening on resistivity. Crystal imperfections, be they vacancies, dislocations, or grain boundaries, introduce deflection sites for electron movements. As is shown in Fig. 14-12, the resistivity is higher in cold-worked metals. A correlation may be made between resistivity and strain hardness, because each depends on the number of dislocations present. Since annealing removes imperfections, it also decreases resistivity.

The total resistivity is additive for the above contributions. As is shown in Fig. 14-12(b),

$$\rho = \rho_T + \rho_x + \rho_s, \tag{14-7}$$

where ρ_s is the resistivity added by the plastic strain.

Example 14-4

A wire must carry a load, F, of 10 lb without plastic deformation, and have a resistance of less than 0.01 ohm/ft.

a) What is the smallest wire that can be used if it is made from 60-40 brass? 80-20 brass? 100% copper?

b) Which wire should be selected if the cost of copper is twice as much per pound as zinc? [Assume that the density varies linearly with the zinc composition.]

Answer

a) Strength:

$$4F/\pi d^2 = \sigma_y \quad \text{(Fig. 11-21b),} \quad \text{or} \quad d = \sqrt{12.7 \text{ lb}/\sigma_y}.$$

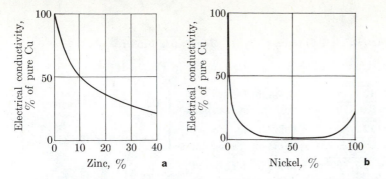

14-10 Electrical conductivity versus solid solution (Cu-Zn and Cu-Ni). Solute atoms reduce the mean free path of the electron movements, thus reducing the conductivity.

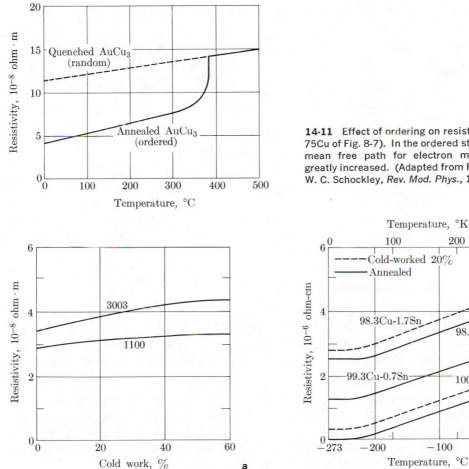

14-11 Effect of ordering on resistivity (25Au-75Cu of Fig. 8-7). In the ordered structure, the mean free path for electron movement is greatly increased. (Adapted from F. C. Nix and W. C. Schockley, *Rev. Mod. Phys.*, **10**, 1938.)

14-12 Effect of strain hardening on resistivity. (a) Resistivity versus cold work (wrought aluminum alloys); 1100 = 99.9% Al; 3003 = 1.2% Mn, balance Al. (b) Additive effects of ρ_T, ρ_r, and ρ_s (Cu-Sn alloys).

Resistance:

$$4\rho l/\pi d^2 = 0.01 \text{ ohm}, \quad \text{or} \quad d = \sqrt{0.001 \text{ in}^2/f},$$

where f is the fraction of conductivity of pure copper in Fig. 14-10(a).

For 60-40: $d = \sqrt{12.7/10,500} = 0.035$ in. for strength
$\phantom{\text{For 60-40: }} d = \sqrt{0.001/0.21} = 0.07$ in. for resistance $\Big\}$ Use 0.07 in. (0.18 cm).

For 80-20: $d = \sqrt{12.7/8500} = 0.039$ in. for strength
$\phantom{\text{For 80-20: }} d = \sqrt{0.001/0.35} = 0.053$ in. for resistance $\Big\}$ Use 0.053 in. (0.135 cm).

For 100-0: $d = \sqrt{12.7/4000} = 0.057$ in. for strength
$\phantom{\text{For 100-0: }} d = \sqrt{0.001/1.0} = 0.032$ in. for resistance $\Big\}$ Use 0.057 in. (0.145 cm).

b) From Appendix B, $\rho_{Cu} = 8.96$ gm/cm^3 and $\rho_{Zn} = 7.13$ gm/cm^3. Therefore,

$$\rho_{60\text{-}40} = (0.6)(8.96) + (0.4)(7.13) = 8.23 \text{ gm/cm}^3$$

and

$$\text{relative cost per cm}^3 = [0.6 + (0.4/2)](8.23)/(8.96) = 0.74.$$

Likewise,

$$\rho_{80\text{-}20} = 8.59 \text{ gm/cm}^3, \quad \text{relative cost per cm}^3 = 0.86,$$
$$\rho_{100\text{-}0} = 8.96 \text{ gm/cm}^3, \quad \text{relative cost per cm}^3 = 1.00.$$

Let C be a constant which includes $\pi/4$ and the copper cost:

$$\text{cost}_{100\text{-}0} = C[(0.145)^2(8.96)(1.00)] = 0.19C$$
$$\text{cost}_{80\text{-}20} = C[(0.135)^2(8.59)(0.86)] = 0.14C \Big\} \text{ Use 80-20.}$$
$$\text{cost}_{60\text{-}40} = C[(0.18)^2(8.23)(0.74)] = 0.20C$$

Note. On the basis of this problem, the cost per pound of an 80Cu-20Zn brass would be 90% that of copper alone. In commerce, the relative costs may vary considerably, depending on the cost of recycled copper scrap. ◄

14-4 THERMAL PROPERTIES OF METALS

Several properties of metals must be interpreted in terms of their electron behavior. The most obvious one is the high thermal conductivity of metals. Also the heat capacity of metals has special interest because it justifies the use of the band model. After discussing these properties, we shall look briefly at several properties described as thermoelectric behavior.

Thermal conductivity. Thermal energy can be transported through solids by either of two conduction mechanisms.* The first is by means of *phonons*, or quantized elastic waves (Section 10-2). The second conduction mechanism is that provided by *electrons*. Both mechanisms are operative in metals (but only phonon conduction occurs in insulators).

---

* It can also be transported through transparent solids by radiation.

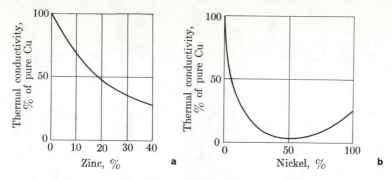

14-13 Thermal conductivity versus solid solution. These data parallel the electrical conductivity (Fig. 14-10) because much of the thermal energy is carried in metals by electrons.

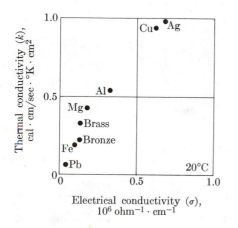

14-14 Thermal versus electrical conductivity. The Wiedemann-Franz ratio, k/σ, increases at higher temperatures, but the Lorentz number remains constant. (See Eq. 14-9.)

Since electrons move rapidly through a metal, we find that a large fraction of the thermal energy is transferred by electrons, and therefore a direct comparison can be made between the thermal conductivity of metals (Fig. 14-13) and their electrical conductivity (Fig. 14-10). This, in addition to the fact that both phonon and electron movements depend on the same structural perfection for long mean free paths, permits one to use the *Wiedemann-Franz ratio*, WF:

$$\text{WF} = k/\sigma \quad (= k\rho), \tag{14-8}$$

where k and σ are the thermal and electrical conductivities. The value of the Wiedemann-Franz ratio is about 1.6×10^{-6} for metals at 20°C when the conductivities are expressed in the dimensions of $(\text{cal} \cdot \text{cm})/(\text{sec} \cdot \text{°K} \cdot \text{cm}^2)$ and $\text{ohm}^{-1} \cdot \text{cm}^{-1}$, respectively (Fig. 14-14).

Because electrical resistivity is nearly proportional to the absolute temperature, we can set up another approximation:

$$\mathcal{L} \cong k\rho/T \quad (= k/\sigma T). \tag{14-9}$$

TABLE 14-3
CONDUCTION OF METALS

Metal	Thermal conductivity,* k, cal $\cdot$ cm/sec $\cdot$ °K $\cdot$ cm^2	Electrical conductivity,* σ, ohm^{-1} $\cdot$ cm^{-1}	Lorentz number, $\mathcal{L}$, cal $\cdot$ ohm/sec $\cdot$ °K^2
Ag	1.0	0.55×10^6	0.61×10^{-8}
Al	0.53	0.34	0.52
Cu	0.95	0.59	0.55
Fe	0.18	0.104	0.59
Pb	0.08	0.049	0.49

* At 20°C.

The constant, $\mathcal{L}$, called the *Lorentz number*, has a value of approximately 0.55×10^{-8} cal $\cdot$ ohm/(sec $\cdot$ °K^2) for a wide variety of metals (Table 14-3).

Example 14-5

Magnesium, although not listed in Table 14-3, is a typical metal. *Estimate* its thermal conductivity at 400°C from its electrical resistivity of 4.4×10^{-6} ohm $\cdot$ cm at 0°C.

Answer. From Eq. (14-5),

$$\rho_{400°C} = (4.4 \times 10^{-6} \text{ ohm} \cdot \text{cm})[1 + (0.005/°K)(673 - 273°K)]$$
$$= 13 \times 10^{-6} \text{ ohm} \cdot \text{cm},$$
$$k_{400°C} = (0.55 \times 10^{-8} \text{ cal} \cdot \text{ohm/sec} \cdot °K^2)(673°K)/(13 \times 10^{-6} \text{ ohm} \cdot \text{cm})$$
$$= 0.29 \text{ (cal} \cdot \text{cm)/(sec} \cdot °K \cdot \text{cm}^2).$$

Note. Experimental results give 0.27 (cal $\cdot$ cm)/(sec $\cdot$ °K $\cdot$ cm^2) at 400°C. ◄

Electronic heat capacity. Prior to the development of the band model for electron energies, the materials scientist was faced with a dilemma. Previously it had been assumed that all the valence electrons could absorb thermal energy. If this were true, each electron would carry the same kinetic energy as an atom or molecule (Eq. 2-12). This would provide an additional $3R/2$ per mole to the heat capacity above the $3R$ per mole (~ 6 cal/mole $\cdot$ °K) for lattice energy, to give a total of $9R/2$ (~ 9 cal/mole $\cdot$ °K) in a metal with one 4s-electron. As shown in Fig. 6-2, this extra heat capacity is not present. The materials scientist could only assume that the majority of the valence electrons are not free to absorb thermal energy. Of course, this is the prediction of the band model, which shows that only those electrons close to the Fermi-energy level may receive thermal energy.

Careful measurements indicate that the electronic heat capacity at room temperature is only about 0.03 cal/mole $\cdot$ °K. Since this is one percent of the previously predicted value, we must assume that only one percent of the electrons are carrying additional energy. As the temperature is increased, the Fermi distribution (Eq. 14-4) permits more electrons to absorb energy, and the *electronic heat capacity*, C_{el}, increases:

$$C_{el} = \gamma T, \qquad (14\text{-}10)$$

where γ is about $10^{-4}/°K$.

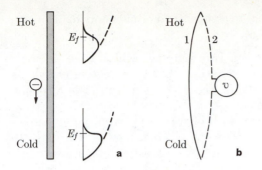

14-15 Thermal emf. (a) The energetic electrons move from the hot to the cold end of the wire more readily than in the opposite direction. (b) Thermocouple. The voltage is equal to $\Delta T(S_1 - S_2)$, where S_1 and S_2 are the thermal emf coefficients for the two metals.

Thermal electromotive force and thermocouples. Figure 14-4(b) showed the electron-density profile of a metal for two different temperatures. Because of the Fermi distribution, there are more high-energy electrons when the temperature is increased. Now consider a metal wire with only one end heated (Fig. 14-15). The energetic electrons at the hot end move more readily to the cold end than vice versa. When this occurs, a voltage difference, $\Delta \upsilon$, is established between the two ends of the wire. This is called a thermal electromotive force, or *thermal emf*. The *thermal emf coefficient*, S, is the ratio of the voltage and temperature differences:

$$S = \Delta \upsilon / \Delta T. \tag{14-11}$$

The thermal emf remains the same no matter how long the wire, or what the intervening temperature profile is.

The thermal emf presents the possibility of measuring temperatures, because if we knew the temperature at one end of a wire, we could calculate the temperature at the other end by means of the voltage difference and the thermal emf coefficient. A difficulty arises, however, for when we attach the voltmeter or potentiometer, a thermal emf is also established in the connecting wires. In fact, if we used the same kind of wire, the emf would be canceled; thus we cannot use simple potential measurements for temperature. Fortunately, different metals have different thermoelectric emf coefficients (because their electron-density profiles are different), and it is possible to measure the difference in the voltage by using a *thermocouple* of *two* kinds of wires (Fig. 14-15b). Since S_1 and S_2 are constants, the voltage read from the thermocouple (called the *Seebeck potential*) is directly related to the temperature difference. Now, if we know the temperature of one metal junction, it is possible to determine the temperature at the other junction.

A second thermoelectric effect, known as the *Thomson effect*, can also be described on the basis of Fig. 14-15(a). If a voltage is placed along a conductor which has a thermal gradient such that electrons are forced from the hot to the cold end, the more energetic electrons lose some of their energy at the lower temperature, thus raising the temperature of the surrounding material. This heating is in addition to the normal I^2R, or joule, heating arising from the electrical resistance. The Thomson effect may be either positive or negative, because, when the voltage is reversed and the electrons move from the cold end to the hot end, they must absorb heat from the surrounding material to assume the Fermi-energy distribution sketched at the upper end of Fig. 14-15(a).

Peltier effect. If a current passes between two materials which have different electron-density profiles, the electrons must either absorb or release energy as they proceed through the contact potential of Fig. 14-7. This will have a cooling or heating effect, called the *Peltier effect.*

Of particular interest is the cooling, because it offers an opportunity for thermo-electric refrigeration, i.e., cooling directly by electric-current flow as contrasted to compressors or related equipment. This presents an interesting materials-system challenge to engineers, because materials must be selected so as to achieve optimum heat absorption at the junction and at the same time avoid I^2R heating and unwanted thermal conductivity.

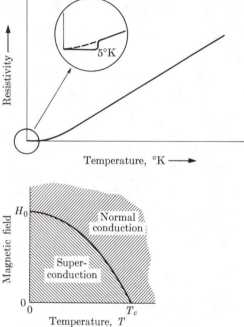

14-16 Zero resistivity (superconductivity). The linear ρ-versus-T relationship changes to ρ versus T^5 below about 100°K, then drops sharply to zero in the last few degrees above absolute zero.

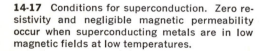

14-17 Conditions for superconduction. Zero resistivity and negligible magnetic permeability occur when superconducting metals are in low magnetic fields at low temperatures.

14-5 SUPERCONDUCTIVITY

At very low temperatures, certain metals and a large number of intermetallic compounds possess *superconductivity*, having zero resistivity (Fig. 14-16) and undetectable magnetic permeability. The origin of superconductivity must be explained in terms of quantum mechanics. Although we will not present the explanation in this text, it will be worthwhile to cite several features of superconductivity because of their obvious engineering interest.

The transition from normal conductivity to superconductivity is abrupt, and occurs as a function of temperature and magnetic field (Fig. 14-17). The critical values of temperature, T_c, and magnetic field, H_0, for several superconductors are

TABLE 14-4

VALUES OF T_c AND H_0 FOR SELECTED SUPERCONDUCTORS
(See Fig. 14-17)

Material	Magnetic field, H_0, at 0°K, oersteds	Transition temperature, T_c, in zero field, °K
Al	106	1.2
Hg	413	4.2
Nb	2000	9.2
Sn	305	3.7
Ti	20	0.4
V	1310	5.0
Nb_3Sn	5000	18.1
V_3Si		17.1
NbN		16.0
MoC		8.0
CuS		1.6

shown in Table 14-4. Since the *H-T* curve of Fig. 14-17 is essentially parabolic for the various superconductors, the data of Table 14-4 permit a calculation of the critical magnetic field, H_c, for various temperatures, T:

$$H_c = H_0[1 - (T/T_c)^2]. \tag{14-12}$$

Twenty-three of the metallic elements are known to be superconductive. Others, including all the alkali metals, the ferromagnetic metals, and the noble metals, have been checked down to less than 0.1°K, and no evidence of a transition to superconductivity has been found. General and empirical rules reveal that superconduction occurs most readily (1) among those metals with low conductivities of the normal type and (2) among those metals with 3, 5, or 7 valence electrons. These generalizations have led to the formulation of many intermetallic compounds with high transition temperatures. Several of these are listed in Table 14-4. The great hope of current research is to develop a material which is superconducting at room temperature. Such a material would bring about radical changes in the field of electrical engineering.

REFERENCES FOR FURTHER READING

American Society for Metals, *Atomic and Electronic Structure of Metals*. Metals Park, O.: American Society for Metals, 1967. Papers presented at an ASM seminar. Graduate-student and professional level.

Azaroff, L. V., and J. J. Brophy, *Electronic Processes in Materials*. New York: McGraw-Hill, 1963. The free-electron and zone theories of metals are discussed in Chapters 6 and 7. Recommended for the advanced undergraduate student who has had "modern physics."

Cadoff, I. B., and E. Miller, *Thermoelectric Materials and Devices*. New York: Reinhold, 1960. For the materials specialist who works with thermoelectricity.

Dekker, A. J., *Electrical Engineering Materials*. Englewood Cliffs, N. J.: Prentice-Hall, 1959. The conductivity of metals is presented in Chapter 5. Supplementary reading for the advanced undergraduate student.

Ehrenreich, H., "The Electrical Properties of Materials," *Scientific American*, **217** [3], September 1967, pp. 194–204. An introduction to electron movements and conductivity in solids. Undergraduate level.

Holden, A., *The Nature of Solids*. New York: Columbia University Press, 1965. Chapters 10 and 11 provide interesting background reading on electron motion in solids.

Hume-Rothery, W., *Atomic Theory for Students of Metallurgy*. London: The Institute of Metals, 1955. A readable presentation of the electron theory of metals. Minimum mathematics.

Hutchinson, T. S., and D. C. Baird, *The Physcis of Engineering Solids*. New York: Wiley, 1963. Electrical and thermal conductivity of metals are presented in Chapter 10. Goes one step further into the physics of conduction than this text. Advanced undergraduate level.

Kittel, C., *Elementary Solid State Physics: A Short Course*. New York: Wiley, 1962. The free-electron model of metals is discussed in Chapters 5 and 6. Advanced undergraduate and graduate level.

Kittel, C., *Introduction to Solid State Physics*, third edition. New York: Wiley, 1966. Free-electron theory and energy bands are presented in Chapters 7 through 9. For the graduate student.

Mitoff, S. P., "Electrical Conduction Mechanism in Oxides," *Progress in Ceramic Science*, Vol. 4 (J. E. Burke, Ed.). Oxford: Pergamon Press, 1966. Ionic conductivity and electronic conductivity are reviewed in terms of defect structures. For the advanced student.

Moore, W. J., *Seven Solid States*. New York: Benjamin, 1967. Paperback. The author uses gold as an example of metallic behavior. This example will be helpful for the reader who wants to study Brillouin zones.

Mott, N. F., and H. Jones, *The Theory of the Properties of Metals and Alloys*. Oxford: Clarendon Press, 1936; and New York: Dover, 1958. Paperback. Although an older text, it is very useful in understanding the electron theory of metals. Advanced undergraduate level.

Rose, R. M., L. A. Shepard, and J. Wulff, *The Structure and Properties of Materials: IV. Electronic Properties*. New York: Wiley, 1966. Paperback. Chapter 3 on thermal behavior and Chapter 4 on electrical conduction provide supplementary reading for this chapter. The student or instructor who wants to explore the electron structure of solids will find a simplified presentation in Chapter 1.

Wert, C. A., and R. M. Thomson, *Physics of Solids*. New York: McGraw-Hill, 1964. Chapters 9, 10, and 11 present the electronic theory of metals at an advanced undergraduate level. Electron transport is discussed.

PROBLEMS

14-1 What fraction of the electron energy states are occupied at $E_f + 0.05$ eV at (a) 100°K? (b) 200°K? (c) 400°K? (d) 800°K?

Answer. a) 0.0032   b) 0.052   c) 0.19   d) 0.33

14-2 Plot the fraction of the electron energy states occupied versus temperature for $E_f - 0.1$ eV.

14-3 At what temperature will 0.65 of the electron energy states be occupied in the $(E_f - 0.04$ eV) energy level?

Answer. 474°C

14-4 Plot the temperatures at which 0.15 of the electron energy states will be occupied versus energy.

14-5 A copper wire has 10^{18} electrons with an energy $= >0.1$ eV in excess of the Fermi energy at 27°C. How many will that wire have above this energy level at 127°C?

Answer. 3.5×10^{18}

14-6 At what temperature must the wire of Problem 14-5 be placed to have only 0.5×10^{18} electrons with an energy level $= >0.1$ eV in excess of the Fermi energy?

14-7 A 0.08-in. diameter copper wire carries a current of 1.1 amp. What is the electron flux?

Answer. $2.1 \times 10^{24}/\text{m}^2 \cdot$ sec

14-8 What is the drift velocity in the wire of Problem 14-7 if 10^{27} electrons/m^3 serve as conductors?

14-9 If one out of every 1000 4s-electrons is a charge carrier, what is the drift velocity in a copper wire where the electron flux is $1.5 \times 10^{24}/\text{m}^2 \cdot$ sec?

Answer. 1.8×10^{-2} m/sec

14-10 A copper wire has a resistance of 0.5 ohm per 100 ft. Consideration is being given to the use of a 75-25 brass wire instead of copper. What would the resistance be if the brass wire were the same size?

14-11 A brass alloy is to be used in an application which will have a tensile strength of more than 40,000 psi and an electrical resistivity of less than 5×10^{-6} ohm $\cdot$ cm (resistivity of Cu $= 1.7 \times 10^{-6}$ ohm $\cdot$ cm). What percent zinc should the brass have?

Answer. 14 to 23% zinc. [Note, however, that the brass having a higher percentage of zinc will be cheaper.]

14-12 A certain application requires a rod of metal having a yield strength greater than 15,000 psi and a thermal conductivity greater than 0.1 cal $\cdot$ cm/$\text{cm}^2 \cdot$ sec $\cdot$ °C. Specify either an annealed brass or an annealed Cu-Ni alloy that could be used.

14-13 A 99.9% pure Al wire 0.10 in. in diameter was annealed before being cold-drawn through a die 0.08 in. in diameter. What electrical conductivity will this wire have after cold-drawing?

Answer. $3.2 \times 10^7/\text{ohm} \cdot$ m

14-14 The electrical resistivity of pure copper is 1.7 micro-ohm $\cdot$ cm at 20°C and 2.4 micro-ohm $\cdot$ cm at 120°C. That of 98Cu-2Ni alloy is 4.1 micro-ohm $\cdot$ cm at 20°C. Estimate the resistivity of (a) pure copper at 45°C; (b) 98Cu-2Ni at 120°C; (c) 96Cu-4Ni at 20°C; and (d) 96Cu-4Ni at 95°C.

15 □ semiconduction in solids

15-1 INTRODUCTION

Metallic conductors used for energy transmission have resistivities of about 2 micro-ohm · cm (2×10^{-8} ohm · m). The resistivities of metallic resistors for heating elements are only one or two orders of magnitude higher. In contrast, a good insulator must have a resistivity of at least 10^8 ohm · m, and preferably higher. Between these extremes is a wide spectrum of materials that fall into the category of *semiconductors*. A significant number of these have assumed substantial engineering importance in the past two decades; hence, we are warranted in devoting a separate chapter to them. Typically they have resistivities between 10^{-2} and 10^{+2} ohm · m. Like insulators and unlike metals (Section 14-3), their conductivities increase with increased temperature and increased impurity levels.

In this chapter we shall limit our consideration to those semiconductors which have electronic conduction. We should be aware, however, that some ionic conductivity values also fall in the resistivity range of 10^{-2} to 10^{+2} ohm · m.

15-2 INTRINSIC SEMICONDUCTION

We observed in Section 14-2 and Fig. 14-2 that the second and third valence energy bands do not overlap. This means that Group IV elements such as carbon, silicon, germanium, and tin (gray) cannot have their Fermi-energy level within a band. The first two bands are exactly filled (Fig. 15-1), since there are four valence electrons per atom and each band may contain twice as many electrons as there are atoms.

An energy gap lies above the second valence band. Thus, vacant states are not immediately available, as they are in metals (Fig. 14-2), to receive electrons for conduction. It is necessary to supply an electron with enough energy to "jump the gap" from the *valence band* to the *conduction band* before it can be accelerated.

Before we consider how electrons enter the conduction band, let us view the electron structure of Group IV materials. The most common semiconductor structure is that of diamond (Fig. 3-12). Each atom is covalently bonded with four neighboring atoms. This is shown schematically in two dimensions in Fig. 15-2(a). There is an average of two electrons per bond. Actually all $4N$ electrons (where N is the number of atoms) can move in a wavelike manner throughout the total

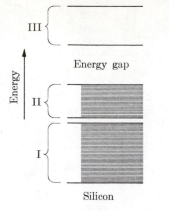

III

Energy

Energy gap

II

I

Silicon

15-1 Energy band of a semiconductor (silicon). Since there are N states per band and $2N$ electrons per band, a Group IV element with N atoms has its first two bands filled. (Cf. Fig. 14-2.)

15-2 Intrinsic semiconductor. (a) Two-dimensional sketch of the diamond structure. (Cf. Fig. 3-12.) Each atom is covalently bonded to four neighbors. (b) Electron-hole pair. If energy is supplied, an electron can be removed from the bond and becomes available for conduction.

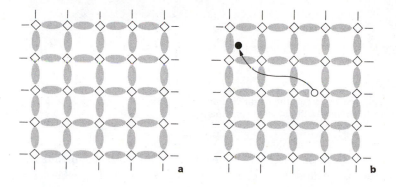

structure, with as many moving in one direction as in another, and at energy levels within the first two bands. Energy is required for an electron to leave this structure and become a conduction electron for acceleration (Fig. 15-2b). This energy may be compared directly with the activation energy required to form a vacancy-interstitial defect of atoms (Section 6-5). In this case, however, the resulting defect is an *electron hole* in the valence band and a charge-carrying electron in the conduction band. In a pure Group IV material there is a one-to-one correspondence between electron holes and conduction electrons. We speak of each pair consisting of an electron hole and a conduction electron as an *electron-hole pair*.

Fermi distribution (intrinsic semiconductors). Thermal energy produces electron-hole pairs just as thermal energy produces more vacancy-interstitial pairs in a crystal. In fact, we can use practically the same equations. If we construct the Fermi distribution across an energy gap, we observe that the "tails" of the distribution curve in both the conduction band and the valence bands are the same size

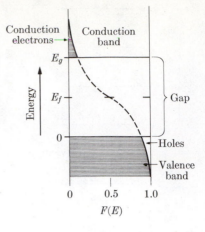

15-3 Fermi distribution (intrinsic semiconductor). Since an intrinsic semiconductor has as many holes as conduction electrons, the distribution curve is symmetrical across the gap and the Fermi energy, E_f, lies in the middle.

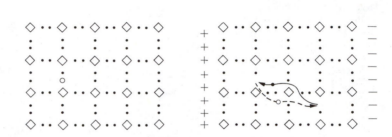

15-4 Hole conduction. The hole, $\bigcirc$, moves toward the negative electrode as if it were a positive charge. In the process, an electron moves in the opposite direction toward the positive electrode.

(Fig. 15-3). The symmetry of the distribution leads us to the conclusion that the Fermi-energy level lies in the middle of the energy gap, since that is where the probability of occupancy is 50%. (Admittedly, no electrons can have energies at this level because of the gap.)

Number of charge carriers. Since there are many available states in the conduction band, each electron in that band is a negative charge carrier, n. Likewise, each electron hole is a positive charge carrier, p, because it represents a vacant state into which other electrons can move. Figure 15-4 provides a convenient two-dimensional illustration of hole conduction. As previously noted, for pure, or *intrinsic semiconductors*,

$$n_n \equiv n_p, \tag{15-1}$$

where n_n is the number of negative charge carriers (conduction electrons) and n_p is the number of positive charge carriers (electron holes) per unit volume.

Since the number and distribution of available states in a conduction band, $N(E)$, are essentially constant, we may compare the number of conduction electrons at various temperatures. Using Eq. (14-4), we find

$$n_n = N(E) \int_{E_g}^{\infty} 1/[1 + e^{(E-E_f)/kT}]\, dE. \tag{15-2a}$$

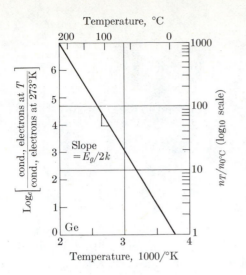

15-5 Conduction electrons versus temperature (intrinsic germanium: $E_g = 0.72$ eV). This Arrhenius plot (Eq. 15-2c) also applies for electron holes, since $n_n = n_p$ in an intrinsic semiconductor.

When $(E - E_f) \gg kT$, .d when $E = 0$ at the bottom of the energy gap (and consequently $E_f = E_g$, then

$$n_n = N(E)kTe^{-E_g/2kT}.$$ (15-2b)

We used ∞ as the upper limit of integration because the top of the conduction band exceeds E_f by a far greater amount than kT:

$$\ln n_n \cong \text{constant} - E_g/2kT,$$ (15-2c)

since $\ln kT$ varies only slightly compared to changes in $1/kT$. Equation (15-2c) may be presented as a straight line on an Arrhenius plot (Fig. 15-5). Since the slope of this plot, $\Delta \ln n/\Delta(1/T)$, is equal to $-E_g/2k$, the size of the energy gap, E_g, may be calculated from n-versus-T data.

TABLE 15-1
PROPERTIES OF SEMICONDUCTORS

Material	Lattice constant, l, A	Energy gap, E_g, eV	Electron mobility, μ_n, m²/sec · volt	Hole mobility, μ_p, m²/sec · volt	Intrinsic conductivity (20°C), σ_i, ohm⁻¹ · m⁻¹
C (diamond)	3.57	5.2	0.17	0.12	10^{-12}
Si	5.42	1.1	0.14	0.05	5×10^{-4}
Ge	5.62	0.72	0.39	0.19	2.5
Sn (gray)	6.49	0.08	0.2	0.1	10^6
GaP	5.45	2.2	0.05	0.002	—
GaAs	5.65	1.4	0.85	0.45	—
InP	5.87	1.3	0.34	0.06	400
InAs	6.04	0.33	2.30	0.01	10^4

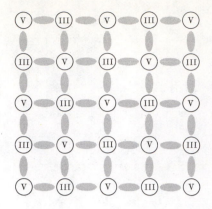

15-6 Semiconducting compounds (III-V). These compounds and certain II-VI compounds have covalent bonds with an average of four valence electrons per atom.

Commonly accepted values for E_g are presented in Table 15-1. It is apparent that the elements carbon (diamond), silicon, germanium, and tin (gray) have successively higher conductivities as a result of their smaller energy gaps.

The III-V and II-VI compounds are important because they offer many varieties of combinations to produce semiconducting materials. They average 4.0 valence electrons per atom (Fig. 15-6), and their crystal structures are very similar to those of the Group IV elements (Section 15-6).

Example 15-1

The energy gap in pure germanium is 0.72 eV. Compare the number of conduction electrons at 20°C and 40°C.

Answer. Using Eq. (15-2b), we get

$$\frac{n_{20°C}}{n_{40°C}} = \frac{N(E)(kT)e^{0.72/2(8.63\times10^{-5}\,\text{eV/°K})(313°\text{K})}}{N(E)(kT)e^{0.72/2(8.63\times10^{-5}\,\text{eV/°K})(293°\text{K})}} = \frac{293°\text{K}}{313°\text{K}}\frac{e^{13.35}}{e^{14.25}} = \frac{1}{2.6}.$$

Note. If we had ignored the kT-term as we did in Eq. (15-2c), the answer would have been 1/2.5. ◀

Conductivity. Equation (14-1) can be adapted to semiconduction where both negative electrons and positive electron holes are charge carriers:

$$\sigma = n_n q\mu_n + n_p q\mu_p. \tag{15-3a}$$

For intrinsic semiconductivity, σ_{in}, where $n_n = n_p = n_{in}$,

$$\sigma_{in} = n_{in}q(\mu_n + \mu_p). \tag{15-3b}$$

The terms μ_n and μ_p are the mobilities of electrons and holes, respectively. In general, $\mu_p < \mu_n$ because the mean free path for hole conduction is shorter than for electrons.

As observed in Example 15-1, the number of charge carriers increases greatly as the temperature increases. Since the mobility is only slightly affected by temperature ($d\mu/dT$ is slightly negative), the net effect is to give positive ($d\sigma/dT$)-

values to materials which exhibit the relationship of Eq. (15-2c):

$$\ln \sigma = \ln \sigma_0 - E_g/2kT, \qquad (15\text{-}4)$$

where σ_0 is a conductivity constant determined by experiment.

Example 15-2

A small rod of pure germanium 1 cm long, 2 mm wide, and 1 mm thick has an electrical resistance of 2160 ohms (lengthwise) at 20°C. Separate data by Hall measurements (Section 15-5) indicate that the mobilities, μ_n and μ_p, are 0.39 m²/sec · volt and 0.19 m²/sec · volt, respectively. How many electrons are in the conduction band?

Answer

$$\sigma = l/RA = (0.01 \text{ m})/(2160 \text{ ohms})(0.002 \text{ m})(0.001 \text{ m})$$
$$= 2.32 \text{ ohm}^{-1} \cdot \text{m}^{-1}.$$

From Eq. (15-3b),

$$n_n = \frac{2.32 \text{ ohm}^{-1} \cdot \text{m}^{-1}}{(1.6 \times 10^{-19} \text{ coul})(0.58 \text{ m}^2/\text{sec} \cdot \text{volt})}$$
$$= 2.5 \times 10^{19}/\text{m}^3.$$

In this rod,

$$n_n = (2.5 \times 10^{19}/\text{m}^3)(2 \times 10^{-8} \text{ m}^3)$$
$$= 5 \times 10^{11}. \quad \blacktriangleleft$$

Photoconduction. Conduction electron-hole pairs may be formed by thermal energy, as discussed in the previous section. There are also other energy sources which can cause an electron to "jump the gap." These include electromagnetic photons, and accelerated electrons and neutrons. Photoconduction arises from photon activation.

An individual photon can transfer all its energy to an electron in the valence band, thus raising the electron to the conduction band. When this happens, both the electron and the hole become charge carriers; thus, an insulator may become a semiconductor in the presence of light, and hence the term *photoconductor*. "Electric eyes" are examples of electrical devices based on this change in electrical conductivity.

Example 15-3

What is the maximum wavelength of light which will produce photoconduction in silicon ($E_g = 1.1$ eV)?

Answer. From Eq. (2-1),

$$E = hc/\lambda,$$

$$(1.1 \text{ eV})(1.6 \times 10^{-19} \text{ joule/eV}) = \frac{(6.62 \times 10^{-34} \text{ joule} \cdot \text{sec})(3 \times 10^8 \text{ m/sec})}{\lambda},$$

$$\lambda = 1.13 \times 10^{-6} \text{ m} \quad (= 11{,}300 \text{ A}).$$

Note. Photons which have longer infrared wavelengths cannot provide the energy necessary for the electron to jump the gap. $\blacktriangleleft$

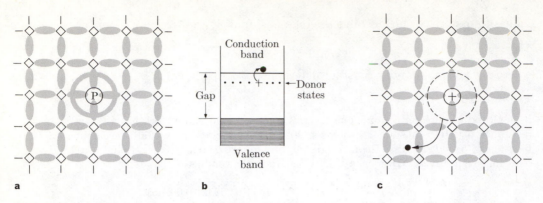

15-7 Extrinsic semiconductor (*n*-type). (a) A Group V (phosphorus) atom has an extra donor electron which does not fit into the covalent bonds. (b) The donor states are near the top of the energy gap. (c) When a donor electron moves to the conduction band, it ionizes the impurity.

15-3 EXTRINSIC SEMICONDUCTION

Impurities markedly alter the number of charge carriers in a semiconductor. We speak of a material altered in this way as an *extrinsic semiconductor*, in contrast to intrinsic semiconductors, which achieve their conductivity by thermal or radiation activation.

n-type semiconduction. If a small amount ($\ll$1 a/o) of phosphorus or some other Group V element is added to silicon, the valence electron-atom ratio increases to slightly above 4.0. We can visualize the consequences as shown in Fig. 15-7(a), where all covalent bonds are satisfied but where there is one additional electron with each phosphorus atom. Although this extra electron is attracted to the phosphorus atom in order to maintain local charge balance, it is held much less firmly than are the covalent electrons. Therefore it is available for electron conduction, and we speak of the material as an *n-type* semiconductor. This electron is called a *donor electron* and its 0°K energy state is a *donor state* (Fig. 15-7b). When raised to the conduction band, the phosphorus atom is ionized (Fig. 15-7c).

Donor exhaustion. The donor states generally lie only a small fraction of an electron volt below the top of the energy gap; that is, $(E_g - E_d) \sim 0.01$ eV. This means that the donor electrons are easily raised to the conduction band, and the extrinsic conductivity has a very low slope in a $\ln \sigma$-versus-$1/T$ plot (Fig. 15-8a). In fact, if the impurity content is controlled, essentially all of the donor electrons will be energized at room temperature where $kT = 0.025$ eV, with the result that donor *exhaustion* occurs as shown in Fig. 15-8(b). Under these conditions, conductivity is essentially constant over a useful temperature range.*

---

* There will be a slight decrease in conductivity with increased temperature, because of decreased electron mobility.

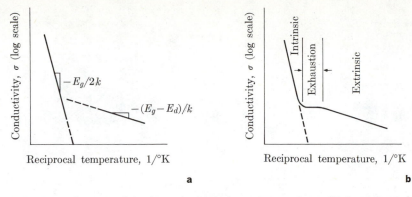

a   b

15-8 Extrinsic conductivity (*n*-type). (a) At lower temperatures (higher values of $1/T$), the extrinsic conductivity is greater than the intrinsic conductivity because the donor states are high in the energy gap ($E_d > E_g/2$). (b) Exhaustion occurs when most of the donor electrons have been moved to the conduction band.

Example 15-4

Silicon contains 0.00001 a/o Sb, or 0.1 ppm.

a) At donor exhaustion, how many donor electrons, n_d, are there in the conduction band for carrying charge?

b) What is the extrinsic conductivity, σ_{ex}, of this *n*-type semiconductor?

Answer. Silicon, like diamond (Fig. 3-12), has eight atoms per unit cell. There is one donor electron per antimony atom, since Sb is in Group V. Other data come from Table 15-1.

a)
$$n_d = \left[\frac{8 \text{ Si atoms}}{(5.42 \times 10^{-10} \text{ m})^3} \right] \left(\frac{10^{-7} \text{ Sb}}{\text{Si}} \right) \left(\frac{1 \text{ carrier}}{\text{Sb}} \right)$$
$$= 5 \times 10^{21} \text{ carriers/m}^3.$$

b)
$$\sigma_{ex} = (5 \times 10^{21}/\text{m}^3)(1.6 \times 10^{-19} \text{ coul})(0.14 \text{ m}^2/\text{sec} \cdot \text{volt})$$
$$= 112 \text{ ohm}^{-1} \cdot \text{m}^{-1} \quad (= 1.12 \text{ ohm}^{-1} \cdot \text{cm}^{-1}).$$

Note. The intrinsic conductivity, σ_{in}, is negligible in comparison, because at room temperature $E_g/2kT = 22$ and the preexponential terms of Eq. (15-2b) equal about $2.5 \times 10^{24}/\text{m}^3$, to give $n_{in} \cong 6 \times 10^{14}/\text{m}^3$, lower than n_d by a factor of 10^7. ◄

Majority and minority carriers. We just noted that silicon *doped* with antimony has many more charge carriers from donor electrons than from intrinsic sources. The electrons are the *majority carriers*, while the electron holes are the *minority carriers* in *n*-type semiconductors.

p-type semiconduction. The addition of Group III elements to silicon, or other semiconduction materials, reduces the valence electron-atom ratio below 4.0. An aluminum atom, for example, provides *acceptor* sites for electrons from elsewhere in the structure (Fig. 15-0a). The energy of these sites is slightly higher than that of other locations in the structure because their presence upsets the local charge bal-

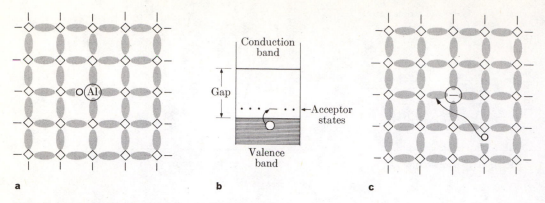

a

b

c

15-9 Extrinsic semiconductor (*p*-type). (a) A Group III (aluminum) atom is an acceptor, since it is short one electron for covalent bonds. (b) The acceptor states are near the bottom of the energy gap. (c) When a valence electron moves to an acceptor state, it ionizes the impurity.

ance; however, with this acceptor state present for receiving electrons (Fig. 15-9b), electron holes can form in the valence band and provide a means of conduction (Fig. 15-9c). Since electron holes, with their positive charges, are responsible for conduction, these materials are called *p-type* semiconductors.

Donor exhaustion in *n*-type semiconductors has a counterpart of *acceptor saturation* in *p*-type semiconductors, and Fig. 15-8(b) could be relabeled accordingly. In a *p*-type semiconductor, the electron holes are the majority charge carriers and any electrons in the conduction band are minority carriers.

Example 15-5

A silicon semiconductor contains 0.000001 a/o aluminum (0.01 ppm) and has a resistivity of 0.25 ohm · m at 20°C.
a) Assuming acceptor saturation, what is the electron-hole concentration, n_a?
b) What is the mobility of the electron holes?

Answer. [As discussed in Example 15-4, the intrinsic semiconduction of silicon is negligible at 20°C.]
a) There is one electron hole per aluminum atom; therefore, there is one electron hole per 10^8 silicon atoms. From Appendix B,

$$\text{Si atoms/m}^3 = \frac{(2.4 \times 10^6 \text{ gm/m}^3)(0.602 \times 10^{24} \text{ atoms/mole})}{(28.1 \text{ gm/mole})}$$

$$= 5.13 \times 10^{28} \text{ atoms/m}^3,$$

$$\text{holes/m}^3 = 5.13 \times 10^{20} \text{ holes/m}^3.$$

b) From Eq. (14-1),

$$\mu_p = \frac{1}{\rho q n_a} = \frac{1}{(0.25 \text{ ohm} \cdot \text{m})(1.6 \times 10^{-19} \text{ coul})(5.13 \times 10^{20}/\text{m}^3)}$$

$$= 0.05 \text{ m}^2/\text{sec} \cdot \text{volt.} \blacktriangleleft$$

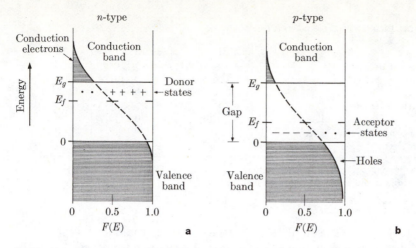

15-10 Fermi distributions (extrinsic semiconductor). (a) *n*-type. There are more conduction electrons than holes as a result of donors. (b) *p*-type. Holes exceed conduction electrons because of acceptor states. Therefore the Fermi levels, where $F(E) = 50\%$, are not at the centers of the gaps.

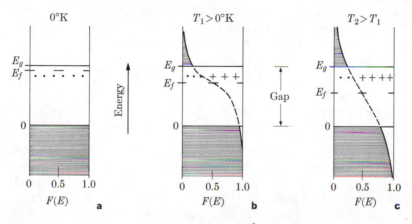

15-11 Fermi energy versus temperature. (a) At $0°K$ all of the donor states and none of the conductor states are occupied. (b) At intermediate temperatures most of the conduction electrons come from the donor states. (c) At high temperatures, valence electrons move across the gap and the Fermi energy drops toward the middle of the energy gap.

Fermi distribution (extrinsic semiconductors). A doped semiconductor has carriers from both intrinsic and extrinsic sources. As a result, the Fermi distribution is not symmetrical across the gap (Fig. 15-10). This means that the Fermi energy does not lie in the middle of the gap, particularly at lower temperatures (Fig. 15-11) where there are very few intrinsic carriers.

The Fermi distribution just described is a result of an equilibrium between the rate of *generation* and *recombination* of conduction electrons and electron holes:

$$\text{valence electrons} \rightleftharpoons \bullet + \bigcirc, \tag{15-5}$$

TABLE 15-2

SEMICONDUCTION PROPERTIES OF n-TYPE GERMANIUM*

Parts of impurity	10^{-10}	10^{-8}	10^{-6}
Donor concentration, N_d, m^{-3}	4.4×10^{18}	4.4×10^{20}	4.4×10^{22}
Conduction electrons, n_n, m^{-3}	2.7×10^{19}	4.4×10^{20}	4.4×10^{22}
Electron holes, n_p, m^{-3}	2.3×10^{19}	1.4×10^{18}	1.4×10^{16}
Equilibrium constant, $(n_n)(n_p)$, m^{-6}	6.2×10^{38}	6.2×10^{38}	6.2×10^{38}
Electron mobility, μ_n, m^2/sec · volt	0.39	0.37	0.29
Hole mobility, μ_p, m^2/sec · volt	0.19	0.18	0.14
Resistivity, ohm^{-1} · m^{-1}	0.42	0.038	0.0005

* Adapted from M. J. Morant, *Introduction to Semiconductor Devices*, London: George Harrap and Co.; and Reading, Mass.: Addison-Wesley Publishing Co., 1964, p. 24.

where the symbols stand for conduction electrons and electron holes, respectively. Like any equilibrium reaction, this one has an equilibrium constant, K:

$$K = n_n n_p. \tag{15-6}$$

As in Eq. (15-1), n_n and n_p are the numbers of conduction electrons and electron holes per unit volume, respectively. The value of K depends on the energy gap, the temperature, and the amount of incident radiation. On the other hand, Eq. (15-6) does not distinguish between intrinsic and extrinsic sources. Thus the equilibrium product of n_n and n_p is the same for pure and for doped semiconductors. If either n_n or n_p is increased, the other must decrease (Table 15-2).

15-4 EXCESS CARRIER LIFETIMES

If the equilibrium constant is suddenly decreased, the semiconductor will have excess carriers. For example, with incident light, the number of conductor electrons (and holes) in a semiconductor is relatively high (Section 15-2). Immediately after the light source is eliminated, there are more conduction electrons and electron holes than the new equilibrium permits, so Eq. (15-5) reacts to the left. This reaction takes time, however; consequently, *excess carriers* are present for a period. The reaction rate is proportional to the number of excess carriers, n:

$$-dn/dt = n/\tau_c \tag{15-7a}$$

or

$$n_t = n_0 e^{-t/\tau_c}. \tag{15-7b}$$

The preexponential term n_0 is the number of excess carriers at $t = 0$, and τ_c is their *lifetime*. [In line with Eqs. (10-10), (12-10), and (13-30), this is also called a relaxation time.]

A second example will be pertinent to transistors, where excess carriers are *injected* into a semiconductor (Section 15-9). Consider n-type germanium from Table 15-2, which has 10^{-8} atom fraction of a Group V element (N, P, As, or Sb). If a current is passed through this material, excess holes will be introduced at the

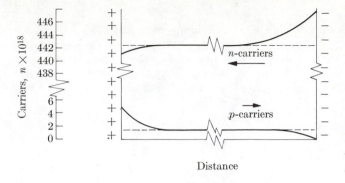

Carriers, $n \times 10^{18}$

446
444
442
440
438

6
4
2
0

n-carriers

p-carriers

Distance

15-12 Excess carriers (n-type germanium). Excess holes (the minority carrier) which are injected into the semiconductor travel a certain distance before recombination with electrons. The $n_n n_p$ product exceeds equilibrium adjacent to the electrodes.

positive electrode and excess conduction electrons at the negative electrode (Fig. 15-12). Thus the actual $n_n n_p$ product will be in excess of the equilibrium constant. The mathematical form of this curve versus distance depends not only on the lifetime of the excess carriers, but also on the diffusion coefficients of the electrons and holes, and on their drift velocity in the electric field (which is related to the total current flux). We shall simply note that the $n_n n_p$ product decreases logarithmically with distance inside the semiconductor.

Electron-hole recombination reactions are nucleated at crystal imperfections, be they vacancies, dislocations, grain boundaries, surfaces, or extraneous impurity atoms. Thus the lifetime of excess carriers depends on the quality of the semiconductor. In very pure germanium, for example, lifetimes as long as 1000 μsec have been obtained, although commercial-quality material usually has a lifetime of 100 to 500 μsec. Excess carriers have somewhat shorter lifetimes in silicon because the energy gap is greater. Since the lifetime depends on the impurity, lifetimes are shorter in extrinsic semiconductors and, in fact, differ between n-type and p-type products because these two types of products have different impurities.

The n_n / n_p ratio does not change in an intrinsic semiconductor by recombination, because the numbers of electrons and holes are identical and the reaction consumes equal numbers of each. In an extrinsic semiconductor, however, recombination consumes a much larger fraction of the minority carriers. For example, consider germanium doped with 10^{-8} atom fraction antimony so that there are 4.4×10^{20} donors/m^3 under equilibrium conditions (Table 15-2). Now if 1.4×10^{18} *excess* electrons/m^3 (and 1.4×10^{18} *excess* holes*) are added, the minority carriers are doubled to 2.8×10^{18} holes/m^3, while the majority carriers have only a fractional increase to 4.414×10^{20} electrons/m^3. Thus equal changes in absolute numbers produce a significant relative change in the number of minority carriers.†

---

* The numbers of *excess* holes and electrons are equal by definition, since the two are generated in pairs in an intrinsic semiconductor.
† In technical terminology one often refers to *minority-carrier lifetimes*, even though the majority carriers recombine in equal numbers and have the same lifetime.

Example 15-6

A semiconductor AX compound with an $n_n n_p$ equilibrium constant of $10^{36}/m^6$ is made in two forms: (a) intrinsic, with an excess-carrier lifetime of 600 μsec, and (b) p-type, with an equilibrium hole concentration of $10^{19}/m^3$ and an excess-carrier lifetime of 200 μsec.

A light source which induced $3 \times 10^{18}/m^3$ excess carriers is turned off. Compare the numbers of excess conduction electrons and holes as a function of time.

Answer

a) Intrinsic semiconductor: At equilibrium,

$$n_n = n_p = \sqrt{K} = 10^{18}/m^3.$$

At t_0,

$$n_n = n_p = 10^{18}/m^3 + 1.5 \times 10^{18}/m^3$$
$$= 2.5 \times 10^{18}/m^3.$$

At 600 μsec,

$$n_n = n_p = 10^{18}/m^3 + (1.5 \times 10^{18}/m^3)e^{-600/600}$$
$$= 1.55 \times 10^{18}/m^3.$$

b) Extrinsic semiconductor (electron holes): At equilibrium,

$$n_p = 10^{19}/m^3.$$

At t_0,

$$n_p = 10^{19}/m^3 + 1.5 \times 10^{18}/m^3$$
$$= 1.15 \times 10^{19}/m^3.$$

At 600 μsec,

$$n_p = 10^{19} + (1.5 \times 10^{18}/m^3)/e^{600/200}$$
$$= 1.007 \times 10^{19}/m^3.$$

c) Extrinsic semiconductor (conduction electrons): At equilibrium,

$$n_n = K/n_p = 10^{17}/m^3.$$

At t_0,

$$n_n = 10^{17}/m^3 + 1.5 \times 10^{18}/m^3$$
$$= 16 \times 10^{17}/m^3.$$

At 600 μsec,

$$n_n = 10^{17}/m^3 + (1.5 \times 10^{18}/m^3)/e^3$$
$$= 1.75 \times 10^{17}/m^3.$$

For other times, see Fig. 15-13. ◄

15-5 HALL EFFECT

As a charge moves across a magnetic field, it is deflected by a 90° force according to the familiar right-hand rule. Thus, a voltage, $\mathcal{E}_H$, is generated as shown in Fig. 15-14 when the charge carriers are electrons. The voltage is in the opposite

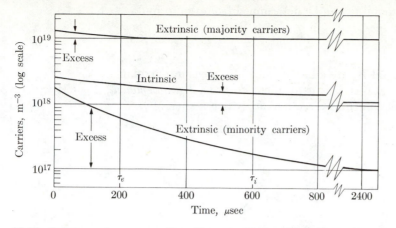

15-13 Excess carriers versus time (Example 15-6). Although equal numbers of minority and majority carriers recombine, the percentage change is more obvious in the former.

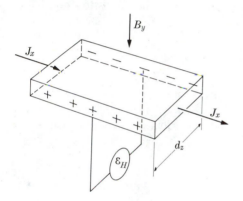

15-14 Hall voltage (*n*-type). As the current flux, J_x, moves through a magnetic flux, B_y, the charge carriers are displaced to produce a Hall field, $\mathcal{E}_H$.

direction when the charge carriers are electron holes. Consequently, it is possible to identify the majority charge carrier in a semiconductor by the sign of the voltage, thus distinguishing between *n*-type and *p*-type semiconductors.

The deflecting force, F, on the electron is

$$F = -qB_yv_x, \tag{15-8a}$$

where B_y is the magnetic flux density in webers/m^2 ($=$ newton/amp $\cdot$ m) and v_x is the drift velocity through the semiconductor. The latter term may be expressed as $\mathcal{E}_x\mu_n$, where $\mathcal{E}_x$ is the voltage gradient for conduction and μ_n is the electron mobility:

$$F = -qB_y\mathcal{E}_x\mu_n. \tag{15-8b}$$

This force is balanced by an equal force, $q\mathcal{E}_H$, on the electron from the induced, or *Hall field*, $\mathcal{E}_H$:

$$F_z = 0 = -qB_y\mathcal{E}_x\mu_n - q\mathcal{E}_H. \tag{15-9}$$

As a result, the mobility can be determined from the formula

$$\mu_n = \frac{-\mathcal{E}_H}{\mathcal{E}_x B_y}, \tag{15-10}$$

where $-\mathcal{E}_H$, $\mathcal{E}_x$, and B_y are all experimentally measurable quantities.

In some calculations, it is convenient to use a *Hall coefficient*, R_H. If we find the current density in the x-direction, J_x, which is given by

$$J_x = n_n q \mathcal{E}_x \mu_n, \tag{15-11}$$

and use Eq. (15-10), we can derive

$$\mathcal{E}_H = \frac{-J_x}{n_n q} B_y = R_H J_x B_y, \tag{15-12}$$

where

$$R_H = \frac{-1}{n_n q}. \tag{15-13a}$$

For p-type semiconductors, where the charge carrier has the opposite sign,

$$R_H = \frac{1}{n_p q}. \tag{15-13b}$$

Example 15-7

An InAs semiconductor is cut into a small bar 1.0 cm × 1.0 mm × 2.0 mm. Its lengthwise resistance is 1.25 ohms. A Hall field of −1.7 volts/m develops when a current of 0.12 amp is carried lengthwise, and the magnetic flux density is 500 gauss (i.e., 0.05 volt · sec/m²).
a) Is the semiconductor n- or p-type?
b) What is the carrier density?
c) What is the mobility of the carriers?

Answer. The current density is

$$J_x = (0.12 \text{ amp})/(2 \times 10^{-6} \text{ m}^2) = 6 \times 10^4 \text{ amp/m}^2.$$

a) From Eq. (15-12),

$$R_H = \frac{(-1.7 \text{ volts/m})}{(6 \times 10^4 \text{ amp/m}^2)(0.05 \text{ volt} \cdot \text{sec/m}^2)}$$

$$= -5.7 \times 10^{-4} \text{ m}^3/\text{coul}.$$

Since the Hall coefficient is negative, the charge carriers are *electrons*.
b) From Eq. (15-13a),

$$n_n = -1/(-5.7 \times 10^{-4} \text{ m}^3/\text{coul})(1.6 \times 10^{-19} \text{ coul})$$

$$= 1.09 \times 10^{22}/\text{m}^3.$$

(c) From Eq. (15-10),

$$\mu_n = \frac{-(-1.7 \text{ volts/m})}{(0.12 \text{ amp} \times 1.25 \text{ ohms}/0.01 \text{ m})(0.05 \text{ volt} \cdot \text{sec/m}^2)}$$

$$= 2.30 \text{ m}^2/\text{volt} \cdot \text{sec.} \blacktriangleleft$$

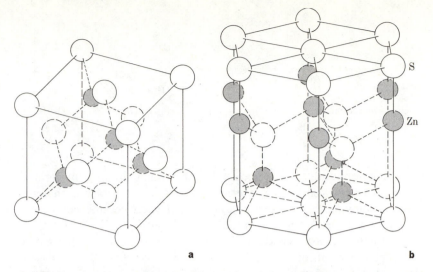

15-15 Crystals of covalent AX compounds. (a) Sphalerite (cf. Fig. 3-12). (b) Würtzite. The majority of III-V and II-VI semiconducting compounds possess one of these two structures.

15-6 SEMICONDUCTING MATERIALS

Materials which are used for semiconductors fall into three general classes: (1) elements, (2) covalent compounds, and (3) nonstoichiometric ionic compounds.

Elements. The Group IV elements have served as the prototype for semiconductors. Silicon and germanium are most widely used. Other covalent elements such as selenium and tellurium (Fig. 4-11a) are also semiconductors, since their covalent electrons fill the third energy band. They have relatively limited use, however, because they are more subject to mechanical and chemical alteration.

It is apparent from previous discussions that the composition must be controlled very closely, because only 10^{-5} a/o of impurities can have a significant effect on conductivity. This requirement has led to the development of a new purification procedure, known as *zone refining* (Section 18-8), as well as a system for growing near-perfect crystals.

Covalent compounds. If the two elements are in exact balance, III-V and II-VI compounds* have an average of four valence electrons per atom. The crystal structure of many such compounds is like diamond, except that the atoms are ordered to give unlike neighbors to each atom. Sphalerite (ZnS) serves as a prototype for many of these compounds (Fig. 15-15a). A second common structure is that of würtzite (also ZnS) as sketched in Fig. 15-15(b).

The coordination number of the atoms is four in both structures; therefore, the atomic packing factor is not high. More highly ionic AX compounds, such as MgO, generally assume higher coordination numbers; and since they have extremely wide energy gaps, they are insulators.

---

* Other possibilities include III$_2$-VI and II-V$_2$ compounds.

$$
\begin{array}{c|ccccccc|c}
- & O^{2-} & Fe^{2+} & O^{2-} & Fe^{2+} & O^{2-} & Fe^{2+} & O^{2-} & + \\
- & Fe^{2+} & O^{2-} & Fe^{2+} & O^{2-} & Fe^{++}\oplus & O^{2-} & Fe^{2+} & + \\
- & O^{2-} & Fe^{2+} & O^{2-} & & O^{2-} & Fe^{2+} & O^{2-} & + \\
- & Fe^{2+} & O^{2-} & Fe^{+++} & O^{2-} & Fe^{2+} & O^{2-} & Fe^{2+} & + \\
- & O^{2-} & Fe^{2+} & O^{2-} & Fe^{2+} & O^{2-} & Fe^{2+} & O^{2-} & + \\
\end{array}
$$

15-16 Defect semiconductor ($Fe_{1-x}O$). The third charge of Fe^{3+} serves as an acceptor to introduce electron holes on other ions.

Nonstoichiometric compounds. The third type of semiconducting material is commonly called a *defect semiconductor* because it is nonstoichiometric and therefore contains anion or cation vacancies. The critical feature in these ceramic compounds is the presence of ions which have more than one valence, e.g., Fe^{2+} and Fe^{3+}. As shown in Fig. 15-16, an electron which moves from one ion to another carries charge just as if by ionic diffusion.

We can also view a defect semiconductor as a material with a wide energy gap, but with acceptor states which will receive electrons from the valence band. In effect, the Fe^{3+} serves as a "trap" for an electron, just as the aluminum atom did in the p-type semiconductor of Fig. 15-9.

15-7 SIMPLE SEMICONDUCTOR APPLICATIONS

A number of simple semiconductor devices depend on the bulk properties of semiconductors. Most, but not all, of these depend on resistivity changes. A few will be given as examples.

Thermistors. The number of charge carriers in a semiconductor is very sensitive to temperature changes (Eq. 15-2b). Since the mobility is less significantly affected and the charge per carrier is an absolute constant, the resistivity generally decreases as the temperature is increased. Some ceramic compounds are available with very high values of dR/dT, with the result that temperature differences as small as $10^{-6}\,°C$ can be detected. These compounds are also stable to relatively high temperatures.

Pressure gauges. Because many semiconductor materials have a low packing factor, they have high compressibility. Under pressure, the volume contracts, reducing the size of the energy gap measurably and thereby significantly increasing the number of charge carriers and conductivity. The pressure can be calibrated against the resistance.

Magnetometers. Since the Hall field, $\mathcal{E}_H$, is directly proportional to the magnetic flux density, the Hall effect may be used to measure magnetic fields (Eq. 15-12).

Electrophotography. Widespread use is currently made of photoconducting materials such as ZnO for electrostatic copying. Paper coated with ZnO is charged electrostatically and then exposed to the light from the image to be reproduced.

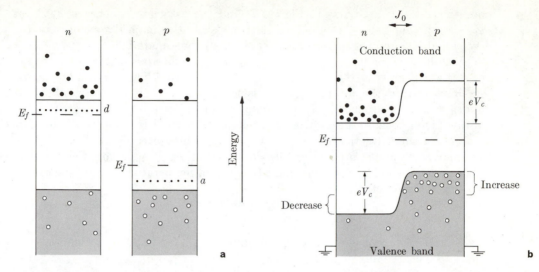

15-17 *n-p* junction. (a) Before contact. (b) After contact. Electrons are shifted from the *valence* band of the *n*-type to the *p*-type semiconductor on contact, so that the two Fermi energies are equal. As a result, the *n*-type material becomes positively charged and the *p*-type negatively charged, producing a potential difference, V_c.

Being photoconductive, the charge is dissipated in the lighted areas but not in the dark areas, which retain the charge. Subsequently the paper is dusted with a thermosetting ink powder which is attracted only to the charged areas. Heating "sets" the ink.

15-8 JUNCTION DEVICES

The better known, and probably more important, semiconductor applications involve a junction between *n*- and *p*-type semiconductors. There are various ways of making a sharp junction ($<10^3$ A). These will be mentioned in a later section. For the time being we will simply assume that we can place the two types of extrinsic conductors in intimate contact with no crystal discontinuity.

Contact potential. The Fermi energy lies above the middle of the energy gap in an *n*-type material and below the middle in a *p*-type material (Fig. 15-17a). Therefore, when these two materials are brought into contact at a junction, the gaps must be shifted so as to equalize the Fermi energy in the two materials (Fig. 15-17b). This means that the top of the valence band must be decreased in the *n*-type material, and increased on the *p*-type side. Of course, the atoms themselves cannot be shifted, but remain as ions within the crystal structure. Consequently, the *n*-type material [with its Group V positive ions (Fig. 15-7c)] becomes positively charged and the *p*-type material [with its Group III negative ions (Fig. 15-9c)] becomes negatively charged. The resulting potential difference is called the contact potential V_c, just as it was in Fig. 14-7, and the associated energy is eV_c* (or $e\phi_1 - e\phi_2$). As before, conduction electrons† are the majority carriers in the n-type material and the minority carriers in the p-type material.

The minority conduction electrons in the p-type materials of Fig. 15-17(b) are not static, but move as they would in any semiconductor. The electron flux across the junction into the n-type material is low because their initial numbers, as minority carriers, are small. In the opposite direction, the numbers are not the limiting factor because the electrons are the majority carriers. But we should note that only a small fraction of them have energy which exceeds eV_c; hence, only this small fraction can produce an electron charge flux, $J_{n \to p}$, across the junction. This fraction is the familiar logarithmic function of temperature, T, and the energy step across the junction, eV_c:

$$\underset{n \to p}{J} = M e^{-eV_c/kT}. \tag{15-14a} ‡$$

The term M is a constant. At equilibrium, $J_{n \to p}$ equals $J_{n \leftarrow p}$, which we shall call J_0, the charge flux of the minority carrier:

$$J_0 = M e^{-eV_c/kT}. \tag{15-14b}$$

Forward bias. Let us now add a forward bias, V_f, to the junction so that its potential is $V_c - V_f$, as shown in Fig. 15-18(a). If we ground the p-side, the electron flux, $n \leftarrow p$, is held constant at J_0, but the flux from $n \to p$ is changed markedly because V_c changes to $V_c - V_f$. Following Eq. (15-14a), we obtain

$$\underset{n \to p}{J} = M e^{-e(V_c - V_f)/kT}. \tag{15-15}$$

The net forward flux, J_f, is the difference:

$$\underset{n \to p}{J} - \underset{n \leftarrow p}{J} = M e^{-e(V_c - V_f)/kT} - J_0,$$

$$J_f = J_0(e^{eV_f/kT} - 1); \tag{15-16a}$$

or, if $eV_f \gg kT$, which is 0.025 eV at room temperature,

$$\ln J_f = eV_f/kT + B, \tag{15-16b}$$

where B is the logarithm of the constant value for J_0, the current carried by the minority carriers in the p-type material. The flux, J_f, is thus a logarithmic function of V_f.

---

* As stated in the last footnote of Section 14-2, we shall use e rather than q to indicate the electron charge when energy is expressed in terms of electron volts. By definition, the electron charge equals 1.0 under these conditions (and not 1.6×10^{-19} coul, which would give energy in joules). Numerically, therefore, one may ignore the e and use voltage values directly.

† From now on we shall direct our attention to electrons. We could discuss electron holes if we used opposite signs and directions.

‡ Since the cross-sectional area is fixed for a given junction, we could replace electron charge flux (coul/cm² · sec) with an opposing current (amp).

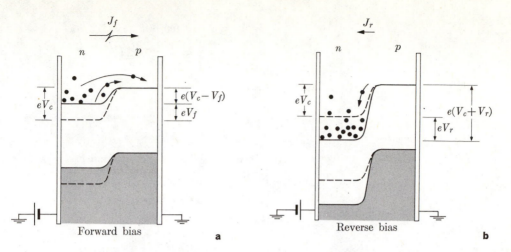

J_f $\quad$ J_r

n $\quad$ p $\qquad$ n $\quad$ p

Forward bias $\qquad$ Reverse bias

a $\qquad$ b

15-18 Biased junctions. (a) Forward bias. The forward electron flux, J_f, is an exponential function of the bias potential, V_f. (See Eq. 15-16.) (b) Reverse bias. The maximum reverse electron flux, J_r, is that of the minority carrier, J_0. (See Eq. 15-17). Everything is inverted for the flux of the electron holes.

Reverse bias. With a reverse voltage, V_r, and by similar arguments (Fig. 15-18b),

$$\underset{n\to p}{J} - \underset{n\leftarrow p}{J} = Me^{-e(V_c+V_r)/kT} - J_0,$$

$$J_r = J_0(e^{-eV_r/kT} - 1); \tag{15-17}$$

or if $eV_r \gg kT$, the net flux approaches J_0, but with opposite flow. The flux, J_r, does not exceed the initial low value.

Rectifiers. According to Eqs. (15-16) and (15-17), the electron charge flux, and therefore the current, varies nonlinearly with the voltage. This agrees with some of the original semiconductor data (Fig. 15-19) obtained by Shockley, one of the chief pioneers in semiconductor development. The practical implications are obvious: an *n-p junction* (or a *p-n junction*) can serve as a rectifier, permitting large quantities of current to pass in one direction and only limited amounts in the opposite direction. This type of rectifier, because of its simplicity, has replaced vacuum tube diodes in most applications.

As current passes in either direction, the carriers move beyond the junction before recombination occurs. The actual number of excess carriers is much greater with a forward bias, however, because the current density is greater. Penetration may occur several microns beyond the junction, and depends on the number of impurities and crystal imperfections present. This fact is significant for transistor applications (Section 15-9).

Zener diodes. If the reverse bias is too large, the rectifying action is eliminated because those few electrons which do pass across the junction are accelerated very

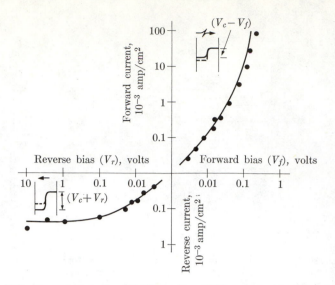

15-19 Rectifier junction characteristics. With forward bias, Eq. (15-16) applies. With reverse bias, Eq. (15-17) applies. (After data by W. Shockley, *Proc. IRE*, **40,** 1952, p. 1289.)

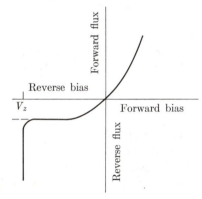

15-20 Zener diode. At high reverse biases, there is a limiting, or Zener, voltage, V_z. An avalanche of current is triggered by those few electrons which enter the junction region.

rapidly in the junction zone. (Recall that the junction zone may be less than a micron wide; if it is, the voltage gradient across the junction becomes extremely high.) The accelerated electrons can dislodge other electrons from the valence band; these other electrons are also accelerated, giving an *avalanche* of current. Thus, an extension of Fig. 15-19 shows a critical limiting voltage. Any attempt to extend the reverse bias beyond this value simply accentuates the breakdown (Fig. 15-20).

Since the critical voltage depends on the impurity levels in the n- and p-materials, it is possible to design junction diodes, called *Zener diodes*, for voltage-limiting devices. Such devices can be used as a bypass if the voltage of a circuit exceeds the desired level.

Photocells. In the discussion of photoconductors (Section 15-2), we saw that photons of light can produce electron-hole pairs. If these pairs are generated in the junction zone, the electrons move into the n-type material and the holes into the p-type material, with the result that the contact voltage is altered just as if a bias were added. The illuminated junction thus serves as a battery, or *solar cell*.

The photocell, unlike a photoconductor, does not need an applied voltage but generates its own. Applications range from exposure meters to power sources. Admittedly, the cost per unit energy is relatively high as compared to most standard power sources; however, there are obvious advantages in isolated locations.

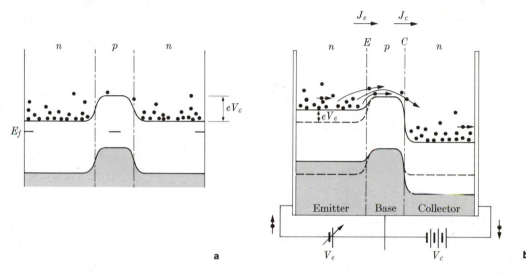

15-21 Transistor action. (a) Unbiased. (b) Forward biased for junction E. If the base is thin to avoid recombination, the electron flux, J_c, through the collector junction, C, is approximately J_e, which in turn is highly sensitive to the emitter voltage, V_e.

15-9 TRANSISTORS

The best-known junction device is a transistor which is a pair of junctions, back-to-back in series, giving an n-p-n (or a p-n-p) sandwich (Fig. 15-21a). When a voltage is placed across the transistor (Fig. 15-21b), junction E between the *emitter* and the *base* is forward biased, and junction C between the *base* and the *collector* is reverse biased.

Electrons pass readily into the base, where they become minority carriers. If the base is thick, the excess minority carriers all disappear by recombination. In practice, however, the junction is made thin enough so that excess conduction electrons reach junction C. Although junction C passes very few electrons to the left ($p \leftarrow n$), there is no barrier for the excess conduction electrons to move to the right ($p \rightarrow n$). Thus, if the base is thin, so that there is little recombination, the electron charge flux density, and therefore the current, I_c, through the collector

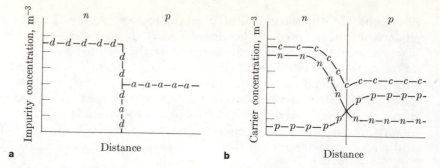

15-22 Concentration profiles (*n-p* junction). (a) ımpurity concentration. Donor—*d*; acceptor—*a*. (b) Carrier concentration. Conduction electrons—*n*; electron holes—*p*. At equilibrium, the product of the minority and majority carriers is a constant. Therefore the total carrier concentration, $c = n + p$, is depleted in the junction zone.

junction is approximately equal to the current at the emitter junction, I_e:

$$I_c \sim I_e. \tag{15-18}$$

But from Eq. (15-16a), the flux density, J_e, and the emitter current, I_c, are very sensitive to the emitter voltage, V_e. This means that a slight voltage change can effect a major *amplification* in current.

The critical feature of the transistor is that the electrons which are injected at junction E must move through the base with minimum recombination.* If this can be done, the I_c/I_e ratio, called the *injection efficiency*, is high.

Transistor structure. The doping levels of donors and acceptors on each side of a junction need not be, and indeed seldom are, identical. This has advantages in producing junctions, and particularly transistors. Consider Fig. 15-22(a), where more donor atoms have been added to the *n*-type material than acceptors to the *p*-type material. The carrier-concentration profiles with zero bias are indicated in Fig. 15-22(b). Because both conduction electrons and holes move across the junction, the carrier profiles are less abrupt than the impurity profiles. There is also a depleted zone with fewer carriers, *c*, in the junction region where recombination has occurred.

Since the impurity level is lower in the *p*-type material than in the *n*-type (in this example), the number of minority carriers must be higher to meet equilibrium requirements (Eq. 15-6). This arrangement has two advantages for the injection efficiency. (1) There are fewer impurities to reduce the lifetime of the excess carriers (Eq. 15-7). (2) The electron charge flux into the *p*-type region is higher, giving more minority carriers for conduction.

---

* In a *p-n-p* transistor, it is the holes that must move through the base with minimum recombination, since they are the minority carriers.

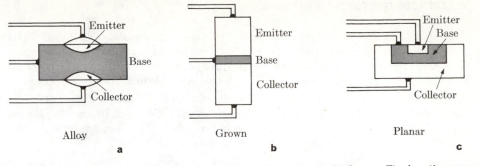

15-23 Transistors. (a) Alloy. The junction is made by alloy melting. (b) Grown. The junctions are made during the original crystal growth. (c) Planar (or diffusion). The junction is made by vapor deposition and diffusion.

Several different methods are available for enabling a single semiconductor crystal to contain both the *n*- and *p*-type regions. These are summarized in Fig. 15-23.

In an *alloy junction* (Fig. 15-23a), a few micrograms of dopant, such as aluminum, is melted on the surface of a single crystal of *n*-type silicon. The melt dissolves some silicon, which on cooling resolidifies with aluminum in solid solution. A similar procedure on the opposite side of a small wafer gives a *p-n-p* junction. An *n-p-n* junction would require an initial *p*-type silicon with a Group V alloying element added later.

For a *grown junction* (Fig. 15-23b), a crystal is solidified from an *n*-type melt. At the desired stage of growth, a Group III dopant is added to the melt. The amount must not only be sufficient to cancel the Group V elements of the *n*-type, but there must be enough extra to give the desired impurity level in the *p*-type material. After the crystal has grown another fraction of a millimeter, the necessary amount of a Group V element is added to the melt to reverse the type from *p* to *n*. As before, a *p-n-p* junction could be made by reversing the dopants. Obviously, processing procedures require close control so that the compositions and dimensions are those required.

The third, and final, process we shall describe involves an *n*-type silicon wafer which is oxidized to give a protective SiO_2 coating. Through a masking procedure, the coating is removed from a selected area before the wafer is heated in a boron-rich vapor. The boron (Group III) diffuses into the silicon to form a *p*-type region. The oxidation, masking, and heating process is repeated again, but with a phosphorus-rich vapor, to form the second *n*-type region, which usually serves as the emitter (Fig. 15-23c). Finally, contacts can be added by repeating the cycle once more, this time with gold or aluminum. A transistor of this type is commonly called a *planar* transistor, simply because all preparation and contacts are on one flat surface. This type of junction has lent itself to mass production because the above steps can be automated.

REFERENCES FOR FURTHER READING

Azaroff, L. V., and J. J. Brophy, *Electronic Processes in Materials*. New York: McGraw-Hill, 1963. The theory of semiconductors, semiconducting materials, and semiconductor devices are discussed in Chapters 8 through 10. For the student who wishes to extend his familiarity of semiconductors beyond the present text. Mathematics is minimized.

Basic Course on Solid-State Electronics. Cleveland: Penton Publishing, 1966–1967. Paperback. Reprints of a series of 12 articles which appeared in *Machine Design*. For the person who wants background reading at a nontechnical level. Also recommended for the instructor who wishes to obtain descriptive analogies for his lectures.

Brophy, J. J., *Semiconductor Devices*. New York: McGraw-Hill, 1964. Paperback. Excellent background for the study of semiconductors. For the freshman student in engineering.

Dekker, A. J., *Electrical Engineering Materials*. Englewood Cliffs, N. J.: Prentice-Hall, 1959. The mechanism of conduction in semiconductors, and junction rectifiers and transistors, are discussed in Chapters 6 and 7, respectively. Supplementary reading for the advanced undergraduate interested in semiconductors.

Hutchinson, T. S., and D. C. Baird, *The Physics of Engineering Solids*. New York: Wiley, 1963. Chapter 11 is alternate reading for this chapter. Same level of presentation.

Kittel, C., *Elementary Solid State Physics: A Short Course*. New York: Wiley, 1962. Semiconductor physics and devices are discussed in Chapter 7. Advanced undergraduate and graduate level.

Leck, J. H., *Theory of Semiconductor Junction Devices*. Oxford: Pergamon, 1967. This textbook for electrical engineers provides a sequel to this chapter. Undergraduate level.

Levine, S. N., *Solid State Microelectronics*. New York: Holt, Rinehart, and Winston, 1963. Chapter 2 amplifies the discussions of this chapter. Subsequent chapters emphasize thin films. A good mathematics background is necessary.

Moore, W. J., *Seven Solid States*. New York: Benjamin, 1967. Paperback. Silicon is used as an example of a semiconductor. This is a good supplement for the present chapter. Undergraduate level.

Morant, M. J., *Introduction to Semiconductor Devices*. London: Harrap; and Reading, Mass.: Addison-Wesley, 1964. A small text which will be useful to the student who wishes to focus his attention on semiconductors. Advanced undergraduate level.

Nussbaum, A., *Electronic and Magnetic Behavior of Materials*. Englewood Cliffs, N. J.: Prentice-Hall, 1967. Paperback. Energy bands and semiconductors are presented in Chapter 3. Part of Chapter 2, beginning at Section 2-3, will provide a useful supplement to this text for students who are interested in semiconductors.

Rose, R. M., L. A. Shepard, and J. Wulff, *The Structure and Properties of Materials: IV. Electronic Properties*. New York: Wiley, 1966. Paperback. There are chapters on semiconductors, semiconducting devices, semiconducting materials, and thermoelectricity. Introductory level. A useful supplement to this book for undergraduates.

Van Vlack, L. H., *Elements of Materials Science*, second edition. Reading, Mass.: Addison-Wesley, 1964. A less technical introduction to semiconductors is given than in this text. Freshman-sophomore level.

Wert, C. A., and R. M. Thomson, *Physics of Solids*. New York: McGraw-Hill, 1964. Semiconductors and their devices are discussed in Chapters 12 and 13, respectively. Advanced undergraduate level.

PROBLEMS

15-1 The resistivity of an intrinsic semiconductor is 4.5 ohm · m at 20°C and 2.0 ohm · m at 32°C. What is its energy gap?

Answer. 1.04 eV   (ln $\sigma_0 = 19.1$)

15-2 At what temperature will the resistivity of the semiconductor in Problem 15-1 be (a) 1 ohm · m? (b) 10 ohm · m?

15-3 Refer to Table 15-1. At what temperature will the intrinsic resistivity of silicon be 500 ohm · m?

Answer. 40°C

15-4 A thermistor has a temperature sensitivity of resistivity of 11%/°C at room temperature. What is its energy gap?

15-5 If intrinsic germanium has a resistivity of 1.0 ohm · m, $\mu_n = 0.39$ m^2/volt · sec, and $\mu_p = 0.19$ m^2/volt · sec, how many electrons serve as carriers per m^3?

Answer. 1.08×10^{19} carriers/m^3

15-6 The temperature is such that there are 10^{20} electrons/m^3 in the conduction band in intrinsic InAs. What is its conductivity?

15-7 Silicon contains antimony as an intentional impurity. There is one Sb atom per 10^8 Si. What is the resistivity of this semiconductor if the electron mobility is 0.14 m^2/volt · sec and the lattice constant is 5.42 A?

Answer: 0.09 ohm · m

15-8 A GaAs semiconductor has the same crystal structure as sphalerite (Fig. 15-15a). This is also the same structure that diamond and silicon have (Fig. 3-12), except that alternate atoms are gallium (and, likewise, alternate atoms are arsenic).

   a) There are ———— Ga atoms and ———— As atoms per unit cell in GaAs.

   b) If there are a few excess Ga atoms as compared to the number of As atoms (1,000,005 Ga per 999,995 As), the electrical conductivity increases tremendously. Will the conduction be p-type or n-type?

   c) In the unbalanced compound of (b), the lattice constant, a, is 5.43 A. As a result, there will be ———— carriers per m^3.

15-9 Silicon has a density of 2.40 gm/cm^3.

   a) What is the concentration of the silicon atoms per m^3?

   b) Phosphorus is added to silicon to make it an n-type semiconductor with an extrinsic conductivity of 100 ohm^{-1} · m^{-1} at donor exhaustion. What is the concentration of the conduction electrons?

Answer. a) 5.15×10^{28} atom/m^3 b) 4.5×10^{21} carrier electrons/m^3

15-10 a) How many silicon atoms are there for each conduction electron in Problem 15-9(b)?

   b) The lattice constant for silicon is 5.42 A, and there are 8 atoms per unit cell. What is the volume associated with each conduction electron?

   c) How many unit cells are there per conduction electron?

15-11 A single crystal of semiconductor-grade germanium has an extrinsic resistivity of 0.02 ohm · m. It contains 1.6×10^{21} aluminum atoms/m^3. Assume complete aluminum ionization, i.e., acceptor saturation.

a) What kind of carriers are present?

b) What is their mobility?

Answer. $\mu_p = 0.19 \text{ m}^2/\text{volt} \cdot \text{sec}$

15-12 Pure InSb is an intrinsic semiconductor with an energy gap of 0.18 eV. Its unit cell is cubic with $a = 6.48$ A, and with four indium and four antimony atoms. An addition is made of 10^{21} zinc atoms/m^3, replacing an equal number of indium atoms.

a) Will the semiconductor now be *n*-type or *p*-type?

b) How many charge carriers will there be per m^3?

15-13 The InSb of Problem 15-12 has the structure of diamond, except that alternate atoms are unlike. Additional properties include the following: $\rho = 5.8 \text{ gm/cm}^3$, $\mu_n = 8.0 \text{ m}^2/\text{volt} \cdot \text{sec}$, $\mu_p = 0.07 \text{ m}^2/\text{volt} \cdot \text{sec}$. These values are relatively constant over the normal temperature range.

a) On the basis of Eq. (15-4) what will the temperature be when the intrinsic conductivity is twice that of 20°C?

b) In each m^3, 10^{21} antimony atoms of pure InSb are replaced by 10^{21} selenium atoms. What is the maximum extrinsic conductivity resulting from this substitution?

c) In another sample (without the selenium doping), 10^{16} pairs/m^3 of indium and antimony atoms are replaced by an equal number of silicon and carbon atom pairs. What is the maximum extrinsic conductivity resulting from this substitution?

Answer. a) 91°C (364°K)   b) 1280 ohm$^{-1} \cdot$ m^{-1} c) zero

15-14 Zinc sulfide is a II-VI semiconductor. It is cubic, with a lattice constant of 5.9 A. As made for our problem, a few sulfur atoms replace zinc atoms so that the Zn/S ratio is 0.9999999/1.0000001. How many charge carriers are there per m^3?

15-15 In light, the conductivity of silicon is increased from 0.0005 ohm$^{-1} \cdot$ m^{-1} to 0.3 ohm$^{-1} \cdot$ m^{-1}. How many electrons have been activated across the energy gap?

Answer. 10^{19}/m^3

15-16 The phosphor of a television tube glows because conduction electrons release light photons as they return across the energy gap. What minimum energy gap is required for a red phosphor ($\lambda = 6700$ A)?

15-17 The lifetime of the conduction-band electrons in Problem 15-15 is 10.0 msec. What concentration of conduction electrons will remain (a) 0.07 sec after the light is removed? (b) 1.0 sec later?

Answer. a) 10^{16}/m^3 b) Nil (except for thermally activated electrons)

15-18 The scanning beam on a television tube covers the screen with 30 frames per sec. What must the lifetime for the electrons in the conduction band of the phosphor be if only 20% of the intensity is to remain when the following frame is scanned?

15-19 The intrinsic conductivity of a III-V compound is 2000 ohm$^{-1} \cdot$ m^{-1} at 30°C and 1200 ohm$^{-1} \cdot$ m^{-1} at 25°C.

a) What is the size of the energy gap?

b) How many atoms of selenium should replace V atoms per m^3 in order to increase the conductivity to 2000 ohm$^{-1} \cdot$ m^{-1} at 20°C ($\mu_p = 0.1 \text{ m}^2/\text{volt} \cdot \text{sec}$, $\mu_n = 0.2 \text{ m}^2/\text{volt} \cdot \text{sec}$)?

Answer. a) 1.59 eV   b) 2.5×10^{22}/m^3

15-20 The sketch in Fig. 15-16 shows the (100) plane of $Fe_{1-x}O$, where x is a small fraction. As indicated, some of the Fe^{2+} ions are replaced by Fe^{3+} ions and vacancies. If there are 4 Fe^{3+} ions per 100 oxygen ions, how many charge carriers are there per cm^3 for electronic conduction? (The lattice constant, a, of FeO is 4.3 A.)

15-21 Some $Fe_{1-x}O$ has an Fe^{3+}/Fe^{2+} ratio of 0.1. What is the mobility of the electron holes if this oxide has a conductivity of 100 $ohm^{-1} \cdot m^{-1}$ and 99% of the charge is carried by the electron holes ($a = 4.3$ A)?

Answer. 1.4×10^{-7} $m^2/volt \cdot sec$

15-22 How many charge carriers, (a) electron holes, (b) cation vacancies, are there per m^3 in the previous problem?

15-23 A *p-n* junction is made by the following steps:
1) 1.000 gm of silicon is melted.
2) 0.0001 w/o of aluminum is added to this 1.000-gm silicon melt.
3) A large single crystal of *p*-type silicon is grown from one-half of the melt.
4) Antimony is added to the remaining melt to produce electron charge carriers.
5) Crystal growth is continued. Thus the initial half of the crystal is *p*-type silicon and the final half is *n*-type silicon.
6) The large crystal is cut into diode junctions.

What percentage antimony must be present in the *n*-type half of the crystal to make the electron charge-carrier concentration equal to the electron-hole carrier concentration in the *p*-type half of the crystal?

Answer. 0.00009 w/o Sb

15-24 A *p-n* junction is made as follows:
1) 100 gm of silicon is melted.
2) 0.001 gm of an alloy is added which contains 0.7 w/o gallium (Ga) and the balance silicon.
3) The total alloy is solidified as a single crystal and cut into small wafers.
4) Antimony (Sb) is vapor-plated onto the surface of the wafer.
5) The coated wafer is heated so that the Sb can diffuse into the silicon.

a) There will be _____ w/o Ga throughout this wafer.
b) There must be _____ w/o Sb in the surface layer to provide it with as many carriers/m^3 from the Sb as there are in the wafer proper from the Ga.
c) Assume (1) that there is 0.000024 w/o Sb on the Sb side of the junction, (2) that the diffusion coefficient for Sb in Si is 10^{-14} cm^2/sec at the heating temperature, and (3) that the junction zone is 100 A wide. How many Sb atoms cross the junction per sec? [*Note.* The density of Si is 2.4 gm/cm^3.]

15-25 A magnetometer is made out of *n*-type InAs with a resistivity of 0.01 $ohm \cdot m$. The Hall field is 97 millivolts per meter when the applied potential is 4.1 volts per meter. What is the magnetic flux density, B?

Answer. 0.01 $volt \cdot sec/m^2$ (0.01 $weber/m^2$)

15-26 The earth's magnetic flux density is approximately 0.5×10^{-4} $volt \cdot sec/m^2$ at mid-latitudes. In designing a magnetometer made out of the InAs in the previous problem, for how much current density should provision be made?

16 □ magnetic behavior of solids

16-1 MAGNETIZATION

We saw in Chapter 13 that an electric field, $\mathcal{E}$, produces an electrical charge density, $\mathfrak{D}$. In a somewhat related way, a magnetic field, H, produces a magnetic flux density, B. Within a vacuum, the magnetic flux density is related to the magnetic field through the *magnetic permeability of a vacuum*, μ_0:

$$\mu_0 = B/H. \tag{16-1a}$$

The *mks units* for μ_0 do not correspond to the units for electric permittivity, however, because different units are used for magnetic and electric fields. Specifically, the mks units for Eq. (16-1a) are

$$(\text{volt} \cdot \text{sec})/(\text{amp} \cdot \text{m}) = (\text{volt} \cdot \text{sec/m}^2)/(\text{amp/m}), \tag{16-1b}$$

or, since a henry is an ohm · sec and a weber is a volt · sec,

$$(\text{henries/m}) = (\text{webers/m}^2)/(\text{amp/m}). \tag{16-1c}$$

The value of μ_0 in mks units* is $4\pi \times 10^{-7}$ henry/m.

Magnetization. With a material present, the magnetic flux density, also called *magnetic induction*, is different from that developed in a vacuum. As with dielectric properties, it is convenient to use a relative value when we compare permeabilities:

$$B = \mu_0\mu_r H, \tag{16-2}$$

where μ_r is called the *relative permeability*. Its value is 1.0 for a vacuum. It is

---

* Although we will not use them, the reader should be aware that practicing engineers make considerable use of *cgs units*. In these units,

$$B = \mu H, \tag{16-1d}$$

where B is in gauss, H is in oersted, and $\mu = 1$ gauss/oersted for a vacuum. Conversions from mks to cgs units are as follows:

H: 1 amp/m (mks) $= 4\pi \times 10^{-3}$ oersted (cgs)
B: 1 weber/m^2 (mks) $= 10^4$ gauss (cgs)
μ: $4\pi \times 10^{-7}$ henry/m (mks) $= 1$ gauss/oersted (cgs).

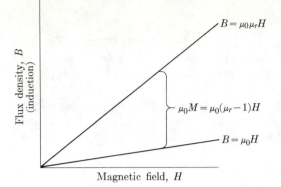

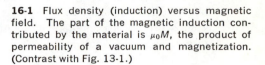

16-1 Flux density (induction) versus magnetic field. The part of the magnetic induction contributed by the material is $\mu_0 M$, the product of permeability of a vacuum and magnetization. (Contrast with Fig. 13-1.)

slightly more than one in paramagnetic materials and slightly less than one in diamagnetic materials (Section 16-2). A few materials, however, have very high permeabilities. These have engineering importance as "magnetic" materials, and will receive the major share of our attention in the last three sections of this chapter.

The part of the magnetic induction which is a consequence of the material will be of specific interest to us. It will be difficult, however, to make a direct analogy between dielectric and magnetic terms, because the two are not handled comparably. In the mks system, which we shall employ, the magnetic flux density (magnetic induction) which develops in the presence of a material is defined as follows:

$$B \equiv \mu_0(H + M) = \mu_0 H + \mu_0 M, \tag{16-3}*$$

where M is the *magnetization* (Fig. 16-1) and has the same units as H, that is, amp/m. Thus, from Eq. (16-2),

$$M = (\mu_r - 1)H; \tag{16-4a}$$

and χ_M, the *magnetic susceptibility*, is related to the other terms as follows:

$$\chi_M = \mu_r - 1 = M/H. \tag{16-4b}$$

Magnetic moments. A bar magnet, or the magnetic behavior of an induction coil, can be visualized as having north and south poles separated by a distance l. Thus they possess a magnetic moment, p_m. In the coil, the magnetic moment can be related to the product of the current and cross-sectional area, amp $\cdot$ m^2.

Since magnetization has units of amps/m, we may consider magnetization as the magnetic moment per unit volume, amp $\cdot$ m^2/m^3. In this respect, magnetization corresponds to dielectric polarization.

The spin magnetic moment of an electron, p_β, called a *Bohr magneton*, will be important to us as we analyze the magnetization of a material. As indicated in

---

* Some authorities define B as $B = \mu_0 H + M$. This, of course, requires appropriate changes in other relationships.

Section 2-2,

$$p_\beta = 9.27 \times 10^{-24} \text{ amp} \cdot \text{m}^2. \tag{16-5}$$

It arises from the spin characteristics of the electron. Magnetic moments also originate from nucleus and from the orbital movements of electrons. These latter two sources make only a minor contribution to the magnetization of most engineering materials.

An atom is given a magnetic moment if its electrons have more spins aligned in one direction than in another. Thus, atoms or ions with unbalanced spins (Fig. 2-5) possess a magnetic moment, and those with completely filled orbitals have no net magnetic moment.

16-2 MAGNETISM IN SOLIDS

Paramagnetism. In a zero applied field, a material containing atoms which have individual magnetic moments exhibits no magnetization, because the atomic magnets are aligned in random directions. In a magnetic field, however, the individual atoms orient their moments to match the direction of the field. A material in which this occurs is called *paramagnetic*.

Since thermal agitation prevents perfect magnetic alignment just as it prevents the perfect alignment of electrical dipoles (Section 13-3), paramagnetization may be expressed as a function of temperature:

$$M_0 = N\mu_0(p_\beta)^2 H/3kT, \tag{16-6}$$

where N is the number of atoms per unit volume with unbalanced spins. Except for the factor μ_0, Eq. (16-6) corresponds closely to orientation polarization (Eq. 13-14).

Since a typical paramagnetic material has about 5×10^{28} atoms per cubic meter, each with a magnetic moment of about one Bohr magneton, Eq. (16-6) indicates that the magnetic susceptibility,

$$\mu_r - 1 = N\mu_0(p_\beta)^2/3kT, \tag{16-7}$$

is less than 10^{-3} at room temperature (and the relative magnetic permeability, μ_r, near 1.001). These values correspond closely to experimental data for paramagnetic materials such as $CrCl_3$, $NiSO_4$, and Fe_2O_3. It is apparent that these materials do not have magnetic applications in the normal sense. They are receiving attention, however, in research studies which make use of paramagnetic resonance.

Diamagnetism. In the absence of any magnetic moment arising from unpaired electrons, an increased magnetic field has a slightly negative effect on a material ($\mu - 1 \sim -10^{-5}$). This effect is called diamagnetism, and occurs in compliance with Lenz's law, according to which an increase in a magnetic field will be opposed by the field arising from the induced emf. Diamagnetism is detected only when other magnetic effects are absent; it has negligible engineering significance.

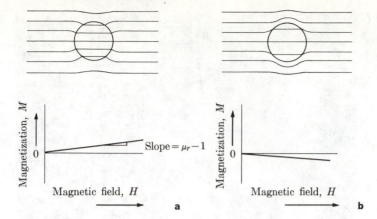

16-2 Flux density and magnetization. (a) Paramagnetism. Since a paramagnetic material has positive susceptibility and magnetization, magnetic flux is concentrated through it. (b) Diamagnetism. The susceptibility and magnetization are negative.

A schematic representation of paramagnetism and diamagnetism is presented in Fig. 16-2.

Ferromagnetism. Iron and certain related materials can possess very high magnetization, and thus high permeabilities and magnetic flux densities. For the sake of comparison, we list the range of magnetic susceptibilities, $\mu_r - 1$, for the three types of materials we have been discussing:

$$\text{diamagnetic:} \quad -0.00005 \text{ to } 0$$
$$\text{paramagnetic:} \quad +0.0001 \text{ to } 0.01$$
$$\text{ferromagnetic:} \quad 10^2 \text{ to } 10^6$$

Although the extremely high magnetic susceptibility of the ferromagnetic materials disappears above the Curie temperature, where they become paramagnetic, it is obvious that the behavior of ferromagnetic materials is markedly different from that of other materials. The remainder of this chapter deals with various aspects of ferromagnetism (and the related ferrimagnetism and antiferromagnetism).

16-3 FERROMAGNETIC PROPERTIES

Atoms with a magnetic moment of several Bohr magnetons and a small $3d$-radius can develop a coupling whereby they spontaneously align their moments with one another to produce a magnetic *domain*. Iron, cobalt, and nickel are the best-known metals which have this spontaneous alignment.

Each domain acts as a magnet which has micron to millimeter dimensions (Fig. 16-3). In pure iron and other *soft* magnetic materials in the absence of an external magnetic field, there are as many domains oriented in one direction as another. Thus the magnetization is not obvious. In a *permanent* magnet, however,

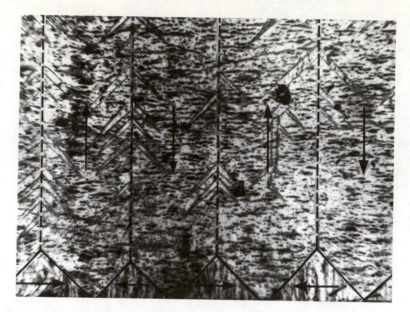

16-3 Ferromagnetic domains (iron, single crystal). The arrows show the direction of magnetization within each domain. If an external field is applied, the domain walls will shift laterally to enlarge those domains with favorable alignment. Photograph by H. J. Williams. (From *Ferromagnetism* by R. M. Bozorth. Copyright 1951 by Litton Educational Publishing, Inc. Reproduced by permission of Van Nostrand Reinhold Company.)

the number of domains having one orientation exceeds the number having other orientations, with the result that a net magnetization is apparent.

Hysteresis loop. Starting with zero flux density (or magnetic induction), the net magnetization of a ferromagnetic solid can be increased if an external magnetic field is applied. Those domains which are favorably oriented grow at the expense of those which are not oriented with the field. This growth occurs by reorientation of electronic spins in atoms adjacent to the domain walls. With domain growth, the net flux density increases rapidly (curve OAC in Fig. 16-4) until most of the domains are aligned in the same direction. This is called *saturation induction*. Additional field strength can only add small amounts of magnetic induction, or flux density, by improving the atomic orientation with the direction of the field.

In a permanent magnet, the induction does not disappear immediately when the field is removed; a residual, or *remanent induction*, B_r, persists. An opposing *coercive field*, $-H_c$, is required to balance the domains and reduce the net induction to zero. Cyclic fields produce a *hysteresis loop*, as indicated by the completed $CDFGC$ path of the B-H curves. The energy consumed per cycle is equal to the BH area within the loop.*

---

* The reader may wish to compare the last two paragraphs with similar paragraphs on ferroelectricity in Section 13-8.

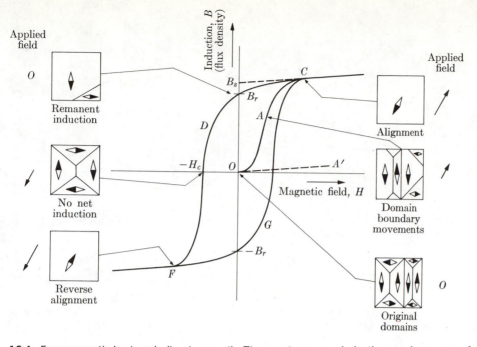

16-4 Ferromagnetic hysteresis (hard magnet). The spontaneous polarization requires energy for reversal. (See text. Also compare with Fig. 13-24.)

Soft and hard magnets. The remanent induction is very low in a soft magnet. In other words, the magnetization *relaxation time* is very short; consequently, demagnetization can take place during the part of the cycle in which the field is being reduced. In contrast, the relaxation time in an acceptable hard magnet appears to be infinite at zero field, with no evidence of a decrease in the magnetic remanence with time. Furthermore, a permanent magnet must have a large coercive field for demagnetization. The permanency of a magnet can also be expressed in terms of the BH energy product required for demagnetization, either as the integrated area of the second quadrant of the hysteresis loop (Fig. 16-5a) or, more commonly, as the *maximum demagnetization product*, BH_{max}, of the points on the curve of this quadrant (Fig. 16-5b).

Any factor which interferes with the movement of domain walls adds to the permanency of the magnet. Perhaps the most important inhibitor is a small particle of a nonmagnetic impurity or inclusion. As is shown in Fig. 16-6, energy is required to move a domain wall beyond an inclusion. There are two reasons for this: (1) The inclusion has eliminated some of the high-energy domain wall, and this must be reintroduced if the wall moves past the inclusion. (2) The magnetic energy across the inclusion is less when the inclusion is at the domain wall, because the average distance between the exposed poles is less. For these reasons, many permanent magnets contain very fine particles of a second phase.

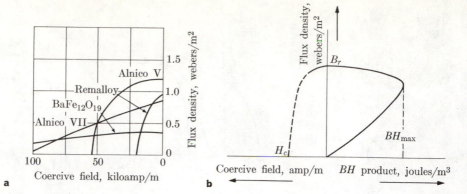

16-5 Demagnetization curves. (a) Hard magnets. (b) *BH* products. The maximum value is commonly used as an index of magnetic permanency.

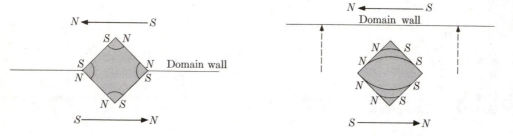

16-6 Domain-wall movements. Energy must be supplied to move the wall from a nonmagnetic inclusion. (See text.)

The movements of the domain wall are also affected by the presence of crystal imperfections. Dislocation networks are particularly effective in this respect. As a result, a strain-hardened metal is also magnetically harder than an annealed metal. In fact, this relationship accounts for the origin of the terms *hard* and *soft* to describe the permanency of magnets.

The properties of chief interest in a permanent magnet are remanence, B_r, coercive field, $-H_c$, and the maximum demagnetizing product, BH_{max}. These are listed for selected hard magnets in Table 16-1. Table 16-2 lists selected data

TABLE 16-1
PROPERTIES OF SELECTED HARD MAGNETS (Various sources)

Magnetic material	Remanence, B_r, webers/m^2	Coercive field, $-H_c$, amp/m	Maximum demagnetizing product, BH_{max}, joules/m^3
Carbon steel	1.0	0.4×10^4	0.1×10^4
Alnico V	1.2	5.5	3.4
Ferroxdur (BaFe$_{12}$O$_{19}$)	0.4	15.	2.0

TABLE 16-2

PROPERTIES OF SELECTED SOFT MAGNETS (Various sources)

Magnetic material	Saturation induction, B_s, webers/m²	Coercive field, $-H_c$, amp/m	Maximum relative permeability, μ_r(max)
Pure iron (bcc)	2.2	80	5,000
Silicon ferrite transformer sheet (oriented)	2.0	40	15,000
Permalloy, Ni-Fe	1.6	10	2,000
Superpermalloy, Ni-Fe-Mo	0.2	0.2	100,000
Ferroxcube A, $(Mn, Zn)Fe_2O_4$	0.4	30	1,200
Ferroxcube B, $(Ni, Zn)Fe_2O_4$	0.3	30	700

of chief interest for soft magnets. These include the coercive field, the *maximum* relative permeability, μ_r(max), and the saturation induction, B_s.

Magnetic permeability. Since the *B-H* curve of a ferromagnetic material is non-linear, there is not a constant $B/\mu_0 H$ ratio. The *initial* relative permeability, μ_r(in), may be defined as the permeability for very low fields. Of greater engineering importance is the *maximum* relative permeability, μ_r(max), as shown in Fig. 16-7. It is useful because it indicates the field necessary to provide high induction values; however, one must be aware that the induction does not continue to increase in proportion to the field. A saturation value is reached.

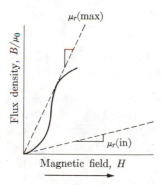

16-7 Ferromagnetic permeabilities. The *B-H* relationship is nonlinear in a ferromagnetic material. Initial relative permeability— μ_r(in); maximum relative permeability— μ_r(max).

Saturation induction. We may assume that at saturation the external field has forced all possible domains, atoms, and spins into alignment with the field. Since the saturation flux density is readily measured experimentally and most of the magnetic induction is from electron spins, it is possible to calculate the number of excess electrons with spins aligned parallel to the field, because, as noted earlier, each electron has a spin magnetic moment of 9.27×10^{-24} amp · m².

TABLE 16-3
FERROMAGNETIC ELEMENTS

Metal	Curie point, °K	Saturation induction, B_s, webers/m²	Saturation magnetization,* M_s, amp/m
Iron	1043	2.1	1.7×10^6
Cobalt	1393	1.7	1.45
Nickel	631	0.6	0.5

* $M_s = B_s/\mu_0$ (approximately).

Example 16-1

The saturation induction, B_s, of iron is 2.1 webers/m². What is N_β, the net number of Bohr magnetons per atom?

Answer. From Appendix B, the density of iron is 7.87 gm/cm³. Thus,

$$n = \frac{(7.87 \times 10^6 \text{ gm/m}^3)(0.602 \times 10^{24} \text{ atoms/mole})}{(55.85 \text{ gm/mole})}$$

$$= 8.5 \times 10^{28} \text{ atoms/m}^3.$$

Since $H \ll M_s$, the *magnetization saturation,*

$$M_s \cong B_s/\mu_0$$

$$= (2.1 \text{ volt} \cdot \sec/\text{m}^2)/(4\pi \times 10^{-7} \text{ volt} \cdot \sec/\text{amp} \cdot \text{m})$$

$$= 1.7 \times 10^6 \text{ amp} \cdot \text{m}^2/\text{m}^3;$$

$$N_\beta = \frac{1.7 \times 10^6 \text{ amp} \cdot \text{m}^2/\text{m}^3}{(9.27 \times 10^{-24} \text{ amp} \cdot \text{m}^2/\text{Bohr magneton})(8.5 \times 10^{28} \text{ atoms/m}^3)}$$

$$= 2.1 \text{ Bohr magnetons/atom.}$$

Note. This means that there is less unbalance in spin alignments per atom in metallic iron than in an individual iron atom which has four Bohr magnetons (Section 2-2). ◄

Curie point. Spontaneous magnetization is reduced at elevated temperatures and disappears entirely at the Curie temperature, or Curie point. This temperature and the saturation values are given for the three most familiar magnetic metals in Table 16-3. These values, of course, show the limit of the useful temperature range of a magnetic material, because above the Curie temperature the material becomes paramagnetic and domains disappear.

16-4 METALLIC MAGNETS

Electron states. We observed in Fig. 2-5 that isolated atoms of transition metal elements obey Hund's rule, according to which the largest possible number of 3d-electrons have parallel spins. This statement also holds for the ions of transition metal elements within compounds. As metallic solids, however, their 3d- and 4s-

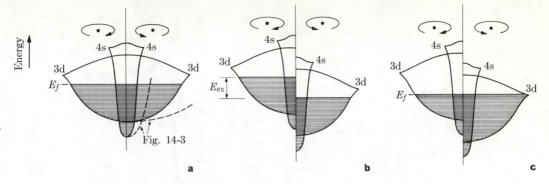

16-8 Exchange energy, E_x. (a) Density of 3d- and 4s-states for the two spin directions. (Both spin directions were included in one curve in Fig. 14-3.) (b) Application of magnetic field. (c) Fermi energy equalization.

electrons form overlapping energy bands with partially filled states. At 0°K, every energy state below the Fermi energy contains two electrons of opposite spins.

Figure 16-8(a) modifies the more conventional sketch of Fig. 14-3 so that the states are divided according to spin direction. However, the exchange force from magnetic coupling changes the energy states; those with unfavorable spin alignments rise to higher energy levels, and those with the favorable spin alignments drop in energy (Fig. 16-8b). The difference is called the *exchange energy*. As might be expected, some electrons invert their spins to occupy states with lower energies; therefore, we find an unbalance in the number of spins (Fig. 16-8c). This unbalance amounts to 2.1 electrons per atom in metallic iron (Example 16-1).

Energy losses. An amount of energy equal to the area of the *B-H* hysteresis loop is lost with each cycle. For this reason, magnets for power and high-frequency applications must have high permeability and low coercive fields. A second and more important loss of energy which occurs in metallic magnetic materials is the joule heating, or I^2R loss, from induced eddy currents. Magnet cores are thus made in sheets (or in some cases powders) so that the induced current is low. Furthermore, alloying elements, such as silicon, are used to increase the resistivity of bcc iron. Even so, losses become sufficient at megacycle frequencies to make heating a problem.

Anisotropy. On at least two occasions in the earlier part of the text, we referred to the anisotropy of magnetic properties in metals. This is shown graphically in Fig. 16-9 for nickel and iron. In each of the three indicated orientations, the saturation induction is the same; but there is a difference in the field required for saturation. This difference has engineering importance in magnetic materials, such as steels for transformer sheets, which are made in tonnage quantities. If the grains of the steel can be aligned so their [100] directions approximate the direction of the magnetization within the transformer, then a greater induction can be obtained without excessive fields and energy losses by induced currents. Two *textures* which have commercial potential are shown in Fig. 16-10.

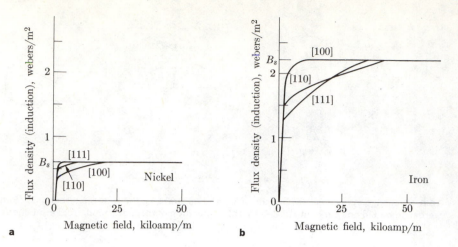

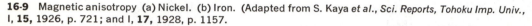

16-9 Magnetic anisotropy (a) Nickel. (b) Iron. (Adapted from S. Kaya *et al.*, *Sci. Reports, Tohoku Imp. Univ.*, I, **15**, 1926, p. 721; and I, **17**, 1928, p. 1157.

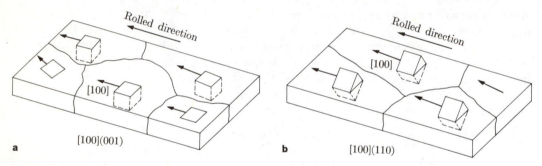

16-10 Preferred orientation (transformer sheet). (a) Cube on face. (b) Cube on edge. Both provide grains with easily magnetized [100] directions in close alignment. The electrical engineer should design the transformer to utilize this orientation.

16-5 CERAMIC MAGNETS

Certain ceramic compounds are magnetic. In fact, nature's original magnet, lodestone, is Fe_3O_4 and is related to many modern-day ceramic oxide magnets.

Ceramic materials have higher resistivities than metals; therefore, they have low energy losses through joule heating. Saturation magnetization and magnetic permeability are generally lower in ceramic magnets than in the widely used metallic magnets. This is due in part to the dilution effect which arises from the fact that the oxygen in ceramic magnets gives fewer ferromagnetic atoms per unit volume.

Ferrimagnetism. Ceramic magnets are usually *ferrimagnetic*. That is, they possess an antiparallel alignment of ferromagnetic atoms.

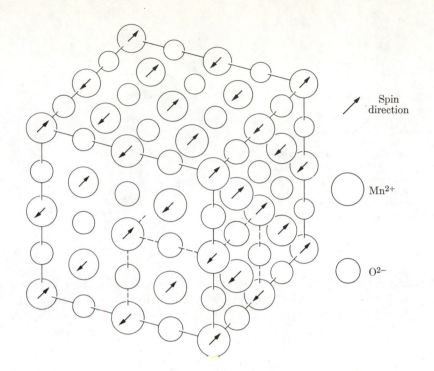

Spin direction

Mn^{2+}

O^{2-}

16-11 Antiferromagnetism (MnO). Although each Mn^{2+} ion has five Bohr magnetons, the net magnetic moment of the unit cell is zero because there are equal numbers of ions with spins aligned in the [$\bar{1}$11] and [1$\bar{1}\bar{1}$] directions.

In the extreme case of ferrimagnetism, the antiparallel moments are equal, and we use the term *antiferromagnetism*. This is illustrated by MnO in Fig. 16-11. Half the Mn^{2+} ions have their spins in each direction, and the O^{2-} ions have no net spin. Although the net magnetic moment is zero, an antiferromagnetic material is unlike a diamagnetic material in that there is a temperature at which this ordered magnetic structure disappears. The magnetic susceptibility of an antiferromagnetic material is large at this temperature.

Example 16-2

How many ions are there *per magnetic unit cell* of MnO? [See Fig. 16-11.]

Answer. Although MnO has an NaCl-type crystal structure, the ions at the corners of the crystal unit cell are not magnetically identical. They have opposite moments. The repeating distance to a magnetically equivalent position is twice that for the crystal unit cell.

Using fractional atoms as in Fig. 4-5, we obtain:

$$\text{Mn}^{2+} \text{ ions/magnetic unit cell} = 8/8 + 12/4 + 30/2 + 13 = 32,$$
$$\text{O}^{2-} \text{ ions/magnetic unit cell} = 24/4 + 24/2 + 14 = 32. \blacktriangleleft$$

16-12 Interstitial sites in oxide structures. Cations are in ordered arrangements between 4-fold sites (CN = 4) and 6-fold sites. Their magnetic spins are in opposite directions. Since there are only three cations for each four anions in a ferrospinel, not all interstitial sites are filled (only one-half of the 6-fold sites and one-eighth of the 4-fold sites). As a result, the lattice constant, a, for equivalent positions in the cubic unit cell is double the dimensions sketched.

Ceramic magnets have an unbalanced ferrimagnetic alignment. This is best illustrated by the *ferrospinels*.* These materials typically have a composition of $[R^{2+}Fe_2^{3+}O_4]_8$, where the iron is trivalent and R is a divalent ion such as Fe^{2+}, Ni^{2+}, Mn^{2+}, or Zn^{2+}. The structure is cubic, with O^{2-} ions at fcc locations (identical to the O^{2-} locations in Fig. 16-11). Two types of interstitial sites lie within the fcc lattice of oxygen ions (Fig. 16-12). Thirty-two interstitial sites have six neighboring O^{2-} ions; 64 sites have four neighboring O^{2-} ions. In the ferrospinel, half the Fe^{3+} ions are in sites where CN = 6, i.e., 6-fold sites, and the remaining half are in 4-fold sites. All the R^{2+} ions are in 6-fold sites.†

Those atoms in 6-fold sites align their magnetic moment in one direction; those in 4-fold sites align their magnetic moments in the opposite direction. It is apparent that the net magnetic moment is equal to the difference between the moments in the two directions.

Partially filled bands are not present in nonmetallic materials. Therefore, Fig. 16-8 does not apply. In fact, following Fig. 2-5, Eq. (2-3), and Eq. (2-4), each Fe^{2+} ion has four, and each Fe^{3+} has five, Bohr magnetons, in compliance with Hund's rule (Section 2-2). Thus, with eight Fe^{2+} ions per unit cell, we should expect 32 Bohr magnetons per unit cell of Fe_3O_4. Experimentally, 34 Bohr magnetons are observed.

Example 16-3

Predict the saturation induction of $[NiFe_2O_4]_8$. It has 32 oxygen ions per unit cell and a lattice constant of 8.34 A.

---

* The term *ferrite* is widely used in place of ferrospinel. Unfortunately, this is also a term used for bcc iron (Section 17-5), which is also ferromagnetic. Likewise, ferrite is a chemical term that includes $PbFe_{12}O_{19}$ and related materials which have other crystal structures and hard magnetic properties. The term *ferrospinel* is therefore recommended to avoid confusion.
† The 4-fold sites are sometimes called *tetrahedral* sites because the four neighboring oxygen ions form a tetrahedron. Likewise, 6-fold sites are sometimes called *octahedral* sites because the six neighboring oxygen ions form an octahedron (see Fig. 16-12).

Answer. Basis: 1 unit cell of 8 Ni^{2+} and 16 Fe^{3+} ions, and 5.8×10^{-28} m³.

$$\text{Spins} \uparrow \begin{cases} Fe^{3+} & 8 \times 5 = +40 \\ Ni^{2+} & 8 \times 2 = +16 \end{cases}$$

$$\text{Spins} \downarrow \quad Fe^{3+} \quad 8 \times 5 = -40$$

$$\text{Bohr magnetons} = +16 = 1.5 \times 10^{-22} \text{ amp} \cdot m^2$$

$$B_s \cong \mu_0 M_s$$

$$= (4\pi \times 10^{-7} \text{ volt} \cdot \text{sec/amp} \cdot m) \frac{1.5 \times 10^{-22} \text{ amp} \cdot m^2}{5.8 \times 10^{-28} \text{ m}^3}$$

$$= 0.33 \text{ volt} \cdot \text{sec/m}^2 \quad (= 0.33 \text{ weber/m}^2).$$

Note. The experimental value is 0.37 weber/m². (See Problem 16–10.)

REFERENCES FOR FURTHER READING

American Society for Metals, *Magnetic Properties of Metals and Alloys*. Metals Park, O.: American Society for Metals, 1959. Seminar on magnetic behavior. Graduate-student and professional level.

Azaroff, L. V., and J. J. Brophy, *Electronic Processes in Materials*. New York: McGraw-Hill, 1963. Recommended as supplementary reading for the student who wishes to expand his knowledge in the paramagnetic and diamagnetic areas. Familiarity with vectors is desirable.

Blasse, G., "Properties of Magnetic Compounds in Connection with Their Crystal Chemistry," *Progress in Ceramic Science*, Vol. 4 (J. E. Burke, Ed.). Oxford: Pergamon, 1966. A thorough review of magnetic compounds. For the materials specialist.

Dekker, A. J., *Electrical Engineering Materials*. Englewood Cliffs, N. J.: Prentice-Hall, 1959. Magnetic properties of materials are discussed in Chapter 4. Advanced undergraduate level.

Guy, A. G., *Physical Metallurgy for Engineers*. Reading, Mass.: Addison-Wesley, 1962. Ferromagnetic properties are summarized in a nonmathematical manner in Chapter 6.

Hutchinson, T. S., and D. C. Baird, *The Physics of Engineering Solids*. New York: Wiley, 1963. Chapter 12 is recommended as supplementary reading for this chapter. Advanced undergraduate level.

Katz, H. W., *Solid State Magnetic and Dielectric Devices*. New York: Wiley, 1959. The origins of magnetic properties are discussed in Chapter 2. Familiarity with field theory is required of the reader.

Keffer, F., "The Magnetic Properties of Materials," *Scientific American*, **217** [3], September 1967, pp. 22–234. Domains receive attention. Undergraduate level.

Kittel, C., *Elementary Solid State Physics: A Short Course*. New York: Wiley, 1962. Magnetism and magnetic resonance are presented in Chapter 8. Graduate level.

Nesbitt, E. A., *Ferromagnetic Domains*. Baltimore: Williams and Wilkins, 1962. Paperback. Available in classroom lots only. Based on high school science. Experiments are described. Excellent background for this chapter.

Nussbaum, A., *Electronic and Magnetic Behavior of Materials*. Englewood Cliffs, N. J.: Prentice-Hall, 1967. Paperback. Chapter 4 presents magnetism in a concise manner, but at an advanced level.

Rose, R. M., L. A. Shepard, and J. Wulff, *The Structure and Properties of Materials: IV. Electronic Properties*. New York: Wiley, 1966. Paperback. The chapters in this reference work on magnetic materials are introductory and can serve as a good supplement to this book.

Stanley, J. K., *Electrical and Magnetic Properties of Metals*. Metals Park, O.: American Society for Metals, 1963. Magnetic properties are presented on the basis of compositions and structure. Advanced undergraduate level.

Wert, C. A., and R. M. Thomson, *Physics of Solids*. New York: McGraw-Hill, 1964. The reader interested in diamagnetism and paramagnetism is referred to Chapter 18. Chapter 19 gives a nonmathematical discussion of ferromagnetism. Chapter 20 presents ferromagnetic domains.

PROBLEMS

16-1 Metallic cobalt has an induction saturation of 1.8 webers/m^2. What is the magnetic moment per atom?

Answer. 1.57×10^{-23} amp $\cdot$ m^2 (or 1.7 Bohr magnetons)

16-2 As observed in Fig. 16-9, the flux density for nickel is saturated at about 0.6 weber/m^2. How many Bohr magnetons are there per atom?

16-3 The rare earth element, gadolinium, is ferromagnetic below 16°C with 7.1 Bohr magnetons per atom.

   a) What is the magnetic moment per *gram* (i.e., amp $\cdot$ m^2/gm)?
   b) What is the saturation magnetization? [The atomic weight is 157.26 amu, and $\rho = 7.8$ gm/cm^3.]

Answer.   a) 0.25 amp $\cdot$ m^2/gm   b) 1.95×10^6 amp $\cdot$ m^2/m^3

16-4 What is the maximum induction possible for gadolinium, assuming that Hund's rule (Section 2-2) applies to the electrons? (There are seven 4f-electrons.)

16-5 (Refer to Fig. 16-5.)

   a) Determine the maximum BH product for Alnico V.
   b) Estimate the total energy required to demagnetize Alnico V.

Answer.   a) 33,500 joules/m^3 b) ~50,000 joules/m^3 (based on the estimated area of the second quadrant)

16-6 How many calories are required to demagnetize an Alnico VII magnet which has a volume of 4.2 cm^3?

16-7 Fe_3O_4 has a unit cell with 8 Fe^{2+} ions, 16 Fe^{3+} ions, and 32 O^{2-} ions. It is cubic and has a lattice constant of 8.37 A. The ferrous and one-half of the ferric ions have their magnetic moments "up"; the remainder of the cations have their moments "down." Estimate the magnetic saturation of Fe_3O_4.

Answer. 0.5×10^6 amp $\cdot$ m^2/m^3 (or 32 Bohr magnetons per unit cell)

16-8 By measurement, the magnetization of $[Li_{0.5}Fe_{2.5}O_4]_8$ is 2×10^{-22} amp $\cdot$ m^2/unit cell of 32 oxygen ions. Is this consistent with its structure, which is like magnetite, but with the 8 ferrous ions in each unit cell replaced with $4 Li^+$ and $4 Fe^{3+}$? (See Problem 16-7).

16-9 Experimental values show 34 Bohr magnetons per Fe_3O_4 unit cell, which is a little higher than determined in Problem 16-7. Assume that this is the result of a few Fe^{2+} ions losing electrons to neighboring Fe^{3+} ions (cf. Fig. 15-16). What fraction of the "up" aligned atoms are now ferric?

Answer. 56%. [*Note.* This difference could also be accounted for by a disordering of the Fe^{2+} and Fe^{3+} ions so that they occupy each other's sites.]

16-10 Rationalize the inconsistency between the two values, 0.33 weber/m^2 and 0.37 weber/m^2, in Example 16-3.

16-11 Estimate the saturation induction if half the Ni^{2+} ions of Example 16-3 are replaced by Mn^{2+} ions.

Answer. ~ 0.6 weber/m^2

part V □ multiphase materials

17 □ phase equilibria

17-1 ONE-COMPONENT SYSTEMS

Temperature. Pure materials, such as tin, water, ethyl alcohol, and SiO_2, have unique melting and boiling temperatures when they are heated under standard pressures. These transformation temperatures are those at which the free energies of the two phases of the same composition are equal (Fig. 3-19). In order to raise or lower the temperature at constant pressure, and at the same time maintain equilibrium, one phase must disappear. This and other conditions of equilibrium may be handled by the *Gibb's phase rule*, where C is the number of *components* and E is the number of *environmental variables:*

$$P + V = C + E. \tag{17-1}$$

On the left side, P is the number of *phases* and V is *variance*.

To illustrate variance, let us take a pure material, such as water, having only one component ($C = 1$). Now assume that pressure is fixed at atmospheric, and consider temperature as the only environmental variable ($E = 1$). With only liquid present, the variance is also 1 according to Eq. (17-1) ($V = 1 + 1 - 1 = 1$). So long as there is only one phase, there is a *choice* of temperatures (in this case, from 0° to 100°C). With two phases, ice and water, $V = 1 + 1 - 2 = 0$; and there is *no choice* of temperature other than 0°C—i.e., the materials system is *invariant*.

Equation (17-1) is called the *phase rule*. Although it seems trivial in a one-component system where temperature is the only variable, it is quite helpful in multicomponent systems where temperature, pressure, and even an electric field may be environmental variables.

Pressure. An increase in pressure as a second environmental variable changes the melting and boiling temperatures in the direction favoring the more dense phase. For example, higher pressures reduce the melting point and increase the boiling point of water because the liquid phase is denser than ice or steam.

In quantitative terms, the increased energy for work, $P \, \Delta V$, must arise from additional transformation energy, $T \, \Delta S$; so

$$\frac{dT}{dP} = \frac{\Delta V}{\Delta S}, \tag{17-2a}$$

Since $\Delta S = \Delta H / T$, where ΔH is the change in enthalpy due to transformation, i.e., the heat of transformation (Fig. 3-19),

$$\frac{dT}{dP} = \frac{T \, \Delta V}{\Delta H}.$$ (17-2b)

This is the well-known *Clausius-Clapeyron* equation. If the change in transformation temperature, ΔT, is small compared with the other terms,

$$\Delta T = \left(\frac{T \, \Delta V}{\Delta H}\right)\Delta P.$$ (17-2c)

In most materials, ΔV on melting is positive, as is the heat of fusion, ΔH (i.e., heat is absorbed during fusion). Therefore, higher pressures increase the melting temperature.*

Example 17-1

Tin changes from its low-temperature, gray form (with the structure of diamond—Fig. 3-12) to the more metallic, white tin (Example 5-8) at 13°C and normal pressure. The densities of these two polymorphic phases are 5.75 gm/cm³ and 7.3 gm/cm³, respectively. At a pressure of 1000 psi, the equilibrium temperature is 9°C. What is the heat of transformation between the two solid phases?

Answer. Based on 1 gm.

$$\Delta V = \frac{1}{7.3 \text{ gm/cm}^3} - \frac{1}{5.75 \text{ gm/cm}^3} = -0.037 \text{ cm}^3/\text{gm},$$

$$\Delta P = 1000 \text{ psi} - 15 \text{ psi} = 985 \text{ psi} = 6.8 \times 10^7 \text{ dynes/cm}^2,$$

$$\Delta H = \frac{(286°\text{K})(-0.037 \text{ cm}^3/\text{gm})(6.8 \times 10^7 \text{ dynes/cm}^2)}{-4°\text{K}}$$

$$= 1.8 \times 10^8 \text{ ergs/gm}$$

$$= 4.3 \text{ cal/gm} \quad \text{(or 500 cal/at wt).} \quad \blacktriangleleft$$

17-1 The **H₂O** system (temperature and pressure). For three phases to be present at the triple point, there is no choice of temperature or pressure. For two phases to coexist, one may choose either temperature *or* pressure; then the other environmental factor is fixed. [The melting temperature drops to 0.0°C when the pressure is increased to 760 mm Hg in accordance with Eq. (17–2c).]

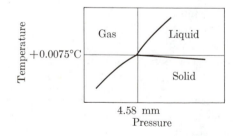

A *phase diagram* of water is shown in Fig. 17-1. Since $\Delta V_{\text{ice}\rightarrow\text{liquid}}$ is negative, dT/dP for the melting point is negative as shown. By similar reasoning, dT/dP for boiling is positive, as it is for sublimation from solid to vapor.

---

* Water is an exception because ΔV is negative.

The curves in Fig. 17-1 are *univariant*, as indicated by the phase rule. With both temperature and pressure as environmental variables, $E = 2$; also, $P = 2$ because two phases are in equilibrium along the curve. Thus, for a one-component system,

$$V = C + E - P = 1.$$

This may be interpreted as follows. With two phases, e.g., ice and liquid, either temperature *or* pressure may be changed without eliminating the second phase. However, if we assign a chosen value to one of the variables, the other becomes fixed, and is no longer subject to choice.

By analogous reasoning, the three regions of Fig. 17-1 are each *bivariant*, with independent variation of temperature and pressure being possible. Finally, the triple point at 0.0075°C and 4.58 mm Hg is *invariant* because there are three phases:

$$V = 1 + 2 - 3 = 0.$$

17-2 TWO-COMPONENT SYSTEMS

Materials with two components are called *binary* systems. They are commonly encountered in engineering, since many commercial products are alloys and/or mixtures (Table 17-1). Note that the components of the material are not necessarily elemental. In the simplest cases, materials are *mixtures* of their components, with no solutions or chemical reactions involved (e.g., glass-reinforced plastic). In other cases, the components react to develop *solutions* (e.g., brass), or even new *compounds* (e.g., Fe_3C in steel).

We can study binary materials most effectively by using *phase diagrams*. The balance of this chapter will be addressed to that subject.

Solubility limits. It was stated earlier (Section 8-1) that unlimited proportions of copper can be dissolved in solid nickel (and vice versa). Because the two atoms are comparable in size and electronic structure, their solution is sufficiently ideal so that a single intermediate solution phase has lower free energy than mixtures of other possible phases (Fig. 17-2a). The same is true for the liquid solutions

TABLE 17-1
COMMON BINARY MATERIALS

Material	Typical prime components
Brass	Cu-Zn
Brick (building)	SiO_2-Al_2O_3
Brine	H_2O-NaCl
Bronze	Cu-Sn
Glass-reinforced plastic	polyester-glass
Rubber (vulcanized)	$(C_5H_8)_n$-S
Semiconductor (*p*-type)	Si-Al
Solder	Pb-Sn
Steel	Fe_3C

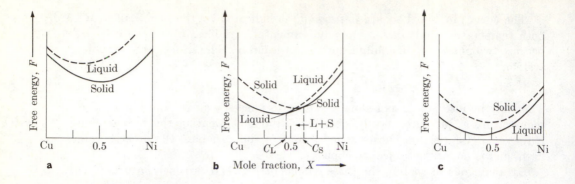

17-2 *F-X* diagram (schematic for Cu-Ni). (a) $T < 1083°C$. (b) $T = 1300°C$. (c) $T > 1455°C$. At 1300°C, the two free-energy curves cross so that liquid is stable below $X_{Ni} = 0.47$. Above $X_{Ni} = 0.63$, solid is stable. When $0.47 < X_{Ni} < 0.63$, a mixture of solid *and* liquid provides lower free energy than does either alone.

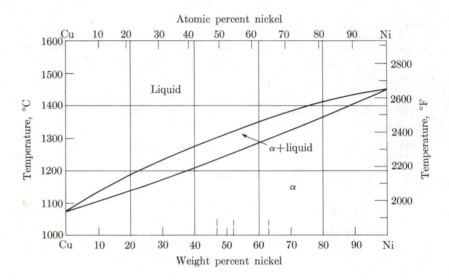

17-3 **Cu-Ni** diagram. The solid phase, α, is fcc. (*ASM Metals Handbook*. Metals Park, O.: American Society for Metals.)

(Fig. 17-2c). However, copper and nickel are not identical, as is evidenced by the difference in their melting points—1083°C for copper and 1455°C for nickel. As a result, we observe a region between these temperatures where a copper-rich liquid *and* a nickel-rich solid are stable. In a 48Cu-52Ni alloy, for example, a mixture of solid and liquid solutions has lower free energy at 1300°C than a single solution (Fig. 17-2b). At this temperature solid nickel cannot contain more than 37 w/o copper at equilibrium. Any excess copper appears in the liquid phase. Also, no more than 47 w/o nickel can dissolve in liquid copper at 1300°C under equilibrium conditions. Figure 17-3 contains a plot of the *solubility limits* just described as a function of temperature. This is a *binary phase diagram* in its simplest form. When a phase diagram like Fig. 17-3 applies to equilibrium conditions, it may also be called an *equilibrium diagram*.

Application of the phase rule. The Cu-Ni phase diagram reveals a two-phase area between the two solubility lines of Fig. 17-3, but only a single phase at higher and lower temperatures. Since $C = 2$ and temperature is the only environmental variable (we take the pressure as being fixed at 1 atm), an application of the phase rule to a one-phase region indicates two degrees of freedom (i.e., the variance is 2), while $V = 1$ in a two-phase region.* In other words, both temperature and composition may be changed independently when only solid (or only liquid) is present; however, in a two-phase region, a choice of one will fix the other. We chose 1300°C in the previous paragraph, and the phase compositions became fixed at the solubility limits of 53Cu-47Ni for the liquid and 37Cu-63Ni for the solid. Had we chosen to have both liquid and solid, with the former containing specifically 75Cu-25Ni, there would be no choice but to go to 1210°C.

Material balances and the lever rule. We can calculate the ratio of solid to liquid in a two-phase material from solubility data. It is possible to write a material balance in the previously discussed 48Cu-52Ni (weight percent) alloy:

$$Cu_{alloy} = Cu_{liquid} + Cu_{solid}.$$

If the alloy contains L lb of liquid with 53 w/o Cu and S lb of solid with 37 w/o Cu, we can write a material balance for copper:

$$0.48(L + S) = 0.53L + 0.37S,$$

or

$$\frac{S}{L} = \frac{53 - 48}{48 - 37} = \frac{5}{11}.$$

A similar calculation can be made in general terms for the weight ratios of any two phases, X and Y, within a material of composition C_0, in which the phase compositions are C_x and C_y:

$$C_0(X + Y) = C_x X + C_y Y,$$

or

$$\frac{X}{Y} = \frac{C_y - C_0}{C_0 - C_x}. \tag{17-3a}$$

This equation is called the *lever rule* (Fig. 17-4). Either component can be used as the basis for calculations, and the lever rule may take on several equivalent alge-

---

* Of course, at the extreme ends of the diagram, where the compositions are 100Cu-0Ni and 100Ni-0Cu, the material has only *one* component. In this case $V = 0$ when there are two phases. Thus the temperature is fixed.

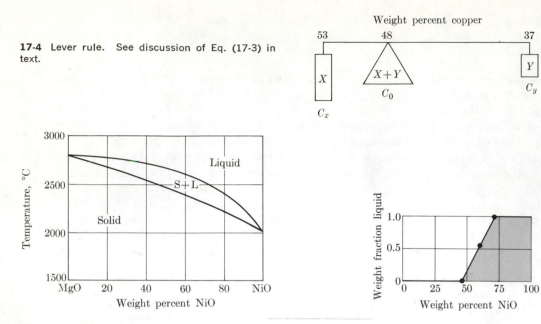

17-4 Lever rule. See discussion of Eq. (17-3) in text.

17-5 **MgO-NiO** diagram. (Adapted from H. Wartenburg and E. Prophet, *Z. Anorg. u. Allgem. Chem.*, **208**, 1932, p. 379.)

17-6 Liquid fraction versus composition (2500°C). See Example 17-2.

braic forms. A common form is

$$\frac{Y}{X+Y} = \frac{C_0 - C_x}{C_y - C_x} = \text{weight fraction.} \tag{17-3b}$$

Example 17-2

Plot the weight fraction of liquid at 2500°C in an MgO-NiO ceramic as a function of the NiO content (Fig. 17-5).

Answer. At 2500°C, the solid can contain up to 46 w/o NiO, and the liquid can contain up to 28 w/o MgO. Therefore, between 0 w/o NiO and 46 w/o NiO there is only solid at 2500°C; between 72 w/o NiO and 100 w/o NiO there is no solid. By using Eq. (17-3b) and letting Y be the liquid, a 60NiO-40MgO (weight percent) composition produces

$$\text{weight fraction liquid} = \frac{60 - 46}{72 - 46} = 0.54.$$

See Fig. 17-6 for other compositions between 46 and 72 w/o NiO. ◀

17-3 INVARIANT REACTIONS IN BINARY SYSTEMS

Eutectics. Commonly, the components of materials are not sufficiently similar to form a complete series of solid solutions. For example, silver and copper atoms are different enough in size so that a 50Ag-50Cu alloy contains two solid phases,

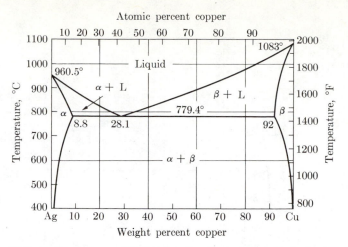

17-7 **Ag-Cu** diagram. (*ASM Metals Handbook.* Metals Park, O.: American Society for Metals.)

one silver-rich and the other copper-rich. The Ag-Cu phase diagram is shown in Fig. 17-7. The curves for the solubility limits of copper in the liquid and of silver in the liquid intersect at 779.4°C and 28.1 w/o copper (71.9 w/o silver). The low-melting liquid at the intersection is called a *eutectic liquid*. The temperature of intersection, 779.4°C, is the *eutectic temperature;* and the 71.9Ag-28.1Cu composition* is the *eutectic composition*.

Although the solid solubility is limited, as much as 8.8 w/o copper can be dissolved in the fcc silver structure at the eutectic temperature; as much as 8.0 w/o silver can be dissolved in the copper structure. In Fig. 17-7, these two solid solutions are called α and β, respectively. Now let us examine a 60Ag-40Cu alloy. Only liquid is stable at high temperatures. During cooling the copper solubility limit in the liquid is reached at 820°C and solid β precipitates. At 780°C (just above the eutectic temperature), the weight ratio of liquid to β will be 0.814/0.186 if equilibrium is attained.† At 775°C (and with equilibrium), there will be α and β but no liquid. As determined by the lever rule, the α/β ratio will be (92 − 40)/(40 − 8.8), or 0.625/0.375.

Precisely at the eutectic temperature, three phases can coexist (α, β, and liquid). The phase rule (for constant pressure) indicates $V = 0$ with three phases; there is no choice in temperature (779.4°C) or compositions (α = 8.8 w/o Cu, β = 92 w/o Cu, and L = 28.1 w/o Cu).

---

* Henceforth we shall follow the standard practice of using weight percents (w/o) when giving solid and liquid compositions, *unless stated otherwise*. The standard practice for gas analyses is to use volume percent (v/o) unless stated otherwise. Volume percent of an ideal gas also equals mole percent (m/o).

† The reader should verify this ratio for himself to ascertain his familiarity with the use of the lever rule.

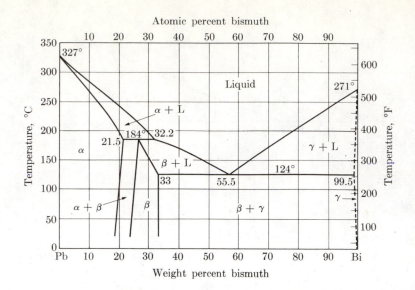

17-8 Pb-Bi diagram. (*ASM Metals Handbook.* Metals Park, O.: American Society for Metals.)

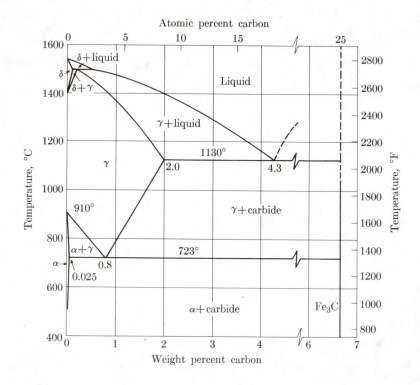

17-9 Fe-C diagram. The eutectoid region is enlarged in Fig. 17-14.

If a liquid of eutectic composition is cooled, it transforms to two solid phases. In general,

$$L_2 \underset{\text{heating}}{\overset{\text{cooling}}{\rightleftarrows}} S_1 + S_3; \qquad (17\text{-}4a)$$

and for the Ag-Cu system,

$$L\ (28.1Cu) \underset{}{\overset{779.4°C}{\rightleftarrows}} \alpha\ (8.8Cu) + \beta\ (92Cu). \qquad (17\text{-}4b)$$

This reaction is called a *eutectic reaction*, and is one of several types of invariant reactions involving three phases in a binary system.

Peritectics. In addition to an eutectic reaction, the Pb-Bi system (Fig. 17-8) also has a *peritectic reaction*. The general form of the peritectic reaction,

$$S_1 + L_3 \underset{\text{heating}}{\overset{\text{cooling}}{\rightleftarrows}} S_2, \qquad (17\text{-}5a)$$

is

$$\alpha\ (21.5Bi) + L\ (32.2Bi) \underset{}{\overset{184°C}{\rightleftarrows}} \beta\ (27Bi) \qquad (17\text{-}5b)$$

for the Pb-Bi system. The phase rule and lever rule operate as before.

In observing the peritectic of Fig. 17-8, it is only necessary to focus attention on the left side of the diagram. This practice of considering *subsystems* permits us to simplify complex diagrams for easier examination.

Eutectoids. The Fe-C system (Fig. 17-9) has a eutectic at 1130°C and 4.3C-95.7Fe. There is also a peritectic at 1500°C and 0.2C-99.8Fe. A third invariant reaction is found at 723°C (1333°F) and 0.8C-99.2Fe:

$$\gamma\ (0.8C) \underset{}{\overset{723°C}{\rightleftarrows}} \alpha\ (0.025C) + \text{carbide}\ (Fe_3C). \qquad (17\text{-}6a)$$

This reaction, like that of a eutectic, occurs when two decreasing solubility lines intersect—the one for carbon in the γ solid solution, and the other for iron in the γ solid solution. It differs from a eutectic reaction only in that all three phases are solids. It is called an *eutectoid* reaction* and has the general form

$$S_2 \underset{\text{heating}}{\overset{\text{cooling}}{\rightleftarrows}} S_1 + S_3. \qquad (17\text{-}6b)$$

The eutectoid reaction of the Fe-C system is of major importance in the heat treatment of steel. As a result, we shall return to it in Section 17-5 and several times in later chapters.

Peritectoids. A peritectic-like reaction which involves only solids is

$$S_1 + S_3 \underset{\text{heating}}{\overset{\text{cooling}}{\rightleftarrows}} S_2. \qquad (17\text{-}7a)$$

---

* Literally, eutectic-like.

An example of this less common *peritectoid* reaction is

$$\gamma_2 \,(21\text{Al}) + \epsilon_2 \,(23\text{Al}) \xrightleftharpoons{690°\text{C}} \delta \,(21.5\text{Al}), \tag{17-7b}$$

which is illustrated in the Cu-Al diagram at the end of the chapter.

Monotectics. The final type of invariant reaction which we shall consider involves two liquids and is called a *monotectic*:

$$L_2 \underset{\text{heating}}{\overset{\text{cooling}}{\xrightleftharpoons{\hspace{1cm}}}} S_1 + L_3. \tag{17-8a}$$

A specific example is found in the $FeO\text{-}SiO_2$ diagram at the end of the chapter, where

$$L \,(4\text{FeO}) \xrightleftharpoons{1690°\text{C}} \text{Crist.}^* \,(\text{nil FeO}) + L \,(42\text{FeO}). \tag{17-8b}$$

Example 17-3

A 70Pb-30Bi alloy is cooled under equilibrium conditions from 300°C to 20°C. Cite the sequence of equilibrium phases and their equilibrium temperatures.

Answer. See Fig. 17-8.

Temperatures	Phases	Phase composition
>196°C	Liquid only	L (30Bi)
196–184(+)	Liquid + α	L (30→32.2Bi), α (20→21.5Bi)
184	β + liquid + α	β (27Bi), L (32.2Bi), α (21.5Bi)
184(−)–155	β + liquid	β (27→30Bi), L (32.2→46 Bi)
<155	β only	β (30Bi) ◀

Example 17-4

Select an isotherm in any binary phase diagram in this book. Formulate a general rule relating the sequential *number* of phases from 100% *A* to 100% *B*.

Answer. For example, select the 150°C isotherm of the Pb-Bi diagram (Fig. 17-8).

Alloy compositions		Phases	Phase composition
0–21	w/o Bi	α only	α (100→79Pb)
21–26	w/o Bi	$\beta + \alpha$	β (74Pb), α (79Pb)
26–30	w/o Bi	β only	β (74→70Pb)
30–47	w/o Bi	Liquid + β	L (53Pb), β (70Pb)
47–64	w/o Bi	Liquid only	L (53→36Pb)
64–99(+)	w/o Bi	γ + liquid	γ (<1Pb), L (36Pb)
99(+)–100	w/o Bi	γ only	γ (<1→0Pb)

The number of phases across an isotherm of a binary diagram alternates 1-2-1-2-1-2-· · · .

Note. This applies only to composition changes at constant temperature, and *not* to temperature changes at constant composition. See Example 17-3. ◀

---

* Cristobalite is the highest-temperature polymorph of silica (SiO_2).

17-4 PHASE SEPARATION (Precipitation)

The invariant reactions of the previous section involve three phases. If equilibrium is maintained, at least one phase has to disappear before the temperature can be either increased or decreased. *Univariant reactions* involve only two phases, with the result that phase changes accompany temperature changes, producing phase separation.

Consider the *solidification* of liquid lead containing 15 w/o bismuth (Fig. 17-8). The solubility limit is reached at about 270°C during cooling, and a two-phase region is entered. At this temperature, the lead-rich phase (α) can contain no more than 11 w/o Bi. At 250°C, the solid and liquid phases contain 13 w/o Bi and 18 w/o Bi, respectively, if equilibrium is maintained. Finally, the last of the equilibrium liquid, with 22 w/o Bi, disappears from an 85Pb-15Bi alloy at about 235°C. Upon equilibrium cooling through the two-phase region, both phases increase in bismuth.* This requires diffusion within both the liquid and the solid during solidification. Thus cooling must be very slow if equilibrium is to be realized. Otherwise, *segregation* can develop in which the first solid to form remains purer than the final solid (Section 18-8).

For a second example of univariant separation, consider *sterling silver*, a 92.5Ag-7.5Cu alloy. According to Fig. 17-7, this alloy of silver and copper contains only a single phase, α, at the eutectic temperature. On cooling from this temperature, however, the solubility limit is reached at about 750°C; at 400°C there is about 0.06 weight fraction of β, a copper-rich phase.

As with solidification, time is required for diffusion to occur. Therefore, it is possible to quench sterling silver to avoid the phase separation. Of course, this produces a supersaturated solid solution which is only metastable. With sufficient time, particularly at intermediate temperatures, copper-rich β will separate from the solution. This separation is called *precipitation* because the second phase usually appears as fine particles† (Section 18-5).

Example 17-5

Plot the equilibrium amount of spinel in a 20MgO-80Al$_2$O$_3$ ceramic as a function of temperature (Fig. 17-10).

Answer. Figure 17-11 is determined from Fig. 17-10 and the lever rule.

Note. This *spinel* has a composition of MgAl$_2$O$_4$ and is the prototype of a large solid-solution family, $R^{2+}R_2^{3+}O_4$, which includes the magnetic ferrospinels. They all have fcc oxygen lattices with various ordered arrangements of the cations (Section 16-5). ◄

17-5 THE Fe-C PHASE DIAGRAM

We shall give more detailed attention to the Fe-C phase diagram (1) because it is basic to steels, one of the most versatile and widely used engineering materials,

---

* This apparent inconsistency is accounted for by the change in relative amounts of the two phases.
† Precipitation in the form of rain and snow also comes from supersaturated solutions, specifically a vapor solution.

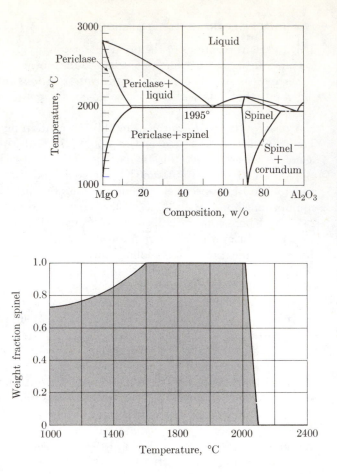

17-10 MgO-Al₂O₃ diagram. (Adapted from E. M. Levin *et al., Phase Diagrams for Ceramists,* 1964 edition. Columbus, O.: American Ceramic Society.)

17-11 Spinel weight fraction versus temperature (20MgO-80Al₂O₃). See Example 17-5.

and (2) because it provides a prototype for microstructural changes and control. If one studies this well-analyzed system, he can understand reactions in most other binary and ternary systems.

On heating, pure iron changes its crystal structure from bcc to fcc at 910°C (1670°F). Iron is unique among materials in that it reverts to bcc at 1400°C (2550°F). This may be explained on the basis of free energies (Fig. 17-12). It is convenient to assign names to these and other phases for ease of discussion.

Ferrite, or α-iron. The structural modification of pure iron at room temperature is called either *α-iron* or *ferrite.* Ferrite is quite soft and ductile; in the purity which is encountered commercially, its tensile strength is less than 45,000 psi. It is a ferromagnetic material at temperatures under 767°C (1414°F).

Since ferrite has a body-centered cubic structure, the interatomic spaces are small and pronouncedly oblate, and cannot readily accommodate even a small carbon atom. Therefore, solubility of carbon in ferrite is very low. The carbon atom is too small for substitutional solid solution, and too large for extensive interstitial solid solution.

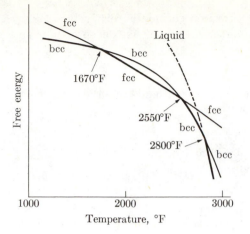

17-12 Free energy of bcc and fcc iron (schematic). The phase with the lowest free energy is the stable phase. (Cf. Fig. 3-19c.)

Austenite, or γ-iron. The face-centered modification of iron is called *austenite*, or *γ-iron*. It is the stable form of pure iron at temperatures between 910°C (1670°F) and 1400°C (2550°F). It is difficult to make a direct comparison between the mechanical properties of austenite and ferrite, because they must be compared at different temperatures. However, at its stable temperatures, austenite is soft and ductile, and consequently is well suited to fabrication processes. Many steel forging and rolling operations are performed at temperatures of 1100°C (2000°F) or above, where the iron is face-centered cubic. Austenite is only paramagnetic.

The face-centered cubic structure of iron has larger interatomic spacings than does ferrite. Even so, in the fcc structure the interstices are barely large enough to accommodate the carbon atoms in solution, and lattice strains are produced. As a result, not all the interstitial sites can be filled at any one time. The maximum solubility is only 2% (8.7 a/o) carbon (Fig. 17-9). By definition, steels contain less than 2% carbon; thus steels may have their carbon completely dissolved in austenite at high temperatures.

δ-iron. Above 1400°C (2550°F) austenite is no longer the most stable form of iron, and the crystal structure changes back to a body-centered cubic phase called *δ-iron*. This iron is the same phase as α-iron except for its temperature range, and so it is commonly called *δ-ferrite*. The solubility of carbon in δ-ferrite is small, but it is appreciably larger than in α-ferrite, because of the higher temperature.

Cementite, or iron carbide. In iron-carbon alloys, carbon in excess of the solubility limit must form a second phase, which is often iron carbide (cementite). Iron carbide has the chemical composition of Fe_3C. This does not mean that iron carbide forms molecules of Fe_3C, but simply that the crystal lattice contains iron and carbon atoms in a three-to-one ratio. The compound Fe_3C has an orthorhombic unit cell with 12 iron atoms and 4 carbon atoms per cell, and thus has a carbon content of 6.67%.

As compared with austenite and ferrite, cementite is very hard and brittle. The presence of iron carbide with ferrite in steel greatly increases the strength of the steel (Section 19-4). However, since pure iron carbide is relatively weak by itself because of its lack of ductility, it cannot adjust to stress concentrations and tends to embrittle the steel.

Example 17-6

Show the phase fractions of ferrite, austenite, carbide, and liquid in an alloy of 0.60% carbon and 99.40% iron as a function of temperature.

Answer. At 724°C (1335°F),

$$\text{percent ferrite} = \frac{0.80 - 0.60}{0.80 - 0.025} = 26\%.$$

At 722°C (1332°F),

$$\text{percent ferrite} = \frac{6.67 - 0.60}{6.67 - 0.025} = 91\%.$$

Fractions for other temperatures and other phases are shown in Fig. 17–13. ◄

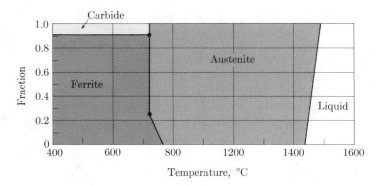

17-13 Phase fractions versus temperature (0.6 carbon, 99.4 iron). See Example 17-6.

The Fe-C eutectoid. The eutectoid region of Fig. 17-9 is expanded in Fig. 17-14. This region will have sufficient use in later chapters so that the reader is urged to become very familiar with it, and with the eutectoid reaction of Eq. (17-6a).

The Fe-C eutectoid reaction on cooling involves the simultaneous formation of ferrite and carbide from austenite of eutectoid composition. There is nearly 12% carbide and slightly more than 88% ferrite in the resulting mixture. Since the carbide and ferrite form simultaneously, they are intimately mixed. Characteristically, the mixture is lamellar; i.e., it is composed of alternate layers of ferrite and carbide (Fig. 17-15). The resulting microstructure, called *pearlite*, is very important in iron and steel technology, because it can be formed in almost all steels by means of suitable heat treatments.

Pearlite is a lamellar two-phase mixture generated by transforming austenite simultaneously to ferrite and carbide. This definition is important, since mixtures of ferrite and carbide may be formed by other reactions; however, the resulting

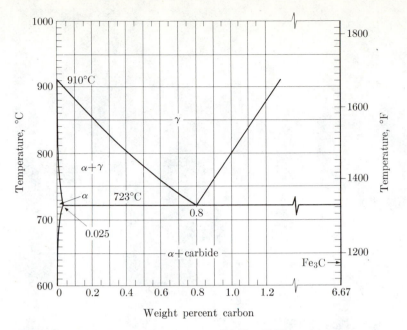

17-14 Eutectoid region of **Fe-C** diagram. (Cf. Fig. 17-9.)

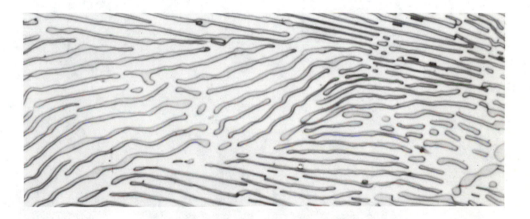

17-15 Pearlite, ×2500. This microstructure is a lamellar mixture of ferrite (lighter matrix) and carbide (darker). Pearlite forms from austenite of eutectoid composition. Therefore the amount and composition of pearlite is the same as the amount and composition of eutectoid. (J. R. Vilella, U.S. Steel Corp.)

microstructures will not be lamellar (compare Figs. 19-2 and 17-15), and consequently the properties of such mixtures will be different (see Section 19-4).

Under conditions of equilibrium cooling, the pearlite comes from austenite of eutectoid composition, and the amount of pearlite present is equal to the amount of eutectoid austenite transformed (Fig. 17-16).

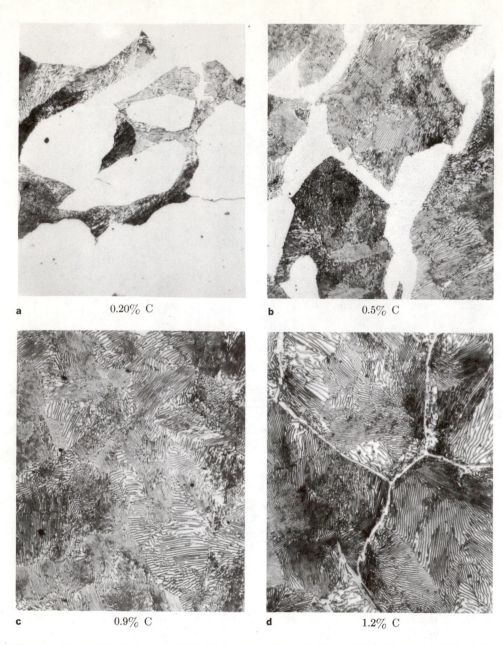

a   0.20% C   b   0.5% C

c   0.9% C   d   1.2% C

17-16   Microstructures of Fe-C alloys (×500). The amount of pearlite is related to the composition of the steel. (United States Steel Corp.)

Example 17-7

Determine the amount of pearlite in a 99.6Fe-0.4C alloy which is cooled slowly from 900°C (1650°F). Basis: 100 lb of alloy.

Answer. From 900°C to 810°C (1650°F to 1490°F), the alloy remains 100 lb austenite with 0.4% C. Between 810°C and 724°C (1490°F and 1335°F), *pro-eutectoid* ferrite separates from austenite and the carbon content of the austenite increases toward the eutectoid composition. At 724°C (1335°F),

$$\text{composition of ferrite} = 0.025\% \text{ C,}$$
$$\text{amount of ferrite} = 51.6 \text{ lb,}$$
$$\text{composition of austenite} = 0.80\% \text{ C,}$$
$$\text{amount of austenite} = 48.4 \text{ lb.}$$

At 722°C (1332°F),

$$\text{amount of pearlite} = 48.4 \text{ lb.}$$

(It came from, and replaced, the austenite of eutectoid composition.) Each of the above calculations assumes sufficient time for equilibrium to be attained. ◄

Example 17-8

From the results of the example above, determine the amount of ferrite and carbide present in the 99.6Fe-0.4C alloy (a) at 722°C (1332°F), (b) at room temperature. Basis: 100 lb of alloy. (Some data come from Example 17-7.)

Answer

a) At 722°C (1332°F), the amount of carbide is

$$48.4 \frac{0.8 - 0.025}{6.67 - 0.025} = 5.7 \frac{\text{lb carbide}}{100 \text{ lb steel}}.$$

The amount of ferrite is

$$48.4 - 5.7 = 42.7 \text{ lb eutectoid ferrite formed with the pearlite}$$
$$\underline{51.6 \text{ lb pro-eutectoid ferrite formed before the pearlite}}$$
$$94.3 \text{ lb ferrite total/100 lb steel.}$$

Alternative calculations: The amount of carbide is

$$\frac{0.4 - 0.025}{6.67 - 0.025} = 5.7 \frac{\text{lb carbide}}{100 \text{ lb carbide}}.$$

The amount of ferrite is

$$\frac{6.67 - 0.4}{6.67 - 0.025} = 94.3 \frac{\text{lb ferrite}}{100 \text{ lb steel}}.$$

See Fig. 17-17.

b) At room temperature (the solubility of carbon in ferrite at room temperature may be considered zero for these calculations),

$$\frac{0.4 - 0}{6.67 - 0} = 6.0 \frac{\text{lb carbide}}{100 \text{ lb steel}}, \qquad \frac{6.67 - 0.4}{6.67 - 0} = 94.0 \frac{\text{lb ferrite}}{100 \text{ lb steel}}.$$

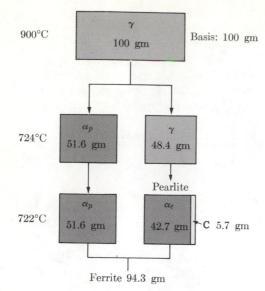

17-17 Phase changes (0.4C-99.6Fe). Equilibrium is assumed; α_p—proeutectoid ferrite; α_e—eutectoid ferrite; **C**—carbide, Fe_3C. (See Examples 17-7 and 17-8.)

17-18 Carbon content versus microstructure (Fe-C alloy, $\times$ 500). See Example 17-9. (United States Steel Corp.)

Additional carbide is precipitated from the ferrite below the eutectoid temperature because the solubility of carbon in ferrite progressively decreases to nearly zero. This additional carbide is not part of the pearlite. (Each of these calculations assumes that equilibrium prevails.) ◄

Example 17-9

Figure 17-18 shows an iron-carbon alloy which was annealed (cooled slowly) from 875°C (1600°F). Estimate its carbon content. (The densities of ferrite and pearlite are 7.87 and 7.84 gm/cm³, respectively.)

Answer. Approximately 40 v/o of the figure is pearlite (60 v/o pro-eutectoid ferrite). The pearlite was formed from austenite of eutectoid composition (0.8 w/o carbon). Basis: 1 cm³ of metal.

$$0.6 \text{ cm}^3 \text{ ferrite} = 4.72 \text{ gm}$$
$$\underline{0.4 \text{ cm}^3 \text{ pearlite} = 3.14 \text{ gm}}$$
$$7.86 \text{ gm alloy}$$

$$(x \text{ w/o})(7.86 \text{ gm}) = (0.8 \text{ w/o})(3.14 \text{ gm}) + (0.02 \text{ w/o})(4.72 \text{ gm}),$$
$$x = 0.33 \text{ w/o}. \quad ◄$$

17-6 POLYCOMPONENT SYSTEMS

Numerous materials have a third component, and some may have several components. With one exception, we shall leave the details of their phase relationships to more advanced courses in materials.

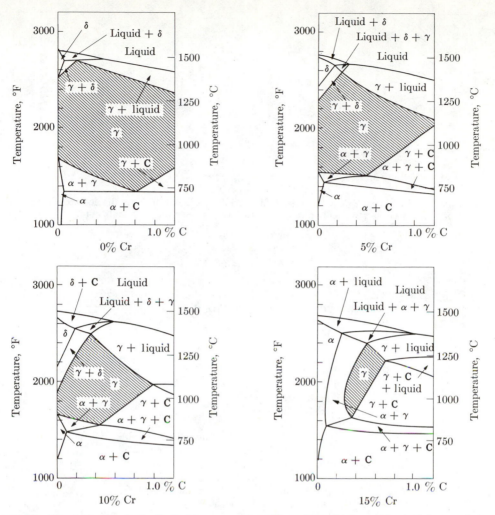

17-19 Pseudobinaries in the Fe-Cr-C ternary system (**C**—carbide which contains both iron and chromium).

It will be desirable, however, to see how a third component affects the Fe-C diagram. This is shown in Fig. 17-19 for various chromium contents. Note that chromium, which has a bcc structure itself, increases the stability range of bcc ferrite by raising the eutectoid temperature and expanding the α field. Also note that the eutectoid composition drops from 0.8 w/o to about 0.4 w/o when 10 w/o Cr is added. These changes due to chromium additions, along with comparable data for other alloying elements, are presented in Fig. 17-20.

Steel nomenclature. The importance of carbon in steel has made it desirable to designate the carbon content when specifying steel types. A four-digit numbering scheme is used, in which the last two digits indicate the number of hundredths of

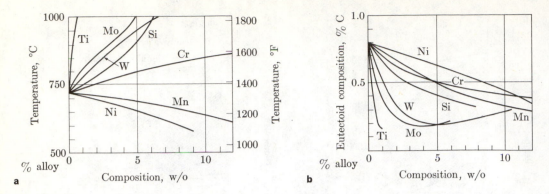

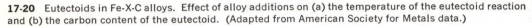

17-20 Eutectoids in Fe-X-C alloys. Effect of alloy additions on (a) the temperature of the eutectoid reaction and (b) the carbon content of the eutectoid. (Adapted from American Society for Metals data.)

TABLE 17-2
NOMENCLATURE FOR AISI AND SAE STEELS

AISI or SAE number	Composition
10xx	Plain carbon steels
11xx	Plain carbon (resulfurized for machinability)
13xx	Manganese (1.5–2.0%)
23xx	Nickel (3.25–3.75%)
25xx	Nickel (4.75–5.25%)
31xx	Nickel (1.10–1.40%), chromium (0.55–0.90%)
33xx	Nickel (3.25–3.75%), chromium (1.40–1.75%)
40xx	Molybdenum (0.20–0.40%)
41xx	Chromium (0.40–1.20%), molybdenum (0.08–0.25%)
43xx	Nickel (1.65–2.00%), chromium (0.40–0.90%), molybdenum (0.20–0.30%)
46xx	Nickel (1.40–2.00%), molybdenum (0.15–0.30%)
48xx	Nickel (3.25–3.75%), molybdenum (0.20–0.30%)
51xx	Chromium (0.70–1.20%)
61xx	Chromium (0.70–1.10%), vanadium (0.10%)
81xx	Nickel (0.20–0.40%), chromium (0.30–0.55%), molybdenum (0.08–0.15%)
86xx	Nickel (0.30–0.70%), chromium (0.40–0.85%), molybdenum (0.08–0.25%)
87xx	Nickel (0.40–0.70%), chromium (0.40–0.60%), molybdenum (0.20–0.30%)
92xx	Silicon (1.80–2.20%)

xx Carbon content, 0.xx%.
Mn All steels contain 0.50% ± manganese.
The letter B is prefixed to show bessemer steel.
The letter C is prefixed to show open-hearth steel.
The letter E is prefixed to show electric furnace steel.

one percent of carbon content (Table 17-2). For example, a 1040 steel has 0.40% carbon (plus or minus a small workable range). The first two digits identify the type of alloying element that has been added to the steel. The classification 10xx is reserved for plain-carbon steels with only a minimum amount of other alloying elements.

The designations for the steels are accepted as standard by both the American Iron and Steel Institute and by the Society of Automotive Engineers (AISI and SAE). Many commercial steels are not included in this classification scheme because of larger additions or more subtle variations in alloy contents.

ADDITIONAL PHASE DIAGRAMS

(See p. 368 for the location of various diagrams.)

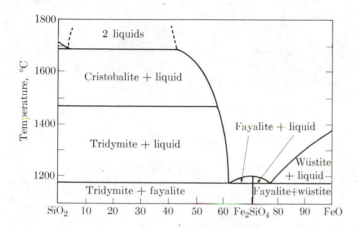

17-21   FeO-SiO$_2$ diagram. (Adapted from N. L. Bowen and J. F. Schairer, *Amer. Journ. Sci.*, 5th series, **24,** 1932, p. 200.)

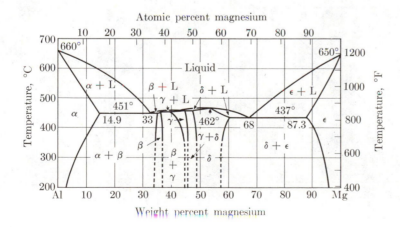

17-22   Al-Mg diagram. (*ASM Metals Handbook.* Metals Park, O.: American Society for Metals.)

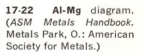

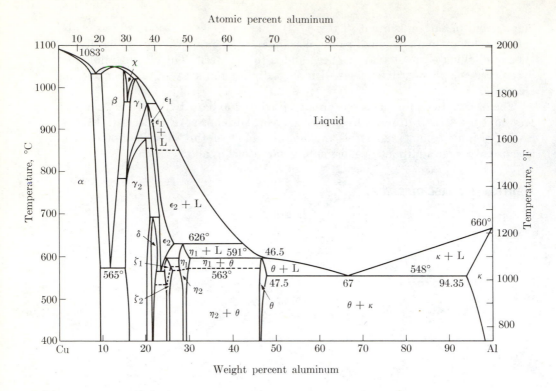

Atomic percent aluminum

Weight percent aluminum

17-23 Al-Cu diagram. The phases of the unlabeled fields may be deduced from the phases of the adjacent single-phase fields. See Example 17-4. (*ASM Metals Handbook.* Metals Park, O.: American Society for Metals.)

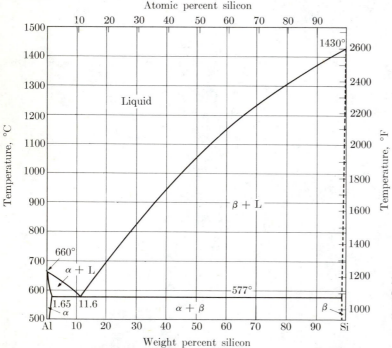

Atomic percent silicon

Weight percent silicon

17-24   Al-Si diagram. (*ASM Metals Handbook.* Metals Park, O.: American Society for Metals.)

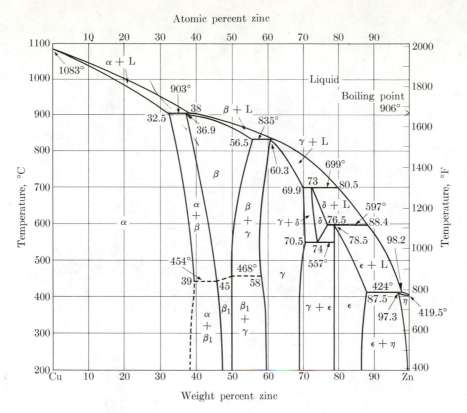

17-25 Cu-Zn diagram. (*ASM Metals Handbook.* Metals Park, O.: American Society for Metals.)

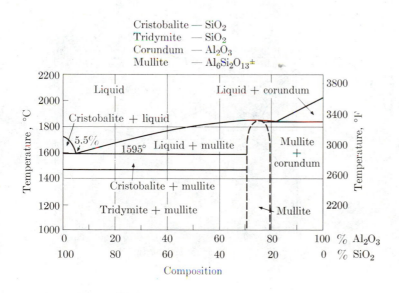

17-26 Al₂O₃-SiO₂ diagram. (Adapted from S. Aramaki and R. Roy, *Journ. Amer. Ceram. Soc.,* **42,** 1959, p. 644.) More recent studies propose that the eutectic temperature is 1547°C (A. J. Majumdar and J. H. Welch, *Trans. Brit. Ceram. Soc.,* **62,** 1963, p. 603).

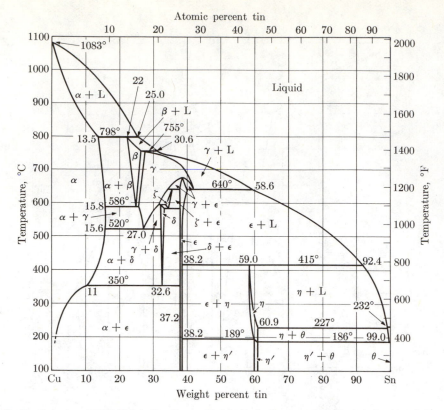

17-27 Cu-Sn diagram. (*ASM Metals Handbook.* Metals Park, O.: American Society for Metals.)

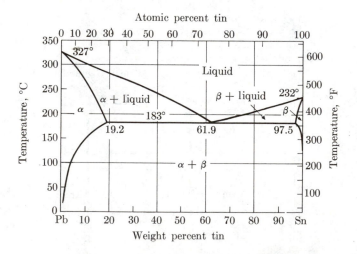

17-28 Pb-Sn diagram. (*ASM Metals Handbook.* Metals Park, O.: American Society for Metals.)

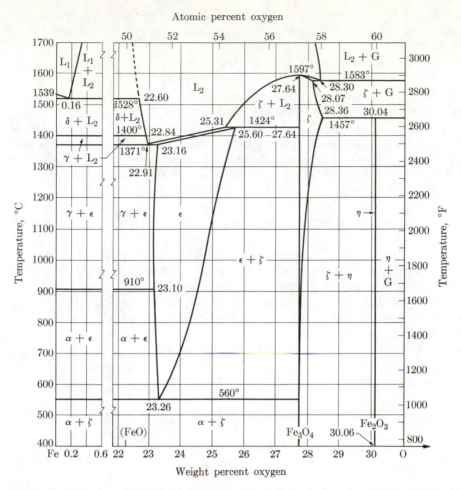

17-29 Fe-O diagram. (*ASM Metals Handbook.* Metals Park, O.: American Society for Metals.)

REFERENCES FOR FURTHER READING

Brophy, J. H., R. M. Rose, and J. Wulff, *The Structure and Properties of Materials: II. Thermodynamics of Structure.* New York: Wiley, 1964. Paperback. Phase equilibria are introduced on the basis of thermodynamics. Advanced undergraduate level.

Gordon, P., *Principles of Phase Diagrams in Materials Systems.* New York: McGraw-Hill, 1968. A very satisfactory text to supplement this chapter for the advanced undergraduate student who has studied thermodynamics.

Guy, A. G., *Physical Metallurgy for Engineers.* Reading, Mass.: Addison-Wesley, 1962. Phase diagrams are presented in Chapter 4. Applications are made in Chapter 5. Alternate reading for this topic.

Hume-Rothery, W., *The Structure of Alloys of Iron*. Oxford: Pergamon, 1966. Paperback. A sequel to this chapter and Chapter 8, with special emphasis on iron alloys. Undergraduate level.

Hurd, P. S., *Metallic Materials*. New York: Holt, Rinehart, and Winston, 1968. Phase diagrams receive a thorough presentation at the introductory level.

Kingery, W. D., *Introduction to Ceramics*. New York: Wiley, 1960. Phase diagrams of ceramic systems are presented in Chapter 9. Undergraduate level.

Levin, E. M., C. R. Robbins, and H. F. McMurdie, *Phase Diagrams for Ceramists*. Columbus, O.: American Ceramic Society, 1964. (Supplement 1969.) The first section presents a general discussion of phase diagrams at the advanced student level. The remainder presents about 4000 phase diagrams of ceramic systems. For easy reference, these diagrams are cross-indexed by components.

Metals Handbook. Metals Park, O.: American Society for Metals, 1948. The last 125 pages of the 1948 edition of this handbook contain the most readily available collection of metallic phase diagrams.

Rhines, F. N., *Phase Diagrams in Metallurgy*. New York: McGraw-Hill, 1956. For the advanced student and the instructor. It is written specially for metallurgists. The illustrations of polycomponent systems are especially good.

Rogers, B. A., *The Nature of Metals*, second edition. Metals Park, O.: American Society for Metals, 1964. Self-education text. This book will be helpful for the student who finds phase diagrams difficult. Freshman level.

Van Vlack, L. H., *Physical Ceramics for Engineers*. Reading, Mass.: Addison-Wesley, 1964. Three-component diagrams are presented in Chapter 6, with special emphasis on the liquidus surface. Undergraduate level.

PROBLEMS

Assume equilibrium in the problems in this chapter unless specifically stated otherwise. Problems use weight fractions unless stated otherwise.

Location of phase diagrams

Ag-Cu	Fig. 17-7	Cu-Sn	Fig. 17-27
Al-Cu	Fig. 17-23	Cu-Zn	Fig. 17-25
Al-Mg	Fig. 17-22	Fe-O	Fig. 17-29
Al-Si	Fig. 17-24	$FeO-SiO_2$	Fig. 17-21
Al_2O_3-MgO	Fig. 17-10	H_2O	Fig. 17-1
Al_2O_3-SiO_2	Fig. 17-26	H_2O-NaCl	Fig. 17-30
Bi-Pb	Fig. 17-8	MgO-NiO	Fig. 17-5
C-Fe	Fig. 17-9 (and 17-14)	Pb-Sn	Fig. 17-28
Cu-Ni	Fig. 17-3		

17-1 The heat of fusion of ice is about 80 cal/gm. On the basis of the data in Fig. 4-14, estimate the pressure necessary to lower the melting point to $-2°C$.

Answer. 3800 psi (2.6×10^8 dynes/cm^2)

17-2 At what temperature will pure lead solidify if it is hydrostatically compressed 1000 psi? ($\Delta H_f = 1200$ cal/at wt; freezing shrinkage = 4 v/o)

17-3 What are the phases present at every 100°C interval for a 92.5Ag-7.5Cu alloy?

Answer. 400, 500, 600, 700°C—$\alpha + \beta$; 800°C—α; 900°C—$\alpha + L$; ≥ 1000°C—L

17-4 What phases are present at every 100°C interval for a 90Cu-10Sn alloy?

17-5 Refer to the Ag-Cu phase diagram (Fig. 17-7). For each of the following alloys, indicate the weight percent silver in the β-phase at 775°C.

a) 90Ag-10Cu   b) 80Ag-20Cu   c) 50Ag-50Cu

d) 20Ag-80Cu   e) 10Ag-90Cu   f) 5Ag-95Cu

Answer. a) through e)   8 w/o Ag   f) 5 w/o Ag

17-6 Refer to Fig. 17-7 and complete the following table.

	Phase	Temperature	Maximum Cu	Maximum Ag
a)	Liquid	800°C	w/o	w/o
b)	Liquid	1000		
c)	α	800		
d)	β	800		
e)	α	600		
f)	β	600		

17-7 Refer to Fig. 17-7 and Problem 17-6. An alloy contains 127.08 gm of copper and 107.87 gm of silver.

a) At what temperature does the liquid of this alloy become saturated with copper?

b) What phases are present in this alloy at 600°C? at 800°C? at 1000°C?

c) On the basis of Problem 17-6, what is the composition of each phase in this alloy at 600°C? at 800°C? at 1000°C?

d) Determine the number of grams of the two phases in this alloy at 800°C and at 600°C.

e) Convert the grams of part (d) to weight fractions.

Answer. a) 875°C   c) at 600°C: α(4Cu-96Ag) and β(98Cu-2Ag)   e) at 800°C: 0.338 β and 0.662 L

17-8 Refer to Fig. 17-7 and Problems 17-6 and 17-7.

a) What is the atomic percent copper in the indicated alloy?

b) Plot the weight percent liquid present in this alloy as a function of temperature.

c) Plot the weight percent β present in this alloy as a function of temperature.

17-9 Refer to Fig. 17-3. At 1300°C the maximum solubility of copper in the solid is 37 w/o. The maximum solubility of nickel in the liquid is 47 w/o. What is the overall composition of an alloy containing (a) 25 w/o liquid and 75 w/o solid? (b) 75 w/o liquid and 25 w/o solid? (c) 100 w/o liquid?

Answer. a) 59Ni-41Cu   b) 51Ni-49Cu   c) <47Ni->53Cu

17-10 An alloy containing 92 lb of magnesium and 8 lb of aluminum is typical of the magnesium alloys used for aircraft parts. What are the phase compositions at 1200°F, 1000°F, 800°F, 600°F, and 400°F?

17-11 Determine the amounts of the phases in a 92Mg-8Al (weight percent) alloy at each temperature cited in the previous problem.

Answer. 1200°F—1.00 L; 1000°F—0.154 L + 0.846 ϵ; 800°F—1.00 ϵ; 600°F—0.054 δ + 0.946 ϵ; 400°F—0.103 δ + 0.897 ϵ

17-12 Calculate the weight fraction liquid in a 25MgO-75NiO ceramic as a function of temperature.

17-13 Calculate the weight fraction of liquid versus the composition of Al_2O_3-SiO_2 materials at 1600°C.

Answer. 0 Al_2O_3, 0.00 L; 5.5 Al_2O_3, 1.00 L; 40 Al_2O_3, 0.50 L; 71–100 Al_2O_3, 0.00 L

17-14 Determine the weight fraction of spinel ($MgAl_2O_4$) in MgO-Al_2O_3 ceramics at 1800°C as a function of composition.

17-15 Refer to the FeO-SiO_2 diagram. Five grams of fayalite (Fe_2SiO_4), 3 gm of quartz (SiO_2) sand, and 7 gm of FeO are intimately mixed and melted at 1400°C.

   a) At what temperature would the first solid form during equilibrium cooling?
   b) How much additional FeO should be added to the liquid at 1400°C to produce a composition which will have only 0.50 weight fraction fayalite at 1100°C?

Answer. a) 1200°C   b) 15 gm

17-16 Perform a material balance on the distribution of lead and tin in a eutectic Pb-Sn solder at 100°C.

17-17 $CuAl_2$, or θ of the Al-Cu system, is tetragonal, with $a = 6.04$ A and $c = 4.86$ A. The atoms are located as follows:

Cu	Al (approximate)	
$\frac{1}{2}$, 0, 0	$\frac{5}{6}$, $\frac{1}{6}$, $\frac{1}{4}$	$\frac{2}{3}$, $\frac{1}{3}$, $\frac{3}{4}$
$\frac{1}{2}$, 0, $\frac{1}{2}$	$\frac{1}{6}$, $\frac{5}{6}$, $\frac{1}{4}$	$\frac{1}{3}$, $\frac{2}{3}$, $\frac{3}{4}$
0, $\frac{1}{2}$, 0	$\frac{1}{3}$, $\frac{1}{3}$, $\frac{1}{4}$	$\frac{1}{6}$, $\frac{1}{6}$, $\frac{3}{4}$
0, $\frac{1}{2}$, $\frac{1}{2}$	$\frac{2}{3}$, $\frac{2}{3}$, $\frac{1}{4}$	$\frac{5}{6}$, $\frac{5}{6}$, $\frac{3}{4}$

What is its density?

Answer. 4.4 gm/cm^3

17-18 Refer to the Cu-Sn diagram for a 60Cu-40Sn alloy.

   a) Plot the amount of liquid versus temperature.
   b) Plot the copper content of the liquid in this alloy versus temperature.

17-19 A 90Cu-10Sn bronze is used for a 57-lb bell. If it is cooled slowly, (a) the first solid will form at ____°C; (b) this solid will contain ____% Cu; (c) the final liquid will disappear at ____°C; (d) this liquid will contain ____% Sn; (e) at 798°C, the fcc phase will contain ____% Cu; (f) the phase(s) at 550°C will be ____, ____. (g) Assume that α contains essentially 0% tin at 20°C. There will be ____ lb of ϵ in the bell.

Answer. g) 15.35

17-20 a) Plot the weight fraction ϵ at 800°C versus oxygen content in Fe-O compositions.
   b) Plot the weight fraction ϵ versus temperature for a 75Fe-25O (weight percent) composition.

17-21 Refer to Fig. 17-30. Sea water (3.5 w/o NaCl) is to be purified by cooling to 0°F so that essentially pure ice is separated from a more salty brine.

   a) How many pounds of NaCl will a cubic foot of sea water contain (specific gravity = 1.023)? How many pounds of H_2O?
   b) At 0°F, the brine will contain ____ w/o H_2O and ____ w/o NaCl.
   c) Essentially all the NaCl will be in the brine; therefore, ____ lb of brine (and ____ lb of ice) can be extracted from each cubic foot of water at 0°F.

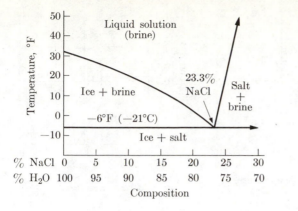

17-30 H$_2$O-NaCl diagram.

The diagram shows axes: Temperature, °F (vertical, values 50, 40, 30, 20, 10, 0, −10). Labels: "Liquid solution (brine)", "Ice + brine", "23.3% NaCl", "Salt + brine", "−6°F (−21°C)", "Ice + salt".

% NaCl	0	5	10	15	20	25	30
% H$_2$O	100	95	90	85	80	75	70

Composition

Answer. a) 2.23 lb NaCl, 61.60 lb H$_2$O   b) 78, 22   c) 10.1, 53.7

17-22 a) An iron-carbon alloy containing 0.67 w/o carbon will contain ___ weight fraction Fe$_3$C at 20°C.

   b) There will be ___ w/o carbon in the pearlite which is in an annealed 0.67C-99.33 Fe alloy.

17-23 An equilibrated Fe-C alloy contained 10 w/o carbide at 722°C. Assuming a new equilibrium, what fraction of the alloy is austenite at 724°C?

Answer. 0.83 γ

17-24 Describe the phase changes which occur on heating a 0.20% carbon steel from room temperature to 1200°C.

17-25 Calculate the weight fraction ferrite, carbide, and pearlite, at room temperature, in iron-carbon alloys containing (a) 0.5% carbon, (b) 0.8% carbon, (c) 1.5% carbon.

Answer. a) 7.5% carbide, 92.5% α, 62% pearlite   b) 12% carbide, 88% α, 100% pearlite   c) 22.5% carbide, 77.5% α, 88% pearlite

17-26 a) Determine the phases present, the composition of each of these phases, and the relative amount of each phase for 1.2% carbon steel at 870°C, 760°C, 700°C.

   b) How much pearlite is present at each of the above temperatures?

17-27 A steel contains 98.5% Fe, 0.5% C, and 1.0% silicon.

   a) What is the eutectoid temperature?

   b) How much pearlite may be formed?

   c) What will be present other than pearlite?

Answer. a) 750°C (1380°F)   b) 0.75 pearlite   c) proeutectoid ferrite (i.e., α formed during cooling before the eutectoid temperature)

17-28 Modify Fig. 7-14 for a steel containing (a) 1% Mn, (b) 1% Cr, (c) 1% W, (d) 1% Ni. (The new solubility curves remain essentially parallel to the previous ones.)

17-29 A steel has the following composition. Give it an AISI number: 0.38 C, 0.75 Mn, 0.87 Cr, 0.18 Mo, 0.03 Ni.

Answer. AISI 4140

17-30 A steel has the following composition. Give it an AISI number: 0.21 C, 0.69 Mn, 0.62 Cr, 0.13 Mo, 0.61 Ni.

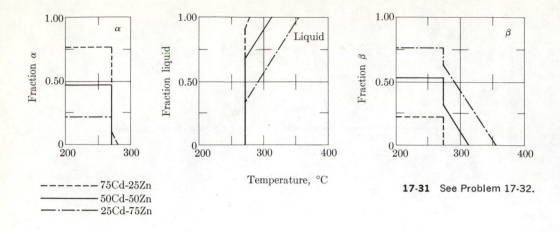

17-31 See Problem 17-32.

75Cd-25Zn
50Cd-50Zn
25Cd-75Zn

Temperature, °C

17-31 The following observations were made in a study of the phase equilibria existing at various temperatures in the system Xm-Yz. On the basis of these observations, draw the equilibrium phase diagram for the Xm-Yz system.

Total composition	Temperature	Structure and composition of phases observed in equilibrium
10 w/o Xm	275°C	Fcc, 8 w/o Xm; L, 13 w/o Xm
	250	Fcc, 10 w/o Xm
	100	Fcc, 10 w/o Xm
20 w/o Xm	225°C	Fcc, 17 w/o Xm; L, 26 w/o Xm
	150	Fcc, 20 w/o Xm
	50	Fcc, 19 w/o Xm; Hex, 25 w/o Xm
30 w/o Xm	200°C	L, 30 w/o Xm
	185	Fcc, 22 w/o Xm; Hex, 27 w/o Xm; L, 32 w/o Xm
	175	Hex, 28 w/o Xm; L, 38 w/o Xm
	100	Hex, 30 w/o Xm
40 w/o Xm	150°C	Hex, 30 w/o Xm; L, 47 w/o Xm
	125	Hex, 33 w/o Xm; L, 56 w/o Xm; Rhomb, 99.5 w/o Xm
	50	Hex, 32.5 w/o Xm; Rhomb, 99.9 w/o Xm
70 w/o Xm	200°C	L, 70 w/o Xm
	150	L, 64 w/o Xm; Rhomb, 99.6 w/o Xm
	125	Hex, 33 w/o Xm; L, 56 w/o Xm; Rhomb, 99.5 w/o Xm
	100	Hex, 32.8 w/o Xm; Rhomb, 99.7 w/o Xm

Pure Xm melts at 271°C; pure Yz melts at 327°C.

Answer. See the Bi-Pb diagram (Fig. 17-8).

17-32 Figure 17-31 shows the fractions α, L, and β in three different Cd-Zn alloys. It is known that the Cd-Zn system has only one invariant reaction other than the 321°C and 419.5°C melting points, respectively. Prepare a phase diagram for the Cd-Zn system.

18 □ phase changes in materials

18-1 REACTION RATES

In Chapter 17 we assumed the existence of phase equilibrium, and no attention was given to the mechanisms by which equilibrium is attained. The mechanisms are important because they affect the phases and their distribution within the material, and therefore the properties of the material (Chapter 19).

The phase diagram shows the equilibrium relationships. Although reactions proceed toward these equilibria, the phase diagram gives no indication of how long it will take to achieve such equilibrium. For example, the solidification of fcc aluminum from a pure liquid occurs so rapidly ($\ll 1$ sec) at 500°C that rates can be determined only by indirect means. In contrast, glass can be used for years without crystallization, even though it is a supercooled liquid with a higher free energy than its crystalline counterpart. In brief, reaction rates vary by many orders of magnitude.

Reaction rates and phase distribution are influenced by the necessity for breaking existing bonds, the requirement for establishing new phase boundaries, and the need for relocating atoms in the new phases or grains. These factors will be discussed in succeeding sections.

18-2 PHASE CHANGES WITHOUT COMPOSITIONAL CHANGES

The simplest phase changes are those in which there is no change in composition. These will be discussed under the headings of congruent transformation, ordering, and martensitic reactions.

Congruent transformations. A single-component material, such as a pure metal, an oxide, or a polymer, crystallizes with no composition change. Even complex substances, such as spinel ($MgAl_2O_4$ of Fig. 17-10), may have identical solid and liquid compositions. Phase changes of this type are termed *congruent*.

Congruent transformations fall into one of two categories, according to whether the atomic coordination is altered or not. The polymorphic transformation from α-Ti (hcp) to β-Ti (bcc) is a *reconstructive transformation*, because the coordination number for the atoms is changed from 12 to 8. A change between two polymorphs of SiO_2, shown in Fig. 18-1, provides a second example of reconstructive transformation and reveals that, for a reaction to occur, there would have to be a significant

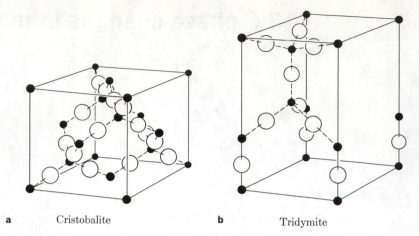

| a | Cristobalite | b | Tridymite |

18-1 Reconstructive transformations (silica). (a) Cristobalite. (b) Tridymite. These two polymorphs have close to the same free energy. However, their transformation at 1470°C is very slow because a high activation energy is required to break bonds and rearrange the atoms.

rearrangement of bonds. Since a relatively high activation energy is required, these reactions are often sluggish, particularly in compounds.

A *displacive transformation* arises from "bond straightening" in which the atoms retain the same neighbors. The cubic-to-tetragonal change in $BaTiO_3$ (Section 13-8) provides one example of this. The low-high temperature inversions of the SiO_2 phases provide a second example (Fig. 18-2). No bonds are broken; therefore, the change is very rapid. These inversions may produce major volume changes which may lead to severe fracturing, particularly if the material is brittle.

18-2 Displacive transformations (silica—schematic). (a) Low-temperature form. (b) High-temperature form. The bond angles across the oxygen atoms straighten to increase the volume and make the crystal structure more symmetric. There is negligible time lag because no bonds are broken.

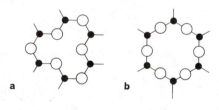

a b

Example 18-1

Assume that the oxygen atoms are placed directly between adjacent silicon atoms in high cristobalite (Figs. 18-1a and 18-2b). At 220°C a displacive transformation occurs during cooling which gives a 1% linear contraction in all dimensions (Fig. 18-2a). What is the new Si-O-Si bond angle, θ?

Answer

$$\sin(\theta/2) = 0.99,$$

$$\theta = 164°. \blacktriangleleft$$

Ordering. The transformation from random β-brass (Fig. 8-6b) to ordered β_1-brass (Fig. 8-6a) does not require a change in coordination number, but does

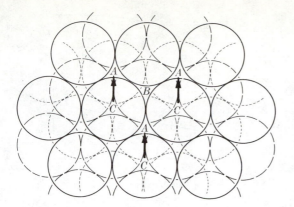

18-3 Shear transformation (Co fcc → Co hcp). The ··· *ABC* ··· stacking of fcc cobalt is transformed by the (111) *C* plane (solid circles) shearing to a location directly overlying the *A* positions (dotted circles). This gives the ··· *ABAB* ··· stacking of hcp cobalt.

require diffusion to relocate atoms. However, the time is so short that it is impossible to quench β-brass from its stable high-temperature form (Fig. 17-25). By contrast, the ordering of an atactic polymer into an isotactic polymer (Section 7-6) is almost infinitely slow, because several strong covalent bonds have to be broken simultaneously for rearrangement to occur.

Shear (martensitic) transformations. Some phase changes occur by the cooperative displacement of the atoms in shearlike fashion. This is most readily described in terms of the reaction

$$\text{Co (fcc)} \xrightarrow{\text{cooling}} \text{Co (hcp)} \tag{18-1}$$

at 1120°C. As shown in Fig. 4-3, the fcc and hcp structures vary mainly in their stacking arrangements. The (111) planes of the former are stacked ··· *ABCABC* ···, while the (0001) planes of the latter are stacked ··· *ABABAB* ···. The planes themselves are identical. By a shear process, the atoms of a *C*-plane of the fcc crystal can move part of an atomic distance to give the hcp stacking (Fig. 18-3). This is called a *shear* transformation for obvious reasons. It is facilitated by supercooling and mechanical stresses. Undoubtedly dislocations are involved in this shearing process, just as in plastic deformation.

Example 18-2

Indicate (a) the shear directions in Eq. (18-1) and (b) the shear strain.

Answer

a) From Fig. 18-3, this direction is observed to bisect two fcc directions of high linear density ⟨110⟩. Therefore, the shear direction is ⟨11$\bar{2}$⟩. [This may be more readily visualized by comparing Fig. 18-3 with the sketch in Fig. 5-18(a).] The corresponding hcp direction is [12$\bar{3}$0].

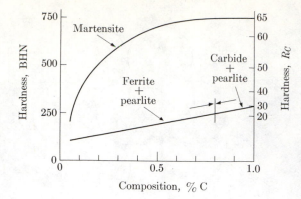

18-4 Hardness of martensite. Quenching produces martensite from austenite by a diffusionless, two-step shear transformation. Martensite is appreciably harder than the equilibrium combination of ferrite plus carbide.

b) From Fig. 18-3,

$$x = R/\cos 30° = 2R/\sqrt{3}.$$

Also, with reference to the fcc lattice,

$$y = 2d_{111} = 2a_{\text{fcc}}/\sqrt{3},$$
$$\gamma = x/y = R/a_{\text{fcc}}.$$

But $a_{\text{fcc}} = 4R/\sqrt{2}$; therefore,

$$\gamma = R\sqrt{2}/4R = 0.35. \quad \blacktriangleleft$$

The best-known shear transformation is that of martensite formation in steel and related alloys. *Martensite* in steel is a body-centered *tetragonal* phase of iron, supersaturated with carbon. It does not appear in the phase diagram because it always has a higher free energy than the phases found in the Fe-C diagram for the same total composition. It appears in steels because it forms by shear (and without diffusion) from quenched austenite, the fcc polymorph. It has tremendous engineering importance because it is extremely hard and strong. In fact, the well-known quenching treatment of steel is used to produce this phase,* in preference to pearlite (Fig. 18-4).

A face-centered crystal can change to a body-centered structure by a shear displacement on {111} planes. These roughly correspond to the {110} planes in the new body-centered structure, but unlike the cobalt transformation of Eq. (18-1), the planes of the two phases are not identical. To complete the transformation requires a two-step shear process which is more complicated than the one sketched in Fig. 18-3. Martensitic transformations can be arrested before completion, because the induced stresses oppose the shearing process. A second consequence of this diffusionless type of transformation is that the carbon dissolved in the fcc austenite becomes trapped in 0, 0, $\frac{1}{2}$ sites of the new body-centered structure,

---

* We will observe later that the quenching process must be followed by a tempering process, because martensite is brittle as well as hard.

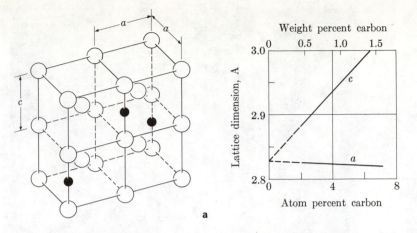

18-5 Body-centered tetragonal (martensite). (a) The carbon atoms are preferentially located in 0, 0, $\frac{1}{2}$ locations to elongate the body-centered lattice into a tetragonal lattice. (b) At low carbon levels, the body-centered lattice becomes cubic (c = a). (Adapted from B. Chalmers, *Physical Metallurgy*. New York: Wiley, 1959, p. 364.)

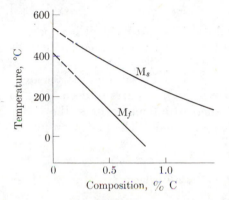

18-6 Martensite transformation temperatures (plain-carbon steels). Transformation is first detected at M_s and is virtually complete at M_f. Between M_s and M_f, austenite is retained as a result of induced stresses.

making it tetragonal (Fig. 18-5). The carbon atoms and the local lattice *distortion*, plus other factors, make slip very difficult, thus giving the extreme hardness and strength cited in the last paragraph.

We observe martensite formation when cooling is fast (i.e., when the material is quenched), so that the more stable ferrite and carbide do not have time to form. (In the latter reaction, carbon atoms must diffuse to form the carbide.) The necessary amount of supercooling varies with carbon content. Figure 18-6 shows an M_s-curve which is the locus of temperatures for the first detection of martensite (~ 1 v/o). Martensite continues to form during further cooling to the final temperature, M_f.[*] Between these two temperatures there is a mixture of martensite

---

[*] It is difficult to detect the last small fraction of retained austenite. Therefore, M_f probably represents about 99% transformation.

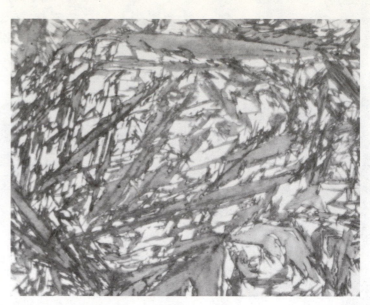

18-7 Martensite (×1000). Since martensite (gray) forms by shear and not by diffusion, it has the same composition as the original austenite.

and austenite (Fig. 18-7), the latter being called *retained austenite* if it is present at room temperature or below. The extent of martensitic transformation depends on the temperature rather than on the time.

Before leaving our discussion of martensite, we should recall that martensite is only metastable; therefore, upon reheating (*tempering*), it starts to decompose to more stable phases. With enough time at increased temperatures, the body-centered tetragonal unit cell changes to ferrite and carbide. The Fe-C transformation reactions may thus be represented by

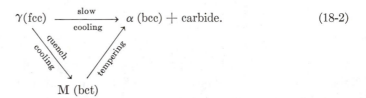

$$\gamma(\text{fcc}) \xrightarrow[\text{cooling}]{\text{slow}} \alpha \, (\text{bcc}) + \text{carbide}. \qquad (18\text{-}2)$$

M (bct)

18-3 NUCLEATION OF PHASE CHANGES

A major barrier to phase transformations is the requirement for nucleation of new phases. It is usually impossible to nucleate a new phase except through the introduction of phase boundaries.* Since atoms along a phase boundary possess more energy than those within the phase itself (Section 6-7), new phases cannot appear automatically at the equilibrium temperature or solubility limit.

---

* The spinodal reaction offers an exception which will be considered separately (Section 18-7).

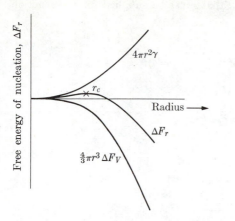

18-8 Critical nucleus radius (homogeneous nucleation). Energy must be supplied to grow a new phase to the critical radius, r_c.

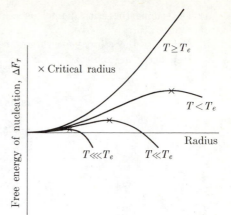

18-9 Effect of supercooling on the critical radius. The critical radius must be infinite at the equilibrium temperature, T_e. Nucleation is more probable with greater supercooling because the critical radius is reduced.

Homogeneous nucleation. Consider a supercooled phase which should transform to a new phase of about the same density, the latter having a lower free energy per unit volume (i.e., ΔF_V is negative). If the new phase is spherical, with radius r, the chemical free-energy change will be $\frac{4}{3}\pi r^3 \Delta F_V$. However, the new phase has a surface of $4\pi r^2$. Its surface energy will be $4\pi r^2 \gamma$, where γ is the energy per unit area of interface. The total free-energy change, ΔF_r, is a function of the radius:

$$\Delta F_r = 4\pi r^2 \gamma + \tfrac{4}{3}\pi r^3 \Delta F_V. \tag{18-3}$$

Although ΔF_V is negative for any spontaneous reaction, γ is always positive; therefore, we can plot Eq. (18-3) as shown in Fig. 18-8. Below a *critical nucleus radius*, r_c, such particles of a new phase will tend to dissolve instead of grow. Cooling beyond the equilibrium temperature, i.e., *supercooling*, makes the value of ΔF_V more negative, and therefore markedly reduces both the critical radius and the energy barrier for nucleation (Fig. 18-9). Thus there is a greater probability that the necessary number of atoms can attain the appropriate energy from some local energy fluctuations. As a result, a reaction which cannot nucleate at the equilibrium temperature occurs more readily, the greater the degree of supercooling. Of course, extremely low temperatures restrict atom movements so that the rate of nucleation then becomes slow again. Hence, the rate of nucleation passes through a maximum as a function of the supercooling temperature.

Example 18-3

Consider the congruent transformation, $\alpha \underset{}{\overset{1025°C}{\rightleftarrows}} \beta$. The interfacial energy between α and β is 500 ergs/cm², and the values of ΔF_V for $\alpha \to \beta$ are -100 cal/cm³ at 1000°C and -500 cal/cm³ at 900°C (1 cal/cm³ = 4.185 × 10⁷ ergs/cm³),

a) Determine the critical nucleus radius, r_c, for the nucleation of β within α at each temperature.

b) Calculate the critical *nucleation energy*, ΔF_c, that must be supplied for the reaction to proceed in each case.

Answer

a)
$$r_c = \text{radius} \quad \text{when} \quad \frac{d\,\Delta F_r}{dr} = 0;$$

$$\Delta F_r = 4\pi r^2 \gamma + \tfrac{4}{3}\pi r^3 \,\Delta F_V,$$

$$\frac{d\,\Delta F_r}{dr} = 0 = 8\pi r \gamma + 4\pi r^2 \,\Delta F_V,$$

$$r_c = \frac{-2\gamma}{\Delta F_V}. \tag{18-4a}$$

At 1000°C,

$$r_c = \frac{-2(500 \text{ ergs/cm}^2)}{-4.185 \times 10^9 \text{ ergs/cm}^3} = 24 \text{ A.}$$

At 900°C,

$$r_c = \frac{-2(500 \text{ ergs/cm}^2)}{-20.925 \times 10^9 \text{ ergs/cm}^3} = 4.8 \text{ A.}$$

b)
$$\Delta F_c = \Delta F_r \quad \text{at} \quad r = r_c.$$

$$\Delta F_c = \frac{16}{3} \frac{\pi \gamma^3}{\Delta F_V^2}. \tag{18-4b}$$

At 1000°C,

$$\Delta F_c = \frac{16\pi}{3} \frac{(500 \text{ ergs/cm}^2)^3}{(-4.185 \times 10^9 \text{ ergs/cm}^3)^2}$$
$$= 1.2 \times 10^{-10} \text{ erg.}$$

At 900°C,

$$\Delta F_c = 0.048 \times 10^{-10} \text{ erg.} \quad \blacktriangleleft$$

Heterogeneous nucleation. Irregularities in crystal structure, such as point defects and dislocations, possess strain energy (Figs. 6-6 and 6-13). Nucleation is facilitated by these imperfections if the transformation reduces the strain energy.* Since this occurs in many cases, the released strain energy can reduce the energy requirements of ΔF_r in Eq. (18-3). Therefore, nucleation proceeds with a smaller critical radius. Heterogeneous precipitation of this type is common.

Nucleation from surfaces. The biggest imperfection is an interface, either as an external surface or as a grain or phase boundary. These interfaces promote heterogeneous nucleation.

Consider the external surface in Fig. 18-10 and a nucleus which has already formed. The equilibrium contact angle, θ, depends on the relative energies, $\gamma_{\alpha/g}$,

---

* Conversely, nucleation will be retarded if the transformation introduces lattice strain energy.

$\gamma_{\beta/g}$, and $\gamma_{\alpha/\beta}$, of the three boundaries:

$$\gamma_{\alpha/g} = \gamma_{\beta/g} + \gamma_{\alpha/\beta} \cos \theta. \qquad (18\text{-}5)$$

Under such conditions the critical radius, r_c, of Eq. (18-4a) is reduced by $\sin \theta$ to $-(2\gamma/\Delta F_V) \sin \theta$. This feature reduces the critical volume, with the result that a new phase has a greater probability of growing than of redissolving. In fact, as θ approaches 0°, the activation energy for nucleation becomes nil and transformation is independent of the energy of the phase boundary.

Inoculants. Cloud-seeding by silver iodide is possible because of the similarity between the unit-cell dimension of silver iodide and that of ice; the two lattices have nearly perfect registry, and therefore negligible phase-boundary energy. Consequently, the ice can crystallize on the silver iodide surface under conditions where it would not be able to undergo homogeneous nucleation. The silver iodide has served as an *inoculant*, and we speak of the process as *heterogeneous nucleation*. The majority of phase transformations are initiated by some type of heterogeneous nucleation—the surface of an inoculant, an impurity, a grain boundary, or simply the container wall.

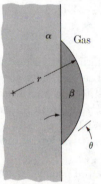

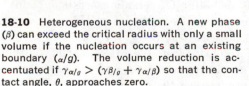

18-10 Heterogeneous nucleation. A new phase (β) can exceed the critical radius with only a small volume if the nucleation occurs at an existing boundary (α/g). The volume reduction is accentuated if $\gamma_{\alpha/g} > (\gamma_{\beta/g} + \gamma_{\alpha/\beta})$ so that the contact angle, θ, approaches zero.

18-11 Diffusional transformation (pearlite formation). Carbon must move from the eutectoid austenite (0.8 w/o carbon) to the carbide areas (>6 w/o carbon) because the ferrite cannot contain more than 0.02 w/o carbon. The carbon may travel through the γ, along the γ-α interface, or through the α.

18-4 DIFFUSION ACCOMPANYING PHASE CHANGES

Although congruent transformations occur with little or no diffusion, this is not the case with the majority of solid-state reactions. This fact is perhaps best illustrated with the Fe-C eutectoid reaction. As austenite (with 0.8 w/o carbon) dissociates into ferrite (with 0.02 w/o carbon) and carbide (with more than 6 w/o carbon), the carbon must diffuse as indicated in Fig. 18-11. It is not known what

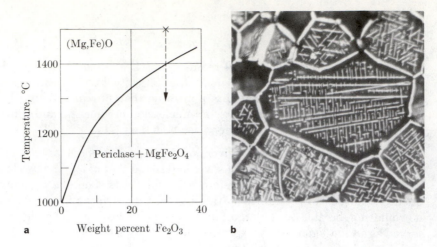

18-12 Diffusional precipitation (MgFe$_2$O$_4$ formation). (a) Solubility limit of Fe$_2$O$_3$ in (Mg, Fe)O. Precipitation produces magnesioferrite (MgFe$_2$O$_4$) and periclase (MgO). (b) Grain-boundary precipitation at higher temperatures plus intragranular precipitation later in the cooling sequence ($\times$1000). (Courtesy of H. McCollister.)

fraction of the carbon atoms move along the three paths: (1) through the austenite, (2) through the ferrite, and (3) along the phase boundary. It is observed, however, that the ferrite and carbide lamellae of pearlite grow edgewise into the austenite from the austenitic grain boundaries. If the cooling rate is fast, the lamellae are thinner and more numerous than when the cooling rate is slow. With slow cooling, (1) more time is available for carbon diffusion, and (2) less supersaturation occurs to nucleate new lamellae.*

A second example of diffusion in solid-state reactions can be seen in precipitation processes. At 1500°C, for example, a 70MgO-30Fe$_2$O$_3$ ceramic consists of a single solid solution (Fig. 18-12a). The material in Fig. 18-12(b) was cooled *continuously* and at a rate such that MgFe$_2$O$_4$ was nucleated at the grain boundaries. The *boundary precipitate* grew by diffusion of iron oxide from the adjacent solid solution. As cooling continued, however, the diffusion rate decreased, and the supersaturation increased, so that nucleation also occurred in the center of the grains. The result was a grain-boundary precipitate, a denuded area as a result of diffusion, and an *intragranular precipitate* which grew along specific planes of the parent crystal.

These two examples reveal that diffusion affects the growth and distribution of reaction products within a solid. In Chapter 19 we shall see how various microstructures control the properties.

---

* We will observe in the next chapter that the finer pearlitic structures produce harder alloys.

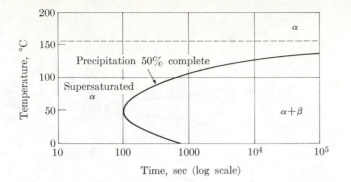

18-13 Isothermal precipitation (90Pb-10Sn). Above 155°C, α can contain all of the tin. The rate of β precipitation varies with the amount of supercooling below the saturation temperature. (Adapted from data by H. K. Hardy and T. J. Heal, *Progress in Metal Physics 5*. London: Pergamon Press, 1954, p. 256.)

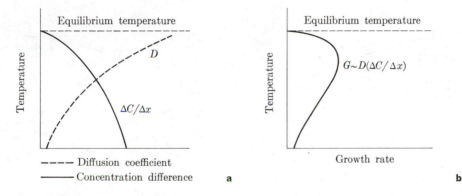

18-14 Particle growth (schematic). (a) Variation of the diffusion coefficient, D, and concentration difference, $\Delta C/\Delta x$, with temperature. (b) Growth rate, G.

18-5 ISOTHERMAL PRECIPITATION

If a solid solution is quenched and held at a temperature below the solubility limit, a long time is usually required (with slight supercooling) to reach the midpoint of the precipitation process. A shorter time is required with intermediate supercooling, and longer times again with severe supercooling. This is indicated in Fig. 18-13 for the precipitation of tin-rich β from a 90Pb-10Sn alloy (Fig. 17-28).

In order to analyze the above behavior, which is characteristic of a wide variety of materials, we will need to consider the diffusion coefficient, the concentration gradient within the matrix phase, and the nucleation rate. In Chapter 9, we saw that the diffusion coefficient increases with temperature (curve D of Fig. 18-14a). The concentration gradient within the matrix phase arises from the difference between the original composition and the solubility limit. This difference typically increases as the supercooling is increased (curve $\Delta C/\Delta x$ of Fig. 18-14a). The flux of atom movements, which makes each precipitate particle *grow*, is the product

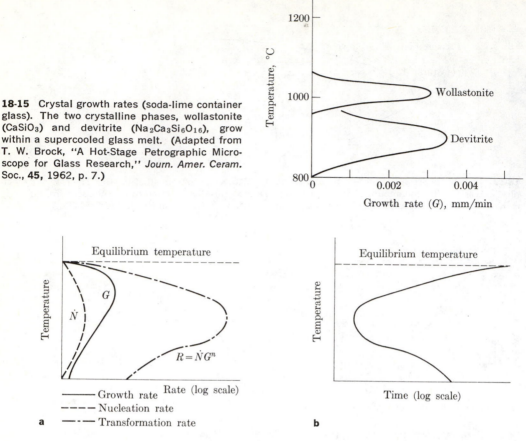

18-15 Crystal growth rates (soda-lime container glass). The two crystalline phases, wollastonite (CaSiO₃) and devitrite (Na₂Ca₃Si₆O₁₆), grow within a supercooled glass melt. (Adapted from T. W. Brock, "A Hot-Stage Petrographic Microscope for Glass Research," *Journ. Amer. Ceram. Soc.*, **45**, 1962, p. 7.)

18-16 Precipitation kinetics (schematic). (a) Transformation rate, *R*. (b) Time (log scale).

$D(\Delta C/\Delta x)$, as shown in Fig. 18-14b, and is proportional to the growth rate, G (Fig. 18-15).

On the basis of Section 18-3, the nucleation rate, $\dot{N}$, like the rate of growth, G, has a maximum with moderate supercooling (curve $\dot{N}$ of Fig. 18-16a). The overall *rate* of precipitation (curve R) is the product $\dot{N}G^n$. Normally, instead of rate, we are interested in the *time* which is required to reach a specified reaction stage, $t = R^{-1}$ (Fig. 18-16b). It is convenient to use a semilog plot for reaction-time data in order to show relative changes in time more clearly (Fig. 18-13).

18-6 ISOTHERMAL TRANSFORMATION OF AUSTENITE

The most thoroughly studied of all reactions occurring in solids is

$$\gamma = \alpha + \text{carbide.} \tag{18-6}$$

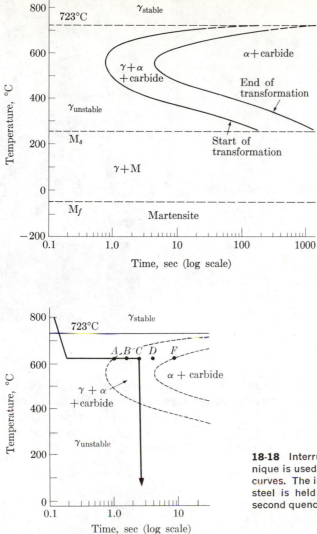

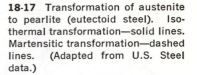

18-17 Transformation of austenite to pearlite (eutectoid steel). Isothermal transformation—solid lines. Martensitic transformation—dashed lines. (Adapted from U.S. Steel data.)

18-18 Interrupted quench (eutectoid steel). This technique is used in establishing isothermal-transformation curves. The initial quench is made into a hot bath. The steel is held there for a prescribed time before the second quench to room temperature.

Of course, it is of major importance in the heat treatment of steel. It also serves as a convenient prototype for other reactions.

Figure 18–17 presents the *isothermal transformation* curves for a eutectoid steel (0.8C–99.2Fe). These may also be called *C-curves* (because of their shape), or *T-T-T curves* (i.e., temperature-time-transformation). The data for the upper part of Fig. 18-17 were obtained as follows. Small samples of steel were heated into the austenite temperature range long enough to ensure complete transformation to austenite. These samples were then quenched to a lower temperature (e.g., 620°C) and held there for varying lengths of time before being quenched a second time to room temperature (Fig. 18-18). The austenite-to-pearlite change

18-19 Start of austenite (γ) → ferrite (α) + carbide transformation at 620°C. P is pearlite, i.e., lamellar (α + carbide) as sketched in Fig. 18-11; M is observed as martensite because it did not have an opportunity to form pearlite before the second quench. (Point *A* of Fig. 18-18.)

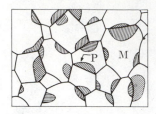

18-20 Transformation 25% complete at 620°C. (Point *B* of Fig. 18-18.)

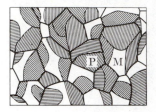

18-21 Transformation 75% complete at 620°C. (Point *D* of Fig. 18-18.)

was not observed in samples held at 620°C for less than 1 sec, and complete transformation to α + carbide was not observed until after more than 10 sec had elapsed (Figs. 18-19 through 18-21). Similar data were obtained at other temperatures to complete the upper part of Fig. 18-17. The M_s- and M_f-data in the lower part of Fig. 18-17 are from Fig. 18-6 for a 0.80 w/o carbon steel.

Example 18-4

A small wire of SAE 1080 steel is subjected to the following treatments as successive steps:

a) heated to 870°C and held there 1 hr;

b) quenched to 550°C and held there 10 sec;

c) quenched to 275°C and held there 10 sec;

d) quenched to 20°C and held.

Indicate the phases and compositions after each step.

Answer

a) Austenite with 0.80 w/o carbon.

b) Ferrite (0.025 w/o carbon) plus carbide (6.67 w/o carbon). (The weight ratio is 0.88 to 0.12.)

c) Same as (b). (Both α and carbide are stable, so *they cannot revert to austenite at this temperature*.)

d) Same as (c).

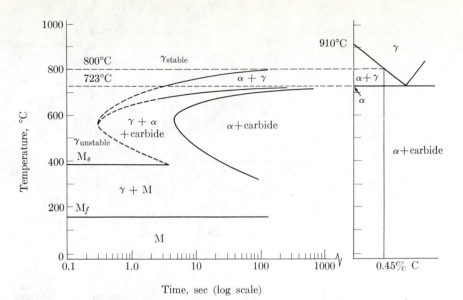

18-22 Transformation of austenite to ferrite plus pearlite (SAE 1045 steel). The equilibrium phases at the right correspond to the long-time product of transformation.

Note. The significance of this example is to emphasize that an isothermal transformation diagram is not an equilibrium diagram. Since equilibrium is attained with time, the phases at the right end of a transformation diagram eventually correspond to the equilibrium phases. ◄

Example 18-5

List factors which make the isothermal transformation for an SAE 1045 steel different from that for an SAE 1080 steel. (See Table 17-2 for steel codes.)

Answer

1) Phase equilibrium: According to Fig. 17-14, the equilibrium phases which appear between 723°C and 800°C after long times are $\alpha + \gamma$.
2) Martensitic temperatures: According to Fig. 18-6, a 0.45 w/o carbon steel has higher M_s and M_f temperatures than a 0.80 w/o carbon steel.
3) Transformation rate: With lower carbon content, the supercooling at any given temperature is greater; therefore, the reaction occurs in less time.

These factors are revealed in Fig. 18-22, which was constructed from experimental data. Note that pro-eutectoid ferrite forms initially at high temperatures (>550°C). ◄

Continuous-cooling transformation of austenite. As shown in Fig. 18-17, reactions occur slowly both at relatively low temperatures and at temperatures close to the equilibrium temperature. At intermediate temperatures the transformation is more rapid. With this information let us consider the progress of transformation with different rates of *continuous* cooling. A severe quench will miss the "knee" of the transformation curve, with the result that the austenite is changed to martensite rather than to pearlite (i.e., α + carbide). Slow cooling

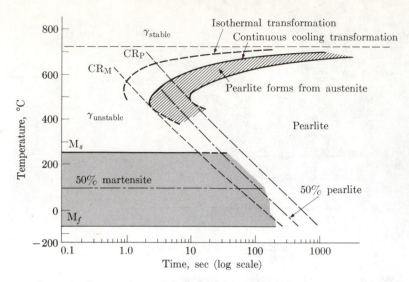

18-23 Continuous-cooling transformation (eutectoid steel). Transformation temperature and times are displaced from the isothermal-transformation curve for the same steel. (Cf. Fig. 18-17.) CR_M = minimum cooling rate for 100% martensite. CR_P = maximum cooling rate for 100% pearlite.

permits pearlite to form; however, the first precipitate occurs after a longer time (and therefore at a lower temperature) than for isothermal transformation, simply because part of the time was spent at higher temperatures where reaction rates are slower. Thus the isothermal-transformation curve is displaced downward and to the right for continuous cooling (Fig. 18-23).*

There are two important cooling rates in the continuous-cooling transformation curve; these are included in Fig. 18-23. The first is the minimum cooling rate, CR_M, which just misses the "knee" of the transformation curve and therefore produces *only* martensite; the second is the maximum cooling rate, CR_P, which produces *no* martensite. *In a eutectoid steel* (0.8C-99.2Fe), these two *critical cooling rates* are about 200°C/sec and 50°C/sec, respectively, through the 750-to-500°C temperature range. These rates are slower in most other steels, as will be discussed later.

Effect of austenitic grain size on transformation. The austenite-to-pearlite reaction is usually nucleated at the austenitic grain boundaries (Fig. 18-11). Thus a steel with a small austenitic grain size, and therefore more grain-boundary area per unit volume, will produce more nuclei per unit volume at any given temperature (curve $\dot{N}$ of Fig. 18-16). The net effect is to increase the reaction rate and decrease the transformation time (Fig. 18-24). This means that the same cooling rate will give more martensite in a coarse-grained steel than in a fine-grained steel.

---

* The underslope of the isothermal curve is meaningless in *continuous* cooling; therefore, it is not shown.

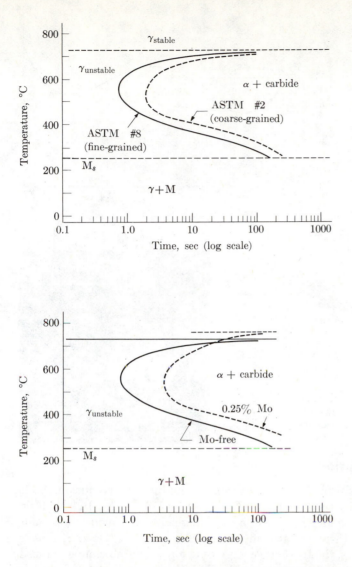

18-24 Grain-boundary nucleation. The effect of grain size on the *start* of austenite transformation is shown for eutectoid steels. The fine-grained steel has more grain-boundary area from which transformation can start.

18-25 Transformation retardation. Molybdenum, like other alloying elements, retards the *start* of transformation of austenite.

Effect of composition on transformation. All the principal alloying elements* in steels retard the transformation of austenite to pearlite, because the solute alloy atoms, as well as the carbon atoms, must redistribute themselves between the growing ferrite and carbide phases. This is shown in Fig. 18-25 for molybdenum. The transformation rate is slower by a factor of 5, which means that the critical cooling rate for obtaining martensite is decreased by a similar factor. One of the chief reasons for adding alloying elements to steels is to give this extra time to obtain the desired martensite for strengthening the steel. (See Section 20-5.)

 Cobalt is the sole exception, and it is rarely used as an alloy addition in tonnage steels.

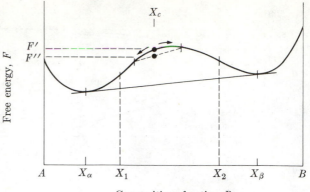

18-26 Spinodal decomposition. In a nonideal solution with negative deviation (Fig. 8-13), a hump, or spinode, occurs between X_1 and X_2, where d^2F/dX^2 is negative. In this range the free energy is lowered by segregation *within* a single solution. Diffusion is required.

Composition, fraction B

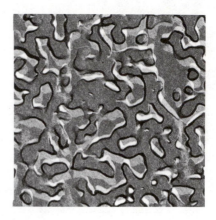

18-27 Spinodal decomposition in glass (Na$_2$O-SiO$_2$). Although originally a single phase, the glass now contains two distinct compositions as a result of spinodal decomposition. The lighter-appearing phase etched more readily (×30,000). (Courtesy of R. Redwine and M. Conrad, Owens-Illinois Glass Co.)

18-7 SPINODAL DECOMPOSITION

Phases can separate without nucleation under special conditions of solution. This reaction is best explained by considering the F-X curve of Fig. 18-26. In a solution with positive deviation from ideality (Section 8-6), there are two reverse inflections in the free-energy curve; these are at compositions X_1 and X_2. On the *spinode* between these compositions, the total free energy of composition X_c can be reduced from F' to F'', or lower, by segregation within the solution, as indicated by the diffusion arrows. In the early, clustering stages of separation, there is no necessity to form a phase boundary, because the structure is a continuous phase with variations in composition. By the time that separation has progressed to the lowest total free energy, the two compositions are markedly different, and correspond to the points of common tangency (Fig. 8-13). Thus, by selective etching, it is possible to reveal the spinodal decomposition into two phases that become progressively more distinct (Fig. 18-27).

One characteristic of spinodal decomposition is the uniform periodicity of the two phases. A second characteristic is the *three-dimensional* continuity of each phase. There are no isolated pockets of either phase.

Glass of the 96% SiO_2 type, familiar to the laboratory worker because of its low thermal expansion, is formed by the spinodal process. A calcium borosilicate glass is melted and formed into the required shape while it is a single phase. Then it is heat-treated to produce spinodal decomposition into a high-silica (96%) and a low-silica phase. The latter is acid soluble and can be leached from the product (thanks to the three-dimensional continuity of the phases). A subsequent sintering process shrinks the high-silica skeleton into a nonporous final product having a very low thermal expansion.

Although long known, spinodal decomposition has only recently been understood. This type of heat treatment possesses significant engineering potential for making ceramic products out of devitrified glasses and for producing certain metallic materials, such as the hard, alnico-type magnets.

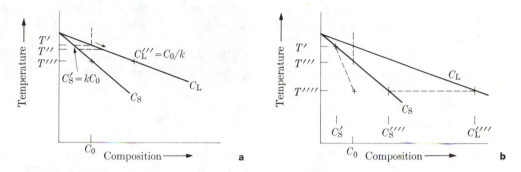

18-28 Segregation. (a) Equilibrium solidification. (b) Non-equilibrium solidification. The first solid has the composition C_S'. With non-equilibrium, the last solid has the composition C_S'''', although the average solid composition is C_0.

18-8 NONEQUILIBRIUM REACTIONS

Compositional *segregation* is a necessary occurrence in those reactions which produce two phases of different compositions. Under equilibrium, compositional segregation is eliminated whenever the product of reaction is a single phase. For phase changes in materials, however, there is not always enough time available to attain equilibrium. Therefore, it is not uncommon to observe nonequilibrium conditions in materials, even if they consist of a single phase. This may or may not be considered desirable, depending on the service requirements.

Solute redistribution. Recall from Section 17-4 that the solidifying phase in a eutectic-type binary system is purer than the liquid. We can express this mathematically in terms of Fig. 18-28(a), where k is the *distribution coefficient:*

$$k = C_S/C_L. \tag{18-7}$$

Since the initial liquid has the composition $C_L' = C_0$, the first solid at T' has the composition $C_S' = kC_0$. Therefore, for each gram of solid formed, there are

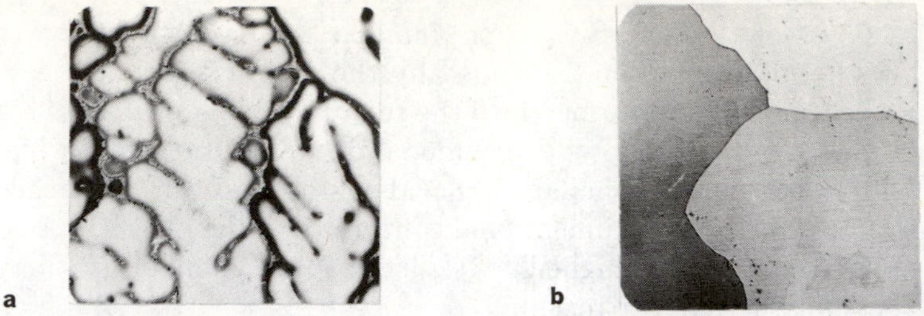

18-29 Solidification segregation (96Al-4 Cu) (×100). The irregular areas of the cast alloy (a) are richer in copper. (b) The same alloy after equilibration. (A. G. Guy, *Elements of Physical Metallurgy,* second edition. Reading, Mass.: Addison-Wesley, 1959. Courtesy of Alcoa Research Laboratories.)

$(1 - k)C_0$ gm of extra solute in the liquid, shifting C_L to the right. With added solute in the liquid, the temperature must be lowered to T'' before further solidification occurs.

Under equilibrium conditions, solute will diffuse into the solid, with the result that C_S follows the lower, or *solidus*, curve, while C_L follows the upper, or *liquidus*, curve. The final liquid disappears at T'''.

With faster cooling and negligible diffusion, the *average* solid composition follows the dashed line below the solidus curve of Fig. 18-28(b), finally crossing C_0 at T'''', where the last liquid solidifies with composition C_L'''', giving an increment of solid, C_S''''. Thus, while the average solid composition is C_0, it ranges from C_S' to C_S''''. This segregation can be eliminated only by the slow process of solid diffusion (Fig. 18-29).

Variations of the segregation just described occur if a eutectic temperature lies between T''' and T''''. Under this condition, a liquid is present at the eutectic temperature even though the phase diagram indicates only solid, e.g., in sterling silver (92.5Ag-7.5Cu in Fig. 17-7).

Example 18-6

A 70Pb-30Bi alloy was thoroughly homogenized as a melt. Microscopic analysis reveals both α and γ (as well as β) in the solid alloy at 20°C. Explain.

Answer. The α which separated between 196°C and 184°C did not have an opportunity to react at the peritectic temperature (Example 17-3). Further nonequilibrium cooling permitted the liquid to follow C_L of Fig. 17-8 to the eutectic, where γ formed.

Note. The phase rule does not apply, since equilibrium does not exist (α and γ should not coexist because they should react to give β). Nevertheless, the phase diagram permits us to interpret what has occurred within a material. ◀

Example 18-7

Seven hundred grams of a 90Cu-10Sn alloy are melted at 1050°C in a long vertical tube, which is then cooled so that all solidification occurs from the bottom upward.

a) Its temperature is dropped quickly to 1000°C and then held until equilibrium occurs (giving solid at the bottom and only liquid at the top).

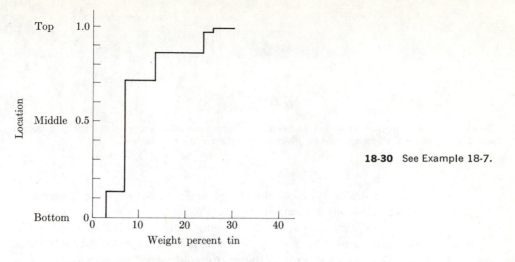

18-30   See Example 18-7.

b) As a second step, the temperature is dropped quickly to 900°C and held.

c) The temperature is dropped to 800°C.

d) The temperature is dropped to 775°C.

e) The temperature is dropped to 750°C.

After each step, additional solid forms from the bottom upward and the remaining liquid collects at the top. Calculate the weight percent tin versus vertical location.

Answer. See Fig. 18-30.

	gm	Percent of height	gm Sn
a) At 1000°C (700 gm): α is 97Cu-3Sn; fraction α = 0.125	88	13	2.6
b) At 900°C (612 gm): α is 93Cu-7Sn; fraction α = 0.667	408	58	28.5
c) At 800°C (204 gm): α is 86.5Cu-13.5Sn; fraction α = 0.522	106	15	14.3
d) At 775°C (98 gm): β = 76Cu-24Sn; fraction β = 0.75	74	11	18.2
e) At 750°C (24 gm): β = 74Cu-26Sn; fraction β = 0.6	14	2	3.5
f) Remaining liquid: 69.4Cu-30.6Sn	10	1	3.0 ◄
	700	100	70.1

Zone refining. The nonequilibrium solidification just described is used intentionally in the zone refining of semiconducting materials. Assume that we have some silicon which contains 0.5 w/o aluminum, and it is desired to reduce the aluminum level to 10^{-8} w/o. According to the Al-Si diagram at the end of Chapter 17, the distribution coefficient, k, for aluminum in silicon is about 0.01. Thus the first solid to separate on solidification will contain only about 0.005 w/o Al.

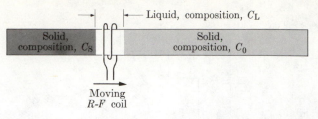

18-31 Zone refining. Each pass of the *R-F* induction coil purifies the material by a factor of *k*, the distribution coefficient. (In calculations, attention must be given to the length of the molten zone.)

As shown in Fig. 18-31, a rod of the initial material is melted by passing an induction heating coil along its length. Since the liquid can contain about 100 times as much Al as the freezing solid, the moving liquid zone carries most of the aluminum to the right end, where it can be discarded. If a narrow molten zone is maintained, a second pass will reduce the solute content by another factor of *k*. With multiple passes, it is possible to reduce the solute content to very low levels, e.g., 1 part in 10^{10}.

REFERENCES FOR FURTHER READING

Brophy, J. H., R. M. Rose, and J. Wulff, *The Structure and Properties of Materials: II. Thermodynamics of Structure.* New York: Wiley, 1964. Paperback. Phase changes and structural changes are discussed in Chapters 6 and 7. Introductory level.

Fine, M. E., *Phase Transformations in Condensed Systems.* New York: Macmillan, 1964. Paperback. Excellent reference for the advanced undergraduate or graduate student who wishes to supplement his study in this area. A thermodynamics background is assumed.

McMillan, P. W., *Glass-Ceramics.* London: Academic Press, 1964. The relatively new process of glass-devitrification is presented. Advanced undergraduate and graduate level.

Parker, E. R., F. R. Davis, and E. L. Langer, *Solid State Structure and Reactions.* Metals Park, O.: American Society for Metals, 1968. Reactions in solids are summarized in an introductory manner.

Smoluchowski, R. (Ed.), *Phase Transformation in Solids.* New York: Wiley, 1951. This is the publication of the symposium which did much to initiate materials science as a discipline. For the instructor and materials engineering major.

Wayman, C. M., *Introduction to the Crystallography of Martensitic Transformations.* New York: Macmillan, 1964. For the advanced undergraduate or graduate student who wishes to learn more about these important transformations.

PROBLEMS

18-1 Refer to Fig. 13-21. Calculate the O^{2-}-Ti^{4+}-O^{2-} angle which develops when $BaTiO_3$ becomes tetragonal below 120°C.

Answer. 173°

18-2 Beta brass orders from β to β_1. What new diffraction lines will appear (although admittedly weak)? [*Hint.* Refer to Table 5-4.]

18-3 The following data apply to the average times and temperatures for a certain solid-state reaction $(\alpha \rightarrow \beta)$ which is diffusion controlled. What temperature would accomplish the same reaction in 1 sec?

Transformation temperature	Time	Rate
380°C	10 sec	0.1 sec^{-1}
315°C	100 sec	0.01 sec^{-1}

Answer. 463°C

18-4 The first noticeable crystallization of glass is found in 12 hr at 2000°F and $2\frac{1}{2}$ days at 1500°F.

   a) Assuming diffusion control, how long would it take for similar crystallization at 1000°F?

   b) At 500°F?

18-5 The ferrite lamellae of pearlite may be as thick as 1 μm. As a 1080 steel transforms, how many carbon atoms must have diffused away from each μm^3 of iron that becomes ferrite?

Answer. $3 \times 10^9/\mu m^3$

18-6 Revise Eq. (18-3) for the situation where each cm^3 of transformation introduces E_ϵ ergs of strain energy.

18-7 (Use the data of Example 18-3 and assume a linear relationship between ΔF_V and temperature.) Plot the values of r_c versus T and ΔF_c versus T at (a) 1025°C, (b) 1000°C, (c) 900°C, and (d) 850°C.

Answer. a) $r_c = \infty$, $\Delta F_c = \infty$ b) $r_c = 24$ A, $\Delta F_c = 1.2 \times 10^{-10}$ erg   c) 4.8 A, 0.048×10^{-10} erg   d) 3.4 A, 0.024×10^{-10} erg

18-8 Revise Eq. (18-3) for the nucleation of a new phase β onto the interface between phase α and phase V, in which the contact angle, θ, equals 90°. Determine r_c and ΔF_c for this special case.

18-9 The surface energy of water is 70 ergs/cm^2, and that for a glass/gas surface is 580 ergs/cm^2. When a drop of water is on the glass, a contact angle of 55° is realized.

   a) What is the energy of the liquid/solid interface?

   b) Assume no change in $\gamma_{L/G}$ or $\gamma_{L/S}$. To what value would the gas/solid energy have to be changed to produce complete wetting?

   c) To what value would the gas/solid energy have to be changed to provide non-spreading (that is, $\theta > 90°$)?

   d) Assume no change in $\gamma_{S/G}$ or $\gamma_{L/G}$. To what value would the liquid/solid energy have to be changed to provide complete wetting?

Answer. a) $\gamma_{L/S} = 540$ ergs/cm^2 b) $\gamma_{S/G} = 610$ ergs/cm^2 c) $\gamma_{S/G} \leq 540$ ergs/cm^2
d) $\gamma_{L/S} = 510$ ergs/cm^2

18-10 Different wire samples of a 1045 steel received one of the following seven heat-treating sequences. Indicate the phases which exist *after* the completion of each sequence.

a) Heated to 825°C,* quenched to 250°C.
b) Heated to 750°C,* quenched to 550°C.
c) Heated to 700°C,* quenched to 250°C.
d) Heated to 825°C,* quenched to 550°C, held 10 sec.
e) Heated to 900°C,* quenched to 550°C, held 10 sec, quenched to 250°C.
f) Heated to 925°C,* quenched to 300°C, held.*
g) Heated to 250°C,* heated to 425°C, held 1 sec.

18-11 One hundred grams of an AISI 1045 steel are heated and quenched in the ways indicated below. Indicate the phase(s) and grams of each phase at the end of each sequence.

a) Heated to 825°C,* quenched to 120°C.
b) Heated to 750°C,* quenched to 120°C.
c) Heated to 825°C,* quenched to 550°C, held 7 sec, quenched to 250°C.
d) Heated to 825°C,* quenched to 550°C, held 10 sec, heated to 750°C.*

Answer. a) 100 gm M b) 65 gm M, 35 gm α c) 93.5 gm α, 6.5 gm carbide d) 65 gm γ, 35 gm α

18-12 A small piece of 1045 steel is heated to 850°C, quenched to 700°C, held at that temperature for 5 sec, and then quenched to 20°C. What phases are present? [*Hint.* Check the method for making isothermal-transformation diagrams.]

18-13 A small piece of 1080 steel is heated to 800°C, quenched to −60°C, reheated to 300°C, and held 10 sec. What phases are present at the end of this time?

Answer. Martensite (with some possible change to α + carbide)

18-14 A small wire of 1045 steel is subjected to the following treatments as *successive* steps:

1) heated to 875°C, held there for 1 hr;
2) quenched to 250°C, held there 2 sec;
3) quenched to 20°C, held there 100 sec;
4) reheated to 550°C, held there 1 hr;
5) quenched to 20°C and held.

Describe the phases or structures present *after each step* of this heat-treatment sequence.

18-15 a) Repeat Problem 18-14 with steps (1), (2), (5).
b) Repeat Problem 18-14 with steps (1), (3), (4), (5).
c) Repeat Problem 18-14 with steps (1), (2), (4), (5).

18-16 Sketch an isothermal-transformation diagram for a steel of 1.0 w/o C and 99.0 w/o Fe.

18-17 Sketch an isothermal-transformation diagram for a 1020 steel.

Answer. $M_s = 450$°C, $M_f = 290$°C; α and γ are stable between 723°C and 860°C after long times; at 550°C the curve is further to the left than in 1045 steel

---

*   An asterisk indicates that equilibrium was attained at this step of the heat treatment before proceeding to the other steps in the sequence.

18-18 Why are Figs. 18-17 and 18-22(a) not equilibrium diagrams?

18-19 A long thin bar of an 80Pb-20Bi alloy is to be enriched in lead by zone melting. Assuming maximum efficiency, what is the minimum number of melting passes required to reduce the bismuth content to less than 5 w/o at the initial end?

Answer. 4

18-20 Repeat Problem 18-19 for an 80Al-20Mg alloy.

19 □ multiphase microstructures

19-1 MICROSTRUCTURAL GEOMETRY

The properties of multiphase materials are even more dependent on microstructures than are the properties of single-phase materials (Section 6-8). After a brief summary of the origins of multiphase microstructures, this chapter presents the factors which alter these microstructures, and then analyzes the dependency of properties on multiphase microstructures.

Multiphase materials are *mixtures* of two or more phases. They may originate through *mechanical mixing* processes, such as those used to add fillers to plastics and rubbers (Fig. 19-1), or they may require solution and subsequent *phase transformations* in heat-treating processes (and, of course, there are combinations of these processes).

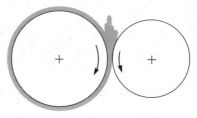

19-1 Rubber mill. Fillers are mixed into plastics and rubbers by rolls which have different surface speeds and provide a kneading action.

For simplicity, we shall direct most of our attention to two-phase microstructures, although various materials may possess several equilibrium phases and additional nonequilibrium phases. As is the case with crystal structures, microstructures are best analyzed through their geometry, which includes the amount, size, shape, orientation, and distribution of the individual phases and grains.

Amount. In a mechanical mixture, the amount of each phase is easily specified in terms of the raw-material additions, if there are no subsequent reactions. Initial compositions also determine the fraction of each phase in materials with transformation microstructures, provided equilibrium is attained. In practice, full equilibrium does not always prevail, and often is not desired. The actual amount of each phase depends on the processing procedure; for example, quenching may markedly alter the fraction of each phase from the equilibrium value.

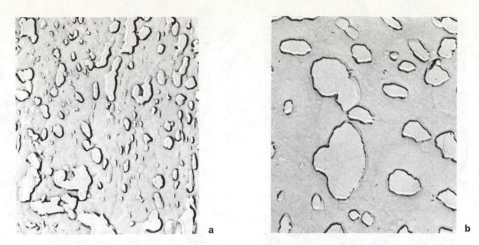

a

b

19-2 Phase size (carbide and ferrite in tempered martensite) (×11,000). The two microstructures are of the same composition. (a) 600°C, 1 hour; hardness, 33 R_C. (b) 675°C, 12 hours; hardness, 20 R_C. Note that the coarser microstructure is softer. (Courtesy of General Motors.)

Size. Since grinding is usually performed separately for each component of a mechanically mixed material, the corresponding particle size depends on the selected raw materials. The size of separate phases within a transformation microstructure is highly sensitive to the processing procedure. Even though the phases may be at chemical equilibrium, geometric changes continue to occur with additional heat treatment (Fig. 19-2). This will be discussed in Section 19-2.

Shape. Particle shapes vary from *equiaxed* to *flake-like* to *fibroid*. In the first, all three of the dimensions are approximately equivalent; in the second, one dimension is markedly smaller than the other two; while the third, one dimension is greatly elongated. Examples are shown in Fig. 19-3. A full choice among these three extremes is seldom available when one is specifying microstructures; however, there is often a choice between two of the three. For example, although gray cast iron has flake-like graphite (Fig. 19-3b), cast iron can also be made with the graphite in a nodular shape. Also, ferrite and carbide mixtures may take on many forms, depending on the heat treatment (Fig. 19-4). Fiber-shaped phases are becoming widely used as reinforcements (Fig. 19-3c). These two-phase structures are usually designed on the basis of a materials system of two distinct, but complementary, materials. They will be considered in Chapter 23.

Orientation. Fibers are often specifically oriented for reinforcement purposes. Likewise, eutectic microstructures may be made with alternate layers of the two eutectic phases, if directional cooling is employed (Fig. 19-5a). Preferential orientation may also be observed within individual grains as *Widmanstätten* structures, where a second phase has precipitated along specific crystallographic planes of the solvent or matrix phase (Fig. 19-5c). In such microstructures, there is a close geometric correspondence between the two lattices.

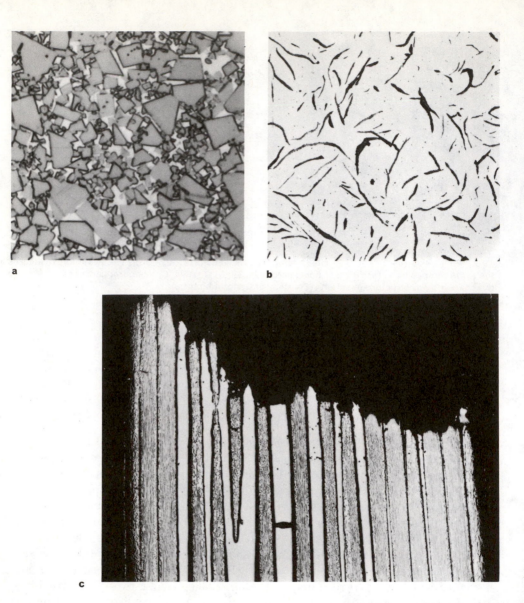

19-3 Phase shape. (a) Equiaxed microstructure (sintered carbides) (×1500). (Courtesy of Metallurgical Products Department, General Electric.) (b) Flake microstructure (graphite in cast iron) (×100). (Courtesy of J. E. Rehder. (c) Fibroid microstructure (tungsten-reinforced copper) (×100). (R. A. Signorelli, D. W. Petrasek, and J. W. Weeton, in *Modern Composite Materials*, L. J. Broutman and R. H. Krock, Eds. Reading, Mass.: Addison-Wesley, 1967.)

Distribution. Heterogeneities of two-phase distributions may arise from incomplete mixing or from segregation during solid-state reactions. This is well illustrated in Fig. 18-12(b). Phase distribution is also influenced by relative interfacial energies, as will be discussed in the following section.

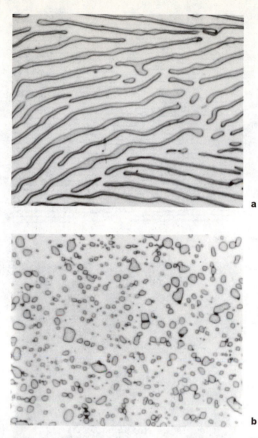

a

b

19-4 Phase shape (carbide and ferrite). (a) Pearlite microstructure (×2500). (b) Spheroidite microstructure (×1000). Each has a composition of 0.8C-99.2Fe. (Courtesy of J. R. Vilella, U.S. Steel Corp.)

19-2 GEOMETRIC EQUILIBRIUM

As presented in the previous two chapters, reactions move toward phase equilibrium in order to minimize the free energy of the system. The phase combinations with the lowest total free energy are the most stable.

Geometric equilibrium arises from a comparable minimization of interfacial energies. There is a "driving force" which directs the microstructure toward the lower-energy geometries. As in the case of phase equilibrium, the time requirement may be very long.

Particle agglomeration. In Fig. 19-2 we observed a two-phase analog to grain growth (Fig. 9-19). If we assume that the grains are spheres, we can calculate the phase-boundary energy per unit particle volume, Γ, as

$$\Gamma = \frac{4\pi r^2 \gamma}{4\pi r^3/3} = \frac{3\gamma}{r}, \tag{19-1}$$

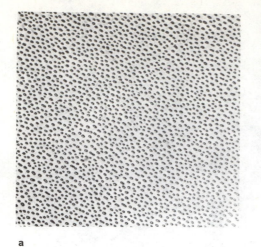

a

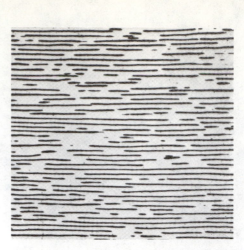

b

c

19-5 Phase orientation. In each microstructure the phases have specific crystallographic relationships. (a and b) Transverse and longitudinal eutectic microstructures of Al-Al$_3$N (X400). [R. W. Hertzberg, F. D. Lemkey, and J. A. Ford, "Mechanical Behavior of Lamellar (Al-CuAl$_2$) and Whisker Type (Al-Al$_3$Ni) Unidirectionally Solidified Eutectic Alloys," *Trans. AIME*, **233**, 1965, p. 342.] (c) Widmanstätten structure of UC$_2$ (bright) in UC (X450). (R. Chang and C. G. Rhodes, "Carbide Powders and Mechanisms of Sintering of Refractory Bodies," *Journ. Amer. Ceram. Soc.*, **45**, 1962, pp. 374–382.

where γ is the interfacial energy per unit boundary area. Thus fewer, but larger, particles possess less total surface energy.

The mechanism of particle agglomeration is more complex than that of grain growth (Fig. 9-18), because the atoms of the particle with a smaller radius go into solution, diffuse through the matrix phase, and finally precipitate onto a particle with a larger radius. This implies that there is a difference in the solubility as a function of particle radius. To show this, let us consider Fig. 8-13 again, where X_α is the normal solubility limit of B in α, the A-rich phase, and X_β is the normal solubility limit of A in β, the B-rich phase. Now consider Fig. 19-6(a), in which a composition, X_0, contains particles of β in a matrix of α. Adjacent to the smallest particles of β, the extra surface per unit volume increases the free energy from curve F_c to F_f, and in turn increases the solubility limit of B in α from X_{α_c} to X_{α_f} (Fig. 19-6b). Thus the finer particles will dissolve into the matrix; but in doing so, the matrix will exceed the solubility limit X_{α_c} adjacent to the coarser particles,

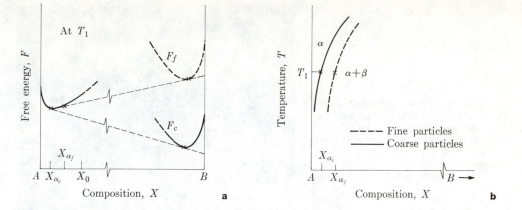

19-6 (a) Free energy versus grain size. The extra surface energy increases the free energy, F_f, of fine particles over the free energy, F_c, of coarse particles. (b) Solubility limits versus particle size. As shown in (a), the solubility limit at T_1 (or any other temperature) is higher for fine particles (X_{α_f}) than for coarse particles (X_{α_c}).

permitting them to grow. The process continually consumes the smaller particles, decreases the total number of particles, and increases the average particle size.

The rate of agglomeration is governed primarily by the rate of diffusion through the matrix. Therefore agglomeration occurs more rapidly at higher temperatures. The rate of agglomeration is also influenced by the distance between particles, the difference in particle sizes, and the interfacial energy.

Equilibrium shapes. Because a sphere has the minimum surface area per unit volume, this will be the equilibrium shape of a minor phase if there is no anisotropy of surface energy with crystal orientation. Thus the spheroidized carbide particles of Fig. 19-4(b) are geometrically more stable than the pearlitic lamellae of Fig. 19-4(a). The lamellar shape was the result of rapid directional growth (Fig. 18-11) rather than conditions of ultimate equilibrium.

Alteration of pearlite into the *spheroidite* microstructure of Fig. 19-4 requires 15 to 20 hr at a temperature just below the eutectoid temperature. Prolonged heating coarsens the carbide particles. The ultimate equilibrium shape for these phases would be a single carbide sphere within a single ferrite crystal; however, the time required to develop this shape would be nearly infinite.

Any variation of the phase-boundary energy with orientation tends to make an included phase adopt a nonspherical shape. It is known that such anisotropies exist; however, the effect is usually small enough so that the nonsphericity is minor, and masked by other factors.

Dihedral angles. We observed in Example 6-6 and Fig. 6-18 that a thermal groove can form a dihedral angle where a grain boundary meets an external surface. In that particular case, the surface between the grain and the vapor had an energy of nearly twice the grain-boundary energy. A similar dihedral angle can develop

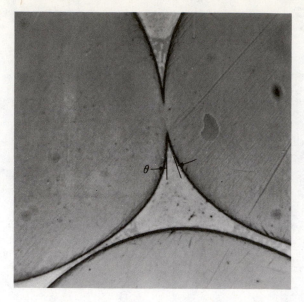

19-7 Dihedral angles (85W-10Ni-5Fe) (×800). At 1460°C the Fe-Ni liquid penetrated the boundary between two tungsten grains with a 15° angle. (R. H. Krock, "Inorganic Particulate Composites," *Modern Composite Materials*, L. J. Broutman and R. H. Krock, Eds. Reading, Mass.: Addison-Wesley, 1967, pp. 455–478.)

within a microstructure when a second phase is present at the grain boundary (Fig. 19-7). Since the energy of a phase boundary is often less than the energy of the grain boundary, the dihedral angle is less than 120°. Although this does not necessarily indicate a minimum boundary area, it does produce minimum boundary energy because higher-energy interfaces are replaced by lower-energy interfaces.

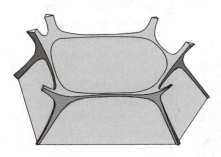

19-8 Phase distribution versus dihedral angle. If the phase-boundary energy is less than half the grain-boundary energy, the grains are separated by a second phase. If the dihedral angle of one phase along the grain boundary of the other phase is less than 60°, the first phase penetrates along grain edges of the second. (After sketches by C. S. Barrett, The University of Chicago.)

If the dihedral angle is less than 60°, the second phase, β, has an equilibrium distribution along the edges of three adjacent grains of phase α (Fig. 19-8). If

$$\gamma_{\alpha/\beta} < 0.5\gamma_{\alpha/\alpha}, \tag{19-2}$$

the minor phase distributes itself as a film between adjacent grains, because the dihedral angle drops to 0°. This distribution has a marked effect on certain properties such as ductility and toughness, particularly if the grain-boundary phase is brittle so that crack propagation can follow a continuous path.

Example 19-1

What is the maximum ratio between the energy of a phase boundary and the energy of a grain boundary if the minor phase is to extend along grain edges?

Answer. As shown in Fig. 19-8, the maximum dihedral angle, θ, is 60°. From Eq. (6-10),

$$\gamma_{\alpha/\alpha} = 2\gamma_{\alpha/\beta} \cos (\theta/2), \qquad (19\text{-}3)$$

or

$$\frac{\gamma_{\alpha/\beta}}{\gamma_{\alpha/\alpha}} = 0.577. \blacktriangleleft$$

19-3 PROPERTIES DEPENDENT ON PHASE AMOUNTS

Two phases never have a completely identical set of properties, because, as we have seen, each is structurally different. This generalization also applies to the properties of multiphase microstructures. A few properties depend only on the amount of each phase and are insensitive to the microstructural geometry. Such properties are exemplified by the two steels shown in Fig. 19-4; although microstructurally different, they have the same density because they have the same weight fraction of ferrite and carbide. Most properties, however, *are* sensitive to the micro-structural geometry. We shall find, for example, that pearlite is harder and stronger, but not so ductile and tough, as spheroidite.

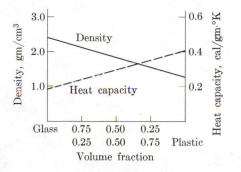

19-9 Additive properties (glass-reinforced plastic). Equation (19-4) applies.

Microstructure-insensitive properties may be determined by suitable weighted averages of the properties of each of the individual phases. *Density* is an additive property which illustrates the statement:

$$\rho_{\text{material}} = f_1\rho_1 + f_2\rho_2 + f_3\rho_3 + \cdots \qquad (19\text{-}4)$$

With only two phases present, the density varies linearly with the *volume fraction, f*, which is present (Fig. 19-9). When there is pore space, the $f\rho$ product is zero for the pore fraction.

The *heat capacity* per unit volume of a multiphase microstructure has a similar additive relationship. The averages of both properties, density and heat capacity, depend on the volume fraction, in that the contribution of one phase does not affect the contribution of the adjacent phase.

Example 19-2

Fifty w/o SiO_2 flour (i.e., very fine quartz powder) is added to a phenol-formaldehyde resin as a filler. What is the density?

Answer. From Appendix C,

$$\rho_{SiO_2} = 2.65 \text{ gm/cm}^3, \qquad \rho_{pf} = 1.3 \text{ gm/cm}^3.$$

Basis: 100 gm.

$$50 \text{ gm } SiO_2 = 18.9 \text{ cm}^3 \text{ } SiO_2 \qquad f_{SiO_2} = 0.33$$
$$50 \text{ gm pf } = \underline{38.4} \text{ cm}^3 \text{ pf} \qquad f_{pf} \quad = \underline{0.67}$$
$$\qquad\qquad\qquad 57.2 \qquad\qquad\qquad\qquad 1.00$$

By Eq. (19-4),

$$\rho_m = (0.33)(2.65) + (0.67)(1.3) = 1.75 \text{ gm/cm}^3. \quad \blacktriangleleft$$

Example 19-3

A cemented-carbide tool is made by impregnating compacted titanium carbide powders with liquid nickel. The carbide compact weighs 3.7 gm/cm^3 before impregnation. What are (a) the weight fraction of nickel present in the final cemented carbide and (b) the density of the cemented carbide tool?

Answer. From Appendixes B and C,

$$\rho_{Ni} = 8.90 \text{ gm/cm}^3, \qquad \rho_{TiC} = 4.5 \text{ gm/cm}^3.$$
$$3.7 \text{ gm/cm}^3 = (1 - x)(4.5 \text{ gm/cm}^3) + (x)(0 \text{ gm/cm}^3),$$
$$x = 0.18 \text{ volume fraction pores (to be filled with nickel).}$$

Basis: 1 cm^3.

a)
$$\text{Weight fraction} = \frac{(0.18)(8.9)}{(0.18)(8.9) + (0.82)(4.5)} = 0.30.$$

b)
$$\rho_m = (0.18)(8.9 \text{ gm/cm}^3) + (0.82)(4.5 \text{ gm/cm}^3) = 5.3 \text{ gm/cm}^3. \quad \blacktriangleleft$$

19-4 MICROSTRUCTURE-SENSITIVE PROPERTIES

Properties which respond to a gradient are sensitive to the geometry of micro-structures. In the simplest case, this is a choice of paths for energy transport. In other cases, the behavior of each phase depends on the characteristics of adjacent grains and phases.

Model microstructures. The thermal and electrical *conductivities* of multiphase microstructures follow *mixture rules*. However, the weighting procedures which are necessary vary with the shape and distribution of the phases. Three simplified

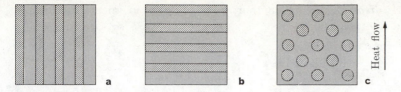

19-10 Conductivity versus phase distribution (model microstructure). (a) Parallel phases (Eq. 19-5). (b) Series phases (Eq. 19-6). (c) Matrix with dispersed phase (Eq. 19-7).

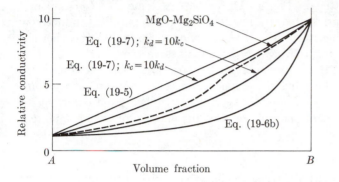

19-11 Conductivity versus volume fractions. (Data for $MgO\text{-}Mg_2SiO_4$ adapted from W. D. Kingery, *Introduction to Ceramics.* New York: Wiley, 1960, p. 502.)

examples are shown in Fig. 19-10. Although they apply equally to either thermal or electrical conductivity, notations of thermal conductivity will be incorporated into the following equations. *Parallel conduction* applies in the first case; therefore,

$$k_m = f_1 k_1 + f_2 k_2 + \cdots , \tag{19-5}$$

and the average conductivity for the material, k_m, is a summation of the volume (or cross-sectional area) contribution of all phases. *Series conduction* applies in Fig. 19-10(b); therefore,

$$\frac{1}{k_m} = \frac{f_A}{k_A} + \frac{f_B}{k_B} + \cdots , \tag{19-6a}*$$

or, for a two-phase microstructure,

$$k_m = \frac{k_A k_B}{f_A k_B + f_B k_A} . \tag{19-6b}$$

In this case the thermal conductivity of the microstructure is smaller than that obtained from a linear interpolation of the volume fractions (Fig. 19-11).

---

* It is convenient with electrical transport to use resistivities (or resistances) with series geometries, so that a reciprocal equation becomes unnecessary.

If one phase is *dispersed* within another, as is common in microstructures, a more complicated interpolation results:

$$k_m = k_c \left[\frac{1 + 2f_d \left[\dfrac{1 - k_c/k_d}{2k_c/k_d + 1} \right]}{1 - f_d \left[\dfrac{1 - k_c/k_d}{2k_c/k_d + 1} \right]} \right]. \tag{19-7}$$

Here, the subscript c refers to the continuous phase and d to the dispersed phase. The above equation is plotted in Fig. 19-11 for two extremes, $k_c/k_d = 10$ and $k_c/k_d = 0.1$. As might be anticipated, the results fall between parallel conductivity (Eq. 19-5) and series conductivity (Eq. 19-6). Kingery* studied the thermal conductivity in periclase-forsterite (MgO-Mg_2SiO_4) mixtures and found substantial agreement with Eq. (19-7) at the two extremes, where one phase may be considered to be the matrix phase and the other the dispersed phase. At intermediate compositions, however, the combined conductivity values switched from one curve of Fig. 19-11 to another, where neither phase was the matrix or was completely dispersed.

Example 19-4

Refer to Example 19-2. Estimate the thermal conductivity of the 50-50 mixture (weight fraction) of silica flour and phenol-formaldehyde resin. (In this material, the silica is the dispersed phase.)

Answer. Conductivity data from Appendix C, volume fraction data from Example 19-2, and Eq. (19-7) give

$$k_m = 0.0004 \left[\frac{1 + 2(0.33) \left[\dfrac{1 - 0.0004/0.03}{2(0.0004/0.03) + 1} \right]}{1 - (0.33) \left[\dfrac{1 - 0.0004/0.03}{2(0.0004/0.03) + 1} \right]} \right]$$

$$= 0.0009 \text{ cal} \cdot \text{cm/°C} \cdot \text{cm}^2 \cdot \text{sec}.$$

Alternative calculation: Since $k_d \gg k_c$,

$$k_m \cong k_c \frac{1 + 2f_d}{1 - f_d} \tag{19-8}$$

$$\cong (0.0004) \frac{1 + 2(0.33)}{1 - 0.33}$$

$$\cong 0.001 \text{ cal} \cdot \text{cm/°C} \cdot \text{cm}^2 \cdot \text{sec}. \blacktriangleleft$$

Young's modulus is also sensitive to microstructure, and can be analyzed on the basis of Fig. 19-10. As shown in the Example 19-5, Eq. (19-6) can be adapted for phases in series if the conductivity is replaced by the modulus, Y. Likewise, Eq. (19-5) serves for an idealized parallel microstructure.†

---

* See the reference for Fig. 19-11.

† Both equations will require a modification if the Poisson ratios of the phases differ appreciably.

Example 19-5

"Explosion-proof" glass often consists of multiple layers of glass and plastic. Which direction, (a) longitudinal (Fig. 19-10a) or (b) transverse (Fig. 19-10b), will have the higher Young's modulus? [The subscripts refer to (1) glass and (2) plastic.]

Answer

a) Longitudinal ($\parallel$) loading: Since the two layers deform together,

$$\epsilon_1 = \epsilon_\parallel = \epsilon_2,$$

or

$$\frac{F_1/f_1 A}{Y_1} = \frac{F/A}{Y_\parallel} = \frac{F_2/f_2 A}{Y_2},$$

where the stress in each type of lamella can be expressed as a load, F, on the appropriate fractional cross-sectional area, fA.

$$F_1 + F_2 = F = \frac{Ff_1 Y_1}{Y_\parallel} + \frac{Ff_2 Y_2}{Y_\parallel}.$$

Thus

$$Y_\parallel = f_1 Y_1 + f_2 Y_2. \tag{19-9}$$

b) Transverse ($\perp$) loading, where $\sigma_1 = \sigma_\perp = \sigma_2$:

$$Y_\perp = \sigma/\epsilon_\perp$$

$$= \frac{\sigma}{\epsilon_1 f_1 + \epsilon_2 f_2} = \frac{1}{f_1/Y_1 + f_2/Y_2},$$

$$\frac{1}{Y_\perp} = \frac{f_1}{Y_1} + \frac{f_2}{Y_2}. \tag{19-10}$$

For any combination of $f_1 + f_2 = 1.0$, $Y_\parallel > Y_\perp$. [Cf. Eqs. (19-5) and (19-6b) in Fig. 19-11.] ◄

Complex microstructures. Equations (19-9) and (19-10) apply to a perfect lamellar structure with compressive (or tensile) stresses either parallel to or across the phase boundaries. Not uncommonly, loads must be transferred from one phase to another by shear. This is illustrated in Fig. 19-12, in which photoelastic techniques were employed to reveal the stress concentrations. The stress within the fiber depends on the distribution of strain of the matrix phase, and vice versa. The overall property depends on the physical interaction of the individual phases. Since these require rather complex mathematical analysis even when the geometries can be characterized, we shall not deal with the quantitative relationships. Qualitative relationships, however, are very useful and are the subject of the next few paragraphs. We shall use ferrite-carbide microstructures as a prototype.

In steels, carbide is harder than ferrite (Section 17-5), and increases the resistance of the steel to deformation. Figure 19-13 shows graphically the hardness values of carbon steels which have been annealed (i.e., cooled slowly from austenitic temperatures to ensure coarse pearlite). The microstructures are given in Fig. 17-16. A 0.40% carbon steel contains approximately 50% pearlite (6% carbide); a 0.80%

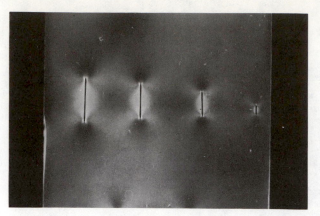

19-12 Stress concentration (boron fibers in plastic) (X27). The vertical tensile load must be transferred from the fibers to the plastic by shear. This is revealed by polarized light. (Courtesy of D. M. Schuster and E. Scala.)

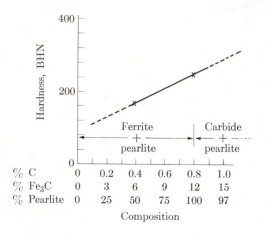

19-13 Hardness versus carbon content of annealed steels. The steels contain mixtures of coarse pearlite and ferrite. The hardness depends on the amount of carbide.

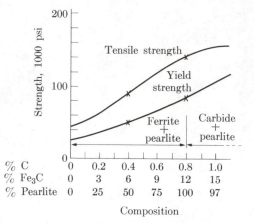

19-14 Strength versus carbon content of annealed steels. (Cf. Fig. 17-16.)

carbon steel is all pearlite (12% carbide). Slip occurs more readily in the former than in the latter.

In addition to being harder, steels with higher carbon contents are stronger. Both the *yield strength* and the *tensile strength* are shown in Fig. 19-14. Note, however, that the curves in this figure cannot be extrapolated to very high values at 100% iron carbide. Because iron carbide is very brittle by itself and fractures before yielding, instead of minimizing stress concentrations by plastic flow, an extension of Fig. 19-14 to 100% iron carbide would show the curves dropping to approximately 5000 psi.

Figure 19-15 illustrates the effect of iron carbide on *ductility*. As expected, the ductility decreases with increased carbon content. Therefore, for automobile

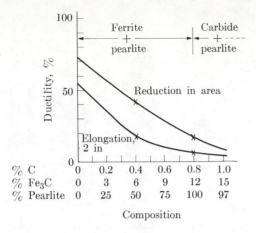

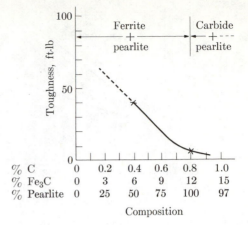

19-15 Ductility versus carbon content of annealed steels. (Cf. Fig. 17-16.)

19-16 Toughness versus carbon content of annealed steels. (Cf. Fig. 17-16.)

fenders the manufacturer chooses a steel which is very low in carbon and thus can be rolled into a thin sheet and deep-drawn to take the sharp curvatures required for styling. By contrast, the dies and shears used in forming and cutting these fenders are made of higher-carbon steels so that they can be made hard and strong.

The *toughness* of a steel is also important if the engineer is designing or using equipment which is subject to impact, since a brittle (the opposite of tough) steel will break under relatively light blows. Fractures are more readily propagated in high-carbon steels because plastic yielding is resisted in these steels; there may also be more carbide paths through which cracks may propagate (Fig. 19-16).

Effects of phase size on mechanical properties. The coarseness of the microstructure of a material directly affects its mechanical properties. The addition of very fine sand to asphalt produces a more viscous mixture than does the addition of an equal weight (or volume) fraction of gravel. Similarly, steel with a very fine microstructure of carbide dispersed in ferrite is harder and stronger than steel of the same carbon content but with a coarser microstructure. A comparison is made in Fig. 19-17. The lower set of data describes a steel which was cooled slowly to produce a coarse pearlitic structure; the upper set is for a steel that was cooled more rapidly, to form a much finer pearlite.

The same variation in properties is observable in martensitic steels tempered at different temperatures. Initial tempering produces a very fine precipitate of carbide particles in a ferritic matrix (Fig. 19-2a). Figure 19-2(b) shows a sample tempered at a higher temperature, which permits the carbide particles to grow. The hardness values in the legend indicate the degree of softening. When the carbide particles are agglomerated, there are larger regions of the soft ferritic matrix in which slip can occur without restriction.

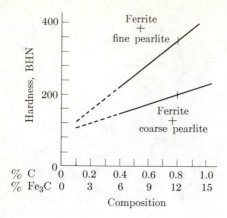

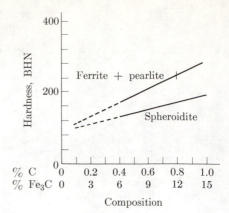

19-17 Effect of particle size on hardness of steel. The finer, harder pearlite was formed by faster cooling.

19-18 Effect of particle shape on hardness. The lamellar pearlite (Fig. 19-4a) provides more resistance to slip than the spheroidite (Fig. 19-4b).

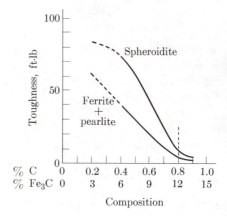

19-19 Effect of particle shape on toughness. In spheroidite, cracks cannot travel very far in the brittle carbide without encountering tougher ferrite. (Cf. Fig. 19-4a and b.)

Effects of phase shape and distribution on mechanical properties. The shape and distribution of microstructural phases greatly affect the properties of a material. For example, in pearlite the carbide is lamellar. However, if the pearlitic steel is held just under the eutectoid temperature for a long period of time, the carbide *spheroidizes* and develops the structure shown in Fig. 19-2(b). The effect of spheroidization on strength is shown in Fig. 19-18; the strength decreases because the spheroidized structure offers less resistance to the movement of dislocations. The shape of the phases also markedly influences the toughness of the mixture (Fig. 19-19). A spherodized structure is tougher because cracks in the carbide cannot propagate very far before entering the more ductile ferrite.

Porosity. Pores play a significant role in the mechanical behavior of a material because they can provide stress concentrations which far exceed the shear or fracture strength of the structure. This will be discussed in connection with failure

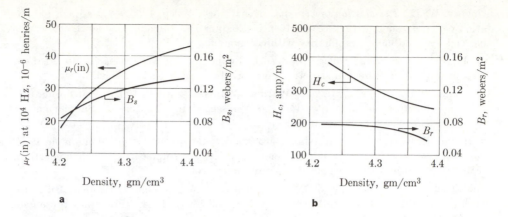

19-20 Magnetic properties versus fired density of magnesioferrite ceramics. (a) Initial relative permeability, μ_r(in), and saturation magnetization, B_s. (b) Coercive force, H_c, and remanent magnetization, B_r. The latter properties decrease because there are fewer pore walls at higher densities to prevent domain-wall movements. (After G. Economos, "Magnetic Ceramics: III. Effect of Fabrication Techniques," *Journ. Amer. Ceram. Soc.*, **38**, 1955, pp. 335-340.)

of a material in Chapter 21. It will be sufficient here to note that none of the linear mixture rules of Eqs. (19-5) through (19-10) are applicable.

Magnetic properties are also affected by porosity, as shown in Fig. 19-20 for a ferrospinel. The initial relative magnetic permeability, μ_r(in), and the saturation induction, B_s, increase with a decrease in porosity (increase in density), because there is then greater magnetic interaction between adjacent grains. In contrast, the remanent magnetization, B_r, and the coercive force, H_c, decrease with increasing density or sintering, because the elimination of pore walls permits domain-wall movements that result in demagnetization.

Space-charge polarization. Microstructures containing electrically conductive phases embedded in insulating phases present a special case of polarization called *space charge*. If small semiconducting particles lie within a dielectric matrix, an applied electric field will displace the electrons in the positive direction (Fig. 13-2d). In its idealized form, this situation may be viewed as a capacitor series in which the spacing is small, giving a large total capacitance and a high apparent dielectric constant. The net effect depends on particle size as well as volume fraction. A very fine dispersion of a semiconducting phase within a glass can produce a dielectric constant well in excess of 1000 simply because of the small particle-to-particle distances and space-charge polarization.

REFERENCES FOR FURTHER READING

Brandon, D. G., *Modern Techniques in Metallography*. Princeton, N. J.: Van Nostrand, 1966. A description of techniques to observe microstructures. Electron microscopy, electron-probe microanalysis, and field-emission microscopy are included. Undergraduate level.

Kingery, W. D., *Introduction to Ceramics*. New York: Wiley, 1960. A thorough presentation of ceramic microstructures is given in Chapter 13. Introductory level. However, the reader should have a familiarity with ceramic phases.

Moore, W. J., *Seven Solid States*. New York: Benjamin, 1967. Paperback. Steel is used as an example of the effect of structure on properties. Introductory level.

Richman, M. H., *Science of Metals*. Waltham, Mass.: Blaisdell, 1967. Transformations and microstructures of binary systems are presented in Chapter 8. Supplementary reading for this chapter. Sophomore level.

Smallman, R. E., and K. H. G. Ashbee, *Modern Metallography*. Oxford: Pergamon, 1966. Paperback. Theory and techniques of metallography are emphasized. Optical, electron, and x-ray procedures are covered. Undergraduate level.

Van Vlack, L. H., "Geometry of Microstructures," *Microstructure of Ceramic Materials*. Washington: U.S. Government Printing Office, 1964, pp. 1–14. Principles presented. Undergraduate level.

Van Vlack, L. H., *Physical Ceramics for Engineers*. Reading, Mass.: Addison-Wesley, 1964. Microstructures of ceramic materials are discussed in Chapter 7. Undergraduate level.

PROBLEMS

19-1 An aluminum alloy (96Al-4Cu) is quenched after solution treatment, then reheated to 212°F until equilibrium is obtained; at equilibrium it contains a θ precipitate (approximately spherical particles) in a κ matrix.

a) If the representative diameter of the θ, or $CuAl_2$, particles is 0.1 μm, how many particles will there be per cm^3? If d $=$ 0.5 μm?

b) What is the average distance between the 0.1-μm particles? The 0.5-μm particles? (ρ_θ = 4.4 gm/cm^3)

Answer. a) 9.0×10^{13}/cm^3, 0.7×10^{12}/cm^3 b) 2200 A, 11,000 A

19-2 The density of 1080 steel is 7.84 gm/cm^3. Estimate the density of Fe_3C. (The density of ferrite is 7.87 gm/cm^3.)

19-3 Calculate the density (lb/in^3) of a glass-reinforced phenol-formaldehyde plastic in which the glass content is 15 w/o. (A borosilicate glass is used for glass fibers.)

Answer. 0.05 lb/in^3

19-4 A gray cast iron contains 96Fe-4C, in which almost all the carbon is present as graphite. What is the density of the cast iron?

19-5 On the basis of Section 6-2, estimate the heat capacity of sterling silver (92.5Ag-7.5Cu) which has been equilibrated at 400°C.

Answer. 0.064 cal/gm · °K

19-6 Repeat Problem 19-5 for a 60Pb-40Sn solder equilibrated at 20°C.

19-7 The energy of a solid-liquid phase boundary is 0.7 of the solid-solid grain boundary. What dihedral angle will result on equilibration?

Answer. 88°

19-8 How much energy is released per cm^3 as dispersed particles grow from 0.1 μm to 1.0 μm (with no change in total volume) if the phase-boundary energy is 250 ergs/cm^2?

19-9 A cube (1 in. along each edge) is made by laminating alternate sheets of aluminum and vulcanized rubber (0.02 and 0.03 cm thick, respectively). What is the thermal conductivity of the laminate (a) parallel to and (b) perpendicular to the sheets? (Use the data of Appendix C.)

Answer. a) 0.21 cal · cm/°C · cm^2 · sec   b) 0.0005 cal · cm/°C · cm^2 · sec

19-10 What are the elastic moduli in the two directions for the laminate of Problem 19-9?

19-11 Estimate the thermal conductivity of the reinforced plastic of Problem 19-3. (Assume that the glass is randomly dispersed.)

Answer. 0.0005 cal · cm/°C · cm^2 · sec

19-12 Assume that the fibers of the previous problem were all in parallel orientation. Derive a mixture rule for conductivity in the parallel direction.

19-13 What is the longitudinal elastic modulus in a rod of borosilicate-reinforced polystyrene in which all the glass fibers are oriented lengthwise? There is 80 w/o glass.

Answer. 6,500,000 psi

part VI □ materials utilization

20 □ strengthening processes

20-1 STRENGTHENING OF MATERIALS

The final four chapters of this text deal with the processing and utilization of materials. The engineer's interest in strengthening processes is readily appreciated. As the result of strengthening processes, (1) engineering structures can be made stronger, (2) engineering products can be manufactured with less material and therefore less weight and cost, and (3) greater reliability and service life can be built into an engineering system.

Useful strength may be limited by either plastic yielding or fracture. Ideally, we would like to increase the resistance to both. Although the two are not fully compatible because greater strength commonly means less ductility and toughness (Fig. 11-15b versus Fig. 11-15c, and Fig. 19-14 versus Fig. 19-16), the engineer must optimize the combination of these properties to meet the service requirements. In this chapter, specific attention will be given to processes which increase the resistance of a material to plastic deformation. The effect of these processes on brittleness will also be cited; however, discussion of the failure mechanisms of brittle fracture, fatigue, and creep will be deferred to Chapter 21.

Since plastic deformation commonly proceeds through dislocation movements, one may strengthen materials by making them dislocation-free, or by making dislocation movements more difficult. Dislocation-free materials, such as oxide or metal whiskers (Fig. 11-7), are strong because there is no possibility of slip except by an all-at-once, total plane movement (Fig. 11-3). It is generally impractical, however, to strengthen materials by making them dislocation-free, because this would preclude all common shaping and fabrication processes.* Therefore, the usual processes for strengthening materials include the development of structures which retard dislocation movement. These are (1) solution treatments, (2) mechanical deformation processes, (3) precipitation processes, and (4) a variety of transformations in the solid state. Each gives a structure which interferes with dislocation movements.

---

* Whiskers can be added to composites as a reinforcing phase (Chapter 23).

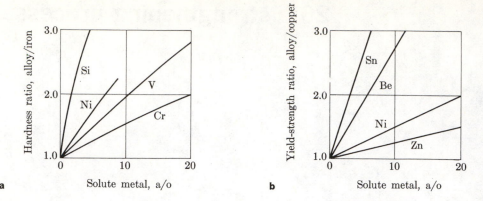

20-1 Solution hardening. (a) Iron. (b) Copper. The hardness and strength ratios for the alloy versus the unalloyed metal depend on the size mismatch of solute versus solvent atoms and the stress field which results. (Adapted from data by E. C. Bain, *Alloying Elements in Steel*, American Society for Metals, 1939; and R. S. French and W. R. Hibbard, "Tensile Deformation of Copper," *Trans. A.I.M.E.*, **188**, 1950, pp. 53–58.)

20-2 SOLUTION TREATMENTS

Solution hardening (Section 11-7) generally requires that the desired compositions be established initially. Most frequently, this is done by a melting operation, although diffusion processes are conceivable. A solution treatment is then introduced (1) to dissolve precipitated phases or (2) to homogenize a single phase. The first procedure can be illustrated by the dissolving of β to give a single α-phase in sterling silver, 92.5Ag-7.5Cu (Fig. 17-7); the second, by the removal of the solidification segregation illustrated in Fig. 18-29 for 70Cu-30Zn brass.

Figure 20-1 shows the effect of solute concentration on the yield strength of ferrite and of copper. The strengthening depends on the atomic percent of solute present. The effectiveness of each atom on solution hardening is a function of the size mismatch, whether larger *or* smaller. Thus the hardening effects of tin ($r = 1.51$ A) and beryllium ($r = 1.1$ A) are significantly greater in copper ($r = 1.28$ A) than are the hardening effects of nickel ($r = 1.24$ A) and zinc ($r = 1.33$ A).*

The figures just cited reveal that the atomic size difference is not the only factor in solution hardening. If it were, zinc would be more effective than nickel in hardening copper because the size mismatches are, respectively, 4% and 2.5%. The second variable affecting solution hardening may be viewed as the deformability of the solute atoms and therefore the resulting distortion it produces. Zinc atoms can be thought of as being elastically softer than nickel (as evidenced by their average shear moduli, 5×10^6 psi and 12×10^6 psi, respectively, for the pure metals). Thus it requires less force to move a dislocation past a zinc atom within copper than past a nickel atom within copper so that slip may proceed.

---

* This mismatch also introduces a limit to solubility; thus, while tin is an effective hardening agent in copper, only a few percent may be contained in solid solution (Fig. 17-27).

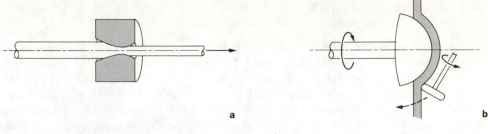

20-2 Mechanical processing. (a) Wire drawing. (b) Spinning of sheet. Strain hardening results when these processes are performed below the recrystallization temperature.

20-3 MECHANICAL DEFORMATION PROCESSES

Strain hardening by cold work is one of the more important means by which engineering materials are strengthened. This can be observed in stress-strain curves (Fig. 1-3), where the true stress for deformation increases logarithmically with strain. Cold work strengthens ductile materials by introducing dislocation networks which interfere with their continued movements, either by pileups or as anchor points (Section 11-5).

Cold-working processes include the *rolling* of sheet and foil between rolls, the *drawing* of rod or wire through circular dies (Fig. 20-2a), the *spinning* of sheet into concave shapes (Fig. 20-2b), *cold forging*, and *extrusion*, as well as a number of other processes. If the deformation is quite nonuniform, as in *stamping* and *deep-drawing* operations for forming automobile parts, the strain hardening will also be nonuniform. In such cases the useful strength may be limited to that induced during the initial cold-rolling of the sheet.

The loss of ductility during strain hardening increases the danger of fracturing during deformation, particularly when the plastic strain is nonuniform (Fig. 20-3). This fact must be considered by the design engineer when he sets his specifications. It also receives considerable attention by the metallurgist and materials engineer,

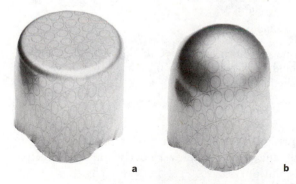

20-3 Drawability (sheet steel). (a) Flat-nose punch. (b) Round-nose punch. (Both have a 2-in. diameter.) The 0.5-in. grid and the 0.2-in. circles show the variation of deformation with position and punch shape. (Courtesy of Bethlehem Steel.)

who can exercise some control over ductility by modifying composition, inclusion content, and the size, shape, and orientation of the grains.

Mechanically induced transformations. Recall from Section 18-2 that a martensitic transformation has a shear origin. Therefore, a martensitic transformation can be induced by mechanical deformation. For example, both nickel and manganese in steel expand the austenitic stability range (Fig. 17-20b) and retard the $\gamma \rightarrow \alpha +$ carbide transformation, making it possible to retain the austenite at room temperature. Such steels can have their strength significantly increased as they are deformed. Practical considerations, however, must take two factors into account. First, the alloy content makes the initial austenite strong from solution hardening. Second, the martensite which forms may itself be exceptionally hard and brittle. As a result, the metal can become difficult to process except as sheet or wire.

Surface hardening of larger sections is possible; for example, consider a crossover for a railroad track. In this application an austenitic 14 w/o manganese steel can be used. The surface which is subjected to the wear and impact of the passing trains changes to a form of martensite by mechanical deformation, thus increasing the hardness in just the region where *wear* is significant. At the same time, the bulk of the material remains in the tougher austenitic form.*

Example 20-1

A copper rod is required to have a diameter of 0.21 in., a tensile strength of more than 40,000 psi, and an elongation of more than 20%. The rod is to be drawn (Fig. 20-2a) from a larger 0.35-in. rod. Specify the final processing steps for making the 0.21-in. rod.

Answer. From Fig. 11-15,

$$CW > 14\% \text{ for TS}, \qquad CW < 24\% \text{ for El.}$$

Use 20% cold work as the last drawing step. By Eq. (11-6),

$$0.20 = \frac{d^2\pi/4 - (0.21)^2\pi/4}{d^2\pi/4},$$

$$d = 0.235 \text{ in.}$$

Cold-draw from 0.35 in. to 0.235 in., and then anneal. Cold-draw 20% to 0.21 in. diameter. (Alternatively, the rod could be hot-worked to 0.235 in. In this case the annealing step is not required before the final cold-working.) ◄

20-4 PRECIPITATION PROCESSES

Age hardening. The first of two precipitation processes that we shall consider is known as age hardening. The hardening occurs in the very early stages of precipitation from a supersaturated solid solution.

---

* This is a relatively expensive steel, (1) because of the alloy requirement and (2) because of its difficult processing.

One of the best known age-hardening materials is 95.5Al-4.5Cu. We shall use it as a prototype, recognizing that age hardening (also called *precipitation hardening*) is found in a wide variety of alloys which exhibit decreasing solubility with decreasing temperature. Furthermore, age hardening is not limited to alloys, but has been found in certain ceramics which have the required solubility characteristics, e.g., $MgO-Fe_2O_3$ of Fig. 18-12(a).

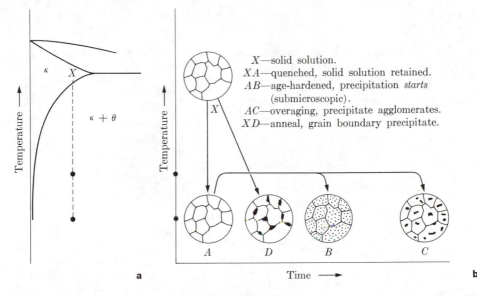

20-4 Age-hardening process (95.5Al–4.5Cu). Maximum hardness occurs when the clusters of $CuAl_2$ which constitute the pre-precipitate are still submicroscopic.

The age-hardening process requires a solution treatment first, followed by a quench to supersaturate the solid solution (*XA* of Fig. 20-4). Usually the quenching medium is at room temperature where the isothermal precipitation time is very long, corresponding to the extreme lower end of the *C*-curve in Fig. 18-16(b). Precipitation is then initiated by reheating to a temperature where the reaction can proceed at a convenient rate for close control. We shall observe in a moment that if the rate is too fast, the material may overage and soften. Of course, if the rate is excessively slow, the process becomes impractical.

The *start* of precipitation has been emphasized above because it is often the first *clustering*, or pre-precipitate, that provides much of the hardening and strengthening. The quenched, supersaturated solid solution (*A* of Fig. 20-4) may be represented by Fig. 20-5(a), in which the solute atoms are randomly distributed. As aging proceeds (*AB* of Fig. 20-4), the solute atoms cluster, but still maintain a matching, or lattice coherency, with the surrounding crystal. This is indicated schematically in Fig. 20-5(b) by showing the continuity of crystal planes from

● Solute atom   o Solvent atom

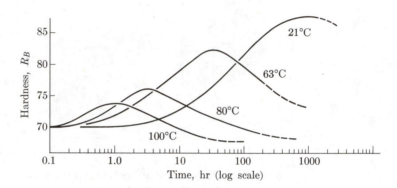

a

b

c

20-5 Age-hardening mechanism. (a) κ solid solution. (b) Age-hardened; the θ precipitation has been initiated. Since the two structures are coherent at this stage, there is a stress field around the precipitate. (c) Overaged. There are two distinct and noncoherent phases, κ and θ. A limited number of solute atoms will provide the greatest interference to dislocation movements in (b). (A. G. Guy, *Elements of Physical Metallurgy.* Reading, Mass.: Addison-Wesley, 1959, p. 448.)

20-6 Overaging (carbide and nitride precipitation in iron). Overaging occurs when the precipitate becomes noncoherent (Fig. 20-5c). Compare with data in Table 20-1. (Adapted from Davenport and Bain, "The Aging of Steel," *Trans. A.S.M., 23,* 1935, p. 1011.)

the matrix and through the cluster, or pre-precipitate.* In this case the boundary energy is low, because of the registry, or *coherency*. Note, however, that the lattice is distorted and strained in the vicinity of the cluster. As a result, anchor points are provided which "lock" dislocation movements and interfere with slip (Chapter 11). The effect of the clustering is magnified because the surrounding strained volume is appreciably greater than the actual volume of the clusters themselves.

Continued precipitation and growth (AC of Fig. 20-4) permits the clusters to form larger but fewer units. The coherency and accompanying strain are progres-

---

* Several stages of the clustering process have been identified in aluminum-copper alloys. The first stage involves clusters of a few hundred copper atoms on [100] planes of the aluminum solute. The maximum hardness occurs when the clusters of copper and aluminum atoms grow to a "pancake" shape—1000 A to 2000 A diameter by 100 A to 200 A thickness. Coherency is maintained on the flat "pancake" surface, but not around the edges. Additional growth destroys the coherency and softens the alloy. Other materials may have different sequences of cluster growth.

TABLE 20-1

PROPERTIES OF AN AGE-HARDENABLE ALLOY (95.5Al-4.5Cu)

Treatment (See Fig. 20-4)		Tensile strength, psi	Yield strength, psi	Ductility, % in 2 in.
A	Solution-treated and quenched	35,000	15,000	40
B	Age-hardened	60,000	45,000	20
C	Overaged	~25,000	~10,000	~20
D	Annealed	25,000	10,000	15

sively lost (Fig. 20-5c) and the spacing between the resulting particles increases, allowing slip to occur more readily. This softening is called *overaging*. Figure 20-6 shows this phenomenon for an iron-nitrogen alloy. Note that both the age hardening and the overaging depend on both temperature and time.

A set of properties which accompany the age hardening of 95.5Al-4.5Cu is summarized in Table 20-1. These changes can be interpreted on the basis of the schematic structures of Figs. 20-4 and 20-5.

Combinations of age hardening and cold-working can give more strengthening than either process alone (Table 20-2). Observe some practical considerations, however. If aging precedes cold work, the strain-hardening process is complicated by higher power requirements for deformation, and less ductility. Conversely, strain hardening can be lost if the aging is performed after cold work, unless time and temperature control are closely maintained to prevent annealing.

Composition-controlled precipitation. In age hardening, the solid solution is supersaturated by cooling rapidly past a solubility limit. By contrast, in composition-controlled precipitation the supersaturation is by diffusion of a solute into the material until the solubility limit is exceeded. Steel which is to be *nitrided*, for example, contains aluminum in solid solution. It is heated in a nitrogen-active atmosphere (usually dissociated ammonia, NH_3) so that nitrogen diffuses into

TABLE 20-2

TENSILE STRENGTHS OF A STRAIN- AND AGE-HARDENED ALLOY (98Cu-2Be)

Annealed (870°C)	35,000 psi
Solution-treated (870°C) and rapidly cooled	72,000
Age-hardened only	175,000
Cold-worked only (37%)	107,000
Age-hardened, then cold-worked	200,000 (cracked)
Cold-worked, then age-hardened	195,000

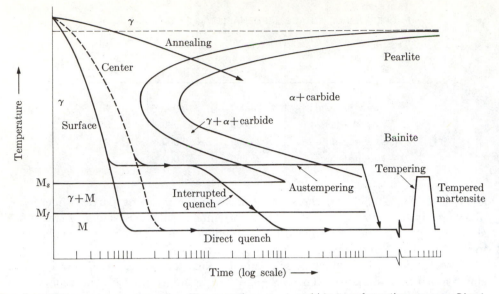

20-7 Transformation processes. *Annealing:* The normal $\gamma \rightarrow \alpha$ + carbide transformation occurs. *Direct quench:* Martensite forms, first in the surface, then in the center. Severe stresses result. *Interrupted quench:* Time is available for the surface and center to transform nearly simultaneously, thus avoiding the quench cracking found in direct quenching. *Tempering:* Both the direct and the interrupted quench must be followed by a tempering process to complete the transformation. *Austempering:* Quenching avoids pearlite formation, but the $\gamma \rightarrow \alpha$ + carbide transformation may still occur above the M_s. The resulting microstructure is bainite.

the ferrite matrix as a second solute. Excess nitrogen causes AlN to precipitate throughout the grains. Because of the resulting fine dispersion of AlN particles, this process can provide extremely high hardness at the surface of aluminum-bearing steels, or other alloy steels containing nitride-forming elements.

Another dispersion-hardening treatment in metals is that of *internal oxidation*. Silver which contains aluminum can be heated in an oxygen-rich atmosphere. As oxygen diffuses into the alloy, the aluminum is oxidized to form small, well-dispersed Al_2O_3 particles. The oxide product is very hard and is probably best known as the emery of abrasive wheels, or as sapphire. Internal oxidation is a particularly valuable method of hardening silver for use in low-resistivity electrical contacts. Other hardening procedures, such as solution treatments or cold work, may increase the resistivity beyond tolerable limits (Section 14-3).

20-5 TRANSFORMATION PROCESSES

The eutectoidal transformation of austenite to ferrite + carbide can proceed in several different manners; consequently, a choice of microstructures is possible. These are summarized in Table 20-3 and Fig. 20-7. Since the microstructures may differ appreciably, the resulting properties also differ significantly.

TABLE 20-3

TRANSFORMATION PROCESSES FOR STEEL

Process	Procedure	Resulting phases
Annealing	Slow cool from γ-stable range	α + carbide
Quenching	Severely rapid cooling from from γ-stable range	martensite*
Interrupted quench	Quench followed by slow cool through M_s and M_f	martensite*
Ausforming	Deformation just above M_s, then slow cool	martensite*
Austempering	Interrupted quench, then isothermal transformation	α + carbide
Tempering	Reheating of martensite	α + carbide

* Steels with martensite must be toughened by a tempering treatment before use. Tempering produces α + carbide.

Heat treatments. *Annealing* (a slow cool from the austenitic temperature range) gives sufficient time for coarse pearlite growth (Fig. 18-11). The properties are summarized in Section 19-4.

Quenching rates faster than the CR_M of Fig. 18-23 give hard (and relatively brittle) martensite by avoiding transformation to ferrite and carbide. Austenite is more dense than either martensite or ferrite plus carbide. This presents a problem with a direct austenite-to-martensite quench, because the center of the steel expands under transformation and *after* the surface has already transformed to martensite. Hence, cracking may be encountered in high-carbon steels (>0.6 w/o C), except in the case of small cross sections. As a result, an *interrupted quench* (also called *martempering* or *marquenching*) has been developed. In this process, the steel is quenched rapidly past the "knee" of the transformation diagram to avoid decomposition into ferrite and carbide, but the cooling is interrupted just above the M_s temperature. Cooling is then continued at a slow rate through the martensitic range to ambient temperatures, so that the surface and center of the steel may transform more or less simultaneously, thus avoiding the quench-cracking. Slow cooling is possible at these lower temperatures because the martensitic transformation is very insensitive to time (Fig. 18-17), and proceeds primarily as a function of the temperature drop below M_s.

Ausforming is a variation of the interrupted quench in which the steel (still in the austenitic condition) is plastically deformed during the quench interruption, and the martensite is formed during subsequent quenching to the ambient temperature. Since the arrest temperature is below the short-time recrystallization temperature (Section 11-6), strain hardening is added to the properties of the quenched material.

Tempering is applied after martensite formation because the martensite is too brittle for most engineering applications. Tempering at 200° to 400°C transforms metastable martensite to ferrite plus carbide (see Eq. 18-2). Carbide nucleation may occur at many sites within the shear-formed martensite (Section 18-2), rather

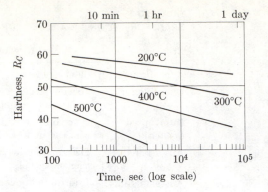

20-8 Hardness after tempering (1080 steel quenched to maximum hardness). The M → α + carbide reaction greatly toughens the steel. It also softens the steel; but with a suitable choice of temperature and time, it is still possible to retain high hardness values.

than simply at the prior austenitic grain boundaries. With a very fine dispersion of carbide particles, the hardness and strength remain high; however, the toughness increases markedly over that of the as-quenched martensite because a fracture path must pass through the more ductile ferrite matrix. This microstructure of ferrite and carbide is called *tempered martensite*. As shown in Fig. 20-8, tempered martensite softens with extended heating times or elevated temperatures, because the carbide particles grow in size and decrease in number (Section 19-2).

The final variation which we shall consider for austenite decomposition is that of *austempering*. In this treatment, the austenite is allowed to transform *isothermally* to ferrite and carbide just above the M_s temperature. This requires a quench to avoid transformation to pearlite at higher temperatures. The advantage of austempering is that transformation occurs by a combination of shear and diffusion, to give a fine dispersion of carbides in ferrite; also, quench-cracking is avoided because the reaction (and volume change) takes place at constant temperature. In many respects the product, *bainite*, is similar to tempered martensite, and the physical properties of the two microstructures are closely related.*

A close examination of the isothermal-transformation process reveals that pearlite formation and bainite formation are different (Fig. 20-9). However, their combination gives the $\gamma \rightarrow \alpha +$ carbide transformation curve of Fig. 18-17 for SAE 1080 steels. In some alloy steels, the two reactions do not blend together because their temperature ranges are separated quite distinctly (Fig. 20-9b). This difference allows bainite to be formed during the continuous cooling of certain steels.

Hardenability. Quenching rates are limited by the volume-to-surface ratio of the specimen and by the fact that the heat from the center must pass through the surface. Figure 20-10(a) plots the cooling rates for water-quenched round bars in which the heat transfer is in the radial direction. Thus, if the minimum cooling rate, CR_M, for 100% martensite is 200°C/sec, we will find hard martensite throughout the 0.5-in. bar but only near the surface of the 2-in. bar (Fig. 20-10b). In alloy steels, where the value of CR_M is only 20°C/sec, martensite may be formed

---

* There is a slight difference in the distribution of the carbides in the two microstructures. This has only subtle effects on the properties.

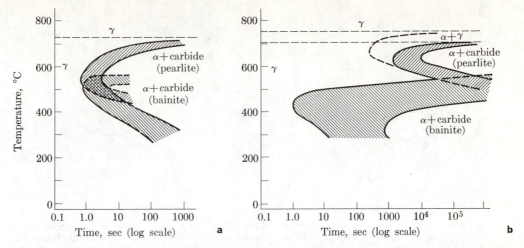

20-9 Pearlite and bainite formation. (a) Eutectoid steel. (b) 4340 steel. Pearlite nucleates at austenite grain boundaries and grows by diffusion. Bainite is formed by a combination of shear and diffusion. Therefore, pearlite and bainite have separate transformation curves even though both products contain α + carbide. The two curves happen to be tangential in a eutectoid steel; in general, they are not.

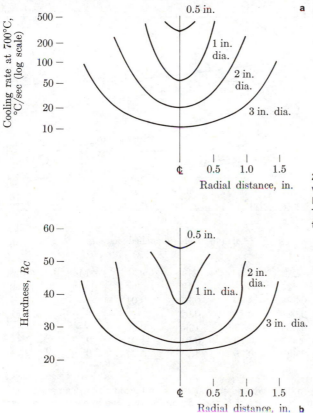

20-10 Quenched steel. (a) Cooling rate versus radius (water-quenched, round bars). (b) Hardness traverses (1040 steel). The cooling rate must be faster than CR_M to produce 100% martensite.

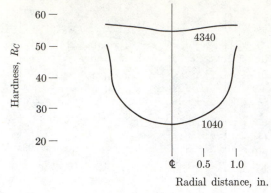

20-11 Hardness traverses (water-quenched 2-in. round bars of 4340 and 1040 steel). Both steels had the same quenching rates; however, the 4340 steel has a lower critical cooling rate for martensite formation than does the 1040 steel. Therefore it formed the harder martensite in the center as well as at the surface.

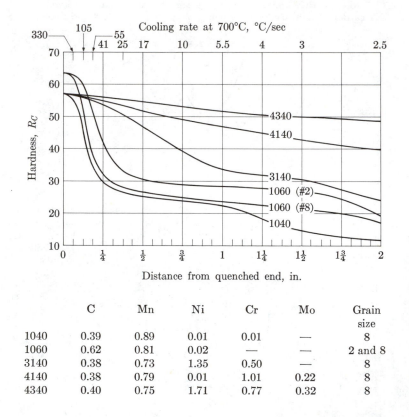

	C	Mn	Ni	Cr	Mo	Grain size
1040	0.39	0.89	0.01	0.01	—	8
1060	0.62	0.81	0.02	—	—	2 and 8
3140	0.38	0.73	1.35	0.50	—	8
4140	0.38	0.79	0.01	1.01	0.22	8
4340	0.40	0.75	1.71	0.77	0.32	8

20-12 Hardenability curves for six steels with the indicated compositions and grain sizes. The steels were end-quenched as shown in Fig. 9-10. In commercial practice, the hardenability curve of each type of steel varies because of small variations in composition. (Adapted from U.S. Steel data.)

throughout a 2-in. bar. Hardness traverses illustrate this contrast in Fig. 20-11 for 2-in. rounds of plain-carbon (1040) and alloy (4340) steels where both steels were quenched in the same manner. The alloy steel has martensite to a greater depth (as shown by the R_C hardness values, which are greater than 40). We speak of the alloy steel as having greater *hardenability;* i.e., it is more readily hardened even at the slower cooling rates found in the interior of large cross sections.

Hardenability curves are shown in Fig. 20-12. These steels were end-quenched as sketched in Fig. 9-10. Note that the hardness profile depends on alloy content (Fig. 18-25), austenitic grain size (Fig. 18-24), and also carbon content. These curves indicate that hardness is a function of the cooling rate (top abscissa of Fig. 20-12), which in turn depends on the location in the end-quenched bar (lower abscissa).

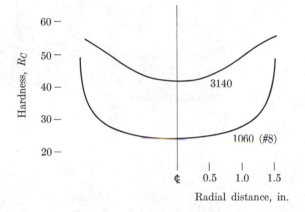

20-13 Hardness traverses (water-quenched 3-in. round bars of 3140 and 1060 steel). (See Example 20-2.)

Example 20-2

Plot a hardness traverse for a 3-in. diameter bar (a) of SAE 3140 steel, (b) of SAE 1060 (grain size No. 8) steel.

Answer. Use the cooling rates of Fig. 20-10 and the hardness-versus-cooling data of Fig. 20-12. For example, from Fig. 20-10, the center of a 3-in. water-quenched bar cools through 700°C at 11°C/sec. From Fig. 20-12, a 3140 steel cooled at this rate develops a hardness of 41 R_C, and a 1060 steel of the same grain size develops a hardness of 25 R_C. (See Fig. 20-13 for other values.) ◄

Example 20-3

Added carbon increases the hardness of both the martensite and the α + carbide (Fig. 18-4). Therefore, the hardenability curves show an overall shift in hardness values with the carbon content, as shown in the nickel-molybdenum (40xx) steels of Fig. 20-14. A 1-in. round bar of 4017 steel is *carburized* so that it has an outer 0.125-in. *case* of 0.6 w/o carbon and its center *core* remains at the original carbon level (cf. Fig. 23-2). What is the hardness traverse after a water quench from the austenitizing temperature?

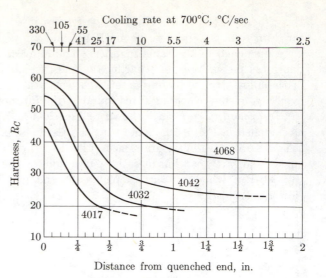

Cooling rate at 700°C, °C/sec

20-14 Hardenability curves for 40xx steels. Except for carbon content, the composition is the same for each. Additional carbon gives harder martensite and harder $\alpha + C$, according to Fig. 18-4.

Distance from quenched end, in.

Answer. From Figs. 20-10 and 20-14, we obtain the following data:

Radial distance	Cooling rate	Carbon	Hardness
0 in.	55°C/sec	0.17	30 R_C
0.25	90	0.17	33
0.35	140	0.17	35
0.4	175	0.6	62
0.5	330	0.6	63 ◀

Example 20-4

Figure 20-15 shows the points in the cross section of a V-bar of SAE 3140 steel in which the following hardness readings were obtained after oil quenching. What hardness values would be expected for an identically shaped bar of SAE 4068 steel?

Answer. For a given steel, the hardness is dependent on the cooling rate.

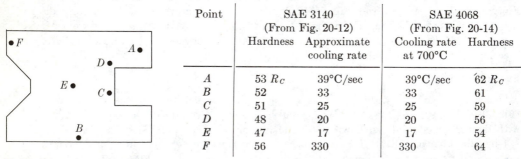

Point	SAE 3140 (From Fig. 20-12)		SAE 4068 (From Fig. 20-14)	
	Hardness	Approximate cooling rate	Cooling rate at 700°C	Hardness
A	53 R_C	39°C/sec	39°C/sec	62 R_C
B	52	33	33	61
C	51	25	25	59
D	48	20	20	56
E	47	17	17	54
F	56	330	330	64

20-15 V-bar cross section. (See Example 20-4.)

REFERENCES FOR FURTHER READING

American Society for Metals, *Strengthening Mechanisms in Solids*. Metals Park, O.: American Society for Metals, 1962. The "Introductory Review" to this book summarizes the principles involved in strengthening mechanisms. Subsequent chapters pay attention to different types of mechanisms. Advanced undergraduate and graduate level.

Brick, R. M., R. B. Gordon, and A. Phillips, *Structure and Properties of Alloys*, third edition. New York: McGraw-Hill, 1965. The theory of steel heat treatment is given in Chapter 9. Applications of steels are discussed in Chapters 10 (structural steels), 11 (tool steels), 12 (stainless steels), and 13 (cast irons). Recommended for the undergraduate who wants to apply the principles.

Burke, J. J., N. L. Reed, and V. Weiss (Eds.), *Strengthening Processes*. Syracuse: Syracuse University Press, 1966. Symposium on strengthening processes in metals and ceramics. For the advanced student and instructor.

Dovey, D. M., E. R. Gadd, E. Mitchell, and W. S. Owen, *Heat Treatment of Metals*. Reading, Mass.: Addison-Wesley, 1962. Lectures delivered at the Institution of Metals (London) refresher course. Applications are emphasized. Suitable as a sequel to this book.

Felbeck, D. K., *Introduction to Strengthening Mechanisms*. Englewood Cliffs, N. J.: Prentice-Hall, 1968. Paperback. Excellent supplementary reading for this chapter.

Guy, A. G., *Elements of Physical Metallurgy*, second edition. Reading, Mass.: Addison-Wesley, 1959. Age hardening and heat treatment of steel are given in Chapters 13 and 14. Undergraduate level.

Sutton, W. H., and J. Chorné, "Factors Affecting the Tensile Strength of Metals Reinforced with Strong Fibers," *Strengthening Mechanisms in Metals and Ceramics*. Syracuse: Syracuse University Press, 1966, pp. 549–577. Presents a good summary of strength theories. For the advanced undergraduate.

Zackay, V. F. (Ed.), *High Strength Materials*. New York: Wiley, 1965. One of the best compilations of papers on strengthening mechanisms from all aspects. Topics are somewhat specialized. For the materials specialist, because of its detail. It is easily read, however, by the advanced undergraduate student.

PROBLEMS

20-1 Specifications call for an iron rod to have a final diameter of 0.15 in., a minimum tensile strength of 50,000 psi, a minimum of 130 BHN, and a ductility of at least 30% elongation. Prescribe cold-work processing steps. One-inch cold-drawn rods are available as starting material.

Answer. Deform to 0.175 in. diameter; anneal; cold-draw 27% to 0.15 in. diameter.

20-2 a) Estimate the yield strength of a 90Cu-10Sn quenched bronze.
  b) Would a slow-cooled bronze be harder or softer?
  c) Is this alloy one that might be age-hardened?

20-3 A copper-base alloy is to be chosen for a brine cooling tank. Specifications call for a tensile strength of more than 40,000 psi, a thermal conductivity of at least 70% of that of pure nickel, and an elongation (% in 2 in.) of at least 40%.

  a) Indicate the best alloy of the compositions shown in Figs. 11-21, 11-22, and 14-13.
  b) Cite the reasons for your choice.

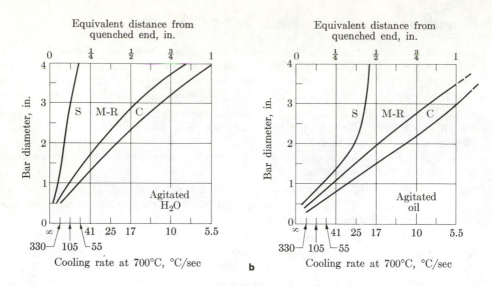

20-16 Cooling rates in round steel bars quenched in (a) agitated water and (b) agitated oil. Bottom abscissa, cooling rates at 700°C; top abscissa, equivalent positions on an end-quench test bar. (C, center; M-R, midradius; S, surface.) The high heat of vaporization of water produces a severe quench in that quenching medium.

Answer. a) 65Cu-35Zn   b) Meets specifications; furthermore, since zinc is cheaper than copper, a brass high in zinc is less expensive than one higher in copper; allow some composition latitude below the 60Cu-40Zn limit.

20-4 A rod is to be made from annealed copper or one of the copper-base alloys in the text.   Its annealed specifications call for $\rho \leq 7 \times 10^{-6}$ ohm · cm, YS $\geq$ 8000 psi, $k \geq 0.40$ cal · cm/cm^2 · sec · °C, El $\geq$ 40%. Select the most appropriate alloy. (See Figs. 11-21, 11-22, 14-10, and 14-13.)

20-5 The temperature dependence of the rate of tempering martensite may be expressed as an activation energy. A particular martensite has an as-quenched hardness, R_C, of 62. After tempering 10 min at 900°F (448°C), the hardness is 45 R_C. It takes 10 times as long to temper to 45 R_C at 800°F (427°C). What is the activation energy for the process?

Answer. 40,000 cal/mole · °K

20-6 a) Write a rate equation (cf. Eq. 9-2) for the tempering of steel in Fig. 20-8.
  b) What is the activation energy for tempering this steel to 50 R_C?
  c) Is it greater or less for tempering to 40 R_C?

20-7 What hardness would you expect at the center of a 2-in. round of 1040 steel if it were quenched in (a) mildly agitated oil? (b) mildly agitated water? (See Fig. 20-16 for cooling rates.)

Answer. a) 23 R_C b) 26 R_C

20-8 A bar of 1040 steel has a surface hardness of 41 R_C and a center hardness of 28 R_C. How rapidly were the surface and center cooled through 700°C (1300°F)?

20-9 A 2.5-in. round of 1040 steel is quenched in agitated oil. Estimate the hardness 1 in. below the surface of the round. (Show your reasoning.)

Answer. 23 R_C

20-10 Exact identification was lost on a round bar of steel (50 ft long × $1\frac{1}{2}$ in. in diameter). It is known only that it may be either AISI 4068 or AISI 4140.

a) It is proposed that the bar be quenched in oil and identified by a hardness traverse. Indicate below the hardness values you would expect.

	4140	4068
Surface	___ R_C	___ R_C
$\frac{3}{8}$ in. below surface	___	___
$\frac{3}{4}$ in. below surface	___	___

b) As a practical matter, a furnace is not available to heat the 50-ft bar. However, since the bar is 2 in. longer than required, it is proposed that 2 in. be cut off for the above identification test. Comment.

c) Suggest and discuss the merits of alternative possibilities for identification.

20-11 Two round bars, each 4 in. in diameter, of the *same* steel were quenched in water and oil, respectively.

Inches below surface	Water	Oil
0.0	54 R_C	48 R_C
0.5	50	47
1.0	36	32
1.5	34	31
2.0	33	31

a) From the above hardness data, plot the hardenability curve for this steel.

b) The hardness traverse for a 3-in. diameter oil-quenched bar of this same steel would show: surface = ___ R_C; mid-radius = ___ R_C; center = ___ R_C.

Answer. a) See AISI 3140. [*Note.* There is only *one* curve.] b) 49; 36; 33

20-12 Two round bars of the same steel composition and grain size have 2-in. and 3-in. diameters, respectively. Their hardness traverses give the following data:

Diameter	2 in.	3 in.
Quench	Water	Oil
Surface hardness	60 R_C	33 R_C
Mid-radius hardness	38	29
Center hardness	34	28

Plot the hardenability curve.

20-13 In commercial practice, an annealed steel is heated one hour at 50°F above the highest temperature of α-ferrite. Indicate the annealing temperature for a 1040, a 1080, and a 1% C steel.

Answer. 1550°F, 1385°F, 1385°F (845°C, 750°C, 750°C)

20-14 An aircraft manufacturer receives a shipment of aluminum-alloy rivets that have already age-hardened. Can they be salvaged? Explain.

21 □ mechanical failure

21-1 FAILURE BY FRACTURE

This chapter is directed toward the mechanisms of failure through applied stresses. Mechanical failure often occurs by *fracture*. We speak of *brittle fracture* if this involves minimal energy absorption and plastic deformation; or of *ductile fracture* if there is significant ductility prior to failure. The mode of brittle fracture may be *cleavage* along crystal planes; or it may be *intergranular*, along grain boundaries. Ductile fracture involves considerable plastic deformation and the possibility of internal rupture from heterogeneities and defects before the actual fracture occurs. The term *fatigue* applies to fracture which arises from cyclic stresses. We shall observe later that mechanical failures can occur which are wholly due to plastic deformation, particularly at higher temperatures; the mechanisms of such failures are *creep* and *stress rupture*. Since attention will not be given to failure by *wear*, the reader interested in this subject is referred to the references at the end of this chapter.

Cleavage. Since crystalline materials have directional properties (Sections 5-7 and 10-3), we should expect that fracture stresses and ductility will be anisotropic. The reader is well aware of the easy splitting, or cleavage, of mica and graphite. Other materials also exhibit cleavage, as indicated in Table 21-1. The resolved

TABLE 21-1
COMMON CLEAVAGE PLANES

Mica	{001}
Rock salt	{100}
MgO	{100}
Bcc metals	{100}
Fcc metals	{111}
Hcp metals	{0001}
Graphite	{0001}
Diamond	{111}
CaF_2	{111}

tensile stress, $\sigma_{(hkl)}$, across a cleavage plane, (hkl), may be calculated in a manner analogous to that used in Schmid's law (Eq. 11-1) for shear stresses. Here, however, only the angle, ϕ, between the force direction, $[h'k'l']$, and the plane normal is

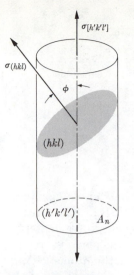

21-1 Resolved tensile stress on a cleavage plane, (*hkl*). Equation (21-1) applies.

required (Fig. 21-1):

$$\sigma_{(hkl)} = \sigma_{[h'k'l']} \cos^2\phi. \tag{21-1a}$$

In cubic crystals the plane normal and the plane itself have the same indices; thus, by replacing ϕ with $[hkl] \not\measuredangle [h'k'l']$, Eq. (21-1a) can be written as follows for easier vector solution:

$$\sigma_{(hkl)} = \sigma_{[h'k'l']} \cos^2 [hkl] \not\measuredangle [h'k'l']. \tag{21-1b}$$

Example 21-1

A cubic crystal is oriented so that its [111] direction is aligned with the applied tensile stress of 1386 psi.

a) What is the resolved tensile stress across the (001) plane?
b) If the critical fracture stress for cleavage is 2200 psi for {110} planes and 1200 psi for {100} planes, which will cleave first under the above [111] loading? (Assume comparable stress irregularities from flaws.)

Answer

a)
$$\sigma_{(001)} = \sigma_{(111)} \cos^2 [111] \not\measuredangle [001]$$
$$= (1386 \text{ psi})(1/\sqrt{3})^2$$
$$= 462 \text{ psi.}$$

b) As oriented, the (001), (010), and (100) planes are symmetric with respect to [111]; likewise, (110), (011), and (101) are, among themselves, equally symmetric to [111]. For $\sigma_{(100)} = 1200 \text{ psi} = \sigma_{(111)} \cos^2 [111] \not\measuredangle [100]$,

$$\sigma_{(111)} = (1200 \text{ psi})/(1/\sqrt{3})^2 = 3600 \text{ psi.}$$

For $\sigma_{(110)} = 2200 \text{ psi} = \sigma_{(111)} \cos^2 [111] \not\measuredangle [110]$,

$$\sigma_{(111)} = (2200 \text{ psi})/(\sqrt{2/3})^2 = 3300 \text{ psi.}$$

Fracture will occur on {110} first if the load is applied in a [111] direction.

Note. Other orientations would commonly favor {100} fracture. ◀

21-2 Cleavage. In cleavage, the fracture path follows crystal planes.

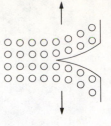

21-2 BRITTLE FRACTURE

For fracture to occur without significant plastic deformation, adjoining atoms must be completely separated, as shown schematically in Fig. 21-2. The force or stress which is required is a function of the sublimation energy of a material. (Recall from Section 3-2 and Fig. 3-1 that the depth of the energy trough, $E_\infty - E_{min}$, is the energy required for complete separation of the atoms.) As a result, the *theoretical fracture stress* is a function of the maximum force, $\times$, of the *net* curve in Fig. 3-2, and has a value which is a major fraction of Young's modulus.

Two noteworthy characteristics of brittle fracture are the following: (1) the fracture stresses are often significantly below the theoretical strengths just cited, and (2) the fracture stress is highly variable, as indicated by the data of Fig. 21-3(a).

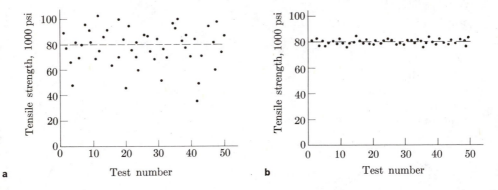

21-3 Variance in strengths. (a) Brittle fracture. (b) Ductile fracture.

Half a century ago, Griffith* proposed that fracture in brittle solids arises from many fine flaws in the surface of the material. As support, he pointed out that pristine glass has high strength, sometimes approaching the theoretical value. Once the glass is exposed to surface deterioration, however, the strength falls rapidly. The surface changes can arise from such subtleties as abrasion by finger

---

* A. A. Griffith, "The Phenomenon of Rupture and Flow in Solids," *Phil. Trans. Royal Soc.* (London), **A221**, 1921, pp. 168–198.

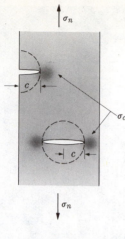

21-4 Stress concentrations in a cracked plate. The stresses at the tip of the flaw depend on the flaw depth, c, and the radius of curvature, r, at the tip of the crack. In brittle materials this radius approximates atomic dimensions.

contact! Or they can originate from reactions with the moisture of the air. In either case, the observed low strength points to surface flaws, often called *Griffith cracks.*

Stress concentrations. A notch concentrates an applied stress at its tip (Fig. 21-4). The effect is much greater than would be anticipated simply on the basis of the restricted cross-sectional area. Griffith concluded that the concentrated stress, σ_c, is related to the nominal stress, σ_n, the depth of the crack, c, and the radius of curvature, r, at the tip of the flaw:

$$\sigma_c = \sigma_n 2\sqrt{c/r}. \tag{21-2}$$

This suggests that the nominal stress may not be the significant stress, particularly in a brittle material, where the tip of a crack may have an effective radius of the order of one atomic radius (say, 1 A). This information, coupled with the existence of flaws which are known to extend from 0.1 to 10 μm (10^{-5} to 10^{-3} cm) below the surface in a material such as glass (Fig. 21-5), shows that the *stress concentration factor*, σ_c/σ_n, may be 10^2 or even 10^3! Thus, when a glass breaks under tension, the localized stress is more than 1,000,000 psi, rather than the nominal 5000 to 15,000 psi. This concentrated stress is more nearly comparable to the theoretical strength of glass.

Crack propagation. When a surface crack propagates, elastic strain energy, ΔU_E, is released. The amount per unit crack length depends on the applied stress, σ_n, Young's modulus, Y, and crack depth, c:

$$\Delta U_E = -\frac{\pi c^2 \sigma_n^2}{2Y}. \tag{21-3}*$$

---

* The volume, $\pi c^2/2$, which releases strain energy appears in Fig. 21-4 as a half cylinder of unit length.

21-5 Surface flaws in container glass. Revealed by special etching and lighting techniques, flaws can develop during service by chemical, mechanical, or thermal environments. (Courtesy of W. C. Levengood and T. S. Vong, The University of Michigan.)

However, energy is consumed in forming the new surfaces of the crack:

$$\Delta U_S = 2c\gamma, \tag{21-4}$$

where γ is the energy per unit area.* Such a crack would spontaneously grow (that is, c would increase) if this growth should lead to a decrease in the overall energy, (i.e., if c is such that $dU/dc = 0$):

$$d(-\pi c^2 \sigma_n^2/2Y + 2c\gamma)/dc = 0,$$

or

$$\pi c \sigma_n^2/Y = 2\gamma,$$

and

$$\sigma_n = [2\gamma Y/\pi c]^{1/2}. \tag{21-5}$$

As expected, the stress necessary for fracture is low when deep flaws, or cracks, are present; it is higher for materials with high elastic moduli and high surface energies.

The above analysis applies to the simplest case, that of a crack which propagates without any deformation across a thin plate by a unidirectional tensile stress.

---

* The internal crack of Fig. 21-4 has identical relationships, except both ΔU_E and ΔU_S are increased by a factor of 2.

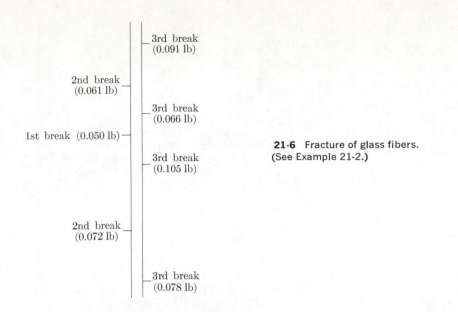

3rd break
(0.091 lb)

2nd break
(0.061 lb)

3rd break
(0.066 lb)

1st break (0.050 lb)

21-6 Fracture of glass fibers.
(See Example 21-2.)

3rd break
(0.105 lb)

2nd break
(0.072 lb)

3rd break
(0.078 lb)

Other geometries require certain modifications, but do not cause a major change in results.

As a crack progresses through a brittle material under sustained loading, the notch effect becomes more pronounced. The stress concentration at the tip of the flaw is increased, making the break catastrophic; in fact, the fracture propagation approaches the speed of an elastic wave, which is related to the speed of sound in the material (Chapter 10).

Example 21-2

A glass has a Young's modulus of 10^7 psi (7×10^{11} dynes/cm^2) and a surface energy of 600 ergs/cm^2. A 20-cm length of this glass in fiber form (diameter $= 0.052$ mm) has been exposed to ambient conditions for 6 hr. After that period it is broken; each part is broken again; and this procedure is repeated once more within a short period of time to give the data of Fig. 21-6.

a) Estimate the crack depth in each case.
b) What was the stress at the tip of *each* crack just prior to the *first break*, assuming the tip radius to be 1.5×10^{-8} cm? Just prior to actual fracture?

Answer. First break:

$$\sigma_n = \frac{(0.050 \text{ lb})(454 \text{ gm/lb})(980 \text{ dynes/gm})}{\pi (0.0026 \text{ cm})^2}$$

$$= 1.05 \times 10^9 \text{ dynes/cm}^2 \quad (= 15{,}000 \text{ psi}).$$

From Eq. (21-5),

$$c = 2\gamma Y/\pi\sigma_n^2$$

$$= \frac{2(600 \text{ ergs/cm}^2)(7 \times 10^{11} \text{ dynes/cm}^2)}{\pi (1.05 \times 10^9 \text{ dynes/cm}^2)^2}$$

$$= 2.5 \times 10^{-4} \text{ cm} \quad (= 2.5 \text{ } \mu\text{m}).$$

From Eq. (21-2),

$$\sigma_c/\sigma_n = 2(2.5 \times 10^{-4}/1.5 \times 10^{-8})^{1/2} = 250,$$

$$\sigma_c = 250(15,000 \text{ psi}) = 3.8 \times 10^6 \text{ psi}.$$

Calculations for each crack provide:

Breaking load, lb	σ_n, psi (at fracture)	Flaw depth, μ	σ_c, psi (at $\sigma_n = 15,000$ psi)	σ_c, psi (at fracture)
1st (0.050)	15,000	2.5	3.8×10^6	$\sim3.8 \times 10^6$
2nd (0.061)	18,300	1.6	3.1	3.8
2nd (0.072)	21,600	1.2	2.6	3.8
3rd (0.091)	27,300	0.7	2.1	3.8
3rd (0.066)	19,800	1.4	2.9	3.8
3rd (0.105)	31,500	0.55	1.8	3.8
3rd (0.078)	23,400	1.0	2.4	3.8

Note 1. Even though calculations in cgs units are more convenient, the more common psi dimensions are used to express the stresses in our answers (1 psi $= 7 \times 10^4$ dynes/cm^2). Since the stress concentration factor, σ_c/σ_n, is dimensionless, we may still use cgs units under the radical of Eq. (21-2) for c/r.

Note 2. Although there is considerable scatter in the nominal stress, σ_n, of the second column of the above table, the stresses at the tips of the cracks which fracture (fifth column) are the same. Since the sites with more severe flaws break first, succeeding tests show higher nominal strengths (provided the test is immediate and no deterioration occurs). This relationship has been verified experimentally by Reinboker. ◄

Deformation during brittle fracture. Most brittle metals (and many nonmetals) realize some local plastic deformation along the fracture path as a result of local shear on adjacent crystal planes. Thus Eq. (21-5) must be rewritten:

$$\sigma = [2(\gamma + \phi_p)Y/\pi c]^{1/2}, \tag{21-6}$$

where ϕ_p is the plastic strain energy per unit area of fracture surface. In many crystalline materials, the value of ϕ_p may significantly exceed γ, the surface energy. The amount of plastic strain energy must be determined experimentally for strength calculations.

Compression strengths of brittle materials. Unlike tensile forces, compressive forces may be transmitted across existing cracks without stress concentrations. As a result, a brittle material is invariably stronger in compression than in tension. Design engineers are fully aware of this. For example, concrete is used under compressive loading and seldom, if ever, in tension (Chapter 23). Furthermore, materials engineers are able to design materials in which the surfaces are under compression with respect to the interior, so that a damaged surface does not lead to premature failure (Section 21-3).

With sufficiently high compressive stresses, however, a brittle material can fail. In a plate, or planar material, failure occurs at the theoretical compressive stress,

21-7 Complex loading. (a) Planar strain. Compression introduces 90° tensile forces in a two-dimensional plate. (b) Tensile failure of concrete compression cylinder. (Courtesy of Frank E. Legg, The University of Michigan.)

which is equal to $8\sqrt{2\gamma Y/\pi c}$, or eight times the theoretical fracture stress in tension shown in Eq. (21-5). When that compressive stress level is reached, a resolved tensile stress is developed in a planar material which is sufficient to initiate tensile fracture at right angles to the compression direction (Fig. 21-7a). A specific example is shown for the concrete test cylinder in Fig. 21-7b. Admittedly, the above factor of 8 must be modified somewhat for geometries which do not give the two-dimensional *plane strain* of Fig. 21-7a. Even so, the very high compressive strengths of nonductile materials are of major significance in engineering design. In brief, the engineer must work toward the high compressive strengths, and around the low tensile strengths, to optimize those designs which must use brittle materials.

21-3 STRENGTHENING BRITTLE MATERIALS

Prestressing is available as a procedure for capitalizing on the high compressive strengths of brittle materials, such as concrete. In prestressing, reinforcing rods place a structure under initial compressive stress before the service load is applied. The bending moment does not build up tensile stresses until the introduced compressive stresses are compensated.

Surface compression. Compressive stresses may be induced in materials by chemical and thermal means. Three examples will be considered. First, consider a plate of Na^+-Ca^{2+}-silicate glass which is suspended in a molten salt bath of K_2SO_4. The *ion exchange* reaction,

$$K_{salt}^+ + Na_{glass}^+ \rightleftarrows K_{glass}^+ + Na_{salt}^+, \qquad (21\text{-}7)$$

can occur. It happens, however, that the incoming K^+ ion is about 30% larger

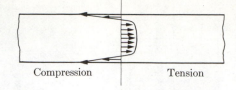

Compression | Tension

21-8 Surface compression. Both chemical and thermal treatments can introduce surface compression (see text). The surface compressive stresses must be overcome before the surface can be broken.

 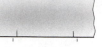 

Hot glass No stresses	Surface cooled quickly Surface contracts Center adjusts Only minor stresses	Center cools Center contracts Surface is compressed Center in tension

21-9 Dimensional changes in "tempered" glass.

than the Na^+ ion that is exchanged and leaves the glass. Thus the glass at the surface of the plate should expand, but cannot because it has a continuous silicate structure within the rigid unaltered interior of the plate. A compressive surface skin forms. Although this skin may be only a fraction of a millimeter thick, it precludes fracture propagation until external tensile stresses are high enough to overcome the induced stresses.

The surface compressive stresses must be balanced by internal tension stresses (Fig. 21-8). The latter are not critical, however, for several reasons. First, the stress level is never high because the tension core constitutes the major fraction of the cross section. Second, the highest flexure stresses are in the outermost "fibers" of the glass; consequently, the core is not so severely stressed by bending moments in service. Third, and most important, the interior is not subject to surface damage which produces flaws or cracks; the stress concentrations of the Eq. (21-2) are thereby minimized. This type of stress distribution provides a considerable improvement in effective strength. Of course, if a deep flaw should penetrate beyond the compressive surface zone, the crack could then proceed spontaneously throughout the tension-stressed core.

As a second example of induced stresses, consider the *temper glass* which is used for glass doors of stores, rear windows of cars, and similar high-strength applications. To produce temper glass, the glass plate is heated above the fictive, or glass-transition, temperature (Fig. 6-26) where the atoms have some freedom of local movement. The glass is then quickly cooled by an air blast or oil quench (Fig. 21-9). The surface contracts because of the drop in temperature and becomes rigid, while the center is still hot and can adjust its dimensions to the surface contraction.

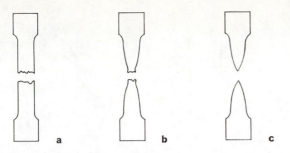

21-10 Tensile failure. (a) Brittle fracture has insignificant plastic deformation. (b) Ductile fracture follows initial necking by plastic deformation. (c) Rupture occurs primarily by plastic deformation.

When the center subsequently cools and contracts, compressive stresses are produced at the surface (and tensile stresses in the center). These residual stresses are diagrammed in Fig. 21-8. A considerable load must be applied to the glass before tensile stresses are developed in the surface.

The third example of surface conditioning is *shot-peening* of metals, which is used to introduce surface compressive stresses, particularly to inhibit fatigue failure (Section 21-6).

21-4 DUCTILE FRACTURE

A tough, nonbrittle material exhibits considerable ductility, either as elongation or reduction in cross-sectional area, before actual fracture occurs. Thus the term *ductile fracture* is appropriate. Ductile fracture is usually preferred to brittle fracture, except in applications which require fragmentation. Hence, the engineer commonly provides both strength and ductility specifications for his structural materials (Example 20-1). There are two reasons for the preference for ductile fracture. First, ductility is desired in some applications to provide a "fail-safe" design and avoid a catastrophic, unpredictable fracture. Second, limited failure by plastic deformation often increases the strength of a material by strain hardening. Also, any plastic flow at the tip of a crack will blunt the crack, reducing the stress concentration (Eq. 21-2); consequently, the stress may not be sufficient to continue the propagation of the crack. Thus the amount of ductility is often highly critical.

Fracture following initial plastic deformation is illustrated in Fig. 21-10(b). We find rupture without fracture (Fig. 21-10c) only at high temperatures, where strain hardening is minimal, or in those crystalline materials where there are very few heterogeneities to nucleate the growth of voids.

A ductile material can initiate subsequent brittle fracture for one of several reasons.

1) Plastic deformation introduces strain hardening, and concurrently reduces the ductility. Hence, a material may initially adapt to stresses, but subsequently lack ductility for further stress relaxation at flaws.

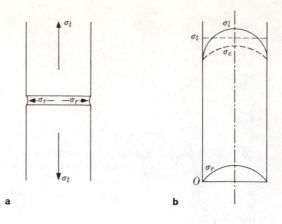

21-11 Plastic constraint (soldered test bar). (a) The rigid steel restrains the solder from necking by introducing a radial stress, σ_r. (b) The radial stresses and the redistributed longitudinal stresses are highest in the center. According to Eq. (21-8) the effective stress, σ_e, is the difference $\sigma_l - \sigma_r$.

2) Nonductile minor phases may serve as an internal stress raiser which eventually produces fracture through the formation and interconnection of voids.

3) Deformation changes the stress pattern so that simple unidirectional loading does not occur.

Plastic constraint. In the unidirectional loading just cited, all the tensile (or compressive) stresses and strains are coaligned. More commonly, stress and strain are planar (bidirectional) or even triaxial (three-directional). We saw in Fig. 21-7(a), for example, a condition of *plane strain* where compression in one direction introduced perpendicular tension. Although axially symmetric, the compressive failure of Fig. 21-7(b) has three-dimensional stress and strain distributions.

The significance of *triaxial stresses* may be revealed by a pair of simple tests. First, a test bar of 50Pb-50Sn solder has a yield strength of 4500 psi and a tensile strength of 6000 psi (based on the original area). A second test bar of steel may be cut in two parts and then rejoined with a thin layer of the above 50Pb-50Sn solder. This test bar will resist a stress of 6000 psi. In fact, if the solder is very thin, failure of the solder will not occur until that value is exceeded by a factor of 5 or more. It would appear that "a chain is *stronger* than its weakest link!"

The above strengthening phenomenon may be explained on the basis of *plastic constraint*. The more rigid steel constrains the ductile solder so that it cannot "neck-down." Thus, one effect is to maintain the full cross-sectional area so that a larger load can be supported before fracturing occurs. There is also another effect, which is shown in Fig. 21-11(a): the constraint of the steel on the solder (a) introduces a radial stress, σ_r, in the soft metal which increases toward the center of the bar, and (b) redistributes the longitudinal stress, σ_l, from the average value of σ_t to that shown in Fig. 21-11(b). One consequence is that the surface

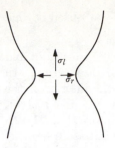

21-12 Radial stresses from necking. The necked area receives plastic constraint from the adjacent metal. The highest longitudinal stresses and radial stresses are in the interior of the necked area. (Cf. Fig. 21-11.)

21-13 Internal fracture. (a) Nucleation of voids on structural heterogeneities. (b) Coalescence of voids. (c) Final rupture to the surface by shear.

stress is reduced below the average applied stress and fracture is less readily initiated. A second, and more significant, consequence arises from the fact that the effective stress, σ_e, in a three-dimensional solid under unidirectional loading is approximately equal to the *difference* between the longitudinal and radial stresses:

$$\sigma_e \cong \sigma_l - \sigma_r. \tag{21-8}*$$

Thus, with plastic constraint, the effective stress is less than the applied tensile stress! When the thickness of the solder joint is 10% of the diameter, the effective stress at the surface is only about one-half the applied stress. Thinner joints reduce the effective stress further, so that still greater applied stresses are possible.

The importance of plastic constraint, as described in the last paragraph, may be found in the data presented in Fig. 19-14. The ferrite of pearlite has a yield strength of no more than 25,000 psi when tested alone. However, as shown in Fig. 19-14, pearlite, with its layers of rigid carbide and ferrite (Fig. 17-15), has a yield strength of 80,000 psi. The basis of this greater strength is the plastic constraint imposed on the ferrite by the rigid carbide. This is also an example of the *interactive properties* cited in Section 19-4.

A further result of plastic constraint is shown in Fig. 19-17, where finer pearlite is harder than coarse pearlite (360 BHN versus 200 BHN). As is the case with thinner solder joints, finer pearlite receives greater plastic constraint. Concurrently with the reduction of plastic deformation, the material becomes more sub-

---

* See McClintock and Argon in the reference list at the end of the chapter.

ject to brittle fracture simply because the yield strength may be raised beyond the fracture stress.

Fracture initiation in ductile materials. As was said earlier, plastic constraint and strain hardening reduce ductility, thus favoring subsequent brittle fracture. In addition, plastic constraint favors the initiation of fracture in the interior of the material as a result of the triaxial stress distributions shown in Fig. 21-11(b). Consider a test bar which has been severely necked just prior to failure (Fig. 21-12). Either the radial stresses, σ_r, or the longitudinal stress, σ_l, may initiate fracturing at internal heterogeneities, such as any preexisting voids, stress concentrations around nonductile phases, or stress concentrations where slip or twinning encounters grain boundaries. Such cracks or voids can interconnect (Fig. 21-13a and b) to increase the notch effect; they may eventually produce rapid fracture from the interior outward by intersecting the surface as a 45° shear failure (Fig. 21-13c).

21-5 DUCTILE-BRITTLE TRANSITION

All materials are more likely to encounter energy-absorbing plastic deformation than brittle failure (1) at low strain rates, (2) at high temperatures, and (3) with low stress concentrations.

Asphalt provides an example of the effect of strain rate. It can be shattered by a hammer blow, but will deform under a small long-time stress. Atoms and molecules are better able to respond to stresses when time is available for their movements.

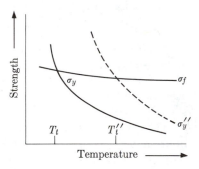

21-14 Transition temperature. Below the transition temperature, T_t, the fracture strength, σ_f, is less than the yield strength, σ_y. Therefore the material fails in a brittle manner. A notch introduces plastic constraint and therefore increases the effective yield strength to σ_y''. As a result, the transition temperature is raised to T_t''.

Most materials* become more brittle at lower temperatures. In simplified terms, the yield stress decreases with increasing temperature while the fracture stress is essentially insensitive to temperature (the solid lines of Fig. 21-14). Thus there is a ductile-brittle transition for many materials, as shown in Fig. 21-15.

---

* Fcc metals are an exception.

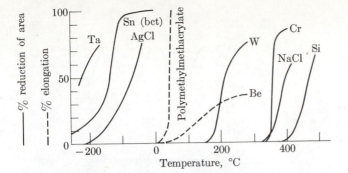

21-15 Ductility versus temperature (tensile tests—schematic). Except for fcc metals, most materials lose ductility at low temperatures. For a given material the transition temperature is higher for higher strain rates, e.g., impact loading. (After data by A. H. Cottrell, *The Mechanical Properties of Matter*. New York: Wiley, 1964, p. 358.)

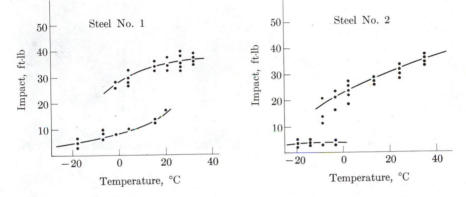

21-16 Brittle-ductile transition (Charpy test). Tough, energy-consuming fractures occur at lower temperatures for Steel No. 2 than for Steel No. 1. (Adapted from N. A. Kahn and E. A. Imbembo, American Society for Testing and Materials.)

The *transition temperature* depends in part on the microstructure of the material; for example, a fine-grained steel has a lower transition temperature than a coarse-grained steel. The transition temperature is also influenced by the stress pattern. To exemplify this, we introduce a notch into the material represented in Fig. 21-14. The resulting plastic constraint raises the yield strength by a factor of 2 or 3 to σ_y'', and thus increases the ductile-to-brittle transition temperature from T_t to T_t''.

A word of caution is in order here. Figure 21-16 implies that Steel No. 2 is tough at temperatures above 0°F and may be used without brittle fracture, whereas Steel No. 1 would undergo brittle failure at most temperatures below freezing. However, since the transition temperature is sensitive to the stress pattern and to the strain rate, the design engineer must realize that, if these factors are altered, the transition temperature is subject to variation. For a given situation we may conclude, however, that Steel No. 2 will have a lower transition temperature than Steel No. 1.

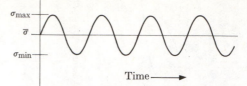

21-17 Cyclic stresses. Typically, but not necessarily, the mean stress, $\bar{\sigma}$, is zero and the stress cycle is approximately sinusoidal.

21-18 Fatigue fracture ($5\frac{1}{2}$-in. steel shaft). Fracture slowly progressed from the set-screw hole at the top through nearly 90% of the cross section before the final rapid fracture (bottom). (Courtesy of H. Mindlin, Battelle Memorial Institute.)

21-6 FATIGUE FAILURE

The term *fatigue* refers to fracture arising from *cyclic stresses* (Fig. 21-17). Commonly these stresses alternate between tension and compression as in a loaded rotating shaft. Fatigue can also occur as the result of fluctuations between cycling stresses of the same sign, as in a leaf spring or similarly loaded component of a car.

Cyclic fatigue is a characteristic of ductile materials.* Even so, the final fracture is rapid (Fig. 21-18). The number of stress cycles prior to fracture is a function of the applied stress (Fig. 21-19). Fortunately, many metals have an *endurance limit* below which the metal is not subject to fatigue failure.

Fatigue cracks usually start at the surface, though in some cases they may be initiated within a material, particularly at high stress levels (the upper part of Fig. 21-19). Surface cracks are more common, however, because (1) bending or torsion will cause the highest stresses to occur at the outer fibers, and (2) surface irregularities introduce stress concentrations as described in Eq. (21-2). As a result, the endurance limit is very sensitive to surface finish (Table 21-2). For this

---

* Glass and other nonductile materials can undergo *static fatigue* from surface reactions or time-dependent internal charges which introduce Griffith flaws. Likewise, brittle materials may undergo *thermal fatigue*, or *spalling* (see Chapter 22).

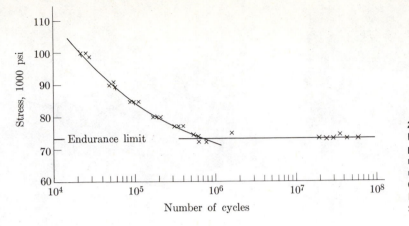

21-19 *S-N* curve (4340 steel, hot-worked bar stock); *S-N* = permissible stress versus the number of cycles before failure. (Adapted from M. F. Garwood, H. H. Zurburg, and M. A. Erickson, American Society for Metals.)

reason, shot-peening and other procedures which introduce surface compression tend to raise the endurance limit. Furthermore, fatigue life can be shortened considerably by even mildly corrosive environments. The endurance limit in dry air will be higher than in humid industrial air, and still higher in a vacuum.

It may seem surprising that stress frequency is not a factor in fatigue failure at normal temperatures. This is fortunate, because it permits accelerated testing at several kilocycles per minute; hence, the endurance limit may be established reasonably easily. An exception to this rule is fatigue failure at elevated temperatures where the fracture is intergranular rather than transgranular.

Lower-stress fatigue failure (with many cycles) commonly starts as surface cracks at 10% to 20% of the fatigue life. The cracks progress slowly from the initiation point to the final rapid failure. The two steps of crack growth and final fracture may be identified on the fracture surface (Fig. 21-18). In addition, a smooth exterior surface will often reveal *extrusions* and *intrusions* prior to crack initiation (Fig. 21-20). These may be interpreted as arising from irreversible slip during successive stress reversals. Apparently orientation, stress levels, imperfections, and other factors favor slip initiation in these localities; however, the

TABLE 21-2

SURFACE FINISH VERSUS ENDURANCE LIMIT
(SAE 4063 steel, quenched and tempered to 44 R_C)*

Type of finish	Surface roughness, μin.	Endurance limit, psi
Circumferential grind	16–25	91,300
Machine lapped	12–20	104,700
Longitudinal grind	8–12	112,000
Superfinished (polished)	3–6	114,000
Superfinished (polished)	0.5–2	116,750

* Adapted from M. F. Garwood, H. H. Zurburg, and M. A. Erickson, "Correlation of Laboratory Tests and Service Performance," *Interpretation of Tests and Correlation with Service,* Metals Park, O.: American Society for Metals, 1951.

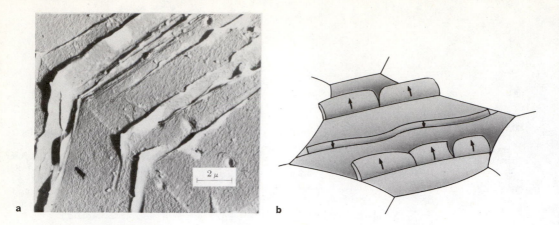

a   b

21-20 Intrusions and extrusions (×4500). (a) Electron microscope replica after cyclic straining. [A. H. Cottrell and D. Hull, *Proc. Royal Soc.* (London), **A242,** 1957, p. 211.] (b) Sketch. Extrusion "tongues" and intrusion channels form at the surface of critically oriented grains.

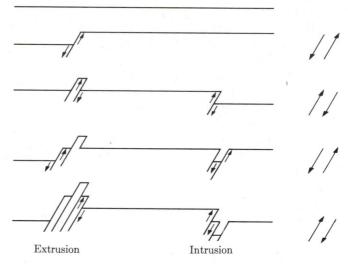

21-21 Formation of intrusions and extrusions. Progressive irreversible slip by cyclic shear stresses can form surface irregularities which produce stress concentrations to nucleate cracks.

Extrusion   Intrusion

strain hardening after each slip movement prevents back-slip on the same plane (Fig. 21-21). Because of this irreversible deformation, even those surfaces which are initially mirror smooth may develop surface irregularities that can initiate cracks after a sufficient number of stress cycles.

Mechanism of fatigue. A decrease in usable strength under cyclic loading is directly attributable to the fact that the material is not an isotropic, homogeneous solid.

a) Fatigue cracks can be initiated under relatively low stresses at surface imperfections. Imperfections from machining marks and surface damage are the most

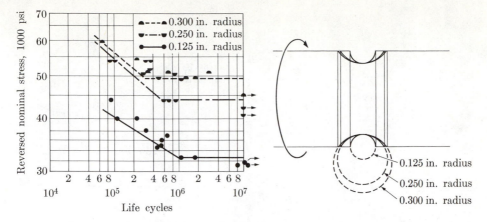

21-22 *S-N* curves of filleted test samples. The smaller radii of curvature permit higher stress concentrations and therefore lower endurance limits. (Adapted from M. F. Garwood, H. H. Zurburg, and M. A. Erickson, American Society for Metals.)

serious types; in their absence, however, imperfections may originate from local slip, as shown in the preceding figures.

b) In normal service, the stress concentrations are initially not severe enough to produce catastrophic failure, particularly since the crack tips are not atomically sharp in a ductile material; however, they are sufficient to cause slow propagation of a crack into the material.

c) Eventually the crack may become sufficiently deep so that the stress concentrations exceed the fracture strength and catastrophic failure occurs.

Steps (b) and (c) also apply to those cracks which are developed internally at the higher stress levels in the upper part of the *S-N* curves.

Design considerations. Any design factor which concentrates stresses can lead to premature fatigue failure. We have already seen from Table 21-2 that surface finish is critical. Keyways and other notches (Fig. 21-18) are also critical, as is revealed by the data of Fig. 21-22, where all three sets of results apply to the same steel. This figure emphasizes the importance of generous fillets. In fact, in many cases a component may be made stronger by *removing* material, *if* that removal increases the radius of curvature at the point of stress concentration.

21-7 CREEP AND STRESS RUPTURE

Mechanical failure of crystalline materials at high temperatures* usually occurs by means of either creep or stress rupture. *Creep* is time-dependent plastic strain. Although the rate of strain may be low, high-temperature service applications are often such that materials are exposed to stresses for long periods of time (e.g.,

---

* Greater than one third to one-half of the melting temperature, T_m.

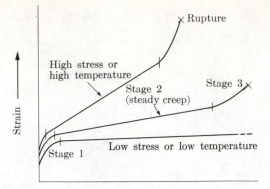

21-23 Creep. The steady rate of creep in the second stage determines the useful life of the material. (Cf. Figs. 10-15 and 12-10.)

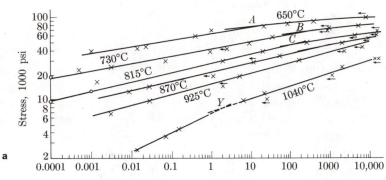

21-24 Creep (a) and stress-rupture (b) properties. The creep rate is the strain rate in stage 2 of Fig. 21-23. Stress rupture shows the time required for final failure. (After N. J. Grant, American Society for Metals.)

boiler tubes in steam power generators). *Stress rupture* is the prolongation of creep to the point where actual failure occurs.

Creep characteristics are shown in Fig. 21-23. Like the deformation behaviors described in Chapters 10 and 12, there is an initial elastic deformation. This is followed in turn by *primary creep* (Stage 1), by secondary or *steady-state creep* (Stage 2), and by *tertiary creep* (Stage 3) before ultimate rupture. The primary creep is partially anelastic and is recoverable if the stress is removed (Section 10-4). The secondary or steady-state creep is of greater engineering importance because it leads to extensive irreversible strain. The *creep rate* is commonly reported as the strain rate, $d\epsilon/dt$, during this stage. The accelerated creep in the tertiary stage is the result of an increase in the true stress, either because of necking (Fig. 21-10c) or because of internal cracking.

Effect of stress and temperature. High stresses and high temperatures have comparable effects, as shown schematically in Fig. 21-23. In quantitative terms, we can observe a logarithmic relationship between stress and creep rate as a function of temperature (Fig. 21-24a). The relationship between stress and rupture time is shown in part (b) of Fig. 21-24.

We will pay specific attention to the steady-state creep, so that we can consider the work hardening from plastic deformation as being balanced by the thermal softening.

During creep, the stress is a function of work hardening from plastic deformation and thermal softening from stress relaxations: $\sigma = \sigma(\epsilon, l)$. At steady-state creep, where the two mechanisms balance each other,

$$d\sigma = 0 = \frac{\partial \sigma}{\partial \epsilon} d\epsilon + \frac{\partial \sigma}{\partial t} dt. \tag{21-9a}$$

Let $\partial \sigma / \partial \epsilon$ equal a *hardening coefficient*, h, and $-\partial \sigma / \partial t$ equal a *thermal-softening rate*, s. At steady-state creep,

$$d\epsilon/dt = s/h = \text{constant}. \tag{21-9b}$$

However, as the temperature is raised, the hardening coefficient usually decreases and the softening rate increases rapidly. Therefore, we find that the creep rate also increases rapidly with temperature. In fact, the *Dorn-Weertman equation*,

$$d\epsilon/dt = C\sigma^n e^{-Q^*/RT}, \tag{21-10}$$

predicts a logarithmic relationship between the creep rate, $d\epsilon/dt$, the activation energy, Q^*, for self-diffusion, and the reciprocal of the absolute temperature, T (see Eq. 9-2). Other terms include C, a proportionality constant which depends on the microstructure, and n, which determines the logarithmic slope in Fig. 21-24.

Creep mechanisms. *Cross slip* and *dislocation climb* are considered to be the softening mechanisms which produce creep.† Dislocation climb is very sensitive

---

† Grain-boundary sliding, vacancy migration, and twinning may also contribute to creep in some materials.

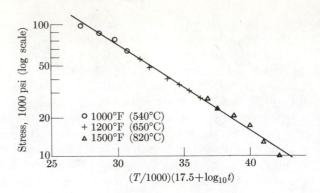

21-25 Larson-Miller parameter (S-590 alloy). The time, *t*, and temperature, *T*, relationships for high-temperature service are empirically related to the applied stress. (Adapted from data by Goldhoff, *Materials in Design Engineering*, **49**, 1949, pp. 93–97; and C. W. Richards, *Engineering Materials Science*, San Francisco: Wadsworth, 1961.)

to temperature, and probably accounts for the rapid increase of the creep rate with temperature. Recall from Section 11-2 that dislocation climb involves the diffusion of vacancies and/or interstitials to or from the dislocation. In effect, these movements are atom movements; thus the self-diffusion activation energy of Eq. (21-10) is applicable.

Materials selection. The choice of high-temperature materials must depend on various factors ranging from microstructural stability to cost. Creep is always an important factor, however, when the high-temperature service involves stresses. One of the most effective ways of ensuring low creep rates is to strengthen the material with a fine, nearly submicroscopic dispersion (Section 20-4) to interfere with dislocation movements. Specifically, the precipitation of carbides or nitrides along dislocations is very effective. To this end, alloying elements such as chromium, vanadium, tungsten, and molybdenum are often added to high-temperature alloys.

Various empirical procedures are available to correlate σ, T, and t in commercial alloys. One of the more useful is the *Larson-Miller parameter*, which expresses stress values for creep as a function of the other two variables:

$$\sigma = f\big(T(C + \log t)\big). \tag{21-11}$$

This relationship is useful because it allows us to present the thermal behavior over a wide range of time-temperature situations as a single curve for each alloy (Fig. 21-25).

REFERENCES FOR FURTHER READING

Brick, R. M., R. B. Gordon, and A. Phillips, *Structure and Properties of Alloys*, third edition. New York: McGraw-Hill, 1965. Metals for high-temperature service and fracture of metals are discussed in the last two chapters of the book. Supplementary reading for this chapter.

Charles, R. J., "A Review of Glass Strength," *Progress in Ceramics Science*, Vol. 1 (J. E. Burke, Ed.). Oxford: Pergamon, 1961. A summary of our understanding of glass fracture. Advanced level.

Cottrell, A. H., *The Mechanical Properties of Matter*. New York: Wiley, 1964. Chapter 11 reviews fracture in solids. May be used as supplementary reading for this text.

Edelglass, S. M., *Engineering Materials Science*. New York: Ronald, 1966. Creep and fracture are discussed in Chapters 12 and 14. Supplementary reading for the undergraduate student.

Ernsberger, F. M., "Current Status of the Griffith Crack Theory of Glass Strength," *Progress in Ceramics Science*, Vol. 3 (J. E. Burke, Ed.). Oxford: Pergamon, 1963. Reviews the theory and applications of the Griffith crack hypothesis to brittle glass. For the materials specialist.

Forrest, P. G., *Fatigue of Metals*. Reading, Mass.: Addison-Wesley, 1962. Reference book on the theory and analysis of fatigue behavior. For the materials specialist and the mechanical engineer. Advanced undergraduate level.

Hall, R. C., "Strengthening Ceramic Materials," *Bulletin of the American Ceramic Society*, **47** [3], pp. 251–254 (1968). The strengthening of a ceramic material depends on avoiding brittle fracture. Upperclass level.

Hayden, H. W., W. G. Moffatt, and J. Wulff, *The Structure and Properties of Materials: III. Mechanical Behavior*. New York: Wiley, 1965. Paperback. An introductory presentation of fracture is given in Chapter 7.

Kingery, W. D., *Introduction to Ceramics*. New York: Wiley, 1964. Fracture is discussed in Chapter 17. Thermal stresses are presented in Chapter 18. For the undergraduate who has previously had an introduction to materials science.

McClintock, F. A., and A. S. Argon, *Mechanical Behavior of Materials*. Reading, Mass.: Addison-Wesley, 1966. Brittle and ductile fracturing are discussed in Chapters 15 and 16. Fatigue and creep are covered in Chapters 18 and 19. Extensive references to other sources. Advanced undergraduate level.

Olcott, J. S. "Chemical Strengthening of Glass," *Science*, **140,** June 14, 1963, pp. 1189–1193. Describes procedures for introducing compressive stresses into glass without heat treatment. Introductory technical level.

Osborn, C. J. (Ed.), *Fracture*. London and Sydney: Butterworth, 1965. The first chapter provides a good summary of fracture mechanisms. A second book (1969) with the same editor, title, and publisher emphasizes the relationship between fracture theory and engineering design. For the materials specialist.

Rosenthal, D., *Introduction to Properties of Materials*. Princeton, N. J.: Van Nostrand, 1964. Fatigue strength is discussed in Chapter 10; creep properties are presented in Chapter 12; brittleness is discussed in Chapter 8. Undergraduate level.

Wulpi, D. J., *How Components Fail*. Metals Park, O.: American Society for Metals, 1967. Written for test engineers and design engineers who have not had a course in materials science.

PROBLEMS

21-1 A stress of 10^7 dynes/cm^2 is applied in the $[2\bar{1}\bar{1}3]$ direction of magnesium ($c/a = 1.624$). What is the normal stress across the (0001) plane?

Answer. 0.725×10^7 dynes/cm^2

21-2 Cleavage occurs on {111} planes of cubic CaF$_2$. Compare the relative stresses in the [110] and [100] directions which are necessary to produce cleavage.

21-3 A tensile stress is applied to a rod of brittle glass with a surface crack $2\,\mu$m deep. [Assume that the radius of curvature at the tip of the crack is 1.5 A. Typically the Young's modulus of glass is 10^7 psi (7×10^{11} dynes/cm^2) and the energy of a fractured surface is 600 ergs/cm^2.] What stress is necessary for the crack to propagate?

Answer. 16,500 psi (1.15×10^9 dynes/cm^2)

21-4 It takes a nominal stress of 3570 psi to fracture a pane of window glass. Estimate the depth of the responsible surface flaws if this glass has the properties cited in Problem 21-3.

21-5 What is the theoretical fracture strength of iron required for a crack to initiate along a cleavage plane of perfect crystal? (The surface energy is approximately 2000 ergs/cm^2. When the crack starts, its depth must equal the interatomic spacing.)

Answer. 3.3×10^{11} dynes/cm^2 (4.7×10^6 psi) [*Note.* This answer assumes an average modulus. The theoretical stress will vary from plane to plane.]

21-6 Repeat Problem 21-5 for copper which has a surface energy of 1800 ergs/cm^2.

21-7 The following data were obtained in a creep-rupture test of Inconel "X" at 1500°F: (a) 1% elongation after 10 hr, (b) 2% elongation after 200 hr, (c) 4% elongation after 2000 hr, (d) 6% elongation after 4000 hr, (e) "neck-down" initiation at 5000 hr and rupture at 5500 hr. What was the creep rate?

Answer. 0.00105%/hr

21-8 Other things being equal, which will have the lowest creep rate: (a) steel in service with a high tensile stress and low temperature, (b) steel in service with a low tensile stress and high temperature, (c) steel in service with a high tensile stress and high temperature, (d) steel in service with a low tensile stress and low temperature? Why?

21-9 On the basis of Example 21-2, explain why larger test samples often exhibit lower strengths.

21-10 Why does a material which is nonductile in tension sometimes exhibit ductility in compression?

22 □ service stability

22-1 SERVICE ALTERATION

An engineering product is made to be used. Service conditions, however, may produce alterations in materials as a result of temperature, irradiation, or chemical environment. Of course, alterations in internal structure lead to property changes, and surface reactions often cause material losses.

Internal alteration requires that energy be introduced within the material to effect structural changes. *Thermal damage* (Section 22-2) and *radiation damage* (Section 22-3) develop when the resulting structural alteration is undesirable from the standpoint of the properties of the material. Much of Materials Engineering is involved with selecting materials which meet the design criteria and at the same time are stable in the anticipated environment.

Surface reactions are of two chief types: (1) reactions with gases and (2) reactions with liquids. In type (1) the gas is usually air, and *oxidation* results (Section 22-4). We should note, however, that other types of surface reactions are possible, e.g., with the sulfur present in some fuels. *Corrosion* is the reaction of major importance between a solid material and a liquid (Section 22-5). Because of its importance in engineering considerations, we will give special attention to corrosion kinetics (Section 22-6).

Although it may be the controlling factor in certain applications, we will not give additional consideration to the *fluxing*, or solution, of materials by solvent liquids. The principles underlying this type of deterioration were covered in the discussions of phase equilibrium in Chapter 17.

22-2 THERMAL DAMAGE

Overaging (Section 20-4) and overtempering at elevated temperatures (Fig. 20-8) are two types of thermal damage which have important effect on strength. To illustrate, an age-hardened aluminum alloy cannot be used at elevated temperatures without encountering the probability of softening. Likewise, the use of a plain-carbon tool steel may be limited because softening can result from the heat generated at the cutting edge (Fig. 22-1). *High-speed* steel tools contain alloying elements such as tungsten, chromium, vanadium, and molybdenum which stabilize the carbide particles and markedly reduce the rate of their coalescence at elevated temperatures.

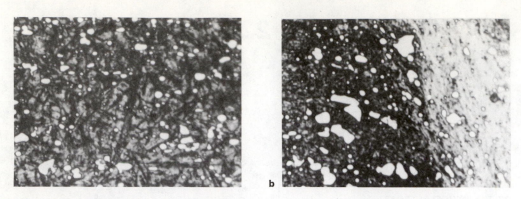

22-1 Cutting tools (high-speed steel) (×500). (a) Microstructure before use. (b) Microstructure after tip failure. (Courtesy of T. Neumeyer, Crucible Steel Co.)

Thermal damage can also arise from stresses that develop during temperature changes. Any cracking which ensues is called *spalling*, and is accentuated in non-ductile materials with high thermal expansion, high elastic moduli, low thermal diffusivity, or low strength. High expansion coefficients produce greater dimensional changes, and high elastic moduli accentuate internal stresses for a given dimensional change. A low thermal diffusivity (Section 9-3) permits steep thermal and stress gradients. Quite expectedly, ductile materials are less subject to spalling than are nonductile materials, because the former can adjust plastically to the thermal strain.

Thermal damage in polymers is of two principal types. (1) Viscoelastic behavior (Chapter 12) produces *creep* above the glass-transistion temperature. Accordingly, very few thermoplastic polymers have dimensional stability above 100° to 130°C. The few exceptions include the silicones and fluorocarbons, some of which can tolerate a temperature of 300°C under favorable conditions. (2) *Degradation* is caused by thermally activated *de*polymerization. The polymerization reactions of Chapter 7 are initiated because energy is released. Elevated temperatures then provide thermal energy which can reverse these reactions to produce *scission* of polymer chains. A reduction of the degree of polymerization, of course, alters the properties, as was discussed in Section 7-3. Degradation can also occur by chemical decomposition into nonmonomeric micromolecules such as H_2O, CH_2O, CH_4, and CH_3OH. *Charring* is the most obvious example of the last stage of decomposition short of complete gasification.

22-3 RADIATION DAMAGE

Unlike elevated temperatures, which energize all the atoms within a solid, radiation processes may focus relatively large amounts of energy locally. As sketched in Fig. 22-2, a *neutron* possessing a number of electron volts may collide with an atom, dislodging it from its crystal site. Typically, the neutron has sufficient energy remaining after the first collision to ricochet to other atoms, dislodging as many as a dozen or more before being finally captured by the nucleus of an atom. In the meantime, considerable damage has occurred within the crystal

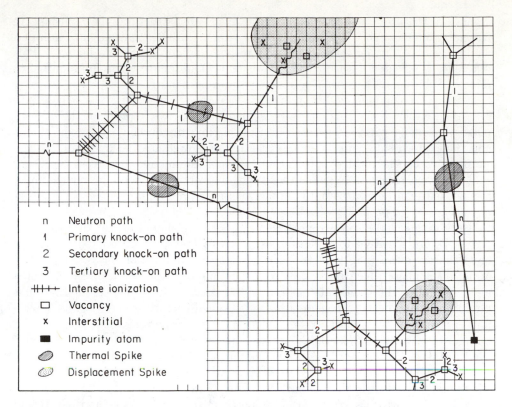

22-2 Radiation effects in a crystal lattice arising from the neutrons entering at the left. (C. O. Smith, *Nuclear Reactor Materials*. Reading, Mass.: Addison-Wesley, 1967, p. 67.)

n   Neutron path
↑   Primary knock-on path
2   Secondary knock-on path
3   Tertiary knock-on path
⫴⊢   Intense ionization
▢   Vacancy
x   Interstitial
◼   Impurity atom
⬭   Thermal Spike
⬭   Displacement Spike

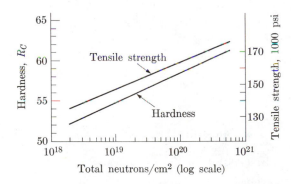

22-3 Radiation hardening (Type 347 stainless steel). Deformation is restricted by the lattice imperfections. (Adapted from C. O. Smith, ORSORT, Oak Ridge, Tenn.)

structure. Slip processes are thereby inhibited, with the result that hardness and strength increase as shown in Fig. 22-3. There is a concurrent decrease in ductility and toughness.

The property changes caused by neutron irradiation of crystalline materials are similar to the changes that arise during cold work (Section 11-5). Similarly, *annealing* will soften and toughen materials which have undergone radiation damage. Despite these comparisons, however, the structural mechanisms are basically different, because point defects are mainly responsible for radiation damage, but linear defects are mainly responsible for strain hardening. This difference is reflected in the fact that lower annealing temperatures are required to remove radiation damage than to remove strain hardening.

Scission is common within polymeric materials when they are exposed to radiation, either of the particulate type (neutrons, α-rays, etc.) or of the electromagnetic type (γ-rays, x-rays, etc.). After scission, the broken ends of the chains and the exposed side radicals possess reactive sites (Sections 3-7 and 7-2). In the simplest case, the reactive sites rejoin. More commonly, however, there are sufficient displacements so that other types of reactions can occur; for example, a chain half can be grafted onto another chain to produce *branching*, with consequent changes in properties (Section 7-6).

By noting several items from earlier chapters, we should expect that *electrical properties* are sensitive to radiation. (1) Recall first that disordered structures have shorter mean free paths for electron movement (Section 14-3). Thus any radiation damage which introduces imperfections into the crystal structure of a metal increases the electrical resistivity. (2) Also recall from Section 15-2 that photoconduction occurs when an electron is excited across the energy gap. Therefore, irradiation by photons in the ultraviolet-to-γ-ray range can increase the number of intrinsic carriers (both electrons and holes) by several orders of magnitude. Hence, the resistivity of semiconductors is decreased. (3) Finally, recall from Section 13-6 that vacancies in some ionic solids may produce color centers which modify dielectric and optical behaviors.

Table 22-1 presents examples of property changes which are attributable to neutron radiation.

Example 22-1

Assume that all the energy required to produce scission in a polyethylene molecule comes from a photon (and that none of the energy is thermal).

a) What is the maximum wavelength which can be used?
b) How many eV are involved?

Answer

a) From Table 3-4,

$$C\text{—}C = \frac{88,000 \text{ cal}}{0.6 \times 10^{24} \text{ bonds}} = 1.46 \times 10^{-19} \text{ cal/bond}$$

$$= (1.46 \times 10^{-19} \text{ cal/bond})(4.185 \times 10^7 \text{ ergs/cal})$$

$$= 6.1 \times 10^{-12} \text{ erg/bond}.$$

From Eq. (2-1),

$$\text{energy} = h\nu = hc/\lambda,$$

$$\lambda = \frac{(6.62 \times 10^{-27} \text{ erg} \cdot \text{sec})(3 \times 10^{10} \text{ cm/sec})}{(6.1 \times 10^{-12} \text{ erg})}$$

$$= 3.2 \times 10^{-5} \text{ cm} = 3700 \text{ A}.$$

b)

$$\text{eV} = (6.1 \times 10^{-12} \text{ erg})(6.24 \times 10^{11} \text{ eV/erg}) = 3.8. \blacktriangleleft$$

TABLE 22-1

EFFECTS OF RADIATION ON VARIOUS MATERIALS*

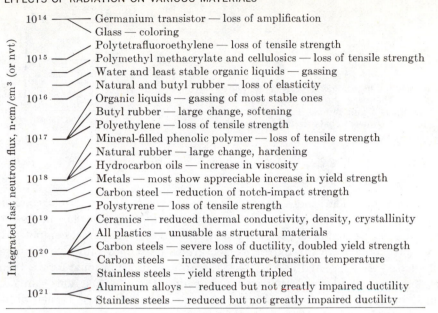

10^{14} — Germanium transistor — loss of amplification
Glass — coloring
Polytetrafluoroethylene — loss of tensile strength
10^{15} — Polymethyl methacrylate and cellulosics — loss of tensile strength
Water and least stable organic liquids — gassing
Natural and butyl rubber — loss of elasticity
10^{16} — Organic liquids — gassing of most stable ones
Butyl rubber — large change, softening
Polyethylene — loss of tensile strength
10^{17} — Mineral-filled phenolic polymer — loss of tensile strength
Natural rubber — large change, hardening
Hydrocarbon oils — increase in viscosity
10^{18} — Metals — most show appreciable increase in yield strength
Carbon steel — reduction of notch-impact strength
Polystyrene — loss of tensile strength
10^{19} — Ceramics — reduced thermal conductivity, density, crystallinity
All plastics — unusable as structural materials
Carbon steels — severe loss of ductility, doubled yield strength
10^{20} — Carbon steels — increased fracture-transition temperature
Stainless steels — yield strength tripled
10^{21} — Aluminum alloys — reduced but not greatly impaired ductility
Stainless steels — reduced but not greatly impaired ductility

Integrated fast neutron flux, n-cm/cm³ (or nvt)

* Indicated exposure levels are approximate. Indicated changes are at least 10%. The table is reprinted from C. O. Smith, *Nuclear Reaction Materials*. Reading, Mass.: Addison-Wesley, 1967, p. 70.

22-4 OXIDATION

In this section we shall consider oxidation in the narrow sense of oxide formation, either

$$\text{metal} + \text{oxygen} \rightarrow \text{metal oxide} \tag{22-1a}$$

or

$$\text{polymer} + \text{oxygen} \rightarrow CO, CO_2, H_2O, \text{etc.} \tag{22-1b}$$

When we study corrosion in the next section, we shall consider oxidation more generally as electron removal (or valence increase):

$$\text{element} \rightarrow \text{cation} + \text{electron} \tag{22-2a}$$

or

$$\text{anion} \rightarrow \text{element} + \text{electron.} \tag{22-2b}$$

Most metals and polymers are not thermodynamically stable in air, but can be used in air as engineering materials because their oxidation rates are tolerably slow, or can be brought under control.* Iron, for example, releases energy as it

* Most ceramic products are stable with respect to oxidation. Many of them, however, would hydrate with the ambient moisture if the reaction rates were significantly increased to achieve equilibrium.

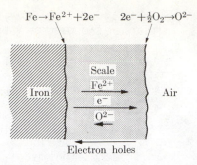

$Fe \rightarrow Fe^{2+} + 2e^-$ $2e^- + \frac{1}{2}O_2 \rightarrow O^{2-}$

22-4 Scale formation. The scale on iron, $Fe_{1-x}O$, has cation vacancies (Fig. 8-9). Therefore, the Fe^{2+} ions diffuse outward more readily than the O^{2-} ions diffuse inward. Electrons and electron holes must also move to balance the charge.

forms iron oxide; likewise, organic materials eventually produce CO_2, H_2O, and other products of combustion.

Scale formation. The oxides of most metals are solids under normal conditions. As a result, diffusion is a very important part of oxidation (*scaling*), because once scale is present the metal and the oxygen of the air are not in direct contact. Metal atoms must move outward and/or oxygen atoms must move inward through the oxide phase. As suggested in Fig. 22-4, the atoms diffuse in ionic form. In most cases, cations will diffuse outward more readily than oxygen ions can diffuse inward, because the cations are commonly smaller and have higher diffusion coefficients. Among the exceptions, however, is the oxidation of uranium. Although larger than the uranium cation, the oxygen anion has a higher diffusion coefficient because this oxide is nonstiochiometric with a deficiency of oxygen ions; i.e., oxygen vacancies are present which greatly increase the anion mobility (Fig. 9-14).

As the scale grows in thickness, the ion flux decreases. In fact, the rate at which the thickness increases, dx/dt, is inversely proportional to the thickness:

$$dx/dt \propto 1/x, \tag{22-3a}$$

or

$$x = k\sqrt{t}. \tag{22-3b}$$

This is known as the *parabolic rate law*, which is applicable to the oxidation of those metals which develop fully dense, nonporous scales. The proportionality constant, k, varies, however, as a direct function of the diffusion coefficient, oxygen pressure, and temperature.

Some metals form porous scales. When this occurs, the oxygen has more direct access to the underlying metal. With perfectly free access, the oxidation rate would be *linear* with time. The oxide may be porous either because the oxide occupies less volume than the initial metal, or because there is sufficient volume increase to force the scale to expand and crack off (literally, "scale off").

Example 22-2

Compare the volume changes of (a) iron and (b) magnesium as they are oxidized to FeO and MgO. [FeO and MgO both have the NaCl structure (Fig. 4-1), with lattice constants of 4.3 A and 4.2 A, respectively.]

Answer. From Appendix B,

$$\rho_{Fe} = 7.87 \text{ gm/cm}^3, \quad \text{and} \quad \rho_{Mg} = 1.74 \text{ gm/cm}^3.$$

a) $\quad \rho_{FeO} = \dfrac{(4 \text{ FeO/unit cell})(55.85 + 16.00 \text{ gm FeO/mole})}{(0.602 \times 10^{24} \text{ FeO/mole})(4.3 \times 10^{-8} \text{ cm})^3/\text{unit cell}} = 5.95 \text{ gm/cm}^3.$

Basis: 1 cm^3 Fe   (= 7.87 gm).

$$(7.87 \text{ gm Fe})(71.85 \text{ gm FeO/mole})/(55.85 \text{ gm Fe/mole}) = 10.1 \text{ gm FeO},$$
$$(10.1 \text{ gm FeO})/(5.95 \text{ gm/cm}^3) = 1.7 \text{ cm}^3 \text{ FeO per cm}^3 \text{ Fe}.$$

b) Basis: 1 cm^3 Mg   (= 1.74 gm).

$$\frac{(1.74 \text{ gm/cm}^3)(0.602 \times 10^{24} \text{ MgO/mole})(4.2 \times 10^{-8} \text{ cm})^3 \text{ MgO/unit cell}}{(24.03 \text{ gm Mg/mole})(4 \text{ MgO/unit cell})} = 0.8 \; \frac{\text{cm}^3 \text{ MgO}}{\text{cm3 Mg}}.$$

Note. In spite of adding oxygen, the volume contracts because the strong coulombic forces between the Mg^{2+} and O^{2-} ions markedly reduce the interatomic distances. ◀

Polymer oxidation. Oxidation presents a special engineering problem for elastomers of the unsaturated, or butadiene, type (Table 7-3), because as oxygen diffuses into the rubber and reacts, it provides additional cross-links. The result is a hardening of the rubber. In addition, oxygen reacts readily in any polymer with reactive sites that are established during chain scission (Section 22-2). This accelerates the depolymerization, and eventually produces micromolecules which escape from the material as a gas.

In applications where oxidation presents a problem, *antioxidant* additives are used to provide competitive reactions, either (a) to consume the oxygen which diffuses in or (b) to preferentially combine with free radicals that result from scission.

22-5 CORROSION

The deterioration of solids by liquid electrolytes is called *corrosion* (Fig. 22-5). The liquids most commonly involved in corrosion are *aqueous solutions*, and the solids with which we will be concerned here are *metallic*.* Since the reactions involve electron transfer, corrosion is an electrochemical process.

Most simply stated, metallic corrosion is the opposite of *electroplating*. In the sketch of Fig. 22-6, the left electrode, called the *anode*, is being corroded and is *supplying electrons to the external circuit*, while the right electrode, or *cathode*, is *receiving electrons from the external circuit*.† Thus the anode is undergoing an

---

* We should be aware, however, that comparable deterioration can occur in nonmetallic systems, e.g., between molten glass and the refractory lining of a glass tank.
† These italicized definitions apply to anodes and cathodes in all electrical circuits—even in radio tubes, which have chemically inert electrodes. In electrochemical reactions, the *anode* is corroded.

22-5 Corrosion. This form of deterioration costs about 10^{10} per year. It involves electrochemical oxidation.

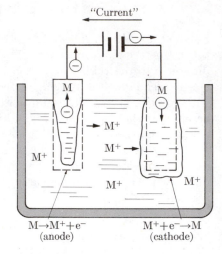

"Current"

M→M⁺+e⁻
(anode)

M⁺+e⁻→M
(cathode)

22-6 Corrosion and electroplating are opposites. The anode reaction is an *oxidation* reaction. The cathode reaction is a *reduction* reaction. Electrons move from the anode through the *external* circuit to the cathode.

oxidation reaction because the valence state is being increased:

$$Cu \rightarrow Cu^{2+} + 2e^-. \tag{22-4a}$$

Meanwhile, *reduction* is occurring at the cathode:

$$Cu^{2+} + 2e^- \rightarrow Cu, \tag{22-4b}$$

where the valence state is being reduced. There are many electrode reactions possible in actual corrosion. As corrosion continues, it is necessary for the amount of oxidation and reduction to be equal in order to balance out the charges.

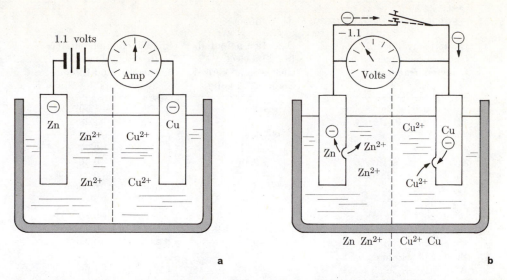

22-7 Electrochemical potential (Zn-Cu in molar nonaqueous solutions). (a) A reverse potential of 1.1 volts is required to prevent electron movements through the external circuit. (b) A voltage difference of 1.1 volts occurs in an open circuit. When closed, corrosion occurs at the zinc anode (Eq. 22-6) and copper is plated at the cathode (Eq. 22-4b).

Example 22-3

The electroplating process of Fig. 22-6 utilizes 25 amp. What is the corrosion rate of copper at the anode (and the plating rate at the cathode, if there are no side reactions)?

Answer. Basis: gm/hr.

Charge transferred = (25 amp)(3600 sec) = 90,000 coul.

Electrons involved = (90,000 coul)/(1.6 × 10^{-19} coul/electron)

$\qquad$ = 5.6 × 10^{23} electrons

$\qquad$ = 2.8 × 10^{23} Cu^{2+} ions,

gm/hr = (2.8 × 10^{23} Cu^{2+})(63.54 gm/mole)/(0.602 × 10^{24} Cu^{2+}/mole)

$\qquad$ = 30 gm/hr.

Alternatively, from Faraday's law, which is taught in general chemistry courses,

$$w = ItM/n\mathfrak{F} \qquad (22\text{-}5)$$

$$= \frac{(25 \text{ amp})(3600 \text{ sec})(63.54 \text{ gm/mole})}{(2 \text{ equiv./mole})(96,500 \text{ coul/equiv.})}$$

$$= 30 \text{ gm}. \quad \blacktriangleleft$$

Electrochemical potentials. For corrosion to occur, the anode and cathode must have electrical contact. Furthermore, to produce a net reaction they must be at different potentials. In Fig. 22-6, the potential was supplied by a battery. Other

TABLE 22-2

ELECTRODE POTENTIALS (25°C; molar solutions)

Oxidation reaction (the arrow is reversed for reduction reactions)	Oxidation potential, ϕ^0 (used by electro-chemists and corrosion engineers*), volts			Reduction potential, ϕ^0 (used by physical chemists and thermo-dynamicists*), volts
$Au \rightarrow Au^{3+} + 3e^-$	$+1.50$			-1.50
$2H_2O \rightarrow O_2 + 4H^+ + 4e^-$	$+1.23$			-1.23
$Ag \rightarrow Ag^+ + e^-$	$+0.80$	cathodic	(noble)	-0.80
$Fe^{2+} \rightarrow Fe^{3+} + e^-$	$+0.77$			-0.77
$4(OH) \rightarrow O_2 + 2H_2O + 4e^-$	$+0.40$			-0.40
$Cu \rightarrow Cu^{2+} + 2e^-$	$+0.34$			-0.34
$H_2 \rightarrow 2H^+ + 2e^-$	0.0000	Reference		0.0000
$Pb \rightarrow Pb^{2+} + 2e^-$	-0.13			$+0.13$
$Ni \rightarrow Ni^{2+} + 2e^-$	-0.25			$+0.25$
$Fe \rightarrow Fe^{2+} + 2e^-$	-0.44			$+0.44$
$Cr \rightarrow Cr^{3+} + 3e^-$	-0.74	anodic	(active)	$+0.74$
$Zn \rightarrow Zn^{2+} + 2e^-$	-0.76			$+0.76$
$Al \rightarrow Al^{3+} + 3e^-$	-1.66			$+1.66$
$Mg \rightarrow Mg^{2+} + 3e^-$	-2.36			$+2.36$
$Li \rightarrow Li^+ + e^-$	-2.96			$+2.96$

* The choice between oxidation potential and reduction potential is arbitrary. Since we are concerned with corrosion, we will follow the middle column (oxidation potential).

wise the reduction reaction (Eq. 22-4b) and the oxidation reaction (Eq. 22-4a), being identical except for their opposite signs, would cancel each other.

When the electrodes are unlike, the oxidation reactions at the two electrodes will not be equal. In fact, it would be necessary to apply 1.1 volts to the Zn-Cu circuit of Fig. 22-7 to *prevent* electron flow. Conversely, a potential of 1.1 volts is established between the electrodes in an open circuit. In effect,

$$Zn \rightarrow Zn^{2+} + 2e^- \tag{22-6}$$

occurs more readily than does the copper oxidation reaction (Eq. 22-4a). When electrical contact is made between the two electrodes, corrosion occurs at the anode, electrons move to the cathode, and (in this example) copper is plated on the cathode (Eq. 22-4b).

Electrode potentials must be based on an arbitrary reference. Because many electrolytes of interest contain hydrogen ions, it has been convenient to establish the reaction

$$H_2 \rightarrow 2H^+ + 2e^- \tag{22-7}$$

as our reference.* The data of Table 22-2 are based on this procedure.

---

* Similarly, elevation measurements are arbitrarily, but conveniently, taken from sea level.

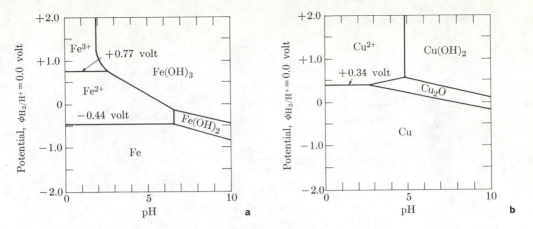

22-8 Potential-pH equilibrium diagram (25°C, 1 molar). (a) Iron. (b) Copper. Called Pourbaix diagrams, these diagrams show the equilibrium oxidation state versus pH. (Simplified from M. Pourbaix, *Atlas of Electrochemical Reactions.* Oxford: Pergamon Press, 1966, pp. 313, 388.)

The products of corrosion may be soluble ions or precipitates, depending on the pH of the electrolyte. This is illustrated for iron in Fig. 22-8(a) through the *Pourbaix diagram.** Although it is not a phase diagram, a *pH-potential diagram* such as this is an *equilibrium diagram* and indicates the ions or precipitates with the lowest free energy. As with a phase diagram, a Pourbaix diagram indicates only the lowest-energy species and *tells nothing about how rapidly* the corrosion reactions will occur.

By comparing Fig. 22-8(a) and (b) we observe that metallic copper and ionic iron are stable in the neutral or low-pH region between -0.44 volt and $+0.34$ volt. Thus if copper ions and metallic iron are in contact within an electrolyte, the iron will oxidize to ions and any Cu^{2+} ions will be reduced to metal. Furthermore, at least $+0.78$ volt, i.e., $[0.34 \text{ volt} - (-0.44 \text{ volt})]$, would have to be applied to the iron before it could become cathodic to the copper in a one-molar solution at 25°C. This conclusion, of course, agrees with the data in Table 22-2.

Cathode reactions. A variety of reactions can consume electrons at the cathode:

Electroplating of (divalent) metals (M): $\quad M^{2+} + 2e^- \rightarrow 2M.$ (22-8)

Hydrogen generation: $\quad 2H^+ + 2e^- \rightarrow H_2\uparrow.$ (22-9)

Water decomposition: $\quad 2H_2O + 2e^- \rightarrow H_2\uparrow + 2(OH)^-.$ (22-10)

Hydroxyl formation: $\quad O_2 + 2H_2O + 4e^- \rightarrow 4(OH)^-.$ (22-11)

Water formation: $\quad O_2 + 4H^+ + 4e^- \rightarrow 2H_2O.$ (22-12)

Which reaction operates depends on the electrolytic environment. Obviously, metal ions, M^+, must be present when plating is the predominant cathodic reaction. Reaction (22-12) requires the presence of oxygen in acid; Eq. (22-11)

---

* Named after its originator.

TABLE 22-3

SUMMARY OF GALVANIC CELLS

	Examples	Anode (oxidation)	Cathode (reduction)
Electrode composition		Baser phase	Nobler phase
	Zn versus Fe	Zn	Fe
	Fe versus H_2	Fe	H_2
	H_2 versus Cu	H_2	Cu
	Pearlite	α	Carbide
Stress cells		Higher energy	Lower energy
	Boundaries	Boundaries	Grain
	Grain size	Fine-grain	Coarse-grain
	Imperfections	Defect	Perfect
	Strains	Cold-worked	Annealed
	Stresses	Loaded areas	Nonloaded areas
Solution concentration		Lower concentration	Higher concentration
	Electrolyte	Dilute solution	Concentrated solution
	Oxidation	Low O_2	High O_2
	Dirt or scale	Covered areas	Clean areas

requires oxygen in basic or neutral environments. Equation (22-9) is predominant in acid, oxygen-free electrolytes.

Galvanic cells. Oxidation-reduction reactions can occur under several conditions, as summarized in Table 22-3. The first situation involves *unlike electrodes* where the nobler phase serves as the cathode and the baser phase as the anode. Note that the couple need not involve massive electrodes, but may be between two phases of a microstructure, such as in pearlite (Fig. 22-9).

The second type of galvanic cell is a *stress cell*. A distorted, higher-energy zone such as a grain boundary, a dislocation line, or cold-worked metal serves as an anode, while the strain- or stress-free region is the cathode. This type of cell is used advantageously in examining microstructures, because grain boundaries (Fig. 6-17) can be revealed by preferential etching (Fig. 22-10). Imperfections serve as anodes in that the adjacent regions possess extra strain energy, and less additional energy is required to oxidize the affected atoms to ions.

The third general type of galvanic cell is called a *concentration cell*, and arises from differences in electrolyte composition. Consider Fig. 22-11, where electrode D of copper is in the more dilute Cu^{2+}-containing electrolyte and electrode C is in the more concentrated solution. Equation (22-4), rewritten as $Cu \rightleftarrows Cu^{2+} + 2e^-$, applies to both electrodes. Reaction (22-4) moves to the left as a cathodic reaction on the electrode in the concentrated Cu^{2+} electrolyte, and corrosion occurs, with more copper being ionized on electrode D.

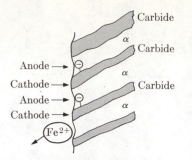

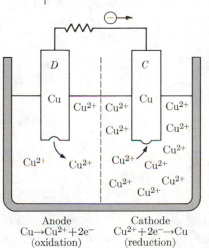

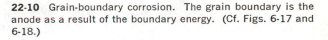

22-9 Galvanic microcell (pearlite). Since the two phases are different in composition, they have different electrode potentials and produce a small galvanic cell.

22-10 Grain-boundary corrosion. The grain boundary is the anode as a result of the boundary energy. (Cf. Figs. 6-17 and 6-18.)

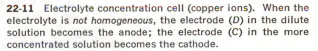

Anode
$Cu \rightarrow Cu^{2+} + 2e^-$
(oxidation)

Cathode
$Cu^{2+} + 2e^- \rightarrow Cu$
(reduction)

22-11 Electrolyte concentration cell (copper ions). When the electrolyte is *not homogeneous*, the electrode (*D*) in the dilute solution becomes the anode; the electrode (*C*) in the more concentrated solution becomes the cathode.

The concentration cell accentuates corrosion by producing an anode where the concentration of the metal ion is lower.

Concentration cells of the above type are frequently encountered in chemical plants, and also under certain erosion-corrosion conditions. However, in general, they are less critical than are *oxidation-type concentration cells*. Consistent with Eqs. (22-11) and (22-12), oxygen consumes electrons as a cathodic reaction, and

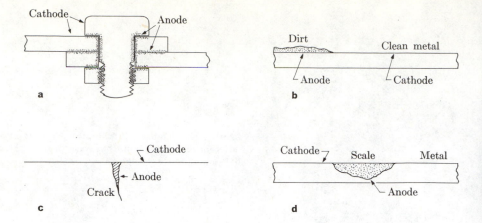

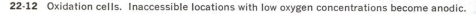

22-12 Oxidation cells. Inaccessible locations with low oxygen concentrations become anodic.

therefore permits the galvanic reaction to proceed. But note in examples like those sketched in Fig. 22-12 that the exposed, oxygen-rich areas are cathodic, while the inaccessible locations serve as anodes and are subjected to corrosion.

The oxidation cell accentuates corrosion by producing an anode where the oxygen concentration is lower. Of course, this is an insidious situation because corrosion will occur in inaccessible locations; and as scale or rust builds up to restrict the supply of oxygen, these areas become even more anodic, while the exposed oxygen-rich areas are cathodic. Rapid pit corrosion can then follow (Fig. 22-12d).

22-6 CORROSION RATES

If zinc is placed alone in a nonaqueous electrolyte, the *oxidation* reaction,

$$\text{Zn} \rightarrow \text{Zn}^{2+} + 2\text{e}^-, \tag{22-13a}$$

and the *reduction* reaction,

$$\text{Zn}^{2+} + 2\text{e}^- \rightarrow \text{Zn}, \tag{22-13b}$$

proceed at equal rates. As a result, equilibrium is developed so that we see *no net* corrosion. It is important, however, to realize that, in this apparently static situation, electrons are being freed and recombined according to Eq. (22-13a and b). It is possible to determine experimentally the number of electrons which are involved, and therefore to calculate the number of amperes per cm^2 at equilibrium. This is called the *exchange current density*, i_0. It varies with the electrode and the electrode reactions as shown in Table 22-4. Of course, at equilibrium this current density must equal the oxidation rate, or anode current density, i_a, and the reduction rate, or cathode current density, i_c*:

$$i_a = i_c = i_0. \tag{22-14}$$

---

* This is based on the assumption of equal anode and cathode areas.

TABLE 22-4

EXCHANGE CURRENT DENSITIES AND TAFEL CONSTANTS
(F. M. Donahue, The University of Michigan)

Reaction	Electrode	Exchange current density, i_0, amp/cm^2*	Tafel constant, B, volts*
Anode:			
$Fe \rightarrow Fe^{2+} + 2e^-$	Iron	10^{-8}	0.02
$Cu \rightarrow Cu^{2+} + 2e^-$	Copper	10^{-5}	0.03
$Zn \rightarrow Zn^{2+} + 2e^-$	Zinc	10^{-5}	0.02
Cathode:			
$Fe^{2+} + 2e^- \rightarrow Fe$	Iron	10^{-8}	0.05
$Cu^{2+} + 2e^- \rightarrow Cu$	Copper	10^{-5}	0.05
$Zn^{2+} + 2e^- \rightarrow Zn$	Zinc	10^{-5}	0.05
$2H^+ + 2e^- \rightarrow H_2$	Iron	10^{-6}	0.05
$2H^+ + 2e^- \rightarrow H_2$	Copper	10^{-7}	0.05
$2H^+ + 2e^- \rightarrow H_2$	Zinc	10^{-8}	0.08
$2H^+ + 2e^- \rightarrow H_2$	Platinum	10^{-3}	0.05
$2H^+ + 2e^- \rightarrow H_2$	Lead	10^{-11}	0.06

* Approximate values, 20°C, for 1-molar solutions.

If an added voltage, or *overpotential*, ϕ, is applied to an electrode and the circuit is completed, the current density, i, will increase according to the equation

$$i = i_0 e^{\phi/B}, \qquad (22\text{-}15)$$

where B, the "Tafel constant," has the same sign as ϕ and is usually between 0.02 and 0.08 volt at normal temperature (Table 22-4).

The effect of overvoltage on the oxidation and reduction of a zinc electrode is shown in Fig. 22-13. In fact, the exchange current density, i_0, is determined by

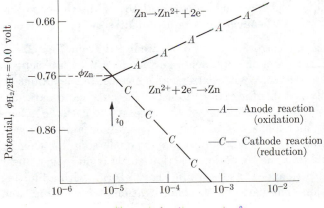

—A— Anode reaction (oxidation)

—C— Cathode reaction (reduction)

22-13 Electrode current density (zinc in 1-molar Zn^{2+} solution). At equilibrium, the anodic current density, i_a, and the cathodic current density, i_c, are equal to the exchange current density, i_0. The two curves are semilogarithmic according to Eq. (22-15). The exchange current density and the Tafel constants, B, which give the slopes of the rate curves must be determined experimentally for each electrode reaction (Table 22-4).

measuring the current densities with positive and negative overvoltages and extrapolating the oxidation and reduction curves to the point where they intersect.

Example 22-4

Assume the metal, M, of Fig. 22-6 is zinc and the cathode area is 23.5 cm². How much zinc is plated per hour in the cathode if the applied voltage is 0.2?

Answer. On the basis of Fig. 22-13, the current density is about 2×10^{-4} amp/cm² when the voltage *difference* between the anode and cathode is 0.2 volt:

$$\frac{(2 \times 10^{-4} \text{ amp/cm}^2)(23.5 \text{ cm}^2)(3600 \text{ sec/hr})(65.37 \text{ gm/mole})}{(2 \times 1.6 \times 10^{-19} \text{ coul/Zn atom})(0.6 \times 10^{24} \text{ Zn atoms/mole})} = 5.74 \times 10^{-3} \text{ gm/hr.} \quad \blacktriangleleft$$

As listed in Table 22-4, the exchange current density for the hydrogen reaction on an inert platinum electrode is 10^{-3} amp/cm². Therefore, when zinc and platinum are connected in an acid solution, we have the two sets of current-density curves shown in Fig. 22-14. The *mixed electrodes* produce a corrosion current density, i_{co}, of 15 amp/cm² where the cathode and anode curves for the *two* electrodes cross each other at about -0.48 volt. This current density determines the *corrosion rate* in a galvanic cell.

Example 22-5

The corrosion current, i_{co}, through an iron anode is 10^{-4} amp/cm². What is the corrosion rate in mm/yr? (Assume uniform corrosion.)

Answer

$$\frac{(10^{-4} \text{ amp/cm}^2)(3.15 \times 10^7 \text{ sec/yr})}{1.6 \times 10^{-19} \text{ coul/electron}} = 2 \times 10^{22} \text{ electrons/cm}^2 \cdot \text{yr}$$

$$= 10^{22} \text{ Fe atoms/cm}^2 \cdot \text{yr,}$$

$$\frac{(10^{22} \text{ Fe/cm}^2 \cdot \text{yr})(55.85 \text{ gm Fe/mole})}{(0.602 \times 10^{24} \text{ Fe/mole})(7.87 \text{ gm Fe/cm}^3)} = 1.18 \text{ mm/yr.} \quad \blacktriangleleft$$

Stress cells. As a result of the stored energy in strained metals, the exchange current density, i_0, is increased as shown in Fig. 22-15, with only slight modification in the equilibrium potential, ϕ, and the reaction constant, B. Thus the corrosion current is increased from i_{co} to i_{co*} whenever the strained metal serves as the anode.

Cathodic polarization. When a corrosion current is predicted from Fig. 22-14, one must assume that ions can leave or approach a surface at a rate sufficient to match the corrosion current. Except with extremely concentrated solutions or isolated surfaces, it is usually possible for cations to leave the anode and enter the electrolyte without diffusional delays. This is not the case at the cathode, where low concentration gradients and molecular oxygen or polyatomic anions are involved (Eqs. 22-10, 22-11, and 22-12). Their supply may be depleted in the vicinity of a cathode, with the result that the reduction reaction curve is no longer linear with a semilog plot; rather it will appear as shown in Fig. 22-16(a), with a limiting rate in the reduction reaction that varies with concentration, temperature, and electrolyte movement. When any of these are high, a sufficient number of cations

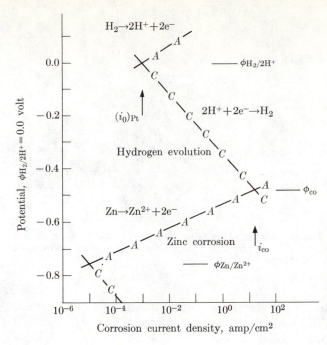

22-14 Mixed electrodes (zinc anode and hydrogen evolution on an inert platinum cathode). With two or more sets of oxidation-reduction reactions, the corrosion current density, i_{co}, is established where the sum of the oxidation current density equals the sum of the reduction current density.

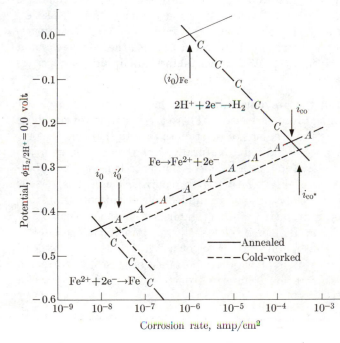

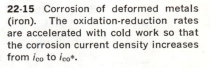

22-15 Corrosion of deformed metals (iron). The oxidation-reduction rates are accelerated with cold work so that the corrosion current density increases from i_{co} to $i_{co}*$.

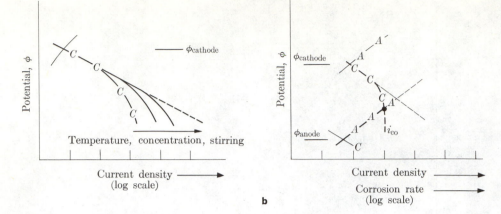

22-16 Cathodic polarization. (a) With diffusion limitations, the reduction current density is restricted. Polarization is removed by higher temperatures, more concentrated electrolytes, and moving electrolytes. (b) A polarized cathode limits the corrosion current density, i_{co}.

can arrive at the cathode surface per second to satisfy the current requirement. When low, the reaction is diffusion-limited and the interception of the cathode reduction curve with the anode oxidation curve is at a lower current density (Fig. 22-16b). Called *polarization*, this corrosion-current decrease is highly desirable for minimizing corrosion.*

Unlike the concentration cell of the previous section, which changed the equilibrium to favor corrosion, we now find that more concentrated electrolytes increase the *rate* of corrosion through decreased polarization and higher current densities. In addition, we expect that a flowing electrolyte, or a rapidly moving pump part, encounters less polarization and more rapid corrosion because the depleted anion layer does not build up at the surface. With other factors equal, higher velocities increase corrosion rates.

Passivation. It was stated in the previous subsection that, except with isolated surfaces, the anode is not subject to polarization. There are cases, however, where the anode surface is isolated by the products of reaction so that the current density, and therefore the corrosion rate, is limited. Naturally, this is desirable from an engineering point of view. Passivation, as anode isolation is called, may be represented by the *passivation curve* of Fig. 22-17. With oxidizing conditions (either because of high oxygen activity, or because of an induced electrical potential from a battery or galvanic source), a protective film is established on anodic surfaces of stainless steels, titanium, silicon, and some related alloys. It has not been determined whether the film is a thin metal oxide layer or a tightly adsorbed layer of oxygen atoms. Whichever it is, it is of major importance because the corrosion rate is almost nil in the *passive* region. Observe from Fig. 22-17 that there is a

---

* It is undesirable in a dry cell or battery, where we utilize corrosion to produce a current.

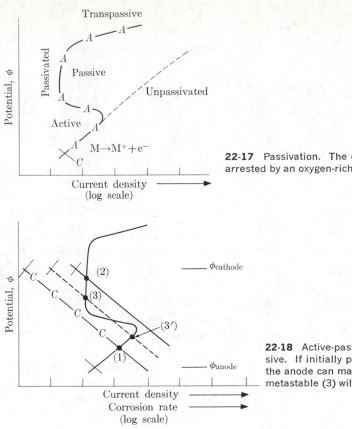

22-17 Passivation. The oxidation reaction on the anode is arrested by an oxygen-rich surface film.

22-18 Active-passive metals. (1) Active. (2) Passive. If initially passivated, and then altered to (3), the anode can maintain protection. If triggered, the metastable (3) will become active (3′).

transpassive region where the potential is sufficient to evolve oxygen and remove the isolating film. The metallurgical engineer must take advantage of the passive region of the anodic reaction.*

Figure 22-18 gives the mixed-potential presentation for a passivated material. We will examine three cases. (1) When the current density of the cathode (or the corresponding voltage) is such that the cathode and anode curves cross in the *active* region, corrosion proceeds as it did in Fig. 22-14. (2) When the current density of the cathode is such that the cathode and anode curves cross in the *passive* region, corrosion is nil. Under oxidizing conditions, stainless steels normally fall in this category. (3) The last case is a special one because the oxidation and reduction curves cross several times. Consider, for example, iron which has been passivated in concentrated nitric acid [point (2)]; this iron is then placed in more dilute, less oxidizing acid so that point (3) is applicable. Because point (3) is

---

* This is another example of why the corrosion mechanism has been confusing. Whereas oxygen can favor the cathode reaction (Eqs. 22-11 and 22-12) and can establish oxidation cells to facilitate corrosion (Fig. 22-12), it can also provide protection for certain metals through passivation. Only as these distinctions have been unraveled has progress been made in understanding corrosion.

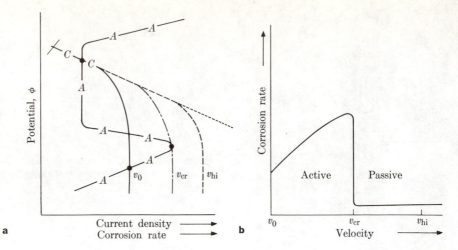

a

b

22-19 Polarization and passivation. (See Example 22-6.)

metastable, the iron can possess passivity, but the corrosion current can be activated to point (3′) if the film is ruptured. Under these conditions corrosion will proceed as rapidly as or more rapidly than if the iron had been immersed directly into the active state of Case 1.

The metallurgical engineer obviously prefers Case 2. Case 3 presents a particular problem because corrosion rates cannot be predicted. A passivated metal may suddenly become activated by various service factors to give a very high corrosion rate.

Example 22-6

Compare the effect of polarization on the corrosion rate (a) in a normal metal and (b) in a passivated metal.

Answer

a) Refer to Fig. 22-16. Stirring, or increased electrolyte velocity, will increase the corrosion rate until the cathode current is not diffusion-limited. Thereafter, the corrosion rate remains constant and high.

b) Refer to Fig. 22-19(a). With stagnant electrolytes (nil velocity, v_0) passivation is only metastable, because the cathode curve crosses the active part of the anode curve. With increased velocities, the corrosion rate increases until the critical velocity, v_{cr}, is exceeded. Thereafter, the passive state becomes stable and the corrosion rate becomes negligible at high velocities, v_{hi} (Fig. 22-19b). ◀

Anode size effect. The previous discussions have assumed equal anode and cathode areas, and therefore equal current densities. In practice, the anode may be very small (e.g., steel screws in a copper plate). In this case, since all of the anode current passes through a small area, the corrosion rate may be excessive. Also,

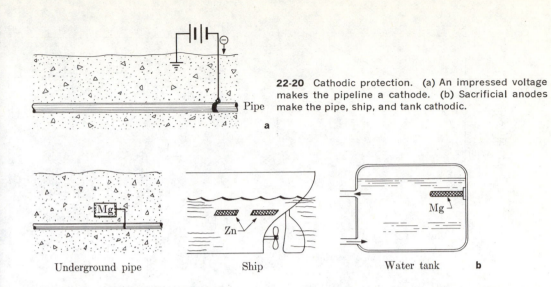

22-20 Cathodic protection. (a) An impressed voltage makes the pipeline a cathode. (b) Sacrificial anodes make the pipe, ship, and tank cathodic.

Pipe

a

Underground pipe   Ship   Water tank   b

if a metastably passivated surface [point (3) of Fig. 22-18] is locally ruptured, that small area will undergo very rapid corrosion [point (3′)]. Such local rupturing can arise from a variety of causes, including mechanical abrasion, surface contamination, and microorganisms. The result is localized pitting, a very common type of corrosion failure of aluminum, stainless steels, and other metals which depend on passivation for protection.

22-7 CORROSION PREVENTION

We can now summarize various methods of corrosion protection. The first is that of *isolation* of the anode from the electrolyte. Paints, tin-plating, and vitreous enamel (i.e., glass coatings) are examples.

The second method of corrosion prevention is *cathodic protection*. In this procedure the metal component which is to be protected is made into a cathode either by an *impressed voltage* (Fig. 22-20a) or by a *sacrificial anode* (Fig. 22-20b). A *galvanized* coating of zinc on iron falls in the latter category.

The third procedure is the use of *inhibitors* which adsorb onto the metal. Rust inhibitors for automobile radiators provide an example.

Finally, corrosion can be minimized (1) through appropriate engineering design, so that *galvanic couples are avoided* (e.g., steel screws should not be used to fasten brass components), and (2) through appropriate materials design, so that *corrosion-sensitive microstructures are avoided*. This latter procedure is illustrated in the use of stainless steel.

If an austenitic* stainless steel containing 0.1% carbon is cooled rapidly from 1000°C, a separate carbide does not form (Fig. 22-21). On the other hand, if the

---

* An 18Cr-8Ni steel remains austenitic at room temperature because the M_s drops below 0°C as a result of the presence of nickel.

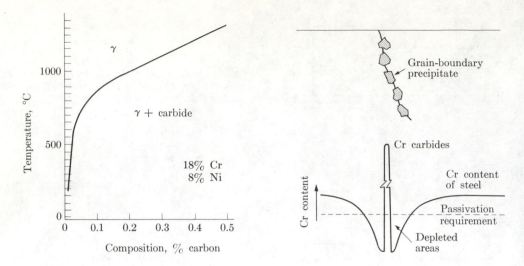

22-21 Carbon solubility in stainless steel. The carbon solubility in an 18-8 type stainless steel decreases markedly with temperature. Consequently, the carbon will precipitate if cooling is not rapid. The precipitated carbide is rich in chromium. (Adapted from E. E. Thum, *Book of Stainless Steels.* Metals Park, O.: American Society for Metals, 1955.)

22-22 Chromium depletion adjacent to the grain boundary. The carbide precipitation consumes nearly 10 times as much chromium as carbon. Since the larger chromium atoms diffuse slowly, the Cr content of the adjacent areas is lowered below protection levels.

same steel is cooled slowly, or held at ~650°C for a short period of time, the carbon precipitates as chromium carbide, usually in the form of a fine precipitate at the grain boundary. In this situation chromium is depleted in the narrow grain-boundary area and its passivation protection is lost (Fig. 22-22). This phenomenon accentuates corrosion at the grain boundaries, and is to be avoided (Fig. 22-23). There are several choices open to the metallurgical engineer for avoiding this microstructural effect. Each draws on his knowledge of materials and their structures.

1) *Quenching to avoid carbide precipitation.* This method is commonly employed unless (a) service conditions require temperatures in the precipitation range or (b) forming, welding, or size prevent such a quenching operation.

2) *Provision for an extremely long anneal in the carbide-separation range.* This technique offers some advantages because of (a) agglomeration of the carbides and (b) homogenization of the chromium content so that there is no deficiency at the grain boundary. However, this procedure is not common, because the improvement in corrosion resistance is relatively small.

3) *Selection of a steel with less than 0.03% carbon.* As indicated in Fig. 22-21, this would virtually eliminate carbide precipitation. However, such a steel is expensive because of the difficulty of removing enough of the carbon to attain this very low level.

22-23 Intergranular corrosion. This type of corrosion becomes severe if stainless steel is heated into the carbide precipitation range. (W. O. Binder, "Corrosion Resistance of Stainless Steels," *Corrosion of Metals.* Metals Park, O.: American Society for Metals.)

4) *Selection of a steel with high chromium content.* A steel which contains 18% chromium corrodes less readily than a plain-carbon steel. The addition of more chromium (and nickel) provides additional protection. This, too, is expensive because of the added alloy costs.

5) *Selection of a steel containing strong carbide formers.* Such elements include titanium, columbium, and tantalum. In these steels, the carbon does not precipitate at the grain boundary during cooling because it is precipitated earlier as titanium carbide, columbium carbide, or tantalum carbide at much higher temperatures. These carbides are innocuous because they neither deplete the chromium from the steel nor localize the galvanic action at the grain boundaries. This technique is used frequently, particularly with stainless steel which must be fabricated by welding.

Although the above examples are somewhat specific, they illustrate the methods which are used to reduce the extent of corrosion in metals. The choice of a particular procedure depends on the alloy and the service conditions involved.

REFERENCES FOR FURTHER READING

Corrosion in Action. New York: International Nickel, 1955. The best single reference on corrosion for the beginning student. Simply written, excellently illustrated; references and simple experiments included. This is the book form of a three-reel motion picture which is available for educational use from the International Nickel Company.

Dorn, J. E., *Mechanical Behavior of Materials at Elevated Temperatures.* New York: Wiley, 1961. A series of articles which discuss the response of materials to stresses at high temperatures. For the advanced student.

Fontana, M. G., and N. D. Greene, *Corrosion Engineering.* New York: McGraw-Hill, 1967. An easily read reference book on the subject.

Guy, A. G., *Physical Metallurgy for Engineers.* Reading, Mass.: Addison-Wesley, 1962. Corrosion is presented in Chapter 11. Same general level as this chapter.

Kopelman, B., *Materials for Nuclear Reactors*. New York: McGraw-Hill, 1959. Useful for the engineer who has some familiarity with the nuclear reactor and has the materials background of this book.

Mahan, B., *University Chemistry*, second edition. Reading, Mass.: Addison-Wesley, 1969. Section 7-4 provides a chemistry background for the study of galvanic cells. Freshman level.

Parker, E. R., *Materials for Missiles and Spacecraft*. New York: McGraw-Hill, 1963. A series of articles on the application of materials to the special environment of space. Advanced undergraduate and graduate level.

Pourbaix, M., *Atlas of Electrochemical Equilibria in Aqueous Solutions*. Oxford: Pergamon, 1966. Thorough presentation of electrochemical reactions in which the roles of pH and concentrations are considered, in addition to emf. For the corrosion engineer.

Primak, W., and M. Bohman, "Radiation Damage," *Progress in Ceramic Science*, Vol. 2 (J. E. Burke, Ed.). Oxford: Pergamon, 1962. Review of radiation damage as it applies to ceramics. For the materials specialist.

Reiss, H., "The Chemical Properties of Materials," *Scientific American*, **217**, [3], September 1967, pp. 210–220. Surface reactions receive attention. Introductory chemistry required.

Scully, J. C., *The Fundamentals of Corrosion*. Oxford: Pergamon, 1966. Paperback. An excellent supplementary book for the study of corrosion. General chemistry background is sufficient.

Smith, C. O., *Nuclear Reactor Materials*. Reading, Mass.: Addison-Wesley, 1967. Types of nuclear reactors, their service environments, and the damage effects on the materials are described. The materials aspects are introductory; a familiarity with the reactors will be helpful.

Uhlig, H. H., *Corrosion and Corrosion Control*. New York: Wiley, 1963. A standard textbook on the subject. Undergraduate level.

PROBLEMS

22-1 Iron oxidizes to FeO above 560°C (Example 22-2 and Fig. 17-29). Magnetite, Fe_3O_4, forms below 560°C ($\rho = 5.18$ gm/cm^3). Calculate the volume change (a) as iron oxidizes to magnetite, (b) as FeO oxidizes to magnetite.

Answer. a) 1 cm^3 Fe → 2.1 cm^3 Fe$_3$O$_4$ b) 1 cm^3 FeO → 1.23 cm^3 Fe$_3$O$_4$

22-2 a) Calculate the density of CaO. It has the NaCl structure.
   b) What is the volume change when metallic calcium oxidizes to CaO?

22-3 a) What frequency and wavelength must a photon have to supply the energy necessary to break an average C—H bond in polyethylene?
   b) Why can some bonds be broken with longer electromagnetic waves?

Answer. a) 1.09×10^{15}/sec, 2750 A   b) Thermal energy is also present.

22-4 A neutron breaks a C—C bond in polystyrene. How many eV were used?

22-5 A photon breaks a C—C bond in polystyrene. What must the minimum energy of the photon be, and what wavelength radiation would supply photons having this energy? (Assume that all the energy came from the photon.)

Answer. 3250 A

22-6 It is desired to set up a small plating process with a capacity of 1 lb Ni/day. What is the minimum amperage requirement?

22-7 Assume that the only cathode reaction in a battery is hydrogen generation. How many cm^3 of H_2 (STP) will be evolved per gm of zinc oxidized at the anode?

Answer. 343 cm^3 H_2/gm Zn

22-8 A 5-lb sacrificial anode of magnesium lasts 3 months on a ship hull. What is the average current during that period?

22-9 The current density for a selected situation of Fig. 22-6 is 3×10^{-6} amp/cm^2 for 0.17 volt, and 4×10^{-7} amp/cm^2 at 0.04 volt.

a) What is the exchange current density, i_0?
b) What is the value of the constant B of Eq. (22-15)?

Answer. a) 2.15×10^{-7} amp/cm^2 b) 0.065 volt

22-10 Under selected conditions the exchange current densities are 10^{-5} amp/cm^2 for each $H_2 = 2H^+ + 2e^-$ and $Cu = Cu^{2+} + 2e^-$. Assuming that B is 0.05 volt for each reaction, estimate the corrosion current for this pair.

22-11 Cold-working increases the corrosion rate over that of the annealed metal by a factor of 2.25. If this increase comes about entirely from a change in the exchange current density, i_0, by what factor has it been changed? (Assume that the values of B for the cathode and anode are equal.)

Answer. $i_0' = 5.1i_0$ (Refer to Fig. 22-15.)

22-12 Local corrosion produces a 0.1-mm diameter "pinhole" through a 0.5-mm thick aluminum pan in a period of 30 days. Assuming that the corrosion was all from one side of the sheet, what was the local current density?

22-13 Two pieces of metal, one copper and the other zinc, are immersed in seawater and connected by a copper wire. Indicate the galvanic cell by writing the half-reaction (a) for the anode, (b) for the cathode; also by indicating (c) the direction of electron flow in the wire and (d) the direction of the "current" flow in the electrolyte. (e) What metal might be used in place of copper so that the zinc changes polarity?

22-14 A zinc-coated, steel nail is cut in half and placed in an electrolyte.

a) What couples must be considered in judging where the anode will be?
b) Cite the location which will be corroded initially.

22-15 Compare and contrast the nature of the protection given to steel by (a) cadmium, (b) zinc, and (c) tin coatings.

22-16 Cite three examples of corrosion from your experience. Describe the nature of the deterioration and account for the corrosion.

22-17 A stainless-steel sheet is welded into a circular duct. After a period of time, rust appears along a band extending about 0.5 in. on each side of the weld. Why did this occur and what could have been done to avoid it?

22-18 An 18Cr-8Ni stainless steel is austenitic at room temperature.

a) Why?
b) Cite distinctive thermal, electrical, and magnetic properties that are consequences of this composition and structure.

23 □ materials systems

23-1 COMPOSITE MATERIALS

In modern-day engineering products, different types of materials are often combined in a single part so as to optimize properties and service behaviors. Consider an automobile tire, a glass-coated metal hot-water tank, reinforced concrete, or a printed circuit (Fig. 23-1). Each of these is a *composite material* in which the contributing materials serve one or more specific functions in the final product. Neither the rubber nor the cord of a tire can perform their respective functions alone. They must be used together as a *materials system*. Likewise, the glass coating in a hot-water tank provides corrosion protection for the easily formed, ductile metal. The combination of these two types of materials permits more life per dollar than does either material alone.

The engineer must design his materials system with due consideration for the properties of the contributing materials. The above-mentioned glass coating must be held under compression by the underlying metal. Otherwise it will be subject to mechanical failure (Section 21-3). Steel reinforcement must be correctly located within a concrete beam so that it carries the tensile load, while the concrete carries the compressive load, provides rigidity, and has the major attribute of "production-in-place." Processing considerations are also important in integrated circuits where resistors, conductors, insulators, p-zones, and n-zones are all incorporated into a materials system, each with its own functions.

For convenience, *composite materials* may be categorized as (1) coated materials, (2) surface-altered materials, (3) agglomerated materials, (4) reinforced materials, and (5) joined materials. We will give some initial attention to these general categories before considering stress distributions (Section 23-2) and bonding mechanisms (Section 23-3) in such materials systems. Finally, we shall also give consideration to some complex materials which are of present and future interest for engineers (Section 23-4). These include cement, soils, and wood.

Surface coatings. Environmental protection is the prime purpose of most surface coatings. As such, they should be resistant to the anticipated service conditions, as well as meet the requirements of production and cost. Table 23-1 makes a comparison of the engineering advantages and disadvantages of the three general types of coatings for environmental protection.

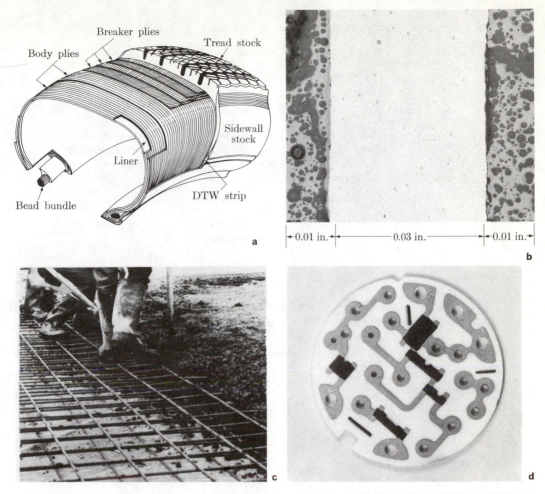

23-1 Materials systems. (a) Fiber and rubber in a tire. (Courtesy of H. Howe, Uniroyal Tire.) (b) Glass-coated sheet steel. (Courtesy of Bethlehem Steel Co.) (c) Steel in a reinforced-concrete floor. (Courtesy of Keystone Consolidated Industries.) (d) Conductors and resistors on an Al_2O_3 substrate of a printed circuit. (Courtesy of A. C. Spark Plug Division.)

Coatings are also used for esthetic purposes in many products. These factors should not be ignored for consumer products or public structures. However, we shall not take the time here to consider pigments, gloss, texture, etc.

Surface-altered materials. The carburized steel of Example 20-3 becomes a composite material because its surface composition is altered from that of the interior (Fig. 23-2). The desired result is a hard, wear-resistant surface on a tough, energy-absorbing core. In this composite, the surface *case* and the interior *core* remain integrally bonded and we do not encounter adherence problems between the two. They do, however, have different properties and each must adapt to the other's dimensional changes during heating or loading.

TABLE 23-1

COMPARISON OF INERT PROTECTIVE COATINGS

Type	Example	Advantages	Disadvantages
Organic	Baked "enamel" paints	Flexible Easily applied Cheap	Oxidizes Soft (relatively) Temperature limitations
Metal	Noble metal electroplates	Deformable Insoluble in organic solutions Thermally conductive	Establishes galvanic cell if ruptured
Ceramic	Vitreous enamel oxide coatings	Temperature resistant Harder Does not produce cell with base	Brittle Thermal insulators

Although normally not categorized as a composite material, an induction-hardened steel assumes two distinct sets of structures, and therefore properties, as a result of surface heating (and rapid quenching). As with the compositional alteration of the surface, this microstructural gradient furnishes a hard case and

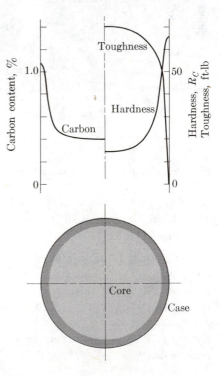

23-2 Carburized steel. The surface has been altered by carbon diffusion to give a high-carbon case around a low-carbon core.

23-3 Surface-hardened steel. The surface of the cam shaft was heated by high-frequency induction. The outer case formed austenite and was then quickly quenched to hard martensite. The center core remained tough with α + carbide. (Courtesy of H. B. Osborn, Jr., Tocco Division, Park Ohio Industries.)

a tough core (Fig. 23-3) to a materials system. Two differences may be cited for *induction hardening* (and flame hardening) as contrasted to normal quenching procedures. First, quenching is much more rapid after induction hardening, since the heat which must be removed from the surface zone only can diffuse inward as well as outward. Second, compressive stresses are set up in the surface during quenching, because martensite occupies a larger volume than its parent austenite and the center does not undergo a $\gamma \rightarrow \alpha$ + carbide expansion after the hard, brittle surface martensite has formed. Thus the surface may be used at a higher hardness level than is permitted in regularly quenched steels.

Agglomerated materials. Concrete, powdered-metal bearings, ceramic magnets, styrofoam insulation, and a number of other materials are made by agglomerating particulate materials. If desired, these products can incorporate markedly different types of materials, e.g., a tough, nickel matrix with a hard, temperature-stable carbide phase for cutting properties.

In producing agglomerated materials with high strength, consideration must be given to the packing factor of the particles—be they submicron Al_2O_3 for a spark plug, or fist-sized gravel for concrete in a dam. The resulting bulk density is usually increased in powdered parts by pressure *compaction*. This is often followed by *sintering* to bond the particles together (Section 23-3).

Higher densities, and therefore high strengths, may also be attained by *sizing*. Consider the concrete of Fig. 23-4, which is an agglomerated material with a coarse gravel aggregate containing interstitial sand and bonded together with a cement-water "paste." The maximum density is realized when the sand is just sufficient to

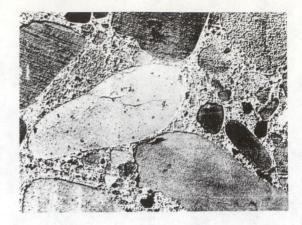

23-4 Agglomerate (concrete) (×1.5). The pores among the gravel are occupied by sand. In turn, the pores among the sand grains were occupied by a cement-water paste. (Portland Cement Association.)

fill the pores among the gravel, and cement fills the pores among the sand grains; finally, water fills the pores among the cement particles.

Example 23-1

Raw materials for a concrete include crushed limestone, quartz sand, and of course cement and water. They have the density properties listed below. What are the optimum rock and sand additions to a unit mix of 1 sack of cement (94 lb) with 6 gal of water so that maximum bulk density is obtained?

	Bulk density*	Apparent density†
Crushed rock	103 lb/ft³	170 lb/ft³
Sand	106 lb/ft³	166 lb/ft³
Cement	94 lb/ft³	197 lb/ft³
Water	62.4 lb/ft³ (8.33 lb/gal)	62.4 lb/ft³

* Mass/total volume (including *all* porosity, open and closed).
† Mass/apparent volume (including *closed* porosity only).

Answer

$$\text{Cement volume} = (94 \text{ lb})/(197 \text{ lb/ft}^3) = 0.48 \text{ ft}^3.$$

$$\text{Water volume} = (6 \text{ gal})(8.33 \text{ lb/gal})/(62.4 \text{ lb/ft}^3) = 0.80 \text{ ft}^3.$$

$$\text{Volume cement-water ``paste''} = 1.28 \text{ ft}^3.$$

$$\text{Open-pore volume of sand} = 1.28 \text{ ft}^3 = \frac{x \text{ lb sand}}{106 \text{ lb/ft}^3} - \frac{x \text{ lb sand}}{166 \text{ lb/ft}^3},$$

$$x = 376 \text{ lb} \quad (= 3.54 \text{ ft}^3 \text{ bulk volume})$$
$$(= 2.26 \text{ ft}^3 \text{ apparent volume}).$$

$$\text{Open-pore volume of crushed rock} = 3.54 \text{ ft}^3 = \frac{y \text{ lb rock}}{103 \text{ lb/ft}^3} - \frac{y \text{ lb rock}}{170 \text{ lb/ft}^3},$$

$$y = 930 \text{ lb} \quad (= 9.0 \text{ ft}^3 \text{ bulk volume})$$
$$(= 5.5 \text{ ft}^3 \text{ apparent volume}).$$

Note. Since mixing cannot be perfect, the amount of aggregate must be less than these figures. A typical mix is 3 ft³ gravel to 2 ft³ sand to 6 gal water per unit (sack) of cement. (Volumes are bulk.) ◄

Reinforced materials. Reinforcement can take the form of steel rods in concrete, glass, or graphite fibers in plastics, boron filaments in aluminum, or nylon cords in tires (Fig. 23-1a). For reinforcement to be effective, the strengthening component must have a higher Young's modulus than the matrix. Otherwise the reinforcement does not carry the load as strain builds up. Second, stresses must be transferred from the matrix to the reinforcement, usually by shear. These stresses will be summarized in Section 23-2.

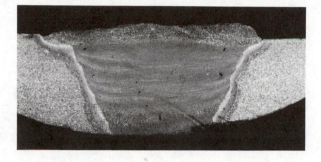

23-5 Joined materials. Weld. The molten weld metal solidifies to form a continuous metal joint. The microstructure of the adjacent metal is also affected by heating. (Courtesy of J. W. Freeman, The University of Michigan.)

Joined materials. Soldering, welding (Fig. 23-5), and organic adhesives are used to join materials into more complex shapes or to combine unlike materials. We will look at the bonds in more detail in the next section. In the meantime, we should remember that any heating process such as welding can introduce thermal damage, as described in Section 22-2. Specifically, overaging, overtempering, and annealing can occur during the heating of materials which have been hardened or strengthened by various cold-working and heat-treatment processes. If small, the welded parts may be heat-treated again. A complete heat treatment is, of course, impossible on bridges or other large structures; and local heating always subjects adjacent areas to intermediate heat treatments. It is obvious that design engineers must be aware of the materials' characteristics before specifying joining processes.

23-2 STRESSES IN MATERIALS SYSTEMS

Commonly the two materials of a composite have different thermal expansions and different Young's moduli, thus producing stress heterogeneities. If the materials system is designed correctly, these differential stresses are utilized advantageously as shown in the following three examples.

Example 23-2

A 1060 steel tire for a locomotive wheel has a 44.97-in. inside diameter at 20°C and a rim cross section of 2.07 in^2.

a) To what temperature must the tire be heated so it will have a 0.04-in. clearance when mounted over a 45.00-in. wheel?

b) What will be the tangential force on the tire after it cools to 20°C? (Assume that the wheel is sufficiently massive so that it has nil strain.)

Answer

a) From Appendix C, $\alpha_{1060} = 11 \times 10^{-6}/°C$. Thus

$$\Delta T = (45.04 \text{ in.} - 44.97 \text{ in.})/(44.97 \text{ in.})(11 \times 10^{-6}/°C)$$
$$= 142°C,$$
$$T = 162°C.$$

b)
$$\sigma = \left(\frac{45.00 \text{ in.} - 44.97 \text{ in.}}{44.97 \text{ in.}} \right) (30 \times 10^6 \text{ psi}) = F/2.07 \text{ in}^2,$$

$$F = 41{,}000 \text{ lb force.} \blacktriangleleft$$

Example 23-3

A glass-reinforced polyvinylidene chloride rod contains 25 w/o borosilicate glass fibers. All the fibers are aligned longitudinally. What fraction of the load is carried by the glass?

Answer. (Data from Appendix C.)

$$\text{v/o glass} = \frac{(0.25 \text{ gm})/(2.4 \text{ gm/cm}^3)}{(0.25/2.4)_{gl} + (0.75)/(1.7)_{pvc}}$$
$$= 19 \text{ v/o} \quad (= 19 \text{ area percent});$$

$$\text{load}_{gl}/A_{gl}Y_{gl} = \epsilon_{gl} = \epsilon_{pvc} = \text{load}_{pvc}/A_{pvc}Y_{pvc},$$

$$\frac{\text{load}_{gl}}{\text{load}_{pvc}} = \frac{(0.19)(10 \times 10^6)}{(0.81)(0.05 \times 10^6)} = \frac{98\%}{2\%}. \quad \blacktriangleleft$$

Example 23-4

A 0.10-in. iron sheet which is to be used in a household oven is coated on *both* sides with a glassy enamel. The final processing occurs above the 500°C fictive temperature, to give a 0.020-in. coating. The glass has a Young's modulus of 10^7 psi and a thermal expansion of $8.0 \times 10^{-6}/°C$.

a) What are the stresses in the glass at 20°C?

b) At 200°C? (Assume no plastic strain.)

Answer. Since $\Delta l/l$ = thermal expansion + elastic strain + plastic strain, and in this case $(\Delta l/l)_{gl} = (\Delta l/l)_{Fe}$, we may write

$$\alpha_{gl} \Delta T + \sigma_{gl}/Y_{gl} = \alpha_{Fe} \Delta T + \sigma_{Fe}/Y_{Fe}.$$

a) By using data from above and from Appendix C,

$$\sigma_{Fe}/30 \times 10^6 \text{ psi} - \sigma_{gl}/10 \times 10^6 \text{ psi} = (8.00 - 11.75)(10^{-6}/°C)(-480°C).$$

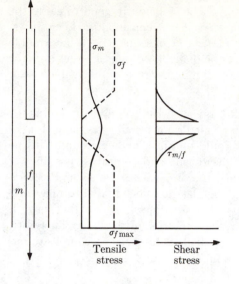

23-6 Stress distributions (fiber composites). The fiber stress, σ_f, drops from its maximum value $\sigma_{f\,max}$ to zero at the end of the fiber; the matrix stress, σ_m, is increased. The shear stress between the matrix and the fiber, $\tau_{m/f}$, is greatest at the ends of the fibers. (Cf. Fig. 19-12.)

But $A_{\mathrm{Fe}} = 2.5A_{\mathrm{gl}}$ and $F_{\mathrm{Fe}} = -F_{\mathrm{gl}}$, so $\sigma_{\mathrm{gl}} = -2.5\sigma_{\mathrm{Fe}}$. Thus

$$\sigma_{\mathrm{Fe}}\left[\frac{1}{30} + \frac{2.5}{10}\right] = (3.75)(+480).$$

Solving, we obtain

$$\sigma_{\mathrm{Fe}} = +6{,}350 \text{ psi} \qquad \text{(tension)},$$

$$\sigma_{\mathrm{gl}} = -15{,}875 \text{ psi} \quad \text{(compression)}.$$

b) By similar calculations for $\Delta T = (500 - 200)°\mathrm{C}$,

$$\sigma_{\mathrm{gl}} = -10{,}000 \text{ psi} \qquad \text{and} \qquad \sigma_{\mathrm{Fe}} = +4000 \text{ psi}.$$

Note. We assumed unidirectional strain. In reality, plane (i.e., two-dimensional) strain occurs. The necessary correction gives a higher stress by the factor of $(1 - \nu)^{-1}$, where ν is Poisson's ratio (Section 10-1). ◄

Interfacial stresses. When stresses are concentrated in a more rigid phase, the forces must be transferred across the interface. We saw an example of this in Section 21-4, where we discussed plastic constraint. Deduced from Fig. 21-11, a shear stress, τ_r, is transferred across the steel-solder interface. Likewise, shear stresses develop at the glass-metal interface of the glass-coated metal described in Fig. 23-1(b) and Example 23-4.

Interfacial stresses are particularly important in *fiber composites*, as illustrated in Fig. 23-6. In this figure, $\sigma_{f\mathrm{max}}$ represents the stress which would be carried by the fiber if there were no end effects (infinite length). This corresponds to the calculation in Example 23-3 and depends on the volume fraction of reinforcement, as well as Young's moduli of the two materials. If the fiber is broken, however, the fiber stress necessarily drops to zero. When this occurs, the load must be transferred to the matrix by shear stresses. Two features stand out in this load transfer.

(1) The bonding between the two materials must be sufficiently good to carry the shear stresses. (2) Reinforcement is most effective if it is continuous. Local interruptions transfer the load into the weaker matrix. Consequently, a deformable matrix has an advantage because the load can be distributed over a larger area for lower maximum stresses. Of course, there is a limit, because an extremely weak matrix will fail completely.

Elastic moduli of composites. Example 19-5 revealed that Young's modulus for a lamellar composite is anisotropic. Similar calculations could be made for an oriented fiber composite, in which case we would obtain a higher modulus in the longitudinal than in the radial direction. The tensile and compressive values of Young's modulus are not necessarily identical in composites. Consider the reinforcement of Example 23-3. Under tension, the glass assumes a load; under compression, however, the small-diameter glass fiber is subject to buckling as it is stressed, and lower-modulus matrix carries a larger fraction of the load. Therefore, the mixture rules of Section 19-4 cannot be applied. This type of composite has a lower compressive modulus than tensile modulus.

23-3 BONDING WITHIN COMPOSITES

A prime production requirement of a materials system is that the components be integrally bonded to one another. The reader is well aware of rivets and several other types of mechanical fasteners which have been ingeniously devised by the mechanical engineer. *Mechanical bonding* also includes shrink-fitting, as discussed in Example 23-2.

Chemical bonding. The greatest coherency between two materials of a composite is attained when chemical bonds are established. In welding, for example, this bond is usually obtained by solidification across the joint to form a continuous metallic structure. *Adhesives* depend on highly polar molecules (Table 7-2) which are strongly attracted to atoms or molecules in the adjacent solid materials. The bonds are complex between glass and metals [e.g., in the glass-coated metals of Fig. 23-1(b), or in glass-metal vacuum seals in electron tubes]. In brief, however, it is necessary for the glass interface to be saturated with the oxide of the underlying metal.

Sintering. During a sintering process where powders are to be agglomerated or bonded, a grain boundary replaces two previous surfaces to reduce the total energy (Fig. 23-7). Sintering thus occurs naturally; however, the rate of sintering is limited by the rate of diffusion. In the absence of plastic deformation, atoms must be removed from the points of contact and moved to the adjacent surface in the pores (Fig. 23-8). Conversely, vacancies diffuse into boundary areas and permit the grains to move together to produce shrinkage. At least two factors are important in this connection. (1) Grain-boundary diffusion is more rapid than diffusion through the crystal (Fig. 6-16). Therefore, intergranular pores are removed more readily than intragranular pores (Fig. 23-8). The materials engineer is careful to

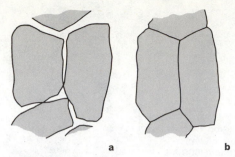

23-7 Solid sintering. The two surfaces of (a) are replaced by one grain boundary in (b) with a reduction of energy.

a   b

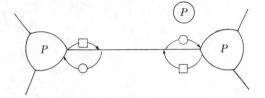

23-8 Sintering mechanism. Densification occurs by atoms (◯) diffusing from grain boundaries to pores (P) at grain corners. There is a counterdiffusion of vacancies (☐) from those pores. The route may be along the boundary or through the crystal. The latter route, however, is sufficiently slow so that intragranular pores are not easily eliminated.

design his sintering process so that grain growth does not precede pore removal; otherwise, entrapped pores significantly delay densification (and in ceramic materials reduce transparency). (2) In compounds, the sintering rate is limited by the diffusion rate of the slowest species; e.g., since the Al^{3+} ions and the O^{2-} ions must maintain a 2-to-3 ratio, sintering cannot proceed more rapidly than is permitted by the slower-moving O^{2-} ions.

Example 23-5

A powder compact for a ceramic magnet has a porosity of 27%. The final product after the sintering process has a porosity of 3%. A magnetic yoke for the cathode tube of an oscilloscope is to have a diameter of 1.75 in. What size die is required for compaction?

Answer. Assume cube equivalents with the required initial D and final d dimensions. Since the amount of material remains unchanged,

$$0.73D^3 = 0.97d^3,$$
$$D = 1.75 \text{ in. } \sqrt[3]{0.97/0.73} = 1.92 \text{ in. } \blacktriangleleft$$

Example 23-6

An iron powder is compacted and sintered. If the initial powder and the final grains may be approximated as 0.01-cm spheres, calculate the reduction in boundary energy

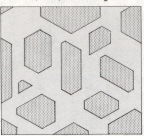

23-9 Epitaxial cadmium on MoS_2 (schematic). The orientation of the graphite-like MoS_2 substrate governs the crystal orientation of vapor-deposited metal. Eventually these small cadmium crystals (about 500 A as sketched) coalesce into a two-dimensional film ($r_{11\bar{2}0}$ of Cd = 2.98 A; $r_{11\bar{2}0}$ of MoS_2 = 3.16 A).

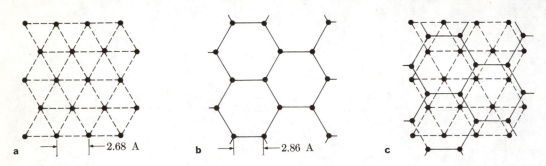

a   2.68 A   b   2.86 A   c

23-10 Coherency (Al_2O_3 and Al). The mismatch is only 6% in lattice spacing between (a) the (111) plane of the metal and (b) the (0001) plane of the oxide. Thus the Al ion layer may stack above the Al atoms of the fcc metal.

per gram during sintering. Surface (external boundary) energy = 2000 ergs/cm². Grain-boundary energy = 800 ergs/cm².

Answer. Basis: 1 gm.

$$\Delta\Gamma = \frac{[\pi(0.01 \text{ cm})^2(0.5)(800 \text{ ergs/cm}^2) + \pi(0.01 \text{ cm})^2(2000 \text{ ergs/cm}^2)]}{(\pi/6)(0.01 \text{ cm})^3(7.87 \text{ gm/cm}^3)}$$

$$= -1.2 \times 10^5 \text{ ergs/gm} \quad (= -0.003 \text{ cal/gm}).$$

Note. The grain-boundary area involves a factor of 0.5, since it is shared by two adjacent grains. Although small, the energy change, $\Delta\Gamma$, is large enough to be of significance in sintering. ◄

Epitaxial bonds. Thin-film composites are important in electrical materials. The film ($\ll 1$ μm) may be deposited on a substrate by *vaporization, electrodeposition* through liquid solutions, and *sputtering*, which is electrodeposition through a vapor phase. A critical feature of these thin films is the bonding which is developed with the substrates. For this to be effective, the structure of the film must be coherent with the substrate; i.e., crystalline films and the substrate must have

structurally related lattice planes so that the depositing film is *epitaxial*. To illustrate, a number of metals such as vapor-deposited cadmium will orient their
structures with an underlying MoS_2 substrate (Fig. 23-9). This means that
relatively large "two-dimensional" crystals can be grown on appropriate substrates.
The atoms which bond across the interface are coherent with each structure;
hence, a low-energy, strong boundary results. This amounts to the substrate
serving as a large nucleus for the depositing metal (Section 18-3). As shown in
Fig. 23-10, the Al-Al spacing in the metal is about 6% less than in the oxide; but
the patterns are compatible, and therefore matching may develop.

Example 23-7

What is the fractional mismatch between the magnesium atoms in the (111) plane of
MgO and the (0001) plane of metallic magnesium ($a_{MgO} = 4.21$ A, $a_{Mg} = 3.21$ A, and
$c_{Mg} = 5.21$ A)? Magnesium ions are located at 0, 0, 0; $\frac{1}{2}, \frac{1}{2}, 0$; $\frac{1}{2}, 0, \frac{1}{2}$; and 0, $\frac{1}{2}, \frac{1}{2}$ in
the fcc oxide. They occur at 0, 0, 0 and $\frac{2}{3}, \frac{1}{3}, \frac{1}{2}$ in the rhombic unit cell of the hexagonal
metal (i.e., the volume enclosed by solid lines in the hexagonal lattice of Fig. 5-3).

Answer. The Mg^{2+}–Mg^{2+} distance in the cubic oxide is $a/\sqrt{2}$, or 2.97 A.

The Mg–Mg distance in the hexagonal metal is equivalent to the translation distance
$\frac{2}{3}, \frac{1}{3}, \frac{1}{2} \rightarrow 1\frac{2}{3}, \frac{1}{3}, \frac{1}{2}$, or 3.21 A.

$$\text{Fractional mismatch} = \frac{3.21 - 2.97}{3.21}$$

$$= 7.5 \text{ l/o.} \blacktriangleleft$$

23-4 COMPLEX MATERIALS

The engineer often must use materials with complex structures which are not
easily characterized but nevertheless important. Structural materials such as
portland cement, soils, and wood are examples. Furthermore, the engineer is
becoming more involved with certain biological materials.

Portland cement. This is the most common commercial cement; it is used in
millions of tons per year for concrete (Section 23-1). Its chief ingredients are
di- and tricalcium silicate and tricalcium aluminate. The hydration reaction for
tricalcium aluminate is relatively simple and rather rapid:

$$Ca_3Al_2O_6 + 6H_2O \rightleftarrows Ca_3Al_2(OH)_{12}. \tag{23-1}$$

The hydrated product is crystalline and subject to dehydration at about 300°C
(570°F), a relatively high temperature for concrete service.

The two silicates of portland cement form appreciably more complex hydration
products:

$$2Ca_2SiO_4 + (5 - y + x)H_2O \rightleftarrows Ca_2[SiO_2(OH)_2]_2 \cdot (CaO)_{y-1} \cdot xH_2O + (3 - y)Ca(OH)_2 \tag{23-2}$$

and

$$2Ca_3SiO_5 + (5 + x)H_2O \rightleftarrows Ca_2[SiO_2(OH)_2]_2 \cdot (CaO) \cdot xH_2O + 3Ca(OH)_2. \quad (23\text{-}3)$$

In these equations, y is approximately 2.3 and the value of x in the hydrated calcium silicate (called *tobermorite*) varies from zero to more than one, depending on the available water.

There are two important factors here: (1) tobermorite has a colloidal structure and is essentially amorphous, and (2) the variability of the water content produces a material which is sensitive to the environment. Because of its amorphous nature, tobermorite forms a stronger, more adaptable bond with various aggregates than does the more crystalline, hydrated tricalcium aluminate. The silicates hydrate slowly and have a wide range of dehydration, as shown in Table 23-2. Also, the hydration-dehydration reaction is a function of the moisture content of the environment. In addition, any change in moisture content produces a volume change. Theoretically, the volume change is reversible; but in practice reversibility is not complete, because volume expansions occur from many colloidal surfaces, and this produces local displacements that are not fully recoverable. As Fig. 23-11 shows, there is a net dimensional growth with humidity cycling. Eventually fatigue may occur.

TABLE 23-2
CEMENT PROPERTIES IN RELATION TO CRYSTALLINITY*

Characteristics	Hydrated tricalcium aluminate	Hydrated dicalcium silicate
Structure	Crystalline	Colloidal and nearly amorphous
Approximate strength of paste (after 1 year)	1000 psi	10,000 psi
Time to achieve 30% strength	12 hr	60 days
Temperature of dehydration	300°C	100–550°C

* Collected from several sources, but primarily from R. H. Bogue and W. Lerch, *P.C.A.F. Reports*, 1936–1945.

Soils. From an engineering point of view, soils are loose or semi-indurated materials of the earth's mantle. They are vital to the structural engineer because he must use them for foundations, or in road beds, retaining walls, etc.

Soils are most readily characterized on the basis of sand, silt, and clay content. To a large extent this is a size classification, *sand* being greater than 0.05 mm, *silt* 0.05 to 0.005 mm, and *clay* less than 0.005 mm (i.e., 5 μm). In practice, how-

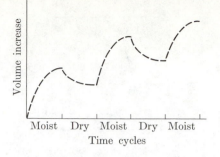

23-11 Humidity expansion (concrete). Dimensional changes are not fully reversible with humidity cycling.

23-12 Clay crystals (kaolinite at ×33,000). These alumino-silicate crystals are micaeous. (W. H. East, *Journ. Amer. Ceram. Soc.*, **33,** 1950, p. 211.)

ever, these size differences also imply compositional differences, particularly for the clays, which are usually hydrated alumino-silicates in contrast to the quartz (SiO_2), calcite ($CaCO_3$), and accompanying minerals found in sands and silts.

The colloidal clays may be platelike (Fig. 23-12) and possess significant surface charges around their edges as a result of their structures (Fig. 23-13). Therefore, the presence of a clay markedly affects the properties of the soils, making them plastic when wet, reasonably strong cements when dry, and impermeable to liquids. Detailed studies of load-bearing capacities of soils and of soil-water relationships must direct considerable attention to the structure and properties of the clays.

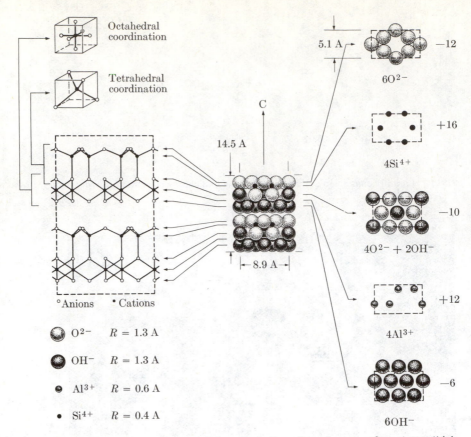

23-13 Structure of kaolinite. The most common of the clays, it contains layers a few atoms thick. Charges are carried by the broken bonds at the edges of the layers. (F. H. Norton, *Elements of Ceramics*. Reading, Mass.: Addison-Wesley, 1952.)

Wood. This familiar but complex material is a natural polymeric composite with highly directional properties. The principal molecules are *cellulose*,

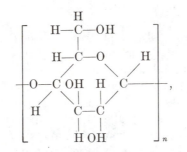

$$\tag{23-4}$$

in which the molecular weights range up to 2,000,000 gm/mole (as many as 15,000 mers). Since cellulose has numerous OH radicals, it develops a fair degree of

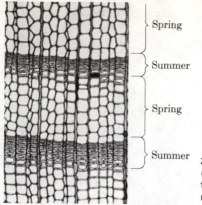

23-14 Structure of wood (cross section). The tubular cells (tracheids) which form with spring growth are larger, but have thinner walls than those which form with summer growth. (The radial vascular rays are not shown.)

Spring

Summer

Spring

Summer

crystallinity. In addition to the more than 50% cellulose, wood contains 10% to 35% *lignin*, an even more complex carbohydrate polymer.

Beyond the polymeric molecules, the next larger structural units of wood are the biological cells. The most extensive of these are called *tracheids*. They are hollow, spindle-shaped cells which are highly elongated in the longitudinal direction of the wood. Finally, the most visible structural unit is the "grain" of the wood, which is made up of *spring* and *summer* layers. The biological cells of the spring wood are larger and have thinner walls than do those of the summer wood (Fig. 23-14). In this respect, biological cells are much more variable in structure than are the unit cells of crystals.

The above description of wood is, of course, oversimplified.* However, it does show the source of the anisotropies in properties that are so characteristic of wood.

In less complex materials, the *density* is a structure-insensitive property (Section 19-3). This is not the case in a complex material such as wood. In the first place, the amount of spring wood and summer wood varies from species to species. Second, the ratio of cellulose to the more dense lignin varies. As a result of these two factors, the density can range from about 0.15 gm/cm³ for *balsa* to 1.3 gm/cm³ for the dense *lignum vitae*, so named because of its abnormally high lignin content (and low tracheid volume). In addition to the above factors, wood is hydroscopic; hence, it will absorb moisture as a function of humidity. In this case, both the mass and the volume increase with a net increase in density in most woods when wet.

It should come as no surprise that *dimensional changes* which accompany variations in temperature, moisture, and mechanical loading are anisotropic. Wood technologists point out that *thermal expansion* is about 40% greater in the tangential direction than in the radial direction, and 6 to 8 times greater in the radial direction than in the logitudinal direction. Thermal expansion in the longitudinal

---

* It does not take into account the wide variety of minor wood chemicals, nor the *vascular rays*, which are rows of single, nearly equidimensional cells that radiate from the center. These become important during deformation.

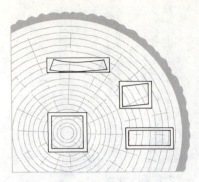

23-15 Dimensional anisotropies (shrinkage). Warpage occurs because tangential shrinkage is greater than radial shrinkage. Longitudinal shrinkage is negligible. (The shrinkage has been exaggerated in this sketch.)

direction is relatively independent of density, while in the other two directions it does depend on density, ρ (in gm/cm^3):

$$\alpha_t \cong 70\rho \times 10^{-6}/°C,$$
$$\alpha_r \cong 50\rho \times 10^{-6}/°C,$$
$$\alpha_l \cong 3 \times 10^{-6}/°C.$$

As an order of magnitude, longitudinal values for *Young's modulus* are between 1,000,000 and 2,000,000 psi when measured in tension. Tangential values are normally between 60,000 and 100,000 psi, while the radial values are commonly in the 80,000- to 150,000-psi range. The much higher longitudinal value is expected. However, from mixture rules (Section 19-4) and from our rather simplified model, one could have predicted the tangential modulus to be higher than the radial values. But in Fig. 23-14 we did not take into account the vascular rays cited in the last footnote. This feature provides an additional rigidity in the radial direction.

Shrinkage is also anisotropic; longitudinal changes are negligible, but tangential shrinkage is very high ($\sim$0.25 l/o per 1 w/o moisture for Douglas fir). Radial shrinkage is intermediate ($\sim$0.15 l/o per 1 w/o moisture for the same wood) because of the restraining effects of the vascular rays. Shrinkage is summarized in Fig. 23-15. The consequences of dimensional distortion on *warpage* are apparent.

The longitudinal tensile *strength*, S_l, is upwards of 20 times the radial tensile strength, S_r, because fracture must occur across the elongated cells. Increased densities (for a given moisture content) reflect an increase in the cell-wall thickness, and therefore give a proportional increase in the longitudinal strength. The transverse strength increases with the square of the density (again for a given moisture content), because there is significantly less opportunity for failure parallel to the hollow tracheid cells in a more dense wood.

Except where designs capitalize on the longitudinal strength, the anisotropies of wood are undesirable. Therefore, much has been done to modify its structure. Examples include (1) *plywood* in which the longitudinal strength is developed in *two* coordinates and (2) *impregnated wood* in which the pores are filled with a polymer such as phenol-formaldehyde to provide better bonding. In effect, the latter product is a plastic which is reinforced with fibers (i.e., elongated tracheid cells).

TABLE 23-3

TYPICAL MECHANICAL PROPERTIES OF HUMAN HARD TISSUES*

	Femur bone	Tooth enamel	Tooth dentin
Modulus of elasticity, 10^6 psi	2.82–2.98	6.7–6.9	1.7–2.0
Yield strength, 10^3 psi		28.2–32.5	18–21.5
Ultimate tensile strength, 10^3 psi	13–17.7		
Ultimate compressive strength, 10^3 psi	18–24	37.8–41.8	10.1–44.2
Shear strength,† 10^3 psi	16.8		
Shear strength,‡ 10^3 psi	5.7–13.3		

* From D. S. Crimmins, "The Selection and Use of Materials for Surgical Implants," *J. of Metals*, **21**, 1969, p. 41.
† Parallel to long axis.
‡ Transverse to long axis.

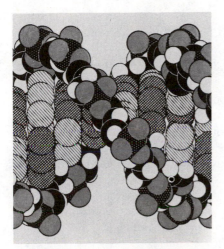

23-16 DNA (Watson-Crick model). When uncoiled by other enzymes, each molecule of the double helix may serve as a primer for a new molecule and new double helix which have the same structure and properties as the original molecule. (Courtesy of Dr. L. D. Hamilton, Brookhaven National Laboratory, and The Upjohn Company, Kalamazoo, Michigan. Not to be reproduced without their permission.)

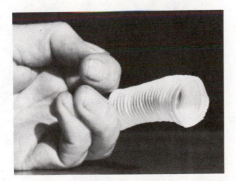

23-17 Prosthetic materials (artificial aorta of polytetrafluoroethylene). The implant must not only be inert to the host environment; it must also be compatible to the adjacent host tissue. (Courtesy of J. W. Freeman, The University of Michigan.)

(3) Finally, the *cellulose* may be extracted from the wood to serve as a raw material for many polymeric products.

Example 23-8

Birchwood veneer is impregnated with phenol-formaldehyde (Fig. 7-16) to ensure water resistance and to increase strength in the final product. Although dry birch weighs only 35 lb/ft^3, the true specific gravity of the cellulose-lignin combination is 1.52.

a) How many pounds of phenol-formaldehyde are required to impregnate 1 ft^3 of dry birchwood?

b) What is the final specific gravity?

Answer. From Appendix C, the density of phenol-formaldehyde is 1.3 gm/cm^3 (or 1.3×62.4 lb/ft^3).

a) Initial solid content of 1 ft^3 birchwood:

$$\frac{35 \text{ lb}}{(1.52)(62.4 \text{ lb/ft}^3)} = 0.37 \text{ ft}^3,$$

$$(1.00 - 0.37 \text{ ft}^3)(1.3)(62.4 \text{ lb/ft}^3) = 51 \text{ lb phenol-formaldehyde}.$$

b)
$$\frac{(51 + 35) \text{ lb/ft}^3}{62.4 \text{ lb H}_2\text{O/ft}^3} = 1.38. \blacktriangleleft$$

Biological materials. We shall end this textbook with a brief comment on biological materials. Whether they are molecular (e.g., amino acids), amorphous, crystalline (e.g., some viruses and certain skeletal materials), or composites (e.g., wood and teeth), they all possess structures which influence their properties (Table 23-3). The structures are often complex; however, once deciphered, they help one to understand the behavior of the materials (Fig. 23-16).

Future achievements in producing artificial materials for prosthetic applications (Fig. 23-17) require a careful characterization of the biological environment and of the structural stability of the material within that environment.

The structure and behavior of materials are always related.

REFERENCES FOR FURTHER READING

Alexander, W. O., "The Competition of Materials," *Scientific American*, **217** [3], September 1967, pp. 254–266. This article is pertinent to the selection of materials for a materials system. General level.

American Society for Metals, *Fiber Composite Materials*. Metals Park, O.: American Society for Metals, 1954. This series of articles will be helpful to the engineer designing composite materials. Advanced undergraduate and graduate level.

Broutman, L. J., and R. H. Krock, *Modern Composite Materials*. Reading, Mass.: Addison-Wesley, 1967. For the materials specialist and structural engineer. Advanced undergraduate level.

Brunauer, S., "Tobermorite Gel—The Heart of Concrete," *Amer. Scientist*, **50**, March 1962, pp. 210–229. Detailed attention is given to portland cement hydration. Professional level.

Kelly, A., "The Nature of Composite Materials," *Scientific American,* **217** [3], September 1967, pp. 160–176. Metals, ceramics, glasses, and polymers can be combined into materials that have properties of their own. Nature uses this principle in wood and bone; man applies it in a new family of supermaterials. Undergraduate level.

Kelly, A., and W. R. Tyson, "Fiber-Strengthened Materials," *High Strength Materials.* New York: Wiley, 1965, pp. 578–602. Failure mechanics in fiber composites. For the advanced undergraduate.

Keyser, C. A., *Materials Science in Engineering.* Columbus, O.: Merrill, 1968. Cement, concrete, and wood are presented in Chapters 11, 12, and 14, respectively, in greater detail than in this book. Introductory undergraduate level.

Moore, W. J., *Seven Solid States.* New York: Benjamin, 1967. Paperback. Anthracene, a relatively complex material, is described. Techniques for structure analysis are reviewed. Interesting supplementary reading for the advanced student.

Rhodin, T. N., and D. Walton, "Nucleation and Growth Processes on Solid Surfaces," *Metal Surfaces.* Metals Park, O.: American Society for Metals, 1963. Epitaxial growth is emphasized. Professional level.

Taylor, H. F. W., "The Chemistry of Cement Hydration," *Progress in Ceramic Science,* Vol. 1 (J. E. Burke, Ed.). Oxford: Pergamon, 1961. Review of the chemical processes in cement. For the materials specialist.

PROBLEMS

23-1 A concrete should have maximum density and low cost. How many tons of sand are required per ton of gravel?

	Specific gravity	Bulk density	Relative cost
Gravel	2.7	114 lb/ft^3	1
Sand	2.6	107	2
Cement	3.15	94	15

Answer. 0.3 ton sand/ton gravel

23-2 A concrete contains a 1:2:3 unit mix (bulk volume) of cement, sand, and gravel with the properties shown in Problem 23-1. If 5.5 gal of water are used with the above unit mix, how many ft^3 of concrete will result?

23-3 Using the data below, compute the apparent density* of a mixture containing 20 ft^3 gravel, 10 ft^3 sand, and 2 ft^3 cement "paste."

	Bulk density, lb/ft^3	True specific gravity
Gravel	100	2.70
Sand	120	2.92
Cement "paste"	80	1.28
Water	62.4	1.00

* Lb/ft^3 of actual material (excluding open pores, but including internal, or closed, pores).

23-4 A certain type of rubber has a true specific gravity of 1.40. This type of rubber is used in the manufacture of a foamed rubber which weighs 0.015 lb/in^3 when dry and 0.025 lb/in^3 when saturated with water.
  a) What is the open porosity of the foamed rubber?
  b) What is the true porosity?

23-5 A polystyrene foam having a bulk density of 4 lb/ft^3 is made of a polymer whose specific gravity is 1.05. One pound of the foam absorbs 1 lb of water. Calculate (a) the percent expansion of the polymer on foaming; (b) the total porosity of the foam; (c) the percent of open pore space; (d) the density of the foam when saturated with water.

Answer. a) 1530 v/o   b) 94 v/o   c) 6.4 v/o   d) 8 lb/ft^3

23-6 A load of 1000 lb is supported in tension by four rods each 0.05 in^2 in cross section and 4.1 in. long, two being copper and two stainless steel (Appendix C).
  a) What fraction of the load is carried by each copper rod at 75°F?
  b) To what temperature should the rods be changed so that each of the four rods would carry 25% of the load?

23-7 An annealed copper wire with a yield strength of 8000 psi is restrained from thermal contraction. How much can its temperature be changed before plastic adjustments are encountered within the wire?

Answer. −56°F

23-8 A 1040 steel wire (cross section 0.1 cm^2) has an aluminum coating, so that the total cross section is 0.12 cm^2.
  a) What fraction of a 1000-lb tensile load will be carried by the iron?
  b) What is the electrical resistance of this wire?

23-9 A "sandwich" of two, 1-in. thick copper plates and one intervening 2-in. aluminum plate is restrained from thermal expansion in its transverse directions. At what temperature will plastic yielding occur if the compressive yield strengths for the copper and aluminum are 12,000 and 8500 psi, respectively? (Assume unidirectional loading.)

Answer. +64°F

23-10 A sintered 70Cu-30Zn brass bearing has a final bulk density of 8.1 gm/cm^3. Its diameter before sintering was 0.317 in.; after sintering, 0.297 in. What was its porosity before sintering?

23-11 A ceramic electrical insulator of Al_2O_3 (specific gravity = 3.85) weighs 1.23 gm, occupies 0.47 cm^3, and is 0.87 cm long before sintering. After sintering, it is 0.77 cm long. What are the porosities (a) before and (b) after sintering?

Answer. a) 32 v/o   b) 2 v/o

23-12 Estimate the surface area per gm for the clay of Figs. 23-12 and 23-13. (The thickness of the platelets averages about 5% of their basal dimensions.)

23-13 What is the minimum amount of water required to hydrate dicalcium silicate in portland cement?

Answer. 0.14 lb H_2O/lb cement (1.6 gal/bulk ft^3)

Appendixes

APPENDIX A
CONSTANTS AND CONVERSIONS

1. Selected constants

Acceleration of gravity	g	$980.7 \ldots$ cm/sec^2
Atomic mass unit	amu	$1.66 \ldots \times 10^{-24}$ gm
Avogadro's number	AN	$0.602 \ldots \times 10^{24}$/mole
Bohr magneton	p_β	$9.27 \ldots \times 10^{-24}$ amp $\cdot$ m^2
Boltzmann's constant	k	$1.38 \ldots \times 10^{-16}$ erg/°K
		$8.63 \ldots \times 10^{-5}$ eV/°K
Electron charge	q	$1.6 \ldots \times 10^{-19}$ coul
		$4.8 \ldots \times 10^{-10}$ esu
Gas constant	R	$1.987 \ldots$ cal/mole $\cdot$ °K
		$0.082 \ldots$ liter $\cdot$ atom/mole $\cdot$ °K
		$82.06 \ldots$ cm^3 $\cdot$ atm/mole $\cdot$ °K
Inductance of free space	μ_0	$4\pi \times 10^{-7}$ henry/m
		$4\pi \times 10^{-7}$ ohm $\cdot$ sec/m
		1 gauss/oersted
Molar volume (0°C, 760 mm Hg)	V_m	22,400 cm^3
Natural logarithm base	e	$2.718 \ldots$
Permittivity of free space	ϵ_0	$8.854 \ldots \times 10^{-12}$ farad/m
		$8.854 \ldots \times 10^{-12}$ coul/volt $\cdot$ m
Planck's constant	h	$6.62 \ldots \times 10^{-27}$ erg $\cdot$ sec
Velocity of light	c	$2.998 \ldots \times 10^{10}$ cm/sec

2. Selected conversions

1 debye = 3.33×10^{-28} coul $\cdot$ cm

1 esu^2 = 1 erg $\cdot$ cm

1 eV = $1.602 \ldots \times 10^{-12}$ erg
    = $0.38 \ldots \times 10^{-19}$ cal

1 eV/atom = 23,100 cal/mole

1 ft^3 H$_2$O = 62.4 lb H$_2$O
       = 8.33 gal H$_2$O

1 gauss = 10^{-4} weber/m^2
       = 10^{-4} volt $\cdot$ sec/m^2

1 gauss/oersted = $4\pi \times 10^{-7}$ henry/m

1 joule = 10^7 ergs
       = 0.239 cal

1 kcal = $4.185 \ldots \times 10^{10}$ ergs

$\ln x = 2.3 \log_{10} x$

1 oersted = $(10^3/4\pi)$ amp/m

1 psi = $6.89 \ldots \times 10^4$ dynes/cm^2

Element	Symbol	Atomic no.	Electrons in shells K	L	M	N	Atomic wt. ($C^{12}=12.000$)	Melting point, °C	Boiling point, °C	Density (g) gm/liter (l) gm/cm³ (s) gm/cm³	Crystal structure of solid	Approx. atomic radius,* A	Valence (most common)	Approx. ionic radius, A (Coord. No. = 6)
Hydrogen	H	1	1				1.008	−259.18	−252.8	(g) 0.0899 (l) 0.070	Hex	0.46	+	Very small
Helium	He	2	2				4.003	−272.2 (26 atm)	−268.9	(g) 0.1785 (l) 0.147	Hcp (?)	1.76	Inert	—
Lithium	Li	3	2	1			6.94	186	1609	(s) 0.534	Bcc	1.519	+	0.78
Beryllium	Be	4	2	2			9.01	1350	1530	(s) 1.85	Hcp	1.12	2+	0.34
Boron	B	5	2	3			10.81	2300	2550	(s) 2.3	Ortho (?)	0.46	3+	~0.25
Carbon	C	6	2	4			12.01	~3500	4200(?)	(s) 2 ± (s) 2.25(gr) (s) 3.51(d)	Amorphous Hex Cubic	0.71 0.77	4+	~0.2
Nitrogen	N	7	2	5			14.007	−209.86	−195.8	(g) 1.2506 (l) 0.808 (s) 1.026	Hex	0.71	3−	
Oxygen	O	8	2	6			15.9994	−218.4	−183.0	(g) 1.429 (l) 1.14 (s) 1.426	Rhombic (?)	0.60	2−	1.32
Fluorine	F	9	2	7			19.00	−223	−188.2	(g) 1.69 (l) 1.108		0.5	−	1.33
Neon	Ne	10	2	8			20.18	−248.67	−245.9	(g) 0.9002 (l) 1.204	Fcc	1.60	Inert	—

* Atomic radii are based on data from *Metals Handbook*. Cleveland: American Society for Metals, 1961.

Element	Symbol	At. No.					At. Wt.	M.P. (°C)	B.P. (°C)	Density	Structure	Atomic radius	Valence	Ionic radius
Sodium	Na	11	2	8	1		22.99	97.5	880	(s) 0.97	Bcc	1.857	+	0.98
Magnesium	Mg	12	2	8	2		24.31	650	1110	(s) 1.74	Hex	1.594	2+	0.78
Aluminum	Al	13	2	8	3		26.98	660.2	2060	(s) 2.699	Fcc	1.431	3+	0.57
Silicon	Si	14	2	8	4		28.09	1430	2300	(s) 2.4	Diamond cubic	1.176	4+	0.41
Phosphorus	P	15	2	8	5		30.97	44.1	280	(s) 1.82	Cubic	—	5+	0.2–0.4
Sulfur	S	16	2	8	6		32.06	119.0	246.2	(s) 2.07 (l) 1.803	Fe ortho	1.06	2− 6+	1.74 0.34
Chlorine	Cl	17	2	8	7		35.45	−101	−34.7	(g) 3.214 (l) 1.557 (s) 1.9	Tetra	0.905	—	1.81
Argon	Ar	18	2	8	8		39.95	−189.4	−185.8	(g) 1.784 (l) 1.40 (s) 1.65			Inert	—
Potassium	K	19	2	8	8	1	39.10	63	770	(s) 0.86	Fcc	1.920	+	1.33
Calcium	Ca	20	2	8	8	2	40.08	850	1440	(s) 1.55	Bcc	2.312	2+	1.06
Scandium	Sc	21	2	8	9	2	44.96	1200		(s) 2.5	Fcc	1.969	3+	0.83
Titanium	Ti	22	2	8	10	2	47.90	1820		(s) 4.54	Fcc	1.605	4+	0.64
Vanadium	V	23	2	8	11	2	50.94	1735	3400	(s) 6.0	Hcp	1.458	3+ 5+	0.65 ~0.4
Chromium	Cr	24	2	8	13	1	52.00	1890	2500	(s) 7.19	Bcc	1.316	3+	0.64
Manganese	Mn	25	2	8	13	2	54.94	1245	2150	(s) 7.43	Cubic comp.	1.249	2+	0.91
Iron	Fe	26	2	8	14	2	55.85	1539	2740	(s) 7.87	Bcc	1.12	2+ 3+	0.83 0.67
Cobalt	Co	27	2	8	15	2	58.93	1495	2900	(s) 8.9	Hcp	1.241	2+	0.82

(Continued)

Element	Symbol	Atomic no.	Electrons in shells						Atomic wt. (C¹²=12.000)	Melting point, °C	Boiling point, °C	Density (g) gm/liter (l) gm/cm³ (s) gm/cm³	Crystal structure of solid	Approx. atomic radius, A	Valence (most common)	Approx. ionic radius, A (Coord. No. = 6)
			K	L	M	N	O	P								
Nickel	Ni	28	2	8	16	2			58.71	1455	2730	(s) 8.90	Fcc	1.245	2+	0.78
Copper	Cu	29	2	8	18	1			63.54	1083	2600	(s) 8.96	Fcc	1.278	+	0.96
Zinc	Zn	30	2	8	18	2			65.37	419.46	906	(s) 7.133	Hcp	1.332	2+	0.83
Gallium	Ga	31	2	8	18	3			69.72	29.78	2070	(s) 5.91	Fc ortho	1.218	3+	0.62
Germanium	Ge	32	2	8	18	4			72.59	958		(s) 5.36	Diamond cubic	1.224	4+	0.44
Arsenic	As	33	2	8	18	5			74.92	814 (36 atm.)	610	(s) 5.73	Rhombic	1.25	3+ 5+	0.69 ~0.4
Selenium	Se	34	2	8	18	6			78.96	220	680	(s) 4.81	Hex	1.16	2−	1.91
Bromine	Br	35	2	8	18	7			79.91	−7.2	19.0	(s) 3.12	Ortho	1.13	−	1.96
Krypton	Kr	36	2	8	18	8			83.80	−157	−152	(g) 3.708 (l) 2.155 (s)	Fcc	2.01	Inert	
Rubidium	Rb	37	2	8	18	8	1		85.47	39	680	(s) 1.53	Bcc	2.44	+	1.49
Strontium	Sr	38	2	8	18	8	2		87.62	770	1380	(s) 2.6	Fcc	2.15	2+	1.27
Yttrium	Y	39	2	8	18	9	2		88.91	1490		(s) 5.51	Hcp	1.79	3+	1.06
Zirconium	Zr	40	2	8	18	10	2		91.22	1750		(s) 6.5	Hcp	1.58	4+	0.87

Element	Symbol	At. No.	2	8	18	N	O	P	At. Weight	M.P.	B.P.	Density (state)	Crystal Structure	Atomic Radius	Valence	Ionic Radius
Niobium (Columbium)	Nb (Cb)	41	2	8	18	12	1		92.91	2415		(s) 8.57	Bcc	1.429	5+	0.69
Molybdenum	Mo	42	2	8	18	13	1		95.94	2625	4800	(s) 10.2	Bcc	1.36	4+	0.68
Technetium	Tc	43	2	8	18	14	1		99	2700	4900		(An artificial element only)			
Ruthenium	Ru	44	2	8	18	15	1		101.07	2500	4900	(s) 12.2	Hcp	1.352	4+	0.65
Rhodium	Rh	45	2	8	18	16	1		102.91	1966	4500	(s) 12.44	Fcc	1.344	3+	0.68
Palladium	Pd	46	2	8	18	18			106.4	1554	4000	(s) 12.0	Fcc	1.375		
Silver	Ag	47	2	8	18	18	1		107.87	960.5	2210	(s) 10.49	Fcc	1.444	+	1.13
Cadmium	Cd	48	2	8	18	18	2		112.40	320.9	765	(s) 8.65	Hcp	1.489	2+	1.03
Indium	In	49	2	8	18	18	3		114.82	156.4		(s) 7.31	Bc tetra	1.625	3+	0.92
Tin	Sn	50	2	8	18	18	4		118.69	231.9	2270	(s) 7.298	Bc tetra	1.509	4+	0.74
Antimony	Sb	51	2	8	18	18	5		121.75	630.5	1440	(s) 6.62	Rhombic	1.452	5+	0.90
Tellurium	Te	52	2	8	18	18	6		127.6	450	1390	(s) 6.24	Hex	1.43	2-	2.11
Iodine	I	53	2	8	18	18	7		126.9	114	183	(s) 4.93	Ortho	1.35	—	2.20
Xenon	Xe	54	2	8	18	18	8		131.3	-112	-108	(g) 5.851 (l) 3.52 (s) 2.7	Fcc		Inert	
Cesium	Cs	55	2	8	18	18	8	1	132.9	28	690	(s) 1.9	Fcc	2.21	+	1.65
Barium	Ba	56	2	8	18	18	8	2	137.3	704	1640	(s) 3.5	Bcc	2.62	2+	1.43
Rare earths	La → Lu	57 → 71	2	8	18	18 → 32	8 → 9	2	138.9 → 175.0				Bcc	2.17	3+	1.22 → 0.99
Hafnium	Hf	72	2	8	18	32	10	2	178.5	1700		(s) 11.4	Hcp	1.59	4+	0.84

(Continued)

Element	Symbol	Atomic no.	K	L	M	N	O	P	Q	Atomic wt. (C^{12} = 12.000)	Melting point, °C	Boiling point, °C	Density (g) gm/liter (l) gm/cm^3 (s) gm/cm^3	Crystal structure of solid	Approx. atomic radius, A	Valence (most common)	Approx. ionic radius, A (Coord. No. = 6)
						Electrons in shells											
Tantalum	Ta	73	2	8	18	32	11	2		180.95	2996	5930	(s) 16.6	Bcc	1.429	5+	0.68
Tungsten	W	74	2	8	18	32	12	2		183.9	3410		(s) 19.3	Bcc	1.369	4+	0.68
Rhenium	Re	75	2	8	18	32	13	2		186.2	3170	—	(s) 20	Hcp	1.370		
Osmium	Os	76	2	8	18	32	14	2		190.2	2700	5500	(s) 22.5	Hcp	1.367	4+	0.67
Iridium	Ir	77	2	8	18	32	17			192.2	2454	5300	(s) 22.5	Fcc	1.357	4+	0.66
Platinum	Pt	78	2	8	18	32	17	1		195.1	1773	4410	(s) 21.45	Fcc	1.387		
Gold	Au	79	2	8	18	32	18	1		197.0	1063	2970	(s) 19.32	Fcc	1.441	+	1.37
Mercury	Hg	80	2	8	18	32	18	2		200.6	−38.87	357	(s) 13.55	Rhombic	1.552	2+	1.12
Thallium	Tl	81	2	8	18	32	18	3		204.4	300	1460	(s) 11.85	Hcp	1.704	3+	1.05
Lead	Pb	82	2	8	18	32	18	4		207.2	327.4	1740	(s) 11.34	Fcc	1.750	2+ 4+	1.32 0.84
Bismuth	Bi	83	2	8	18	32	18	5		209.0	271.3	1420	(s) 9.80	Rhombic	1.556		
Polonium	Po	84	2	8	18	32	18	6		210	600			Monoclinic	1.7		
Astatine	At	85	2	8	18	32	18	7		210							
Radon	Rn	86	2	8	18	32	18	8		222	−71	−61.8				Inert	

Element	Symbol	At. No.								Mass	M.P.	Density	Crystal		Charge	
Francium	Fa	87	2	8	18	32	18	8	1	223					+	
Radium	Ra	88	2	8	18	32	18	8	2	226	700	(s) 5.0				
Actinium	Ac	89	2	8	18	32	18	9	2	227	1600					
Thorium	Th	90	2	8	18	32	18	10	2	232	1800	(s) 11.5	Fcc	1.800	4+	1.10
Protactinium	Pa	91	2	8	18	32	20	9	2	231	3000					
Uranium	U	92	2	8	18	32	21	9	2	238	1130	(s) 18.7	Ortho	1.38	4+	1.05
Neptunium	Np	93	2	8	18	32	22	9	2	237						
Plutonium	Pu	94	2	8	18	32	23	9	2	239						
Americium	Am	95	2	8	18	32	24	9	2	241						
Curium	Cm	96	2	8	18	32	25	9	2	242						
Berkelium	Bk	97	2	8	18	32	26	9	2	249						
Californium	Cf	98	2	8	18	32	27	9	2	252						
Einsteinium	E	99	2	8	18	32	28	9	2	254						
Fermium	Fm	100	2	8	18	32	29	9	2	253						
Mendelevium	Md	101	2	8	18	32	30	9	2	256						

PROPERTIES OF SELECTED ENGINEERING MATERIALS

Part 1 — Metals (Taken from numerous sources)

Material	Specific gravity	Thermal conductivity, $\dfrac{\text{cal·cm}}{\text{°C·cm}^2\text{·sec}}$, at 20°C*	Thermal expansion, in./in.°F, at 68°F†	Electrical resistivity, ohm·cm, at 20°C‡	Average modulus of elasticity, psi, at 68°F
Aluminum (99.9+)	2.7	0.53	12.5×10^{-6}	2.9×10^{-6}	10×10^{6}
Aluminum alloys	2.7(+)	0.4(±)	12×10^{-6}	$3.5 \times 10^{-6}(+)$	10×10^{6}
Brass (70Cu–30Zn)	8.5	0.3	11×10^{-6}	6.2×10^{-6}	16×10^{6}
Bronze (95Cu–5Sn)	8.8	0.2	10×10^{-6}	9.6×10^{-6}	16×10^{6}
Cast iron (gray)	7.15	—	5.8×10^{-6}	—	$12–25 \times 10^{6}$
Cast iron (white)	7.7	—	5×10^{-6}	—	30×10^{6}
Copper (99.9+)	8.9	0.95	9×10^{-6}	1.7×10^{-6}	16×10^{6}
Iron (99.9+)	7.87	0.18	6.53×10^{-6}	9.7×10^{-6}	30×10^{6}
Lead (99+)	11.34	0.08	16×10^{-6}	20.65×10^{-6}	2×10^{6}
Magnesium (99+)	1.74	0.38	14×10^{-6}	4.5×10^{-6}	6.5×10^{6}
Monel (70Ni–30Cu)	8.8	0.06	8×10^{-6}	48.2×10^{-6}	26×10^{6}
Silver (sterling)	10.4	1.0	10×10^{-6}	1.8×10^{-6}	11×10^{6}
Steel (1020)	7.86	0.12	6.5×10^{-6}	16.9×10^{-6}	30×10^{6}
Steel (1040)	7.85	0.115	6.3×10^{-6}	17.1×10^{-6}	30×10^{6}
Steel (1080)	7.84	0.11	6.0×10^{-6}	18.0×10^{-6}	30×10^{6}
Steel (18Cr–8Ni stainless)	7.93	0.035	5×10^{-6}	70×10^{-6}	30×10^{6}

* Multiply by 0.806 to get Btu·in./°F·ft²·sec.   † Multiply by 1.8 to get cm/cm/°C.   ‡ Divide by 2.54 to get ohm·in.

Part 2 — Ceramics (Taken from numerous sources)

Material	Specific gravity	Thermal conductivity, $\frac{cal \cdot cm}{°C \cdot cm^2 \cdot sec}$, at 20°C*	Thermal expansion, in./in·°F, at 68°F†	Electrical resistivity, ohm·cm, at 20°C‡	Average modulus of elasticity, psi, at 68°F
Al_2O_3	3.8	0.07	5×10^{-6}	—	50×10^6
Brick					
Building	2.3(±)	0.0015	5×10^{-6}		—
Fireclay	2.1	0.002	2.5×10^{-6}	1.4×10^8	—
Graphite	1.5	—	3×10^{-6}	—	—
Paving	2.5	—	2×10^{-6}	—	—
Silica	1.75	0.002		1.2×10^8	—
Concrete	2.4(±)	0.0025	7×10^{-6}	—	2×10^6
Glass					
Plate	2.5	0.0018	5×10^{-6}	10^{14}	10×10^6
Borosilicate	2.4	0.0025	1.5×10^{-6}	—	10×10^6
Silica	2.2	0.003	0.3×10^{-6}	10^{20}	—
Vycor	2.2	0.003	0.35×10^{-6}	—	—
Wool	0.05	0.0006		—	—
Graphite (bulk)	1.9	—	3×10^{-6}	10^{-3}	1×10^6
MgO	3.6	—	5×10^{-6}	10^5 (2000°F)	30×10^6
Quartz (SiO_2)	2.65	0.03	7×10^{-6}	—	45×10^6
SiC	3.17	0.029	2.5×10^{-6}	2.5 (2000°F)	—
TiC	4.5	0.07	4×10^{-6}	50×10^{-6}	50×10^6

(Continued)

* Multiply by 0.806 to get Btu·in/°F·ft²·sec. † Multiply by 1.8 to get cm/cm/°C. ‡ Divide by 2.54 to get ohm·in.

Part 3 — Organic Materials (Taken from numerous sources)

Material	Specific gravity	Thermal conductivity, $\frac{cal \cdot cm}{°C \cdot cm^2 \cdot sec}$, at 20°C*	Thermal expansion, in./in./°F, at 68°F†	Electrical resistivity, ohm·cm, at 20°C‡	Average modulus of elasticity, psi, at 68°F
Melamine-formaldehyde	1.5	0.0007	15×10^{-6}	10^{13}	1.3×10^6
Phenol-formaldehyde	1.3	0.0004	40×10^{-6}	10^{12}	0.5×10^6
Urea-formaldehyde	1.5	0.0007	15×10^{-6}	10^{12}	1.5×10^6
Rubbers (synthetic)	1.5	0.0003	—	—	500–10,000
Rubber (vulcanized)	1.2	0.0003	45×10^{-6}	10^{14}	0.5×10^6
Polyethylene	0.9	0.0008	100×10^{-6}	10^{13}	—
Polystyrene	1.05	0.0002	35×10^{-6}	10^{18}	0.4×10^6
Polyvinylidene chloride	1.7	0.0003	105×10^{-6}	10^{13}	0.05×10^6
Polytetrafluoroethylene	2.2	0.0005	55×10^{-6}	10^{16}	—
Polymethyl methacrylate	1.2	0.0005	50×10^{-6}	10^{16}	0.5×10^6
Nylon	1.15	0.0006	55×10^{-6}	10^{14}	0.4×10^6

* Multiply by 0.806 to get Btu·in./°F·ft²·sec.   † Multiply by 1.8 to get cm/cm/°C.   ‡ Divide by 2.54 to get ohm·in.

REFERENCES FOR ADDITIONAL INFORMATION (APPENDIX C)

Ceramic Data Book. Chicago: Cahners. Published yearly. Includes many property data on ceramics.

Materials Selector Issue of Materials Engineering. New York: Chapman-Reinhold. Published yearly. Includes many property data on all types of materials.

Metals Handbook, eighth edition. Metals Park, O.: American Society for Metals, 1961ff. A widely available reference book. Volume 1 considers properties and selection of materials. The remaining volumes cover metals processing.

Modern Plastics Encyclopedia. New York: McGraw-Hill. Published yearly. Includes many property data on plastics.

Parker, E. R., *Materials Data Book for Engineers and Scientists.* New York: McGraw-Hill, 1967. A convenient source of engineering data.

Smithells, C. J., *Metals Reference Book,* third edition. London: Butterworths, 1962. A standard two-volume reference for metal properties.

APPENDIX D
SELECTED ORGANIC STRUCTURES

Acetic acid:

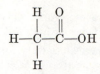

Acetone:

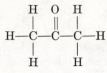

Acrylonitrile:

Adipic acid:

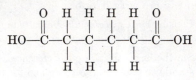

Butadiene:

Butylene (iso): see isobutylene.

Chloroprene:

Dimethylsilanediol:

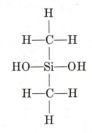

Divinyl benzene:

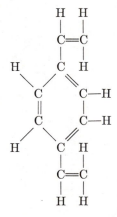

Ethylene glycol:

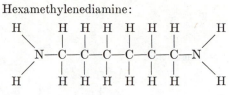

Formaldehyde:

Hexamethylenediamine:

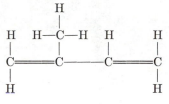

Isobutylene:

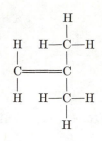

Isoprene:

Isopropyl alcohol:

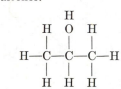

Melamine:

Methyl methacrylate:

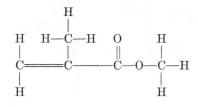

Phenol:

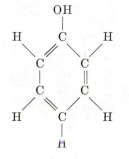

Phthalic acid: see terephthalic acid.

Propanol:

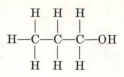

Propylene:

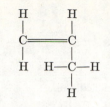

Styrene:

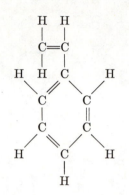

Terephthalic acid:

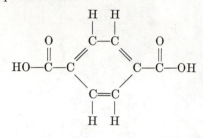

Urea:

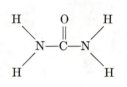

Vinyl acetate:

H H
| |
C═C   H
| | |
H O—C—C—H
   ‖ |
   O H

Vinyl alcohol:

Vinyl benzene: see styrene.

Vinyl chloride:

Vinylidene chloride:

H Cl
| |
C═C
| |
H Cl

APPENDIX E

SYMBOLS, DIMENSIONS, AND UNITS†

Symbol	Meaning	Dimensions	Common units
A	area	l^2	cm^2
A	cation	—	—
APF	atomic packing factor	—	—
a	interatomic distance	l	A
a	unit-cell dimension	l	A
B	induction (magnetic flux density)	Vtl^{-2}	webers/m^2
B_r	remanent induction	Vtl^{-2}	webers/m^2
B_s	saturation induction	Vtl^{-2}	webers/m^2
$B(E)$	Boltzmann distribution	$t^2m^{-1}l^{-2}$	per energy increment
b	unit-cell dimensions	l	A
b	Burgers vector	l	A
C	capacitance	CV^{-1}	farads
C	composition	—	weight percent
C	concentration	m/l^3	gm/cm^3
C	coulomb	C	coul
C	molar heat capacity	$l^2t^{-2}T^{-1}$	cal/mol wt · °K
C	number of components	—	—
CN	coordination number	—	—
CW	cold work	—	percent
c	mass heat capacity	$l^2t^{-2}T^{-1}$	cal/gm · °K
c	unit-cell dimension	l	A
c	velocity of light	lt^{-1}	cm/sec
D	diffusion coefficient	l^2t^{-1}	cm^2/sec
DP	degree of polymerization	—	mers/mole
$\mathfrak{D}$	charge density	Cl^{-2}	coul/m^2
d	distance	l	cm
d_{hkl}	interplanar spacing	l	A
E	energy	ml^2t^{-2} (or CV)	ergs eV
E	(see Y for Young's modulus)		
E^*	activation energy	ml^2t^{-2} (or CV)	cal/mol wt eV
E_f	Fermi energy	CV	eV
E_g	energy gap	CV	eV
$\mathcal{E}$	electric field	Vl^{-1}	volts/cm
$\mathcal{E}_H$	Hall voltage	Vl^{-1}	volts/cm
e	electron	—	(see q)

† Greek symbols follow the roman symbols.

(Continued)

Symbol	Meaning	Dimensions	Common units
F	force	mlt^{-2}	dynes (or lb_f)
F	molar free energy	l^2t^{-2}	cal/mol wt
F	shape factor	—	—
$F(E)$	Fermi distribution	—	fraction
f	farad	CV^{-1}	coul/volt
f	fluidity	ltm^{-1}	cm^2/sec · dyne
G	shear modulus	$ml^{-1}t^{-2}$	lb_f/in^2 (dynes/cm^2)
H	enthalpy (heat content)	l^2t^{-2}	cal/mol wt
H	magnetic field	$Ct^{-1}l^{-1}$	amp/m
h	hardening coefficient	$ml^{-1}t^{-2}$	lb_f/in^2
h	Planck's constant	ml^2t^{-1}	erg · sec
h	reciprocal of x-intercept	—	—
h	thermal diffusivity	l^2t^{-1}	cm^2/sec
hkl	Miller indices	—	—
I	current	Ct^{-1}	amp
i	current density	$Ct^{-1}l^{-2}$	amp/cm^2
J	flux	$l^{-2}t^{-1}$	per cm^2/sec
K	bulk modulus	$ml^{-1}t^{-2}$	lb_f/in^2 (dynes/cm^2)
K	equilibrium constant	—	—
KE	kinetic energy	ml^2t^{-2}	ergs
k	Boltzmann's constant	$ml^2t^{-2}T^{-1}$	ergs/°K
k	reciprocal of y-intercept	—	—
k	thermal conductivity	$mlt^{-3}T^{-1}$	cal · cm/cm^2 · sec · °K
L	length	l	cm
LRO	long-range order	—	—
$\mathcal{L}$	Lorentz number	V^2/T^2	cal · ohm/sec · °K^2
l	length	l	cm
l	reciprocal of z-intercept	—	—
M	magnetization	$Ct^{-1}l^{-1}$	amp/m
M	metallic atom	—	—
$\overline{M}_n$	number average molecular weight	m	gm/mole
M_{ve}	viscoelastic modulus	$ml^{-1}t^{-2}$	dynes/cm^2
$\overline{M}_w$	weight average molecular weight	m	gm/mole
Me	methyl group, CH_3	—	—
MW	molecular weight	m	gm/mole
m	mass	m	gm

SYMBOLS, DIMENSIONS, AND UNITS (Continued)

Symbol	Meaning	Dimensions	Common units
N	number	—	—
N_β	number of Bohr magnetons per atom	—	—
$N(E)$	density of states	$t^2 m^{-1} l^{-2}$	per energy increment
n	density of charge carriers	l^{-3}	number/m^3
n	integer	—	—
n	number	—	—
n	index of refraction	—	—
P	intercept points	l^{-1}	points/cm
P	number of phases	—	—
P	pressure	$m l^{-1} t^{-2}$	dynes/cm^2
PE	potential energy	$m l^2 t^{-2}$	ergs
PL	power loss	$V C t^{-1} l^{-3}$	watts/m^3
$\mathcal{P}$	polarization	$C l^{-2}$	coul/m^2
$\mathcal{P}_r$	remanent polarization	$C l^{-2}$	coul/m^2
p	electron hole	—	—
p	momentum	$m l t^{-1}$	dyne · sec
p	probability	—	fraction
p	dipole moment	$C l$	coul · m
p_m	magnetic moment	$C l^2 t^{-1}$	amp · m^2
p_β	Bohr magneton	$C l^2 t^{-1}$	amp · m^2/electron
Q	charge	C	coulomb
Q	molar activation energy	$l^2 t^{-2}$	cal/mol wt
q	electronic charge	C	coul/electron
R	gas constant	$l^2 t^{-2} T^{-1}$	cal/mol wt · °K
R	radical	—	—
R	radius	l	A
R	resistance	$V t C^{-1}$	ohm
R_H	Hall coefficient	$l^3 C^{-1}$	m^3/coul
r	radius	l	cm
$\mathbf{r}_{uvw}$	lattice vector	l	A
S	entropy per mol wt	$l^2 t^{-2} T^{-1}$	cal/mol wt · °K
S	surface (boundary area) per unit volume	l^{-1}	cm^2/cm^3
S	thermal emf coefficient	$V T^{-1}$	volts/°K
s	softening rate	$m l^{-1} t^{-3}$	lb$_f$/in^2 · sec
T	temperature	T	°K
T	torque	$m l^2 t^{-2}$	ergs
t	time	t	sec
t	transport number	—	fraction

(Continued)

Symbol	Meaning	Dimensions	Common units
U	elastic energy per unit volume	$ml^{-1}t^{-2}$	ergs/cm^3
u	vector coefficient (x-direction)	—	—
V	variance	—	—
V	voltage	V	volts
V	volume	l^3	cm^3
$\mho$	thermal emf	V	volts
v	vector coefficient (y-direction)	—	—
v	velocity	lt^{-1}	cm/sec
WF	Wiedemann-Franz ratio	V^2/T	cal/°K · sec · mho
w	vector coefficient (z-direction)	—	—
X	anion	—	—
$X\bullet$	free radical	—	—
X	mole fraction	—	—
X	number of molecules	—	—
x	coordinate direction	—	—
x	distance	l	cm
Y	Young's modulus	$ml^{-1}t^{-2}$	lb$_f$/in^2 (dynes/cm^2)
y	coordinate direction	—	—
y_T	thermal-resistivity coefficient	T^{-1}	per °C
y_x	solution resistivity	$VtlC^{-1}$	ohm · cm/percent
Z	valence number	—	—
z	coordinate direction	—	—
α	polarizability	Cl^2V^{-1}	coul · m^2/volt
α	short-range order	—	—
α	thermal-expansion coefficient	T^{-1}	per degree
β	absorption coefficient	l^{-1}	per cm
β	compressibility	lt^2m^{-1}	cm^2/dyne
Γ	surface energy	ml^2t^{-2}	ergs
γ	displacement	—	—
γ	shear strain	—	—
γ	surface-energy density	mt^{-2}	ergs/cm^2
δ	grain dimension	l	μm
δ	loss angle	—	degrees
ϵ	electric permittivity	$CV^{-1}l^{-1}$	farads/m
ϵ	strain	—	in./in.

Symbol	Meaning	Dimensions	Common units
η	viscosity	$ml^{-1}t^{-1}$	dyne $\cdot$ sec/cm^2
κ	relative dielectric constant	—	—
Λ	mean free path	l	A
λ	relaxation time	t	sec
λ	wavelength	l	cm
μ	magnetic permeability	$Vt^2C^{-1}l^{-1}$	henries/m
μ	mobility	$l^2V^{-1}t^{-1}$	cm^2/volt $\cdot$ sec
ν	frequency	t^{-1}	per sec
ν	Poisson's ratio	—	—
ρ	density	ml^{-3}	gm/cm^3
ρ	resistivity	$VtlC^{-1}$	ohm $\cdot$ cm
σ	conductivity	$CV^{-1}t^{-1}l^{-1}$	mho/cm
σ	stress	$mt^{-2}l^{-1}$	lb$_f$/in^2 (dynes/cm^2)
τ	relaxation time	t	sec
τ	shear stress	$mt^{-2}l^{-1}$	lb$_f$/in^2 (dynes/cm^2)
ϕ	phase angle	—	degrees
ϕ	voltage	V	volts
ϕ_p	plastic-strain energy	mt^{-2}	ergs/cm^2
χ_e	electric susceptibility	—	—
χ_m	magnetic susceptibility	—	—
ω	angular velocity	t^{-1}	radians/sec

index

BROPHY, J. J., 275, 293, 320, 337
BROUTMAN, L. J., 400, 404, 502
BRUNAUER, S., 502
Bubble-raft, 109
Bulk modulus, 191
   units, 520
Buna-S rubbers, 152
BUNN, C. W., 137, 139
Burgers vector, 110
BURKE, J. E., 183, 184, 223, 228, 294, 337, 456,
   457, 482, 503
BURKE, J. J., 433
Butadiene, 142, 516
By-product molecules, 129

γ-iron, 355
C-curves, 385
CADOFF, I. B., 293
Cancellation, diffraction, 94
Capacitor, 11
Carbon content
   vs. ductility, 411
   vs. hardness, 410
   vs. strength, 410
   vs. toughness, 411
Carbide
   iron, 355
   spheroidized, 403
Carburized steel, 486
Carburizing, 431
Carriers
   majority, 303
   minority, 303
Case, 431
CASTON, R. H., 199
Cathode, 465
Cathode reactions, 469
Cathodic polarization, 474
Cathodic protection, 479
Cations, 26
Cell
   concentration, 470
   galvanic, 470
   oxidation, 472
   photo-, 317
   solar, 317
   stress, 470, 474
   unit, 61
Cellulose, 498
Cement, portland, 495
Cement properties, 496
Cemented-carbide tools, 406
Cementite, 355
Center of symmetry, 77
Ceramic Data Book, 515

Ceramic dielectrics, 266
Ceramic magnets, 334
Ceramics, properties of, 513
Ceramics: Stone Age to Space Age, 15
Chain, folded, 138
Chain reaction, 49, 127
Chain-reaction polymerization, 127
CHALMERS, B., 377
CHANG, R., 402
CHAO, H. C., 110
Charge, electron, 20
Charge carriers, number of, 298
Charge density, 249
   units, 521
CHARLES, R. J., 122, 456
Charpy test, 449
Charring, 460
Chemical Bonding of Molecules, 32, 53
Chemical Crystallography, 137
Chloroprene, 516
CHORNE, J., 433
Circuits, electronic, 3
Cis-isomers, 142
Clausius-Clapeyron, 344
Clay, 496
Cleavage, 436
Climb, dislocation, 214, 415
Cloud-seeding, 381
Close-packed coordinations, 61
Close-packed crystal, 62
Close-packed solids, heat capacities of, 101
Close-packed unit cell, 62
Clustering, 156, 423
Coating
   galvanized, 479
   surface, 484
   protective, 486
Coefficient
   thermal expansion, 103
   vector, 80
Coercive field
   electric, 273
   magnetic, 328, 413
Coherency, 424
Cold forging, 421
Cold work, 225, 422
Collector, 317
Color centers, 108, 267
Combination, termination by, 129
Components, 343
Composite materials, 484
Composites
   bonding within, 492
   elastic moduli of, 492
Composition, eutectic, 349

Current density, exchange, 472
Curtis, H. L., 262
Curves, hardenability, 431
Cyclic fatigue, 450
Cyclic stresses, 204, 450

δ-ferrite, 355
δ-iron, 355
Dacron, 131
Damping, 203
Damping capacity, 205
Darken, L. S., 172, 185
Davis, F. R., 53, 72, 97, 394
Debye, 505
Decomposition, spinodal, 390
Deep drawing, 421
Defect
    Frenkel, 106
    linear, 108
    point, 105
    two-dimensional, 112
Defect semiconductor, 312
Defect structures, 157
Deformation
    elastic, 188
    plastic, 6, 208
    reversible, 188
    viscoelastic, 231
Deformation in crystals, permanent, 208
Deformation of rubber, 241
Deformation by twinning, 216
Degradation, 460
Degree of polymerization, 132
DeHoff, R. T., 123
Dekker, A. J., 275, 294, 320, 337
Demagnetization product, 329
Density, 405
    electron charge, 10
Density of states, 279
Depolarization, 273
Depolymerization, 460
Design considerations, 453
Devices, junction, 313
Diagram
    Ag-Cu, 349
    Al-Cu, 364
    Al-Mg, 363
    Al-Si, 364
    Al_2O_3-SiO_2, 365
    Cu-Ni, 346
    Cu-Sn, 366
    Cu-Zn, 365
    Fe-C, 350, 357
    Fe-O, 367

FeO-SiO_2, 363
H_2O-NaCl, 371
MgO-Al_2O_3, 354
Pb-Bi, 350
Pb-Sn, 366
Diamagnetism, 326
Diamond, 68
Diamond structure, 44
DiBenedetto, A. T., 53, 146, 245
Dielectric behavior of materials, 249
Dielectric constant, 10, 249
    relative, 8, 11, 249
Dielectric constants of amorphous solids, 260
Dielectric constants of liquids, 258
Dielectric constants of polymers, 260
Dielectric constants of solids, 258
Dielectric constant vs. frequence, 261
Dielectric constant vs. temperature, 259
Dielectric losses, 262
Dielectric properties of electrical insulators, 266
Dielectric relaxation, 273
Dielectrics
    ceramic, 266
    polymer, 265
Dieter, G. E., 205, 228
Diffraction, 89
    Bragg's law of, 89
    second-order, 95
Diffraction cancellation, 94
Diffraction lines, 94
Diffraction patterns, 91
Diffusion
    accompanying phase changes, 381
    activation energy for, 168
    inter-, 175
    ring, 167
    self-, 165
Diffusion coefficient, 168
    factors affecting, 170
Diffusion data, 169
Diffusion in compounds, 176
Diffusion in polymers, 180
    factors affecting, 181
Diffusion in Solids, 169, 185
Diffusion mechanisms, 166
Diffusion profiles, 171
Diffusivity, 168
    thermal, 174
Dihedral angle, 115, 403
Dimensions, 521 ff.
Dimethylsilanediol, 516
Diode, Zener, 315
Dipole
    electric, 10

induced, 252
permanent, 250, 256
Dipole moment, 10, 250, 256
Directionality, 40
Directions, angles between, 82
Directions, crystal, 81
Directions, family of, 83
Directions, lattice, 80
Dislocation, 108
edge, 109
energy of, 111
screw, 109
Dislocation climb, 214, 455
Dislocation density, 222
Dislocation generation, 213
Dislocation loop, 109
Dislocation movement, slip by, 211
Dislocation network, 110
Dislocation pile-up, 221
Dislocation strains, 110
Dislocation source, Frank-Read, 214
Dislocations, 228
Disorder, 52
structural, 100
thermal, 100
Dispersed phases, 408
Dispersion effects, 45
Displacement models, 235
Displacive transformation, 374
Disproportionation, 129
Dissipation factor, 263
Distance
interatomic, 35, 64
repetitive, 80
Distortion of metal crystals, 209
Distribution, phase, 400
Distribution coefficient, 391
Divinyl benzene, 516
Dodecahedron, 75
Domain
ferroelectric, 271, 272
magnetic, 327
Domain walls, 329
Domain-wall movements, 330
DONAHUE, F. M., 473
Donor exhaustion, 302
Donor state, 302
DORN, J. E., 481
Dorn-Weertman equation, 455
Dot product, 82, 89
Double bonds, 41
Double refraction, 268
DOVEY, D. M., 433
Drawability, 421

Drawing, 421
deep, 421
Drift velocity, 284
Ductile-brittle transition, 448
Ductile fracture, 445
Ductile materials, fracture in, 448
Ductility, 5, 410
vs. carbon content, 411

EAST, W. H., 497
ECONOMOS, G., 413
EDELGLASS, S. M., 228, 457
Edge dislocation, 109
EHRENREICH, H., 294
ELAM, C., 209
Elastic behavior, 188
Elastic deformation, 188
Elastic modulus, 5, 6
variations in, 196
Elastic waves, 194
Elasticity
thermo-, 202
time-dependent, 200
Elastomers, 143, 241
Electric charge density, 10
Electric dipoles, 10
Electric eye, 301
Electric field, 249
Electric insulators, 265
Electric moment, 29
Electric permittivity, 249
units, 522
Electric properties, 8
Electric susceptibility, 250, 255, 257
vs. temperature, 260
Electrical and Magnetic Properties of Metals, 338
Electrical conductivity, 8, 9, 283, 407
Electrical Engineering Materials, 275, 294, 320, 337
Electrical insulators, dielectric properties of, 266
Electrical processes in solids, 247
Electrical properties, 9
Electricity
ferro-, 272
piezo-, 270
Electrochemical potentials, 467
Electrode potential, 468
Electrodeposition, 494
Electromagnetic and Quantum Properties of Materials, 275
Electromotive force, thermal, 291
Electron
mass of, 18
3d-, 24

NORTON, F. H., 498
n-p junction, 315
n-p-n, 317
n-type semiconduction, 302
Nuclear Reactor Materials, 461, 463, 482
Nucleation, 378
  grain-boundary, 389
  heterogeneous, 380
  homogeneous, 379
Number, atomic, 18
Number-average molecular weight, 133
NUSSBAUM, A., 275, 320, 337
NUTT, M., 96
Nylon, 131

Octahedral sites, 336
Octahedron, 75
O'DRISCOLL, K. F., 146
Oersted, 324, 505
OLCOTT, J. S., 457
Optical properties, 267
Orbital, 22
Order, 52
  diffraction, 95
  long-range, 61, 154
  molecular, 156
  short-range, 60, 156
Order-disorder transitions, 153
Order parameters, 154
Ordered solid solutions, 286
Ordering, 153, 374
  stress-induced, 199
Organic materials, properties of, 514
Organic polymers, 146, 245
Orientation, preferred, 118, 399
Orientation polarization, 250
OROWAN, E., 245
Orthorhombic, 75
OSBORN, C. J., 457
OSBORN, H. B., 487
Overaging, 425
Overpotential, 473
OWEN, W. S., 433
Oxidation, 463, 466
  polymer, 465
Oxidation cell, 472
Oxidation potential, 468
Oxidation reaction, 468

Packing factor, atomic, 62
Parabolic rate law, 464
Paraelectric, 274
Paramagnetism, 326
PARKER, E. R., 53, 72, 97, 394, 482, 515

Particle agglomeration, 401
Particle shapes, 399
Passivation, 476
Passivation curve, 476
Pattern
  diffraction, 91
  Laue, 91
  powder, 93
  repetition, 77
Pauli exclusion principle, 21, 35, 277
Pb-Bi diagram, 350
Pb-Sn diagram, 366
Pearlite, 356
PEARSALL, G. W., 32, 71, 96, 123
Peltier effect, 292
Permanent dipole, 45, 250, 256
Permanent magnet, 327
Permeability
  initial, 331, 413
  magnetic, 324, 331
  maximum, 331
  relative, 324
  units, 523
Periodic table, 19
Peritectic, 351
Peritectic reaction, 351
Peritectoid, 351, 352
Peritectoid reaction, 352
Permittivity, 249
  electric (units), 522
Permittivity of free space, 505
PETCH, N. J., 227
PETRASEK, D. W., 400
Phase, 4, 59, 149, 343
  amount of, 398
  condensed, 49
  crystalline, 59
  dispersed, 408
  molecular, 126
  size of, 399
  stable, 51
  structure of, 57
Phase angle, 204
Phase changes
  diffusion accompanying, 381
  nucleation of, 378
Phase changes in materials, 373
Phase diagram
  binary, 347
  Fe-C, 353
  location of, 368
Phase diagram of water, 344
Phase Diagrams for Ceramists, 354, 368
Phase Diagrams in Metallurgy, 368
Phase distribution, 400